ADVANCES IN BLADDER RESEARCH

ADVANCES IN EXPERIMENTAL MEDICINE AND BIOLOGY

Recent Volumes in this Series

Volume 454
OXYGEN TRANSPORT TO TISSUE XX
Edited by Antal G. Hudetz and Duane F. Bruley

Volume 455
RHEUMADERM: Current Issues in Rheumatology and Dermatology
Edited by Carmel Mallia and Jouni Uitto

Volume 456
RESOLVING THE ANTIBIOTIC PARADOX:
Progress in Understanding Drug Resistance and Development of New Antibiotics
Edited by Barry P. Rosen and Shahriar Mobashery

Volume 457
DRUG RESISTANCE IN LEUKEMIA AND LYMPHOMA III
Edited by G. J. L. Kaspers, R. Pieters, and A. J. P. Veerman

Volume 458
ANTIVIRAL CHEMOTHERAPY 5: New Directions for Clinical Application and Research
Edited by John Mills, Paul A. Volberding, and Lawrence Corey

Volume 459
IMPACT OF PROCESSING ON FOOD SAFETY
Edited by Lauren S. Jackson, Mark G. Knize, and Jeffrey N. Morgan

Volume 460
MELATONIN AFTER FOUR DECADES
Edited by James Olcese

Volume 461
CYTOKINES, STRESS, AND DEPRESSION
Edited by Robert Dantzer, Emmanuele Wollman, and Raz Yirmiya

Volume 462
ADVANCES IN BLADDER RESEARCH
Edited by Laurence S. Baskin and Simon W. Hayward

Volume 463
ENZYMOLOGY AND MOLECULAR BIOLOGY OF CARBONYL METABOLISM 7
Edited by Henry Weiner, Edmund Maser, David W. Crabb, and Ronald Lindahl

Volume 464
CHEMICALS VIA HIGHER PLANT BIOENGINEERING
Edited by Fereidoon Shahidi, Paul Kolodziejczyk, John R. Whitaker, Agustin Lopez Munguia, and Glenn Fuller

ADVANCES IN BLADDER RESEARCH

Edited by

Laurence S. Baskin

and

Simon W. Hayward

Department of Urology
University of California, San Francisco

Kluwer Academic / Plenum Publishers
New York, Boston, Dordrecht, London, Moscow

Proceedings of the International Bladder Research Congress, held April 23–25, 1998, in San Francisco, California

ISBN 0-306-46112-9

233 Spring Street, New York, N.Y. 10013

10 9 8 7 6 5 4 3 2 1

A C.I.P. record for this book is available from the Library of Congress

Printed in the United States of America

SAN FRANCISCO · Golden Gate Bridge from the Presidio

This book is dedicated to my parents, Fred and Cynthia Baskin

PREFACE

The aim of Bladder Research Congress, San Francisco, California, April 23–25, 1998, was to provide a forum for authoritative investigators who are actively involved in the various disciplines which define the leading edges of bladder research. It is important for such investigators to continue to meet regularly for the purpose of discussing the latest developments in their individual fields, to analyze the significance of current research, to discuss new tactics for unresolved problems, to critically evaluate current theories, and to develop new theories and approaches as needed.

The two and a half day meeting was organized into five half day sessions, with each session encompassing one of five topics: (1) Epithelial-Mesenchymal Interactions; (2) Extracellular Matrix and Muscle; (3) Nerves and Pharmacology; (4) Infection and Immunology; and (5) Oncology. Each session was introduced by a moderator followed by five to six invited expert speakers with time for extensive interaction from the participants. Two late-afternoon poster sessions allowed further interactions between investigators.

This book documents the proceedings of the Bladder Research Congress. It is organized into the five half-day sessions of the meeting with moderators overview and an edited transcription of discussions that followed each presentation.

I would like to thank Sarah Burke and the Office of Continuing Medical Education, USCF; Joanne Hayward, Editorial Assistant; and Miriam Escamilla, Administrative Assistant.

I hope you find this resource useful.

Laurence S. Baskin, M.D.
Program Chair
Chief, Pediatric Urology
UCSF

ACKNOWLEDGMENTS

Support from the following organisations is gratefully acknowledged; without this funding, the meeting would not have taken place.

The National Institutes of Health

UCSF Department of Urology

* * *

Alza Pharmecuticals

American Foundation of Urologic Disease

Astra Tech, Inc.

Bayer Corporation

C.R. Bard, Inc.

Circon ACMI

Interstitial Cystitis Association

Laborie Medical Technologies

Matritech

Mentor Urology

National Kidney Foundation of Northern California

Phone-Poulenc Rorer

SmithKline Beecham

CONTENTS

Section V. Cancer of the Bladder

Section I

EPITHELIAL-MESENCHYMAL INTERACTIONS IN THE BLADDER

OVERVIEW OF EPITHELIAL-MESENCHYMAL INTERACTIONS IN THE BLADDER

Gerald R. Cunha

Department of Anatomy and Reproductive Endocrinology Center
University of California, San Francisco
California, 94143-0452

Like many other internal organs, the bladder is composed of epithelium and fibromuscular tissue and owes its development to reciprocal interactions between mesenchyme and epithelium. The epithelium of the bladder is endodermal in origin being derived from the embryonic cloaca, the caudal terminus of the hindgut, and its ventral diverticulum the allantois. The cloaca is subsequently divided by the urorectal septum into the primitive urogenital sinus and rectum. Epithelium of the upper urogenital sinus and the proximal portion of the allantois give rise to the lining of the bladder, a highly specialized urothelium.[1] Epithelium of the embryonic bladder is in turn surrounded by mesenchyme whose origin has not been adequately examined. Bladder mesenchyme differentiates into the cells that comprise the fibromuscular wall of the bladder: fibroblasts of the mucosa, inter-muscular fibroblasts, sub-serosal fibroblasts, vasculature, and smooth muscle cells of the muscularis mucosa and detrusor.

Initially it is impossible to perceive organ-specific boundaries in the regions of the endodermal epithelium of the embryonic urogenital sinus. The developmental repertoire of urogenital sinus epithelium is to differentiate into the urothelia of the bladder and urethra, the glandular epithelia of the prostate, urethral glands, and bulbourethral glands, and the stratified squamous epithelium of the lower vagina.[2–4] Significantly, adult bladder epithelium is capable of expressing under certain spontaneous and experimental conditions the urothelial phenotype, various glandular phenotypes, and the stratified squamous phenotype.[5]

The unique differentiation program of urothelium is not unique to endoderm-derived epithelium. The ureter, whose epithelial lining is mesodermal in origin, also undergoes urothelial differentiation. Differences in the biological properties of endoderm-versus mesoderm-derived urothelium have not been adequately explored, even though ureteric epithelium is much easier to establish in culture than bladder urothelium. Urothelium is a highly differentiated epithelium specialized to contain urine and to accommodate vast changes in the overall size of the bladder. To achieve these functions a specialized apical plasma mem-

Advances in Bladder Research, edited by Baskin and Hayward.
Kluwer Academic / Plenum Publishers, New York, 1999.

brane has evolved which is asymmetric and about one and a half times thicker than typical cell membranes. Chapters by Drs. Sun and Southgate describe the unique membrane proteins expressed in urothelial cells. Apical membrane plaques express three major proteins, uroplakins I, II, III, that have been extensively characterized at the structural and molecular levels by Dr. Sun and colleagues. In the inter plaque regions of the apical plasma membrane urothelial membrane antigen and UGP-85 are expressed (See chapter by Southgate et al). The cytokeratin profile of urothelium has been characterized, and surprisingly, many of the cytokeratins normally restricted to simple epithelia (#s 7, 8, 18) are expressed in urothelium which is multilayered. Taken together, the chapters by Sun et al and Southgate et al provide a comprehensive overview of urothelial differentiation.

Based upon an extensive literature on instructive mesenchymal-epithelial interactions in development, it is assumed that the urothelial differentiation program is induced by mesenchyme. However, evidence for such an instructive mesenchymal-epithelial interactions is lacking. This has important implications in regard to the potential mechanisms of maintenance of adult urothelial differentiation. The chapter by Southgate reviews the experiments pro and con whether adult urothelial differentiation is maintained via instructive versus permissive interactions between urothelium and its cellular and extracellular microenvironment. This issue has practical ramifications in regard to the design of therapeutic strategies for bladder repair, augmentation and tissue engineering, which are treated in considerable depth in the chapters by Southgate and Atala.

For most internal organs, the epithelial element is clearly the functional component of the organ. In the case of the bladder, the smooth muscle detrusor is equally important in the overall functioning of the organ. Indeed, the unique ability of the bladder to maintain low pressure through the filling phase, and the ability to empty during micturation is a function of relaxation of the smooth muscle during filling and contraction of the detrusor during voiding. Studies on the ontogeny of bladder smooth muscle indicate that the epithelium induces the undifferentiated embryonic mesenchyme to differentiate into smooth muscle.[6] Clearly, bladder development involves reciprocal epithelial-mesenchymal interactions in which mesenchyme plays a key role in epithelial differentiation, and epithelium induces differentiation of smooth muscle in the mesenchyme as reviewed in the chapter by Baskin. Such reciprocal signaling is indicated by paracrine acting growth factors such as keratinocyte growth factor, although other growth factors are also involved. Growth factors are expressed during normal bladder development, in wound healing and in response to bladder neck obstruction as discussed by Baskin.[7] Such studies also have practical application in the reconstruction of the bladder.

Practical aspects of bladder repair are discussed in the chapter by Atala. Strategies for the propagation of bladder epithelial and smooth muscle cells were described. Artificial matrices containing bladder epithelial cells alone or in combination with bladder smooth muscle cells can now be used in the repair and augmentation of the bladder. The result is the attainment of reasonably normal function as described by Atala. The biological underpinning of these clinical studies of bladder reconstruction are derived from basic science studies on the underlying cellular and molecular mechanisms of bladder development which will continue to provide important insights into the biology of the bladder.

REFERENCES

1. Felix, W.: The development of the urogenital organs. In: Manual of Human Embryology. Edited by Kiebel, R. and Mall, F.P. Philadelphia: Lippincott, chapt., 869, 1912.

2. Cunha, G.R., Lung, B. and Reese, B.: Glandular epithelial induction by embryonic mesenchyme in adult bladder epithelium of Balb/c mice. Invest. Urol., **17:** 302,1980.
3. Kinbara, H., Cunha, G.R., Boutin, E., Hayashi, N. and Kawamura, J.: Evidence of stem cells in the adult prostatic epithelium based upon responsiveness to mesenchymal inductors. Prostate, **29:** 107,1996.
4. Boutin, E.L., Battle, E. and Cunha, G.R.: The response of female urogenital tract epithelia to mesenchymal inductors is restricted by the germ layer origin of the epithelium: prostatic inductions. Differentiation, **48:** 99,1991.
5. Cunha, G.R., Fujii, H., Neubauer, B.L., Shannon, J.M., Sawyer, L.M. and Reese, B.A.: Epithelial-mesenchymal interactions in prostatic development. I. Morphological observations of prostatic induction by urogenital sinus mesenchyme in epithelium of the adult rodent urinary bladder. J. Cell Biol., **96:** 1662,1983.
6. Baskin, L.S., Hayward, S.W., Young, P. and Cunha, G.: Role of mesenchymal-epithelial interactions in bladder development. J. Urol, **156:** 1820,1996.
7. Baskin, L.S., Sutherland, R.S., Thomson, A.A., Hayward, S.W. and Cunha, G.R.: Growth factors and receptors in bladder development and obstruction. Lab Invest, **75:** 157,1996.

1

UROPLAKINS AS MARKERS OF UROTHELIAL DIFFERENTIATION

Tung-Tien Sun,[1,2,3,5] Feng-Xia Liang,[1] and Xue-Ru Wu[3,4]

[1]Epithelial Biology Unit, Ronald Perelman Department of Dermatology
[2]Department of Pharmacology
[3]Department of Urology
[4]Department of Microbiology
[5]Kaplan Comprehensive Cancer Center
New York University Medical School
New York, New York 10016

1. INTRODUCTION

In studying the physiology and diseases of a given organ, the identification of the major differentiation products of that particular organ is a crucially important step. Detailed knowledge on the structure and function of these molecules, which are frequently accumulated in large quantities during cellular differentiation, help elucidate the molecular basis of important physiological processes, such as muscle contraction. Mutations of many of the tissue-specific products have been found to be the underlying causes for many organ-restricted diseases. The availability of promoters of some of these tissue-specific genes enables one to drive various transgenes including mutated differentiation products, specific oncogenes, dominant negative mutants of various receptors, to express uniquely in the target organ. The resultant pathology and abnormal physiology can yield great insights on the functions of the differentiation products per se and various test molecules, as well as the mechanisms of many human diseases. It is in this context that the recent identification of uroplakins as the major differentiation products of mammalian urothelium is important, as this has made it possible to employ many approaches, that have proven to be fruitful in other organ systems, for elucidating the biochemical mechanisms of urothelial growth and differentiation. In this paper, we will discuss what has been learned recently about the uroplakin molecules and their possible role in urothelial differentiation.

Advances in Bladder Research, edited by Baskin and Hayward.
Kluwer Academic / Plenum Publishers, New York, 1999.

2. EARLIER STUDIES ON UROTHELIAL DIFFERENTIATION

Keith Porter first described in 1963 that the apical surface of urinary epithelium is covered by numerous rigid-looking, concave-shaped plaques interconnected by regular-looking membranous hinges.[1] In the ensuing two decades, pioneering work by several investigators established mainly by electron microscopy that the outer leaflet of these plaques is almost twice as thick as the inner one hence the term asymmetrical unit membrane (AUM[2]); that a tremendous number of Golgi-derived cytoplasmic vesicles are accumulated underneath the apical surface of the umbrella cells and are most likely involved in delivering these plaques to the surface;[3–6] that these plaques consist of 16-nm protein particles arranged hexagonally in two dimensional crystalline arrays (Figure 1 [7,8]); that the

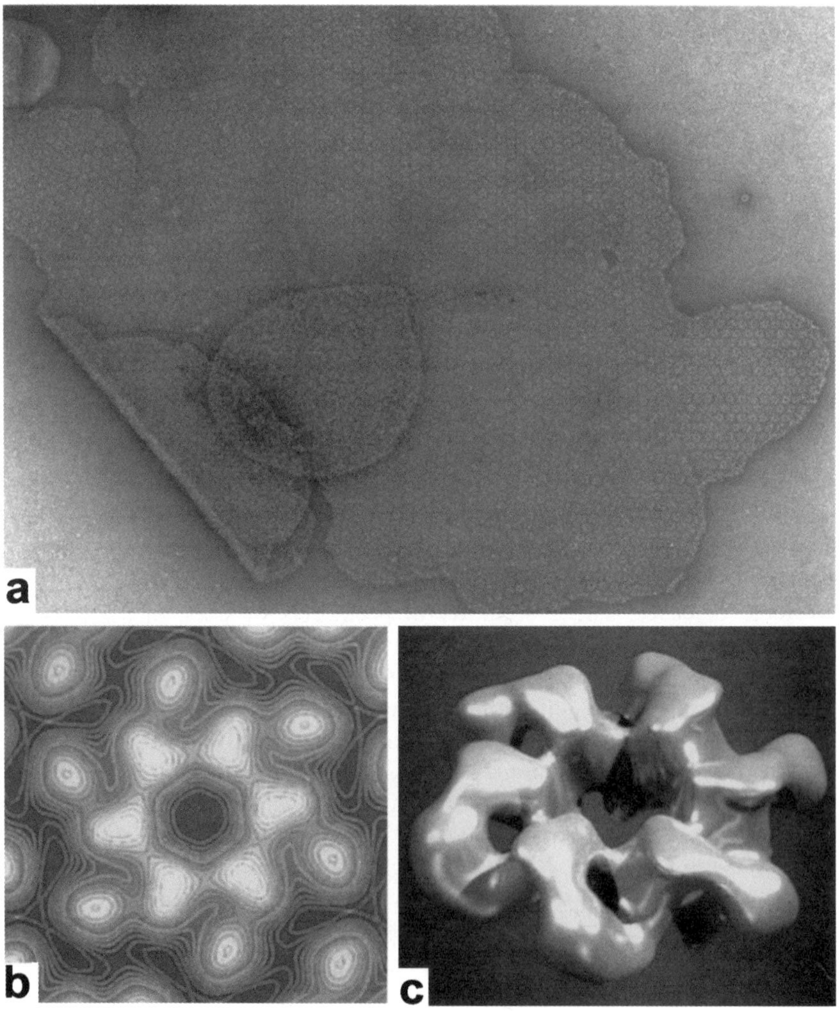

Figure 1. Ultrastructure of the urothelial plaques. Panel (a) EM of a negative-stained bovine AUM exhibiting a crystalline array of 16-nm protein particles.[21,29] (b) A computer-enhanced image of a 16-nm particle showing six inner and six outer subdomains.[29] (c) A three-dimensional model of the extracellular (luminal) portion of a 16-nm protein particle.[29]

cytoplasmic surface of these plaques seem to interact with a cytoplasmic network of filaments suggesting that the plaques may function as a physical stabilizer of the apical surface thus preventing it from rupturing during bladder distention;[9] that some of the cytoplasmic plaque-containing vesicles may fuse with the apical surface upon bladder distention thus contributing to an increased apical surface area,[1,2,5,9,10] and, finally and remarkably, that upon bladder contraction, some of the luminal plaques may actually be "retrieved" as vesicles back into the cytoplasm thus accomplishing a reversible modulation of the apical surface area.[2,10–14] These results, together with the finding that such crystalline plaques are unique to urothelial umbrella cells, clearly established AUM plaques as highly specialized products of urothelial differentiation. The fact that these plaques represent the major specialized subcellular structure of urothelium, and are thus potentially available in large quantities, offers unique opportunities for studying the detailed structure and function of membrane proteins. Finally, from a membrane dynamics point of view, the targeting of the AUM plaques to the luminal surface and their possible retrieval into the cytoplasm is both intriguing and fascinating.

While excellent progress was made on ultrastructural studies, the earlier biochemical characterization of AUM plaques was met with only limited success. SDS-PAGE analyses of partially purified AUM plaques from cattle, pig and sheep revealed multiple protein components in the MW range of 12 to 100-Kd.[15–18] The field reached an impasse at this stage, however, as no one seemed to be able to produce antibodies with proven specificity for any of the putative AUM protein subunits. Hence important questions were left open as to whether some higher MW AUM components are oligomers of the smaller subunits, as suggested by Stubbs et al.,[17] or conversely whether some of the smaller components are degradation products of the larger ones. Even more crucial was the open question as to whether any of the putative AUM subunits are truely AUM-associated in vivo; without this information, one cannot be sure whether some of these proteins are merely contaminants. Without understanding the basic protein composition of the major differentiation products of mammalian urothelium, i.e., the urothelial plaques, the field of urothelial differentiation slowed down in the mid-80's.

The situation improved in the late 80's due to several advances. The first came when we succeeded in raising a panel of monoclonal antibodies (against crude bovine urothelial plaques) that selectively stained the urothelial umbrella cells. One of these antibodies, AE31, was shown by EM to decorate the outer leaflet of the apical AUM plaques, and by immunoblotting to recognize specifically a 27-Kd urothelial integral membrane protein.[19] Moreover, this monoclonal antibody can affinity-purify, in one step, not only the 27-Kd protein, but also two other associated proteins of 15-Kd and 47-Kd. While these data gave us the first glimpse of the AUM composition, the yield of the AE31-affinity purification was so low ((10-μg per bovine bladder) that we were unable to obtain enough material to raise specific antisera or to determine the partial amino acid sequences of the individual protein components. Another significant development was therefore the large scale purification of urothelial plaques. By modifying and improving an existing protocol,[16,18,20] we were able to isolate in two days 10 to 20-mg of highly purified bovine urothelial plaques (yield, 0.5 to 1-mg per bladder) which contained, like the AE31-purified material, three major proteins of 15-, 27- and 47-Kd.[21] The availability of such a large quantity of highly purified bovine urothelial plaques made it feasible for us to embark on a systematic study of their biochemical and molecular features.

3. IDENTIFICATION OF UROPLAKINS AS MAJOR DIFFERENTIATION PRODUCTS OF UROTHELIUM

Although we obtained sufficient AUM proteins, it was still difficult for us to produce antisera to the individual AUM components. While the electrophoretically purified 47-kDa protein proved to be quite immunogenic, similarly purified 27-kDa and 15-kDa proteins were found to be extremely non-immunogenic as repeated immunization of these antigens in rabbits and chickens elicited little imunological response.[19,21] As an alternative approach, we immunized the animals with purified bovine urothelial plaques. The rationale of this approach was that the highly regular, polyvalent, hexagonal arrays of protein epitopes exposed on the plaque surface may be more favorably presented to the immune cells and hence may be more immunogenic.[22] The validity of this rationale was supported by our own earlier experience in raising rabbit antibodies to keratins. In this case we found that intact (reconstituted) keratin filaments were much more immunogenic than the individual keratins purified by preparative SDS-PAGE.[23] The immunization of intact urothelial plaques indeed worked well yielding rabbit and chicken antibodies against all three uroplakins.[21] Immunoblotting showed that these antisera against total uroplakins reacted strongly with the 47-kDa UPIII, moderately with the 27-kDa UPI, and weakly with the 15-kDa UPII.[21] From these antisera, we affinity-purified antibodies monospecific for individual uroplakins that we isolated by preparative SDS-PAGE.[21] The availability of these antiboies enable us to show by immunofluorescence staining that the three major AUM proteins are associated with urothelial umbrella cells indicating that they are markers for an advanced stage of urothelial differentiation.[19,21,24] EM-localization showed that these three proteins are indeed AUM-associated in vivo.[19,21] Based on their proven association with the urothelial plaques, we named these proteins uroplakins I (27-Kd), II (15-Kd) and III (47-Kd,[19,21]).

cDNA cloning showed that the 27-Kd component contains two closely related isoforms that we named uroplakins Ia (27-Kd) and Ib (28-Kd) thus establishing the existence of four uroplakins.[25] Northern blot analyses showed that the mRNAs for all these uroplakins are accumulated in large quantities in urothelium, but undetectable in other epithelia including the epidermis, esophageal epithelium, the liver and a large number of other epithelial and mesenchymal cell types.[24–27] This finding, in conjunction with the above-mentioned immunofluorescence staining data, clearly established uroplakins as markers for an advanced stage of urothelial differentiation.[25–27]

4. MOLECULAR STRUCTURE OF UROPLAKINS AND THE 16NM AUM PARTICLES

4.1. Two Isoforms of UPI Are Members of the Tetraspan Family

Our cDNA data revealed the existence of two UPI isomers, i.e., UPIa and UPIb. These two isoforms share 39% of their amino acid sequences, and both belong to a novel family of integral membrane proteins all having 4 transmembrane domains (the "TM4 family"). The two major hydrophilic loops, interconnecting the transmembrane domains (TMDs) 1/2 and 3/4, face the lumen since they become protected from protease digestion once an in vitro translated UPI is inserted into pancreatic microsomes. The major loop of each UPI has a single potential N-glycosylation site, which according to deglycosylation data harbors 2 to 3-Kd equivalents of high mannose-type of sugars. As one can see from the schematic drawing in Figure 2, the uroplakin I's have relatively short cytoplasmic domains.[25]

4.2. UPII Is Synthesized with a PrePro Sequence

The cDNA of the 15-Kd uroplakin II[27]) revealed a precursor structure containing a hydrophobic N-terminal signal peptide, and a long and heavily glycosylated pro-sequence of 59 amino acid residues terminating in RGRRña consensus substrate for furin, a trans-

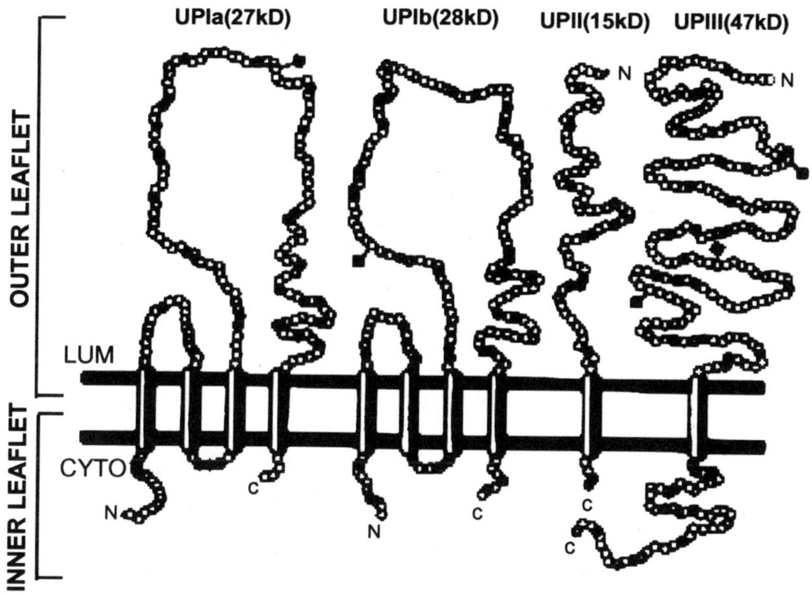

Figure 2. The proposed transmembrane topology of the four major bovine uroplakins showing that the bulk of most uroplakins are luminal, possibly involved in interacting with one another to form the 16-nm protein particles, and that UPIII is the only uroplakin possessing a significant cytoplasmic domain, possibly involved in AUM:cytoskeletal interactions.[25]

Golgi-network-associated processing enzyme. The mature UPII per se is not glycosylated. Unlike the UPI's, UPII has only a single TMD located at its C-terminus possibly serving as a membrane anchorage (Figure 2).[27]

4.3. UPIII May Play a Role in Mediating AUM/Cytoskeletal Interaction

Molecular cloning of bovine UPIII cDNA showed that it, like UPII, has a single TMD, which however in this case is not located at the end but rather it divides the protein into a long N-terminal luminal domain (189 amino acids) and a short C-terminal cytoplasmic domain (52 residues; Figure 2). Consistent with this designated transmembrane topology, the N-terminal domain is highly glycosylated with 20-Kd equivalents of complex type of sugars, while the cytoplasmic domain harbors several potential phosphorylation sites. Interestingly, UPIII shares a stretch of about 12 amino acid sequence, located on the N-terminal side of the single TMD, with a similar domain of UPII suggesting that these two uroplakins are related (see below). Since UPIII is the only uroplakin that possesses a significant cytoplasmic domain, this raises the interesting possibility that UPIII may be involved in anchoring the AUMs to the underlying cytoskeleton[9] and that phosphorylation may be involved in regulating such an association.[26]

4.4. Most of the Hydrophilic Domains of Uroplakins Are Extracellular

Our results thus indicate that UPI's and UPII have almost no cytoplasmic domain. Even for UPIII, the mass of its extracellular domain (189 amino acids plus 20-Kd of carbohydrates) greatly exceeds its cytoplasmic domain (52 amino acids). Thus the bulk of all the uroplakins projects into the luminal space most likely interacting tightly with one another to form the 16-nm AUM particles that can be seen by EM to protrude into the lumen. Although it was previously thought that AUM particles are "floating" on the luminal leaflet without any of its parts penetrating into the inner leaflet,[28] our data strongly suggest that all the major AUM subunits are integral membrane proteins with their particle-forming domains anchored to AUM via their transmembrane domains that penetrate through the lipid bilayer. This model suggests that the asymmetry of uroplakin mass distribution across the lipid bilayer may account in part for the EM observation that the outer leaflet of AUM is much thicker than the inner one (Figure 2[25–27]).

4.5. The 16-nm AUM Particle as a Twisted Ribbon Structure

In collaboration with Ueli Aebi, we have refined the 3-D structure of the 16-nm protein particles of bovine urothelial plaques using negative staining coupled with image enhancement, metal shadowing and scanning transmission electron microscopy.[29] Our results led to a three dimensional model, at about 2-nm resolution, depicting each AUM particle as a continuous, "twisted ribbon" embracing a central cavity (Figure 1c).

4.6. Uroplakins Are Highly Organized with Clearly Defined Partners

To better understand how the extracellular domains of uroplakins interact to form the 16-nm particles that are arranged in two-dimensional crystalline arrays, we analyzed the nearest neighbor relationship among uroplakins by chemical crosslinking and by identifying the crosslinked species by immunoblotting using monospecific antibodies to individual uroplakins.[30] Our results indicate that uroplakins Ia and Ib, the two four transmembrane-domained AUM subunits, were crosslinked to different single transmembrane-domained partners, i.e., uroplakins II and III, respectively. Our data also established that UPII could be readily crosslinked to form homodimers, that were sensitive, however, to octyl-glucoside that can solubilize the AUMs. These data raised the interesting possibility that there exist two types of 16-nm particles containing two different pairs of uroplakins, i.e., UPIa and UPII or UPIb and UPIII. In the case of UPIa/UPII particle, our data support a model in which the UPIa and UPII occupy the inner and outer domains, respectively.[30] This model can account for the apparent "redundancy" of the uroplakins, as the structurally related UPIa and UPIb, by interacting with different partners, may play different roles in AUM formation. This model also suggests that AUM plaques with different uroplakin composition may differ in the regulation of their assembly, and in their ability to interact with the underlying cytoskeleton and overlying glycocalyx components.

4.7. The AUM Structure Is Highly Conserved

We have shown that the ultrastructure, protein composition and immunological properties of AUM plaques are highly conserved during mammalian evolution. Sequencing data further showed that the primary sequences of bovine and mouse UPII are highly similar sharing about 90% of identity.[31]

5. UROPLAKINS 1A AND 1B CAN SERVE AS THE UROTHELIAL RECEPTORS FOR THE TYPE 1-FIMBRIATED UROPATHOGENIC E. COLI

A great majority of urinary tract infection is caused by type 1-fimbriated E. coli. Although it is known that the attachment of the bacteria to urothelial surface is a crucial first

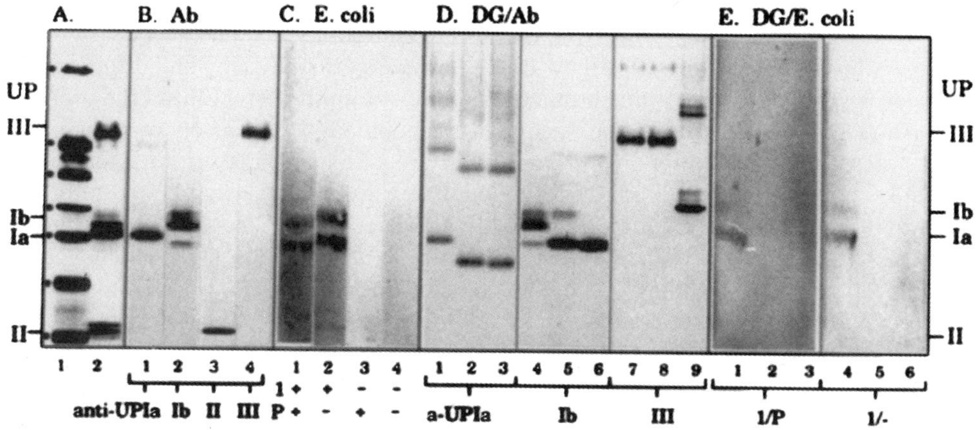

Figure 3. Type 1 fimbriated E. coli bind specifically to uroplakins Ia and Ib.[32]

step for UTI, the urothelial receptors for the bacteria were not identified. Given that uroplakins are the major subunits of the urothelial plaques that cover over 75% of the urothelial surface, these proteins represent the predominant apical membrane proteins in the entire urinary tract starting from the urethra all the way to renal pelvis. We therefore tested the possibility that some of the uroplakin molecules may serve as the elusive urothelial receptors of the uropathogenic type 1-fimbriated E. coli. In this study, we immobilized purified AUMs onto microtiter wells and incubated them with ^{35}S-labeled E. coli that produce type 1, P, or both fimbriae.[32] The results indicate that only E. coli strains expressing the type 1 fimbriae bound to the AUMs (Figure 3). This fimbria-AUM binding has several notable characteristics. First, the binding is specific for the type 1 fimbriae because nonfimbriated strains or strains expressing P fimbriae did not show any significant binding. Second, the binding is saturable, being linearly proportional to the amount of AUMs and bacteria. Third, the binding is mediated by mannose-containing receptors on the AUM luminal surface because free mannose could specifically block the binding. Fourth, the binding is highly conserved across mammalian species, in that AUMs from not only cattle, but also from human, monkey, and mouse bladders showed similar binding. Finally, by using gel overlay assay, we demonstrated that the type 1-fimbriated E. coli selectively bind to uroplakins Ia and Ib, two major mannosylated glycoproteins of AUM (Figure 3, panel C[32]). These results, along with the fact that uroplakins Ia and Ib are the major glycoproteins of AUM which covers the bulk of the urothelium surface, strongly suggest that these two proteins can serve as the major urothelial receptors for type 1-fimbriated E. coli.[32]

6. ALTERED UROPLAKIN EXPRESSION AND ASSEMBLY IN CULTURED UROTHELIAL CELLS

Although we were able to grow rabbit bladder epithelial cells in the presence of lethally irradiated 3T3 feeder cells, which can support the growth of a number of stratified squamous epithelia,[33–37] urothelial cells apparently differentiated “too well” in this culture system as they soon formed a superficial layer of umbrella-like cells with tight junctions that effectively sealed the lower cell layers from the nutrient supply; these cells therefore formed only small colonies of <200–300 cells before they died (Sun, unpublished observation, 1982–83). By slightly modifying the culture conditions, we later succeeded in serially cultivating bovine urothelial cells with the formation of stratifying and differentiating colonies.[38] A unique feature of this urothelial culture system is that it allows the urothelial cells to undergo clonal growth thus achieving a population expansion of greater than 10^4–10^5 folds. Such a differentiating urothelial culture system offers exciting opportunities for studying urothelial growth and differentiation. We have shown that bovine urothelial cells, when plated in the presence of lethally irradiated 3T3 feeder cells, can undergo clonal growth forming stratified epithelial colonies. Northern blot analysis showed that cultured urothelial cells synthesize large amounts of mRNAs of UPIa, Ib, II and III.[27] Immunoprecipitation of ^{35}S-methionine-labeled urothelial cells showed that they indeed synthesize and process all the uroplakins to their mature sizes (T.-T. Chen, G. Kreibich, X.-R. Wu and T.-T. Sun, in preparation).

7. SUPPRESSION OF UROPLAKIN EXPRESSION IN BLADDER SQUAMOUS CELL CARCINOMA SUGGESTS ALTERNATIVE PATHWAYS OF UROTHELIAL DIFFERENTIATION

Since in all the normal bovine and human tissues that we have surveyed we found uroplakins only in the urothelia, we tested in collaboration with Roland Moll (Univ. of Halle, Germany) the possible use of anti-uroplakins as a diagnostic tool for identifying metastatic transitional cell carcinomas. Our results indicate that 66% of the metastatic TCC are UPIII-positive indicating that, although there are false negatives, positive UP-staining can be used as a reliable marker for the identification of metastatic TCC.[39] More recently, we showed, in collaboration with Carlos Cordon-Cardo of Sloan Kettering, that human uroplakin genes are expressed in transitional cell carcinoma, but not in bilharzia related squamous cell carcinoma. These results suggest that urothelium can undergo several alternative pathways of differentiation. Urothelial cells undergoing these different pathways express different differentiation markers, and yield different types of cancers.[40]

8. UROPLAKIN PROMOTER CAN DRIVE THE UROTHELIAL EXPRESSION OF A HUMAN GROWTH GENE—A NEW BIOPHARMING STRATEGY

We have isolated the genes of mouse UPIa, UPII and UPIII.[41] We showed that a 3.6-Kb 5'-upstream sequence of UPII gene can drive a bacterial LacZ reporter gene to express efficiently in suprabasal urothelial cells (Figure 4[41]), indicating that this sequence must contain most of the cis-elements that are required to confer the bladder-specific and differentiation-dependent expression of the gene.[41] The ectopic expression in the hypothalamus may be due to the LacZ gene per se as this expression is shared by many other unrelated promoters.[41] We showed that a 4-Kb 5'-upstream sequence of mouse UPIa gene can also serve as a bladder-specific and differentiation-dependent promoter (H.P. Zhao et al, in preparation). More recently, we used the 3.6-Kb UPII promoter to drive human growth hormone gene to express in the bladder epithelium of transgenic mice. Human growth hormone was indeed synthesized in urothelium and, moreover, secreted into the urine at up to 500 ng/ml. These findings lay the foundation for a new strategy of biopharming in which

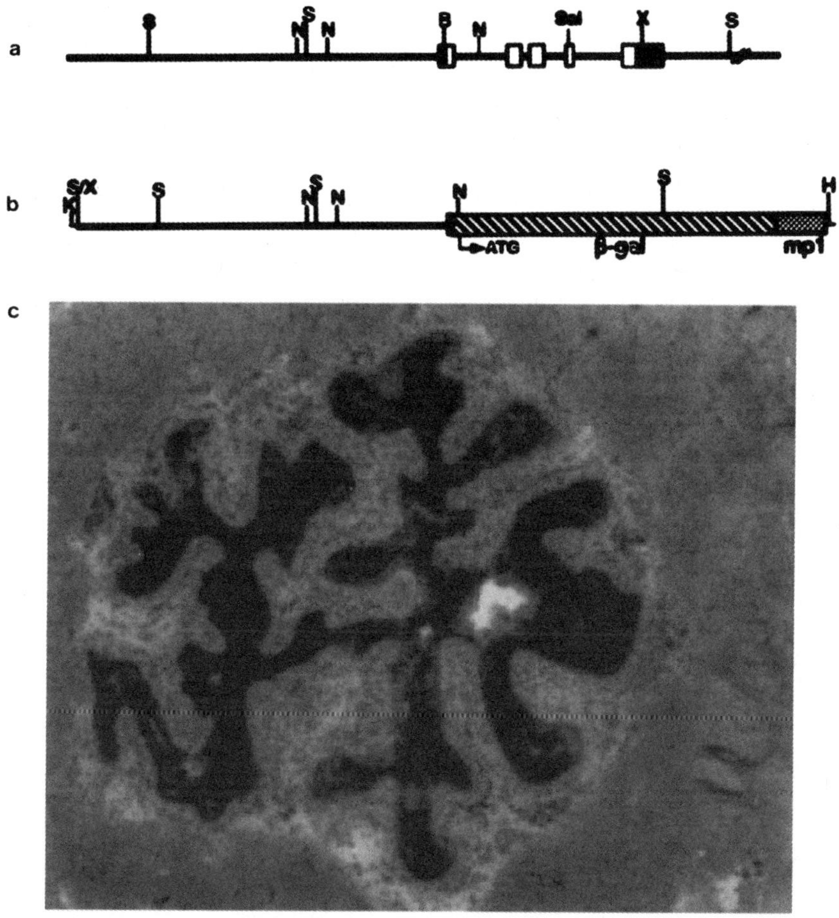

Figure 4. The 3.6-Kb 5'-upstream sequence of mouse UPII gene contains sufficient information to drive the urothelial expression of foreign genes. (a) The exon-intron structure of mouse UPII gene. (b) shows the construction of a fusion gene in which a β-galactosidase reporter gene is driven by a 3.6-Kb 5'-upstream sequence of the mouse UPII gene. (c) The urothelial expression of bacterial B-galactosidase gene.[41]

the bladder of a transgenic animal can be made into a bioreactor to produce pharmaceutically important recombinant products that can then be purified from the urine of such an animal.[42]

9. PERSPECTIVES AND CONCLUDING REMARKS

Recent studies have therefore established uroplakins as the main protein subunits of asymmetric unit membrane, their transmembrane topology and possible nearest neighbor relationship, their urothelial-specific and differentiation-dependent expression, and their altered expression in cultured urothelial cells and some forms of bladder cancers. The results also indicate that a 3.6-kb promoter of uroplakin II gene can account for much of the urothelium-specific and differentiation-dependent expression pattern of the intact gene. With these as a foundation, we can now entertain many interesting questions in the next phase of the investigation. These include: How do the 4 uroplakins interact with one another to form the twisted ribbon shaped 16-nm particle protruding from the luminal surface of the urothelial plaques? What is the mechanism that is responsible for targeting most of the uroplakins to the apical urothelial surface? What is the mechanism for regulating the assembly of uroplakins which form 16-nm particles that become progressively more organized during intracellular vesicular maturation? Urothelial plaques are thought to play a role in forming a permeability barrier, in enhancing the physical stability of the urothelial apical surface via membrane-cytoskeletal interaction, and in the reversible adjustment of the apical cell surface area by pinching off and re-fusing with the apical membrane. What is the relative importance of these proposed functions, and how do the uroplakin molecules accomplish these tasks? What is the physiological role of the uroplakin receptors for the uropathogenic bacteria, and can one design new therapies to prevent bacterial colonization? How are the uroplakin genes regulated during wound healing, metaplasia, and bladder tumorigenesis? Can one generate various types and stages of bladder cancers if one uses the uroplakin promoters to drive the urothelial expression of selected oncogenes in transgenic mice? We hope that answers to these questions will lead to an improved understanding of how the normal bladder functions and how a perturbation of urothelial differentiation leads to some of the bladder diseases.

ACKNOWLEDGMENTS

This work was supported by National Institutes of Health grants DK49469 and DK39753 (T.-T.S.) and a Merit Review Grant from the Verterans Administration (X.-R. W.).

REFERENCES

1. Porter, K. R., and M. A. Bonneville., An introduction to the fine structure of cells and Tissues, New York, Lea and Febiger. 1963.
2. Hicks, R. M.: The fine structure of the transitional epithelium of rat ureter. J. Cell Biol., **26:** 25, 1965.
3. Alroy, J., Merk, F. B., Morre, J., and Weinstein, R. S.: Membrane differentiation in the Golgi apparatus of mammalian urinary bladder epithelium. Anat. Rec., **203:** 429, 1982.
4. Hicks, R. M.: The permeability of rat transitional epithelium. Keratinization and the barrier to water. J Cell. Biol., **28:** 21, 1966.
5. Koss, L. G.: The asymmetric unit membranes of the epithelium of the urinary bladder of the rat. An electron microscopic study of a mechanism of epithelial maturation and function. Lab. Invest., **21:** 154, 1969.
6. Severs, N. J., and Hicks, R. M.: Analysis of membrane structure in the transitional epithelium of rat urinary bladder. 2. The discoidal vesicles and Golgi apparatus their role in luminal membrane biogenesis. J. Ultrastruct. Res., **69:** 279, 1979.
7. Hicks, R. M., and Ketterer, B.: Hexagonal lattice of subunits in the thick luminal membrane of the rat urinary bladder. Nature (Lond.), **224:** 1304, 1969.
8. Vergara, J. A., Longley, W., and Robertson, J. D.: A hexagonal arrangement of subunits in membrane of mouse urinary bladder. J. Mol. Biol., **46:** 593, 1969.
9. Staehelin, L. A., Chlapowski, F. J., and Bonneville, M. A.: Lumenal plasma membrane of the urinary bladder. I. Three-dimensional reconstruction from freeze-etch images. Journal of Cell Biology, **53:** 73, 1972.
10. Porter, K. R., Kenyon, K., and Badenhausen, S.: Specifications of the unit membrane. Protoplasma, **63**: 263, 1967.
11. Lewis, S. A., and de Moura, J. L. D.: Incorporation of cytoplasmic vesicles into apical membrane of mammalian urinary bladder epithelium. Nature, **297:** 685, 1982.
12. Lewis, S. A., and deMoura, J. L. C.: Apical membrane area of rabbit urinary bladder increases by fusion of intracellular vesicles: an electrophysiological study. J. Membr. Biol., **82:** 123, 1984.

13. Minsky, B. D., and Chlapowski, F. J.: Morphometric analysis of the translocation of lumenal membrane between cytoplasm and cell surface of transitional epithelial cells during the expansion-contraction cycles of mammalian urinary bladder. Journal of Cell Biology, **77:** 685, 1978.
14. Sarikas, S. N., and Chlapowski, F. J.: Effect of ATP inhibitors on the translocation of luminal membrane between cytoplasm and cell surface of transitional epithelial cells during the expansion-contraction cycle of the rat urinary bladder. Cell Tissue Res., **246:** 109, 1986.
15. Caruthers, J. S., and Bonneville, M. A.: Isolation and characterization of the urothelial lumenal plasma membrane. Journal of Cell Biology, **73:** 382, 1977.
16. Ketterer, B., Hicks, R. M., Christodoulides, L., and Beale, D.: Studies of the chemistry of the luminal plasma membrane of rat bladder epithelial cells. Biochimica et Biophysica Acta, **311:** 180, 1973.
17. Stubbs, C. D., Ketterer, B., and Hicks, R. M.: The isolation and analysis of the luminal plasma membrane of calf urinary bladder epithelium. Biochim. Biophys. Acta, **558:** 58, 1979.
18. Vergara, J., Zambrano, F., Robertson, J. D., and Elrod, H.: Isolation and characterization of luminal membranes from urinary bladder. J. Cell Biol., **61:** 83, 1974.
19. Yu, J., Manabe, M., Wu, X. R., Xu, C., Surya, B., and Sun, T. T.: Uroplakin I: a 27-kD protein associated with the asymmetric unit membrane of mammalian urothelium. Journal of Cell Biology, **111:** 1207, 1990.
20. Chlapowski, F. J., Bonneville, M. A., and Staehelin, L. A.: Lumenal plasma membrane of the urinary bladder. II. Isolation and structure of membrane components. Journal of Cell Biology, **53:** 92, 1972.
21. Wu, X. R., Manabe, M., Yu, J., and Sun, T. T.: Large scale purification and immunolocalization of bovine uroplakins I, II, and III. Molecular markers of urothelial differentiation. Journal of Biological Chemistry, **265:** 19170, 1990.
22. Dintzis, H. M., and Dintzis, R. Z.: A molecular basis for immune regulation: the immunon hypothesis. In: Book. Edited by Perelson, A. S. (ed.). Reading, Mass.: Addison-Wesley Publishing Co., vol. Part I, Vol. II, p. 283, 1988.
23. Sun, T. T., and Green, H.: Keratin filaments of cultured human epidermal cells. Formation of intermolecular disulfide bonds during terminal differentiation. Journal of Biological Chemistry, **253:** 2053, 1978.
24. Sun, T. T., Zhao, H., Provet, J., Aebi, U., and Wu, X. R.: Formation of asymmetric unit membrane during urothelial differentiation. Molecular Biology Reports, **23:** 3, 1996.
25. Yu, J., Lin, J. H., Wu, X. R., and Sun, T. T.: Uroplakins Ia and Ib, two major differentiation products of bladder epithelium, belong to a family of four transmembrane domain (4TM) proteins. Journal of Cell Biology, **125:** 171, 1994.
26. Wu, X. R., and Sun, T. T.: Molecular cloning of a 47 kDa tissue-specific and differentiation-dependent urothelial cell surface glycoprotein. Journal of Cell Science, **106:** 31, 1993.
27. Lin, J. H., Wu, X. R., Kreibich, G., and Sun, T. T.: Precursor sequence, processing, and urothelium-specific expression of a major 15-kDa protein subunit of asymmetric unit membrane. Journal of Biological Chemistry, **269:** 1775, 1994.
28. Brisson, A., and Wade, R. H.: Three-dimensional structure of luminal plasma membrane protein from urinary bladder. Journal of Molecular Biology, **166:** 21, 1983.
29. Walz, T., Haner, M., Wu, X. R., Henn, C., Engel, A., Sun, T. T., and Aebi, U.: Towards the molecular architecture of the asymmetric unit membrane of the mammalian urinary bladder epithelium: a closed "twisted ribbon" structure. Journal of Molecular Biology, **248:** 887, 1995.
30. Wu, X.-R., Medina, J. J., and Sun, T.-T.: Selective interactions of UPIa and UPIb, two members of the transmembrane 4 superfamily, with distinct single-tansmembrane-domained proteins in differentiated urothelial cells. J. Biol. Chem., **270:** 29752, 1995.
31. Wu, X. R., Lin, J. H., Walz, T., Haner, M., Yu, J., Aebi, U., and Sun, T. T.: Mammalian uroplakins. A group of highly conserved urothelial differentiation-related membrane proteins. Journal of Biological Chemistry, **269:** 13716, 1994.
32. Wu, X.-R., Sun, T.-T., and Medina, J. J.: In vitro binding of type-1 fimbriated E. coli to uroplakin I's: relation to urinary tract infections. Proc. Natl. Acad. Sci. USA, **93:** 9630, 1996.
33. Rheinwald, J. G., and Green, H.: Serial cultivation of strains of human epidermal keatinocytes: the formation of keratinizing colonies from single cells. Cell, **6:** 331, 1980.
34. Doran, T. I., Vidrich, A., and Sun, T. T.: Intrinsic and extrinsic regulation of the differentiation of skin, corneal and esophageal epithelial cells. Cell, **22:** 17, 1980.
35. Pang, Y. Y., Schermer, A., Yu, J., and Sun, T. T.: Suprabasal change and subsequent formation of disulfide-stabilized homo- and hetero-dimers of keratins during esophageal epithelial differentiation. Journal of Cell Science, **104:** 727, 1993.
36. Schermer, A., Galvin, S., and Sun, T. T.: Differentiation-related expression of a major 64K corneal keratin in vivo and in culture suggests limbal location of corneal epithelial stem cells. Journal of Cell Biology, **103:** 49, 1986.

37. Sun, T. T., Bonitz, P., and Burns, W. H.: Cell culture of mammalian thymic epithelial cells: growth, structural, and antigenic properties. Cellular Immunology, **83:** 1, 1984.
38. Surya, B., Yu, J., Manabe, M., and Sun, T. T.: Assessing the differentiation state of cultured bovine urothelial cells: elevated synthesis of stratification-related K5 and K6 keratins and persistent expression of uroplakin I. Journal of Cell Science, **97:** 419, 1990.
39. Moll, R., Wu, X.-R., and Sun, T.-T.: Uroplakins as biological markers of metastatic transtitional cell carcinomas. Am. J. Pathol., **147:** 1383, 1995.
40. Wu, R.-L., Osman, I., Wu, X.-R., Lu, M.-L., Zhang, Z.-F., Liang, F.-X., Hamza, R., Scher, H., Cordon-Cardo, C., and Sun, T.-T.: Human uroplakin II gene is expressed in transitional cell carcinoma but not in bilharzial-related squamous cell carcinoma of the urinary bladder: alternative pathways of bladder epithelial differentiation and tumor formation. Cancer Res., 1998. (In Press)
41. Lin, J. H., Zhao, H., and Sun, T. T.: A tissue-specific promoter that can drive a foreign gene to express in the suprabasal urothelial cells of transgenic mice. Proceedings of the National Academy of Sciences of the United States of America, **92:** 679, 1995.
42. Kerr, D. E., Liang, F. X., Bondioli, K. R., Zhao, H. P., Kreibich, G., Wall, R. J., and Sun, T.-T.: The bladder as a bioreactor: Urothelium production and secretion of growth hormone into urine. Nature Biotechnology, **16:** 75, 1997.

2

UROTHELIAL TISSUE REGULATION

Unraveling the Role of the Stroma

Jennifer Southgate,[1] Patricia Harnden,[2] Peter J. Selby,[1] David F. M. Thomas,[3] and Ludwik K. Trejdosiewicz[1]

[1]Imperial Cancer Research Fund Cancer Medicine Research Unit
[2]Department of Pathology
[3]Department of Paediatric Urology
St James's University Hospital
Leeds, LS9 7TF, United Kingdom

1. EPITHELIAL TISSUE ORGANIZATION AND HOMEOSTASIS

The internal and external surfaces of the body are lined by epithelial tissues. As such, epithelial tissues have a common role in forming the interface with the external environment. This commonality of function is apparent in the basic polarized structure of epithelia, with the apical cell surface abutting onto a luminal or external free space and the basolateral surface anchored to a basement membrane with an underlying mesenchymal stroma. A number of commonly-expressed epithelial gene products play a role in epithelial tissue organization and can be considered as markers of epithelial cell lineage. Such markers include the cytokeratins, E-cadherin and various junctional proteins. Different epithelial tissues also show a diverse range of highly specialized functions which are reflected in their tissue organization, cellular morphology and the expression of specific gene products.

The co-ordination of proliferation, cytodifferentiation and histioarchitecture in epithelial tissues is mediated through interactions with the associated stroma. These interactions are essential during development and tissue repair, serve to maintain tissue homeostasis in adult epithelial organs and are implicated in neoplastic disease progression. The molecular basis for regulation by the stroma is poorly understood, but is likely to be mediated at the epigenetic level through the specific ligation of cytokine/growth factor receptors and adhesion molecules involved in cell:cell and cell:matrix interactions. Cell:matrix interactions are primarily co-ordinated through the expression, localization and activation of integrins interacting with basement membrane proteins.[1,2] A large number of factors have been identified which affect epithelial cell behavior. Nevertheless, it is unclear how these are co-ordinated to achieve the structural integrity and specialized

Advances in Bladder Research, edited by Baskin and Hayward.
Kluwer Academic / Plenum Publishers, New York, 1999.

differentiated function of normal epithelia, or to what extent these processes are involved in pathogenesis.

2. THE UROTHELIUM

The function of the urinary bladder as an impermeable, low pressure reservoir for urine is reflected in the specialized characteristics of the lining urothelium. Urothelium is a stratified transitional epithelium consisting of basal, intermediate and superficial cell zones which move relative to each other to accommodate changes in bladder volume, with no loss of permeability barrier function.[3,4] The barrier function of the urothelium is dependent on the integrity of cell:cell interactions involving adhesion molecules and junctional complexes. A unique feature of the terminally-differentiated superficial "umbrella" cells is the molecular composition and structural organization of the plasma membrane. Whereas basal and intermediate cells have a normal trilaminar plasma membrane, the luminal surface of umbrella cells is composed of thickened plaques of asymmetric unit membrane (AUM), arranged in characteristic hexagonal subunits on the cell surface, bounded by "interplaque" or "hinge" regions of normal symmetrical membrane. The AUM probably functions to reduce the luminal surface area during bladder contraction by invagination, decrease the permeability of the urothelium to urine and to provide structural support. AUM plaques are formed in the vicinity of the Golgi complex by incorporation of specific proteins into the membranes of Golgi-derived fusiform vesicles which move to the surface of maturing cells where they fuse with and insert into the plasma membrane.[5] Insertion of the AUM plaques into the free luminal membrane is generally considered to be the final stage of urothelial cytodifferentiation.

Urothelium represents an excellent model for the study of normal and neoplastic epithelial cell regulation. Although normally a slow turnover epithelium[3,6–8], urothelium has a high proliferative potential during fetal development.[3,9] This potential is maintained in the adult epithelium as seen during tissue regeneration[3,10] and in a range of benign proliferative conditions.[11] In addition, the AUM of the superficial cell serves as an unequivocal ultrastructural marker of terminal urothelial cytodifferentiation.[12] However, the differentiation pathway of urothelial cells appears to be susceptible to re-direction, as shown by the propensity for human bladder epithelium to undergo glandular and squamous metaplasia.[13] There is also evidence from experimental rodent studies that retinoid-deficiency can induce squamous metaplasia[14–18] and that, under the influence of urinogenital sinus mesenchyme, urothelial cells can transdifferentiate into prostatic epithelium.[19,20]

Urothelial cells also exhibit a propensity for neoplastic transformation and are the precursor for transitional cell carcinoma (TCC), the most common form of bladder carcinoma, although other types are recognized.[13,21] TCC shows a spectrum of differentiation- and invasion-related stages which correlate with clinical prognosis[22], although the natural succession of genetic and phenotypic changes associated with disease progression remains unclear.[23–25] A panel of well-characterized bladder cancer cell lines is available, which show a range of phenotypes confirming the urothelial origin and reflecting the different grades of cancer from which they were derived.[26–29]

3. MARKERS OF UROTHELIAL CELL PHENOTYPE

A prerequisite for studies of epithelial cell regulation is the availability of reliable markers of cell phenotype. In the urothelium, a number of differentiation-associated antigens have been identified. The best characterized moieties are the four major integral membrane proteins, known collectively as uroplakins (UP), which are species-conserved and constitute the specialized AUM plaques of umbrella cells.[30,31] The 27,000 M_r UPIa and the 28,000 M_r UPIb proteins are members of the family of 4 transmembrane domain proteins[32,33] and interact respectively with the unrelated 15,000 M_r UPII[34] and 47,000 M_r UPIII[35] proteins. A range of anti-uroplakin antibody reagents have been developed as markers of urothelial cytodifferentiation, including anti-AUM heteroantisera which recognize multiple uroplakin species.[36] Other urothelium-specific antigens have been identified which are associated with the interplaque regions of umbrella cells. The 53,000 M_r "Urothelial Membrane Antigen" is expressed in a differentiation-associated manner[37] and localizes to the interplaque regions of umbrella cells.[38] The UGP85 glycoprotein antigen also shows specific localization to the interplaque regions.[39]

In addition to urothelium-specific antigens, several cytokeratin (CK) polypeptides also show differentiation related expression. In accordance with its classification as a "complex" epithelium,[40,41] urothelium expresses CKs characteristic of "simple" epithelia (CK7, CK8, CK18 and CK20), as well as CKs associated with stratified epithelia (CK13 and CK17, with small amounts of CK5 and CK4). The consensus is that in normal human urothelium CK7, CK8, CK18 and CK19 are expressed throughout all urothelial cell layers, CK17 is basally expressed, CK13 is present in all but the superficial cell layer whereas CK20 is associated with umbrella cells.[28,42–47] It is worth noting that some antibodies show differential labeling patterns in urothelium during urothelial cytodifferentiation. This phenomenon is commonly observed with antibodies recognizing CK8 or CK18; some antibodies react with all urothelial cell layers, whereas others react exclusively with superficial cells.[42,48–50] This phenomenon may reflect the CK constitution of tonofilaments in different cells resulting in altered polypeptide conformations within particular heterotypic complexes (epitope masking/unmasking) or structural processing or other post-translational modifications of the polypeptides (eg phosphorylation). A CK19 epitope has also been described which is uniquely unmasked during terminal urothelial cytodifferentiation.[51,52]

4. IN VITRO UROTHELIAL CELL SYSTEMS

One approach to understanding regulatory mechanisms in human urothelium has been to establish in vitro systems.[53–59] Such systems permit study of genetic and epigenetic factors influencing urothelial cells during normal homeostasis, pathogenesis, neoplasia and malignant progression. The basis for our in vitro work has been the finding that the urothelium can be detached as a complete sheet from the stroma, leaving the stroma with an intact basement membrane.[58,59] From this, we have developed several models.

4.1. Monocultures

Normal human urothelial (NHU) cells can be very reliably maintained in a serum-free medium designed for keratinocyte cell growth without stromal "feeder" cells, thereby eliminating a source of potentially confounding factors.[58,59] In the standard (low Ca^{++}) medium, NHU cells grow as non-stratified cell monolayers and adopt a high proliferation rate, with a rapid (15 hour) doubling time, which is sustainable through up to 12 serial passages. The cells show a basal/intermediate urothelial cell phenotype and do not express markers associated with late urothelial cytodifferentiation such as uroplakins, UMA or CK20.[58,59] Stratification can be induced by increasing exogenous [Ca^{++}] and results in the formation of desmosomes, relocalization of α6β4 integrin and increased cell surface expression of E-cadherin, α3β1 and α5β1 integrins.[59,60] However, Ca^{++} alone is insufficient to induce late or terminal cytodifferentiation, as judged by ultrastructure and antigen expression.[58,59] This observation suggests that in urothelium, the processes which regulate tissue histioarchitecture, such as homotypic cell adhesion and stratification, are independent of the induction of differentiation. This is an important distinction from the keratinocyte system, where differentiation occurs spontaneously in culture[61] and where commitment to terminal differentiation is associated with inhibition of integrin function and expression ensuring selective migration from the basal cell layer.[62]

4.2. Intact Organ Cultures

As a model for normal tissue homeostasis and epithelial:stromal interactions, an organ culture system has been developed in which normal human urothelial tissues can be maintained at an air-liquid interface in organ culture at for at least 18 weeks.[63] In organ culture, we have found that the urothelium retains a transitional epithelial morphology with normal antigenic characteristics, including expression of CD44, integrins, cytokeratins and anti-AUM reactive antigens. The only notable difference was a reduction in the number of CK20-positive superficial cells, suggesting some loss of tissue maturation with time.[63]

4.3. Organoids

Reconstituted tissues or organoids have been formed by seeding NHU cells from monocultures onto de-epithelialized bladder stroma.[63] Within 24 hours of reconstitution, macromolecules involved in epithelial cell:cell and cell:matrix interactions (i.e., E-cadherin, CD44 and α1-integrins) were relocated to intercellular junctions, indicating rapid formation of an intact neo-epithelium which preceded the acquisition of a more differentiated phenotype. After 2 weeks, the NHU cells had formed a multilayered transitional neo-epithelium, with a histioarchitecture essentially indistinguishable from urothelium in situ. The high rate of urothelial cell proliferation seen in vitro was suppressed, suggesting reinstatement of normal growth regulatory controls. CK20 was not expressed. However, analysis of other CKs showed a normal differentiation-associated pattern of antigen expression and there was some superficial reactivity with an anti-AUM antibody.[63]

4.4. Model of TCC Progression

The reconstituted organ culture model has been used as the basis for studying tumor:stromal interactions in TCC.[64] De-epithelialized stromas were seeded with the established TCC-derived cell lines, RT4, RT112 and EJ, which differ in phenotype from well- and moderately- to poorly-differentiated, respectively, reflecting the grades of tumor from which the cell lines were derived.[26,27] On a normal urothelial stroma, the degree of differentiation shown by the TCC cell lines was related to the grade of the original tumor and the proliferation rate was reduced relative to the same cells grown in tissue culture. RT4 cells formed a stratified epithelium with low proliferation, some characteristics of urothelial differentiation and no evidence of stromal invasion over an 8 week culture period. RT112 cells formed a stratified epithelium with limited differentiation and by 2 weeks showed intravasation of the sub-epithelial capillary bed. Anaplastic EJ cells did not stratify, but diffusely invaded the stromal matrix and the superficial vasculature. With time, the EJ cells also extravasated from vessels in the deeper layers of the stroma. The patterns of behavior shown by the three cell lines correlate with different patterns of TCC behavior seen in vivo, which suggests that this system might be invaluable for studying mechanisms and interventions in the processes of tumor progression.

5. THE UROTHELIAL CELL PHENOTYPE AND RESPONSE REPERTOIRE

Recent developments in techniques to support normal human epithelial cells in vitro has meant that a number of groups have been able to sustain normal human urothelial cells through multiple passages in monoculture[59,65–67] and even in the absence of exogenous growth factors.[54,68] The lack of requirement for feeder cells in these systems implies that the stroma is not essential for epithelial cell survival per se, but may be required to integrate different aspects of epithelial cell behavior, such as proliferation and cytodifferentiation, within a structured, functional and renewable framework. It is likely that stromal regulation of the urothelium occurs at the level of signaling via cell surface receptors (for soluble factors) and adhesion molecules (for cell surface and extracellular matrix proteins). Clearly, urothelial cells can only respond to exogenous signals if they express cognate receptors and the ultimate cellular response will be dependent upon downstream signaling response elements and the interaction of different signaling pathways.

The monoculture system is an important model as it can be used to a) identify the factors to which urothelial cells are responsive and b) characterize the full genetically-determined response repertoire of the normal urothelial cell. The use of purified growth factors and other defined soluble mediators to modulate cell behavior in vitro is a synthetic approach which aims to dissociate epithelial cell control into its constituent parts by defining the signals responsible for eliciting particular cellular responses. Using a defined system, it is also possible to study the interaction of mediators eliciting agonistic and antagonistic effects on cell behavior. "Mapping" of the phenotypic responses of a cell in terms of the molecular mechanisms of known signal transduction pathways should give information as to the cellular response repertoire, mechanisms of signal-response and the interaction and control of different response pathways. The eventual aim of such an approach would be to correlate individual effector and response pathways as a basis for the dissection of multifactorial signaling as occurs in tissue interactive systems.

It is important that the culture system be defined, as the culture conditions themselves may modify the cellular and response phenotype. Nevertheless, such adaptive changes are themselves clearly within the response repertoire and hence can help interpret aspects of biological behavior. An interesting addendum is that very few superficial TCCs yield urothelial cells which can be maintained in monoculture,[69] suggesting a reduced adaptability and response repertoire compared to specimens from normal individuals, which routinely yield viable cultures.[59]

6. CYTODIFFERENTIATION

The pertinence of the NHU monoculture system for investigating normal human urothelial cell behavior is attested by the ability of NHU cells to reconstitute a stratified transitional epithelium when transplanted onto a stroma.[63] However, there are limitations to the model in that some loss of tissue maturation was seen with time in intact organ cultures and the expression of terminally-differentiated urothelial markers was limited in tissues reconstituted with cultured NHU cells.[63] This suggests that other factors, such as urine-derived soluble factors and/or mechanical forces reflecting normal bladder contraction and expansion[70] may also be required to induce and maintain full tissue maturation.

An intriguing question is whether the bladder stroma provides any instructive signaling for urothelial cytodifferentiation. This question is important in view of the propensity shown by urothelium for metaplastic change in situ and the trans-differentiation of adult urothelium to prostatic epithelium in response to induction by urogenital sinus embryonic mesenchyme.[19,20] In rodent[71] and porcine[69,72] systems, morphological urothelial differentiation has been achieved after seeding cells onto fibroblast-containing collagen matrices, implying permissive stromal signaling. Nevertheless, there may be inter-species differences in the regulation of differentiation as, unlike NHU cells, bovine[69,73] and rodent[37] urothelial cells have been shown to express urothelial differentiation antigens in monoculture. The permissive versus instructive nature of stromal signaling in human urothelial cytodifferentiation remains to be elucidated, perhaps in a heterotypic recombinant organ culture system.

The relevance of in vitro models for studying the physiology and phenotypic response of normal urothelium is illustrated by the propensity for urothelium to undergo squamous redifferentiation in inflammatory, regenerative and neoplastic states, as this propensity is also seen in NHU cells in monoculture[59,67] and can be prevented or reversed by the use of exogenous retinoids.[59] In rodents, urothelium undergoes squamous metaplasia to a hyperproliferative cornifying epithelium in response to a dietary deficiency of retinoids.[14–18] Although NHU cells showed a transitional rather than squamous phenotype when recombined with a de-epithelialized stroma, TCC cell lines derived from tumors which contained some squamous elements developed areas of squamous differentiation in the surface epithelium of composite tissues.[64] This suggests that whereas squamous differentiation may occur as a physiological response in normal urothelium, the switch may be more permanent in neoplastic urothelium. In TCC, squamous differentiation has been associated with poorer prognosis[74] and may signify a lack of response to radiotherapy.[75]

7. PROLIFERATION

The urothelium has one of the lowest turnover rates among mammalian epithelia, estimated at once every 365 days.[3,6] In normal urothelium, mitotic figures are rare and only 0.1–2% of urothelial cells are found in the cell cycle.[7,8] This contrasts with the rapid regeneration rate during development and repair.[3,10] This high proliferative capacity is reiterated in NHU cell monocultures from pediatric and adult sources[59] and in the intact and NHU-reconstituted organ cultures, where a period of regeneration characterized by high proliferative activity preceded reversion to a quiescent homeostatic state.[63] The growth kinetics of NHU cells demonstrate that proliferation is inhibited when cells reach confluence but recurs when cells are reseeded at lower density.[59] Thus, cell proliferation in isolated and organotypic cultures may be regulated through cell:cell contact, potentially as in other cell systems, acting through the p21 or p27 cyclin-dependent kinase inhibitors.[76,77] If this is the case then the function of the stroma in urothelial proliferation/turnover might be to provide the correct physical framework or histioarchitecture, rather than having any active regulatory role.

Established TCC-derived cell lines have been used extensively as models of epithelial and carcinoma behavior. However, it is only with the more recent availability of normal urothelial cell systems that it has been possible to directly compare the normal and malignant cell counterparts and hence define normal from abnormal behavior in vitro.[78,79]

We have found that, like NHU cells, TCC cells show the same growth kinetics of lag, exponential and plateau phase growth in tissue culture and show a reduced number of cells in the mitotic cycle in organotypic cultures.[64] This suggests that the rapid growth rate of established TCC-derived cell lines in vitro does not necessarily reflect a neoplastically-derived attribute per se and further implies that even aggressive tumor-derived cells remain, at least partially, responsive to growth regulatory controls.

The implication from the organ culture model is that proliferation in urothelial tissues is suppressed during homeostasis, but activated as required for tissue repair. The importance of mitogenic paracrine and autocrine factors in epithelial growth regulation is being recognized increasingly. Several studies have demonstrated the secretion of autocrine factors by TCC cell lines[80–82] and it has been suggested that members of the EGF family may act as autocrine factors for normal urothelium.[68,83,84] As yet no clear picture has emerged of mitogenic regulation in urothelial cells and the specific role(s) of the cell cycle regulatory p15/p16 and p19^{ARF} gene products needs to be considered in view of the association between deletion of these genes and urothelial tumorigenesis.[25,85]

8. REGENERATION AND WOUND HEALING

Characterization of NHU cells has revealed expression of some antigens which are not usually associated with urothelium in situ. This includes expression of the $\alpha5$ integrin subunit.[60] In the epidermis, $\alpha5\beta1$ is expressed as part of the normal wound healing process and keratinocytes cultured in similar low calcium conditions to NHU cells express the $\alpha5\beta1$ integrin as part of a "wound response phenotype."[86] Little is known about urothelial tissue repair processes. However, it seems likely that some of the phenotypic changes associated with NHU cells in vitro might reflect a re-epithelialization response. In TCC, expression of $\alpha5$ integrin subunit has been associated with aggressive behavior.[87] Thus, tumor cells may appropriate the orchestrated inducible wound response programs of normal cells during tumor progression, rather than progressing on the basis of uncoordinated ectopic gene expression.

In concert with expression of the $\alpha5\beta1$ integrin, NHU cells also secrete fibronectin.[59] During wound repair in the epidermis, keratinocyte migration is thought to be mediated through $\alpha5\beta1$ binding to fibronectin.[88] The expression of $\alpha5\beta1$ and fibronectin by NHU cells in monoculture is suggestive of a migratory cell phenotype and is supported by observations in vitro and in intact organ cultures, where cut edges of the stroma rapidly become re-epithelialized.[63] The urothelium appears active in the process of reconstituting an intact basement membrane, as laminin is expressed in the cytoplasm of basal urothelial cells juxtaposed to developing basement membranes.[63] Thus, in keeping with their primary function as a covering/lining tissue, urothelial cells have a propensity to cover surfaces, constructing a basement membrane at the point of contact. Although urothelial monocultures express laminin, they do not appear to express collagen IV [59] and it appears likely that the urothelial cells are induced to express collagen IV through direct stromal contact, as NHU cells plated onto cell-free extracellular matrix gels will construct a laminin- and collagen IV-rich basement membrane.[69] Interestingly, we have found that, following isolation of urothelial cell sheets from the basement membrane, initial plating of primary NHU cultures is enhanced if the cells are pretreated with collagenase.[59] One possibility is that proteolytic cleavage of collagen from the hemidesmosome complex may serve to initiate the wound response phenotype, mimicking the activation of proteolytic cascades in the urothelium and stroma as a result of tissue injury or infection.

9. CONCLUSIONS

In summary, it appears that the stroma provides a structural framework for urothelial cells to cover and that the urothelium self-regulates cellular renewal and maintenance of the basement membrane during homeostasis. Following injury, there is a rapid wound healing response which requires the interaction of stroma and urothelium, with activation of protease cascades being a critical step in initiating a urothelial response. The urothelial cells adopt a "wound response phenotype" characterized by rapid migration and proliferation. This appears to be equivalent to the phenotype of NHU cells in monoculture. It is worth noting that, in many respects, the behavior of TCC cells suggests that they may appropriate the wound healing response during tumor progression.

In vitro models can provide important insights into the mechanisms regulating normal urothelium and pathogenesis. The potential for in vitro manipulation and phenotypic modulation makes them ideal subjects for identifying differential gene expression using techniques such as differential display RT-PCR, an approach being pursued by ourselves and others.[57,69] Further characterization of the physiological, phenotypic and gene expression responses of NHU cells in monoculture will help identify the mechanistic pathways of tumor progression and hence identify candidate targets for future intervention therapies. In vitro approaches can also lead to the development of tools for diagnostic and prognostic uses[47,89,90] and several groups are attempting to exploit the high regenerative capacity of urothelium in vitro for bioengineering/tissue reconstructive purposes.[58,63,68] Another potential development is the introduction of genes for therapeutic or pathogenetic use. The genetic manipulation of normal urothelial cells in vitro and reimplantation in vivo has been proposed as a gene delivery system[91] and the identification of elements responsible for the control of transcription of urothelium-specific genes has implications for controlling the expression of transfected genes in a tissue and/or differentiation-restricted manner.[69,34] Thus, the recent advances in urothelial in vitro systems should permit many outstanding questions to be resolved and presents the potential for powerful applications in the future.

ACKNOWLEDGMENTS

Work in our laboratory is or has been supported by the Imperial Cancer Research Fund; Action Research; the Association for International Cancer Research; the Ministry of Agriculture, Fisheries and Food; the Medical Research Council; the Biotechnology and Biological Sciences Research Council and Ethicon Ltd.

REFERENCES

1. Sheppard D.: Epithelial integrins. Bioessays., **18:** 655, 1996.
2. Tozer E.C., Hughes P.E., Loftus J.C.: Ligand binding and affinity modulation of integrins. Biochem Cell Biol., **74:** 785, 1996.
3. Hicks R.M. The mammalian urinary bladder: an accommodating organ.: Biol. Rev. Cambridge Phil. Soc., **50:** 215, 1975.
4. Jost S.P., Gosling J.A., Dixon J.S.: The morphology of normal human bladder urothelium. J. Anat., **167**: 103, 1989.
5. Hicks R.M.: The function of the Golgi complex in transitional epithelium. Synthesis of the thick cell membrane. J. Cell Biol., **30:** 623, 1966.

6. Marceau N.: Cell lineages and differentiation programs in epidermal, urothelial and hepatic tissues and their neoplasms. Lab. Invest., **63:** 4, 1990.
7. Limas C., Bigler A., Bair R., Bernhart P., Reddy P.: Proliferative activity of urothelial neoplasms: comparison of BrdU incorporation, Ki67 expression, and nucleolar organiser regions. J. Clin. Pathol., **46:** 159, 1993.
8. Limas C., Bair R., Bernhart P., Reddy P.: Proliferative activity of normal and neoplastic urothelium and its relation to epidermal growth factor and transferrin receptors. J. Clin. Pathol., **46:** 810, 1993.
9. Jost S.P.: Cell cycle of normal bladder urothelium in developing and adult mice. Virchows Archiv., **57:** 27, 1993.
10. Stewart F.A.: Mechanism of bladder damage and repair after treatment with radiation and cytostatic drugs. Br. J. Cancer., **7:** 280, 1986.
11. Limas C.: Proliferative state of the urothelium with benign and atypical changes. Correlation with transferrin and epidermal growth factor receptors and blood group antigens. J. Pathol., **171:** 39, 1993.
12. Hodges G.M.: The urinary system - bladder. In: "Biomedical Research Applications of Scanning Electron Microscopy.", Hodges G.M. & Hallowes R.C. (eds), Academic Press, London:: pp. 307, 1979.
13. Eble J.N., Young R.H.: Carcinoma of the urinary bladder: a review of its diverse morphology. Sem Diagnostic Pathol., **14:** 98, 1997.
14. Wolbach S.B., Howe P.R.: Tissue changes following deprivation of fat soluble A vitamin. J. Exp. Med., **42:** 753, 1925.
15. Capurro P., Angrist A., Black J., Moumgis B.: Studies in squamous metaplasia in rat bladder. I. Effects of hypovitaminosis A, foreign bodies and methylcholanthrene. Cancer Res., **20:** 563, 1960.
16. Hicks R.M.: Hyperplasia and cornification of the transitional epithelium in vitamin A-deficient rat. J. Ultrastruct. Res., **22:** 206, 1968.
17. Molloy C.J., Laskin J.D.: Effect of retinoid deficiency on keratin expression in mouse bladder. Exp. Mol. Pathol., **49:** 128, 1988.
18. Gijbels M.J., van der Ham F., van Bennekum A.M., Hendriks H.F., Roholl P.J.: Alterations in cytokeratin expression precede histological changes in epithelia of vitamin A-deficient rats. Cell Tissue Res., **268:** 197, 1992.
19. Cunha G.R., Lung B., Reese B.: Glandular epithelial induction by embryonic mesenchyme in adult bladder epithelium of BALB/c mice. Invest. Urol., **17:** 302, 1980.
20. Neubauer B.L., Chung L.W., McCormick K.A., Taguchi O., Thompson T.C., Cunha G.R.: Epithelial-mesenchymal interactions in prostatic development. II. Biochemical observations of prostatic induction by urogenital sinus mesenchyme in epithelium of the adult rodent urinary bladder. J. Cell Biol., **96:** 1671, 1983.
21. Young R.H., Eble J.N.: Unusual forms of carcinoma of the urinary bladder. Human Patholl., **22:** 948, 1991.
22. Harnden P., Parkinson M.C.: Transitional cell carcinoma of the bladder: diagnosis and prognosis. Curr. Diagnostic Pathol., **3:** 109, 1996.
23. Spruck C.H., 3rd, Ohneseit P.F., Gonzalez-Zulueta M., Esrig D., Miyao N., Tsai Y.C., Lerner, P., Schmutte C., Yang A.S., Cote R., et al.: Two molecular pathways to transitional cell carcinoma of the bladder. Cancer Res., **54:** 784, 1994.
24. Knowles M.A. Molecular genetics of bladder cancer.: Br. J. Urol., **75:** Suppl 1:57, 1995.
25. Reznikoff C.A., Belair C.D., Yeager T.R., Savelieva E., Blelloch R.H., Puthenveettil J.A., Cuthill S.: A molecular genetic model of human bladder cancer pathogenesis. Sem. Oncol., **23:** 571, 1996.
26. Trejdosiewicz L.K., Southgate J., Donald J.A., Masters J.R., Hepburn P.J., Hodges G.M.: Monoclonal antibodies to human urothelial cell lines and hybrids: production and characterization. J. Urol., **133:** 533, 1985.
27. Masters J.R.W., Hepburn P.J., Walker L., Highman W.J., Trejdosiewicz L.K., Povey S., Parkar M., Hill B.T., Riddle P.R., Franks L.M.: Tissue culture model of transitional cell carcinoma: characterization of twenty-two human urothelial cell lines. Cancer Res., **46:** 3630, 1986.
28. Moll R., Achtstatter T., Becht E., Balcarova-Stander J., Ittensohn M., Franke W.W.: Cytokeratins in normal and malignant transitional epithelium. Maintenance of expression of urothelial differentiation features in transitional cell carcinomas and bladder carcinoma cell culture lines. Am. J. Pathol., **132:** 123, 1988.
29. Rieger K.M., Little A.R., Swart J.M., Kastrinakis W.V., Fitzgerald J.M., Hess D.T., Libertino J.A., Summerhayes I.C.: Human bladder carcinoma cell lines as indicators of oncogenic change relevant to urothelial neoplastic progression. Br. J. Cancer., **72:** 683, 1995.
30. Wu X.R., Lin J.H., Walz T., Haner M., Yu J., Aebi U., Sun T.T.: Mammalian uroplakins. A group of highly conserved urothelial differentiation-related membrane proteins. J. Biol. Chem., **269:** 13716, 1994.
31. Sun T.T., Zhao H.P., Provet J., Aebi U., Wu X.R.: Formation of asymmetric unit membrane during urothelial differentiation. Mol. Biol. Rep., **23:** 3, 1996.

32. Yu J., Manabe M., Wu X.R., Xu C., Surya B., Sun T.T.: Uroplakin I: a 27-kD protein associated with the asymmetric unit membrane of mammalian urothelium. J. Cell Biol., **111:** 1207, 1990.
33. Yu J., Lin J.H., Wu X.R., Sun T.T.: Uroplakins Ia and Ib, two major differentiation products of bladder epithelium, belong to a family of four transmembrane domain (4TM) proteins. J. Cell Biol., **125:** 171, 1994.
34. Lin J.H., Wu X.R., Kreibich G., Sun T.T.: Precursor sequence, processing, and urothelium-specific expression of a major 15-kDa protein subunit of asymmetric unit membrane. J. Biol. Chem., **269:** 1775, 1994.
35. Wu X.R., Sun T.T.: Molecular cloning of a 47 kDa tissue-specific and differentiation-dependent urothelial cell surface glycoprotein. J. Cell Sci., **106:** 31, 1993.
36. Wu X.R., Manabe M., Yu J., Sun T.T.: Large scale purification and immunolocalization of bovine uroplakins I, II, and III. Molecular markers of urothelial differentiation. J. Biol. Chem., **265:** 19170, 1990.
37. Trejdosiewicz L.K., Wilson P.D., Hodges G.M.: Species Cross-reactive membrane-associated urothelial differentiation antigen. J. Natl. Cancer Inst., **72:** 355, 1984.
38. Hodges G.M., Smolira M.A., Trejdosiewicz L.K.: Urothelium-specific antibody and lectin surface mapping of bladder urothelium. Histochem. J., **14:** 755, 1982.
39. Yu J., Manabe M., Sun T.T.: Identification of an 85–100 kDa glycoprotein as a cell surface marker for an advanced stage of urothelial differentiation: association with the inter-plaque ('hinge') area. Epithelial Cell Biol., **1:** 4, 1992.
40. Sun T.T., Eichner R., Schermer A., Cooper D., Nelson W.G., Weiss R.A.: Classification, expression, and possible mechanisms of evolution of mammalian epithelial keratins: a unifying model. In: "Cancer Cells, the Transformed Phenotype.", Levine A.J., van den Woude G.F., Topp W.C. & Watson J.D. (eds), Cold Spring Harbor Laboratory, New York:: pp. 169, 1984.
41. Cooper D., Schermer A., Sun T.T.: Classification of human epithelia and their neoplasms using monoclonal antibodies to keratins: strategies, applications, and limitations. Lab. Invest., **52:** 243, 1985.
42. Achtstatter T., Moll R., Moore B., Franke W.W.: Cytokeratin polypeptide patterns of different epithelia of the human male urogenital tract: immunofluorescence and gel electrophoretic studies. J. Histochem. Cytochem., **33:** 415, 1985.
43. van Muijen G.N., Ruiter D.J., Franke W.W., Achtstatter T., Haasnoot W.H., Ponec M., Warnaar S.O.: Cell type heterogeneity of cytokeratin expression in complex epithelia and carcinomas as demonstrated by monoclonal antibodies specific for cytokeratins nos. 4 and 13. Exp. Cell Res., **162:** 97, 1986.
44. Ramaekers F., Huysmans A., Schaart G., Moesker O., Vooijs P.: Tissue distribution of keratin 7 as monitored by a monoclonal antibody. Exp. Cell Res., **170:** 235, 1987.
45. Schaafsma H.E., Ramaekers F.C., van Muijen G.N., Ooms E.C., Ruiter D.J.: Distribution of cytokeratin polypeptides in epithelia of the adult human urinary tract. Histochem., **91:** 151, 1989.
46. Troyanovsky S.M., Guelstein V.I., Tchipysheva T.A., Krutovskikh V.A., Bannikov G.A.: Patterns of expression of keratin 17 in human epithelia: dependency on cell position. J. Cell Sci., **93:** 419, 1989.
47. Harnden P., Allam A., Joyce A.D., Patel A., Selby P., Southgate J.: Cytokeratin 20 expression by non-invasive transitional cell carcinomas: potential for distinguishing recurrent from non-recurrent disease. Histopathol., **27:** 169, 1995.
48. Cintorino M., Del Vecchio M.T., Bugnoli M., Petracca R., Leoncini P.: Cytokeratin pattern in normal and pathological bladder urothelium: immunohistochemical investigation using monoclonal antibodies. J. Urol., **139:** 428, 1988.
49. Ramaekers F.C., Beck J.L., Feitz W.F., Debruyne F.M., Vooijs G.P.: Keratin expression during neoplastic progression of bladder cancer. Eur. Urol., **14:** Suppl 1:7, 1988.
50. Schaafsma H.E., Ramaekers F.C., van Muijen G.N., Lane E.B., Leigh I.M., Robben H., Huijsmans A., Ooms E.C., Ruiter D.J.: Distribution of cytokeratin polypeptides in human transitional cell carcinomas, with special emphasis on changing expression patterns during tumor progression. Am. J. Pathol., **136:** 329, 1990.
51. von Koskull H., Aula P., Trejdosiewicz L.K., Virtanen I.: Identification of cells from fetal bladder epithelium in human amniotic fluid. Human Gen., **65:** 262, 1984.
52. Trejdosiewicz L.K., Southgate J., Hodges G.M.: Unmasking of a novel lineage-specific keratin epitope during urothelial cytodifferentiation. Struct. Funct. Cytoskeleton., **171:** 265, 1988.
53. Knowles M.A., Finesilver A., Harvey A.E., Berry R.J., Hicks R.M.: Long-term organ culture of normal human bladder. Cancer Res., **43:** 374, 1983.
54. Reznikoff C.A., Loretz L.J., Pesciotta D.M., Oberley T.D., Ignjatovic M.M.: Growth kinetics and differentiation in vitro of normal human uroepithelial cells on collagen gel substrates in defined medium. J. Cell Physiol., **131:** 285, 1987.
55. Mothersill C., Seymour C.B., O'Brien A., Hennessy T.: Proliferation of normal and malignant human epithelial cells post irradiation. Acta Oncol., **30:** 851, 1991.

56. Wu S.Q., Storer B.E., Bookland E.A., Klingelhutz A.J., Gilchrist K.W., Meisner L.F., Oyasu R., Reznikoff C.A.: Nonrandom chromosome losses in stepwise neoplastic transformation in vitro of human uroepithelial cells. Cancer Res 1., **51:** 3323, 199.
57. Liebert M., Hubbel A., Chung M., Wedemeyer G., Lomax M.I., Hegeman A., Yuan T.Y., Brozovich M., Wheelock M.J., Grossman H.B.: Expression of mal is associated with urothelial differentiation in vitro: Identification by differential display reverse-transcriptase polymerase chain reaction. Differentiation., **61:** 177, 1997.
58. Hutton K.A., Trejdosiewicz L.K., Thomas D.F., Southgate J.: Urothelial tissue culture for bladder reconstruction: an experimental study. J. Urol., **150:** 721, 1993.
59. Southgate J., Hutton K.A., Thomas D.F., Trejdosiewicz L.K.: Normal human urothelial cells in vitro: proliferation and induction of stratification. Lab. Invest., **71:** 583, 1994.
60. Southgate J., Kennedy W., Hutton K.A., Trejdosiewicz L.K.: Expression and in vitro regulation of integrins by normal human urothelial cells. Cell Adhes. Commun., **3:** 231, 1995.
61. Boyce S.T., Ham R.G.: Calcium-regulated differentiation of normal human epidermal keratinocytes in chemically defined clonal culture and serum-free serial culture. J. Invest. Dermatol., **81:** 33s, 1983.
62. Adams J.C., Watt F.M.: Changes in keratinocyte adhesion during terminal differentiation: reduction in fibronectin binding precedes alpha 5 beta 1 integrin loss from the cell surface. Cell, **63:** 425, 1990.
63. Scriven S.D., Booth C., Thomas D.F., Trejdosiewicz L.K., Southgate J.: Reconstitution of human urothelium from monolayer cultures. J. Urol., **158:** 1147, 1997.
64. Booth C., Harnden P., Trejdosiewicz L.K., Scriven S., Selby P.J., Southgate J.: Stromal and vascular invasion in an human in vitro bladder cancer model. Lab. Invest., **76:** 843, 1997.
65. Reznikoff C.A., Johnson M.D., Norback D.H., Bryan G.T. Growth and characterization of normal human urothelium in vitro. In Vitro., **19:** 326, 1983.
66. Liebert M., Wedemeyer G., Chang J.H., Stein J.A., McKeever P.E., Carey T.E., Flint A., Steplewski Z., Buchsbaum D.J., Wahl R.L., et al.: Comparison of antigen expression on normal urothelial cells in tissue section and tissue culture. J. Urol., **144:** 1288, 1990.
67. Petzoldt J.L., Leigh I.M., Duffy P.G., Masters J.R.: Culture and characterisation of human urothelium in vivo and in vitro. Urol. Res., **22:** 67, 1994.
68. Cilento B.G., Freeman M.R., Schneck F.X., Retik A.B., Atala A.: Phenotypic and cytogenetic characterization of human bladder urothelia expanded in vitro. J. Urol., **152:** 665, 1994.
69. Manuscript in preparation.
70. Baskin L., Howard P.S., Macarak E.: Effect of physical forces on bladder smooth muscle and urothelium. J. Urol., **150**: 601, 1993.
71. Howlett A.R., Hodges G.M., Rowlatt C.: Epithelial-stromal interactions in the adult bladder: urothelial growth, differentiation, and maturation on culture facsimiles of bladder stroma. Develop. Biol., **118:** 403, 1986.
72. Fujiyama C., Masaki Z., Sugihara H.: Reconstruction of the urinary bladder mucosa in three-dimensional collagen gel culture: Fibroblast-extracellular matrix interactions on the differentiation of transitional epithelial cells. J. Urol., **153:** 2060, 1995.
73. Surya B., Yu J., Manabe M., Sun T.T.: Assessing the differentiation state of cultured bovine urothelial cells: elevated synthesis of stratification-related K5 and K6 keratins and persistent expression of uroplakin I. J. Cell Sci., **97:** 419, 1990.
74. Tannenbaum S.I., Carson C.C., Tatum A., Paulson D.F.: Squamous carcinoma of urinary bladder. Urol., **22:** 597, 1983.
75. Jenkins B.J., Martin J.E., Baithun S.I., Zuk R.J., Oliver R.T., Blandy J.P.: Prediction of response to radiotherapy in invasive bladder cancer. Br. J. Urol., **65:** 345, 1990.
76. Noda A., Ning Y., Venable S.F., Pereira-Smith O.M., Smith J.R.: Cloning of senescent cell-derived inhibitors of DNA synthesis using an expression screen. Exp. Cell Res., **211:** 90, 1994.
77. Polyak K., Kato J.Y., Solomon M.J., Sherr C.J., Massague J., Roberts J.M., Koff A.: p27Kip1, a cyclin-Cdk inhibitor, links transforming growth factor-beta and contact inhibition to cell cycle arrest. Gene Devel., **8:** 9, 1994.
78. Southgate J., Trejdosiewicz L.K., Smith B., Selby P.J.: Patterns of splice variant CD44 expression by normal human urothelium in situ and in vitro and by bladder-carcinoma cell lines. Int. J. Cancer., **62:** 449, 1995.
79. Southgate J., Pitt E., Trejdosiewicz L.K.: The effects of dietary fatty acids on the proliferation of normal human urothelial cells in vitro. Br J Cancer., **74:** 728, 1996.
80. De Boer W.I., Houtsmuller A.B., Izadifar V., Muscatelli-Groux B., Van der Kwast T.H., Chopin D.K.: Expression and functions of EGF, FGF and TGFbeta-growth-factor family members and their receptors in invasive human transitional-cell-carcinoma cells. Int. J. Cancer., **71:** 284, 1997.

81. Ruck A., Paulie S.: The epidermal growth factor receptor is involved in autocrine growth of human bladder carcinoma cell lines. Anticancer Res., **17:** 1925, 1997.
82. Tachibana M., Miyakawa A., Tazaki H., Nakamura K., Kubo A., Hata J.I., Nishi T., Amano Y.: Autocrine growth of transitional cell carcinoma of the bladder induced by granulocyte-colony stimulating factor. Cancer Res., **55:** 3438, 1995.
83. Baskin L.S., Sutherland R.S., Thomson A.A., Nguyen H.T., Morgan D.M., Hayward S.W., Hom Y.K., DiSandro M., Cunha G.R.: Growth factors in bladder wound healing. J. Urol., **157:** 2388, 1997.
84. Freeman M.R., Yoo J.J., Raab G., Soker S., Adam R.M., Schneck F.X., Renshaw A.A., Klagsbrun M., Atala A.: Heparin-binding EGF-like growth factor is an autocrine growth factor for human urothelial cells and is synthesized by epithelial and smooth muscle cells in the human bladder. J. Clin. Invest., **99:** 1028, 1997.
85. Southgate J, Proffitt J, Roberts P, Smith B, Selby P.: Loss of cyclin-dependent kinase inhibitor genes and chromosome 9 karyotypic abnormalities in human bladder cancer cell lines. Br. J. Cancer., **72:** 1214, 1995.
86. Clark R.A.F.: Wound repair: overview and general considerations. In: "The Molecular and Cellular Biology of Wound Repair.", Clark R.A.F. (ed), 2nd ed. Plenum Press, New York: pp. 3, 1996.
87. Saito T., Kimura M., Kawasaki T., Sato S., Tomita Y.: Correlation between integrin alpha 5 expression and the malignant phenotype of transitional cell carcinoma. Br. J. Cancer., **73:** 327, 1996.
88. Kim J.P., Zhang K., Chen J.D., Wynn K.C., Kramer R.H., Woodley D.T.: Mechanism of human keratinocyte migration on fibronectin: unique roles of RGD site and integrins. J. Cell Physiol., **151:** 443, 1992.
89. Harnden P., Eardley I., Joyce A.D., Southgate J.: Cytokeratin 20 as an objective marker of urothelial dysplasia. Br. J. Urol., **78:** 870, 1996.
90. Harnden P., Southgate J.: Cytokeratin 14 as a marker of squamous differentiation in transitional cell carcinomas. J. Clin. Pathol 1998 (in press)
91. Yoo J.J., Atala A.: A novel gene delivery system using urothelial tissue engineered neo-organs. J. Urol., **158:** 1066, 1997.

3

CREATION OF BLADDER TISSUE IN VITRO AND IN VIVO

A System for Organ Replacement

Anthony Atala

Director
Laboratory for Cellular Therapeutics and Tissue Engineering
Department of Urology, Children's Hospital, and
Harvard Medical School
300 Longwood Ave., Boston, Massachusetts 02115

1. INTRODUCTION

The bladder is exposed to a variety of possible injuries from the time the fetus develops. Aside from congenital abnormalities, individuals may also suffer from other disorders such as cancer, trauma, infection, inflammation, iatrogenic injuries or other conditions which may lead to bladder damage or loss, requiring eventual reconstruction.

Currently, gastrointestinal segments are used for bladder repair or replacement. The use of gastrointestinal segments for bladder reconstruction is associated with a variety of complications.[1] These include metabolic abnormalities, infection, perforation, urolithiasis, increased mucous production and malignancy. Alternative approaches are constantly being explored in order to overcome the problems associated with the incorporation of non-urologic segments into the bladder.

Natural tissues and synthetic materials which have been tried previously in experimental and clinical settings for bladder replacement include omentum, peritoneum, seromuscular grafts, de-epithelialized segments of bowel and polyvinyl sponge.[1] These attempts have usually failed. It is evident that urothelial to urothelial anastomoses are preferable functionally. However, the limited amount of autologous urothelial tissue for reconstruction generally precludes this option. There is a critical need for tissues to replace lost and functionally-deficient bladder tissues. Engineering bladder tissue using selective cell transplantation may provide a means to create functional new genitourinary tissues.[2]

Advances in Bladder Research, edited by Baskin and Hayward.
Kluwer Academic / Plenum Publishers, New York, 1999.

2. ENGINEERING BLADDER TISSUE: STRATEGY

In tissue engineering, donor tissue is dissociated into individual cells or small tissue fragments and either implanted directly into the autologous host, or attached to a support matrix, expanded in culture, and re-implanted after expansion. Ideally, this approach might allow lost tissue function to be restored or replaced in toto and with limited complications.

The success of using cell transplantation strategies for bladder reconstruction depends on the ability to use donor tissue efficiently and to provide the right conditions for long term survival, differentiation and growth. We have achieved an approach to bladder tissue regeneration by patching isolated cells to a support structure which would have suitable surface chemistry for guiding the reorganization and growth of the cells.[3,4] The supporting matrix is composed of either biodegradable artificial or natural polymers which can allow cell survival by diffusion of nutrients across short distances once the cell-support matrix is implanted. Ideally, the cell-support matrix would become vascularized in concert with expansion of the cell mass following implantation.

3. CELL EXPANSION AND CHARACTERIZATION

One of the initial limitations of applying tissue engineering techniques to the bladder, had been the previously encountered inherent difficulty of growing genitourinary associated cells in large quantities. Even as recently as 10 years ago, it was believed that urothelial cells had a natural senescence which was hard to overcome. Normal urothelial cells could be grown in the laboratory setting, but without any demonstrable expansion.

We have shown that normal human bladder epithelial cells can be efficiently harvested from surgical material, extensively expanded in culture in serum-free conditions, and their differentiation characteristics, growth requirements and other biological properties studied.[5–7] Using our methods of cell culture we estimate that it would be theoretically possible to expand a urothelial strain from a single specimen which initially covers a surface area of 1 cm^2 to one covering a surface area of 4202 m^2 (the equivalent area of one football field) within 8 weeks even if it assumed that 50% of the cells would be lost with each passage.[5] This indicates that it should be possible to collect autologous urothelial cells from human patients, expand them in culture, and return them to the human donor in sufficient quantities for reconstructive purposes. Human bladder smooth muscle can be isolated in a similar fashion and studied transiently under serum-free conditions. The ability to greatly expand primary populations of human urothelial-associated cells in serum-free conditions is important because it indicates that frequent or continuous replenishment of in vitro stocks of normal cells from fresh surgical material, a requirement that would significantly inhibit the use of these cells for tissue engineering and research purposes, is not needed.

Whether the cells retain a normal program of differentiation is an issue of great importance. In our studies, immunofluorescent cell staining with broadly-reacting anti-cytokeratin antibodies (AE1/AE3) and an antibody which reacts specifically to cytokeratin 7, present in bladder mucosa in vivo, confirmed the epithelial phenotype of the cells and the cells' bladder origin. The epithelial cells also stained positively with an antibody to vimentin, suggesting that the cells did not exhibit a barrier-forming epithelial phenotype.[5] To examine this possibility further, the urothelial cells were stained with an antibody to E-cadherin, a ubiquitous epithelial cell surface adhesion protein. The barrier-forming kidney

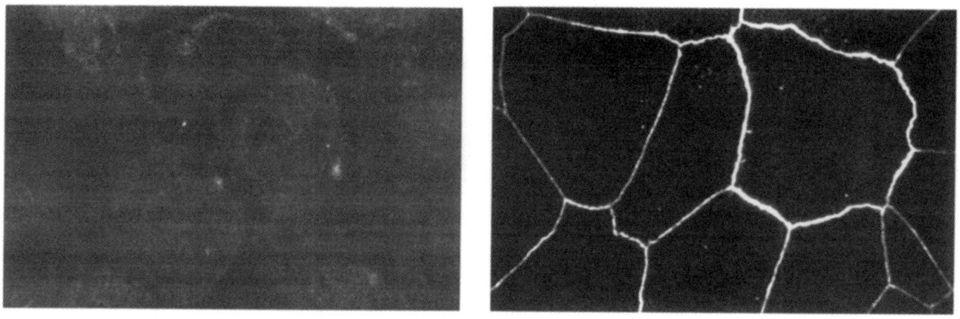

Figure 1. Confocal microscopy of normal human urothelial cells evaluated for ZO-1 expression. Urothelial cells expanded in vitro do not exhibit a barrier-forming epithelial phenotype (left). Epithelial differentiation of these cells can be induced by serum, as evidenced by the localization of the tight junction associated protein ZO-1 to the cell periphery (right).

cell line MDCK, which forms intercellular epithelial adhesion complexes, stained positively with this antibody, however the urothelial cells stained very weakly. This suggested that the cells lacked a ubiquitous protein found in adherens junctions in barrier-forming epithelia. Northern blot hybridization with an anti-E-cadherin cDNA probe confirmed that the urothelial cells expressed low levels of E-cadherin mRNA when compared to a barrier-forming prostate epithelial cell line. These results indicate that the urothelial cells grown on plastic do not form functional epithelial monolayers under these culture conditions. Urothelial cells grown on collagen type I matrix were positive for the expression E-cadherin as detected by indirect immunofluorescence. Cells grown on Matrigel surfaces formed complex cell aggregates in sheet- and tube-like structures. Confocal microscopy demonstrated that cells grown on Matrigel or with the addition of serum localized E-cadherin to interepithelial junctions. Therefore epithelial differentiation of these cells can be induced by some extracellular matrices/growth conditions.[6] Similar findings were evident with the localization of the tight junction associated protein zona occludens-1 (ZO1), to the cell periphery (Figure 1). Also of concern is whether the cells retain a normal cytogenetic component. Chromosomal analysis was performed on urothelial cells at the third and sixth passages. A normal chromosomal complement with no detectable abnormalities was observed by Giemsa staining of metaphase chromosomes.[5]

Most of the growth factor biology in human urothelial-associated cells is entirely unknown. This area of study is important in order to further understand the regulatory mechanisms involved with the growth and development of these cells and how it ultimately affects the entire genitourinary system. Studies performed at our laboratory have demonstrated that unlike keratinocytes and a variety of other epithelia, human bladder urothelial cells do not require EGF for growth.[5] Growth kinetics using multiple independently-derived urothelial strains were observed to be essentially identical under (+)EGF and (-)EGF conditions. These data highlight an important potential difference between the keratinocyte and bladder in vitro epithelial cell models. One possible explanation for these observations is that EGF may not play a role in normal bladder epithelial cell growth. However, epithelial cells in the germinal layer of the bladder epithelium, which are thought to repopulate the suprabasal epithelial layers after injury, express the epidermal growth factor receptor (HER1). Despite the apparent absence of a growth-promoting effect of EGF, these observations suggest a role for a regulatory pathway involving HER1 in urothelial cell growth and/or differentiation.[7] Recently it has become evident that autocrine or paracrine regulation through the HER1

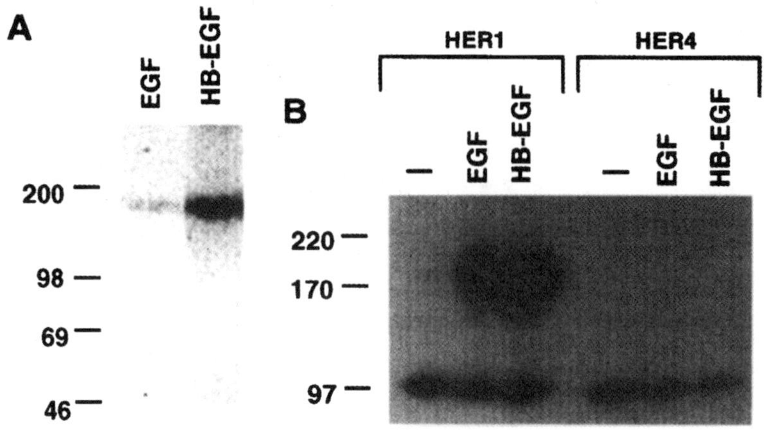

Figure 2. (A) Affinity cross-linking of ^{125}I-EGF and ^{125}I-HB-EGF to human urothelial cell monolayers. Both affinity probes recognize a ca. 170 kDa binding protein. In the experiment shown, the specific activity of ^{125}I-EGF was approximately 10-fold higher than that of ^{125}I-HB-EGF. (B) Detection of tyrosine phosphorylation of HER1 and HER4 in to human urothelial cells in response to no growth factor (-), 100 ng/ml EGF or 100 ng/ml HB-EGF. Immunoprecipitation was with anti-HER1 and anti-HER4 Abs, detected by anti-phosphotyrosine Western blot. MW markers in kDa.

does not require binding of the receptor to EGF. Multiple high-affinity ligands for HER1, which are capable of activating EGF receptor-mediated signal transduction pathways in a fashion similar to EGF have recently been identified.[8] The discovery of these distinct EGF-like molecules indicates that HER1 is a degenerate receptor capable of binding different activating ligands with high specificity in different contexts. HER1 has been implicated in regenerative growth and proliferative diseases of the human bladder epithelium, however a cognate HER1 ligand that can act as a growth factor for normal human urothelial cells has not been identified. We have shown that heparin-binding epidermal growth factor-like growth factor (HB-EGF), an activating HER1 ligand, is an autocrine regulator of human urothelial cell growth. This conclusion is based on demonstration of HB-EGF synthesis and secretion by primary culture human urothelial cells, identification of HER1 as an activatable HB-EGF receptor on human urothelial cell surfaces, stimulation of human urothelial cell clonal growth by HB-EGF, inhibition of HB-EGF-stimulated growth by heparin and of log-phase growth by CRM 197, a specific inhibitor of HB-EGF/HER1 interaction, and identification of human urothelium as a site of proHB-EGF synthesis in vivo. ProHB-EGF expression was also detected in the vascular and detrusor smooth muscle of the human bladder. These data suggest a physiologic role for HB-EGF in the regulation of urothelial proliferation and regeneration subsequent to mucosal injury. Expression of proHB-EGF is also a feature of differentiated vascular and detrusor smooth muscle in the bladder. Because proHB-EGF is known to be the high affinity diphtheria toxin (DT) receptor in human cells, synthesis of the HB-EGF precursor by human urothelium also suggests the possibility of using the DT-binding sites of proHB-EGF as an in vivo target for the intraluminal treatment of urothelial diseases (Figures 2 and 3).[7]

4. CELL DELIVERY VEHICLES

The support structure chosen for cell delivery is of utmost importance. It is known from previous studies that artificial permanent support structures are lithogenic (Teflon, sili-

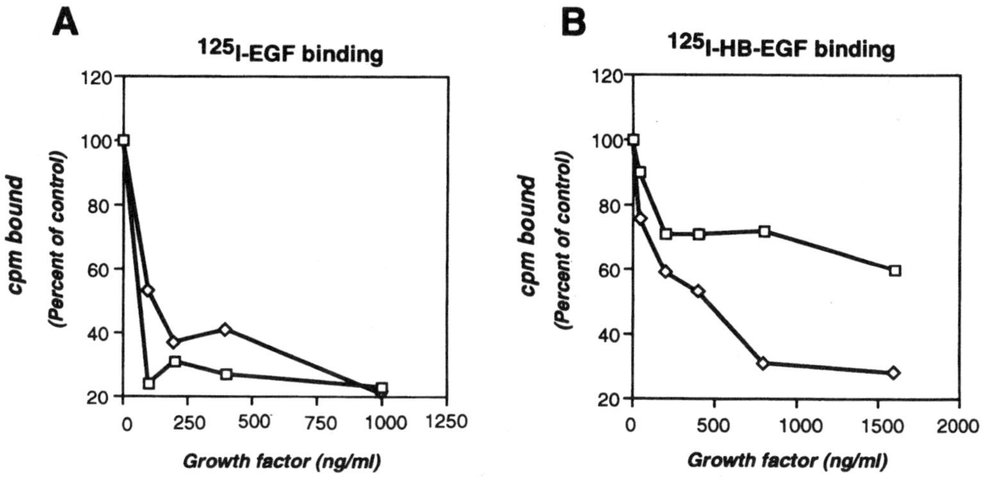

Figure 3. Binding of ^{125}I-EGF and ^{125}I-HB-EGF to human urothelial cell monolayers in the presence of excess unlabeled EGF and HB-EGF. (A) Displacement of ^{125}EGF was similar for EGF and HB-EGF, however (B) EGF only displaced approximately 35% of ^{125}HB-EGF binding.

cone).[1] Other investigators have tried permanent homograft or heterograft support structures such as dura, however these contract with time and are problematic in a clinical setting. Natural permanent support structures, such as denuded bowel retain their inherent properties and mucosal regrowth invariably occurs with time. A variety of synthetic polymers, both degradable and non-degradable, have been utilized to fabricate tissue engineering matrices.[9–12] An early attempt to engineer urologic structures involved implanting collagen sponge tubes as ureteral replacements. It was hoped that the implanted matrix would induce migration of epithelial cells from the adjacent tissue and the formation of an epithelial cell-lined tubular tissue. However, salt deposits onto the collagen matrix were noted following exposure to urine. Bladder submucosa was proposed as a matrix for tissue regeneration in 1961, and there has been a recent resurgence of interest in this material for bladder replacement.[9–11] Intestinal submucosa has also been proposed as a scaffold for the regeneration of urologic tissue.[12] All of the materials used until recently in the urinary tract, both synthetic and natural, were applied without the use of cells. A common finding with these materials was an adequate histological result, but with subsequent graft contraction and shrinkage.

Synthetic polymers can be manufactured reproducibly and can be designed to exhibit the necessary mechanical properties.[12] Among synthetic materials, resorbable polymers are preferable because permanent polymers carry the risk of infection, calcification, and unfavorable connective tissue response. Polymers of lactic and glycolic acid have been extensively utilized to fabricate tissue engineering matrices.[3,4,13] These polymers have many desirable features; they are biocompatible, processable, and biodegradable. Degradation occurs by hydrolysis and the time sequence can be varied from weeks to over a year by manipulating the ratio of monomers and by varying the processing conditions. These polymers can be readily formed into a variety of structures, including small diameter fibers and porous films.

The porosity, pore size distribution and continuity dictate the interaction of the biomaterials and transplanted cells with the host tissue. Fibrovascular tissue will invade a device if the pores are larger than approximately 10 μm, and the rate of invasion will increase with the pore size and total porosity of a device.[14,15] This process results in the formation of a capillary network in the developing tissue.[15] Vascularization of the engineered tissue may be required to meet the metabolic requirements of the tissue and to integrate it with the surrounding host. In urologic applications it may also be desirable to have a non-porous luminal surface (e.g., to prevent leakage of urine from the tissue).

5. FORMATION OF BLADDER TISSUE

The direction which we have followed to engineer bladder tissue involves the use of both synthetic and natural biodegradable materials which act as cell delivery vehicles. Urothelial and muscle cells can be expanded in vitro, seeded onto the polymer scaffold, and allowed to attach and form sheets of cells (Figure 4). The cell-polymer scaffold can then be implanted in vivo. We have performed a series of in vivo urologic associated cell-polymer experiments. Histologic analysis of human urothelial, bladder muscle, and composite urothelial and bladder muscle-polymer scaffolds, implanted in athymic mice and retrieved at different time points, indicated that viable cells were evident in all three experimental groups.[4] Implanted cells oriented themselves spatially along the polymer surfaces. The cell populations appeared to expand from one layer to several layers of thickness with progressive cell organization with extended implantation times. Polymers alone evoked an angiogenic response by 5 days, which increased with time. Polymer fiber degradation was evident after 20 days. An inflammatory response was also evident at 5 days, and its resolution correlated with the biodegradation sequence. Cell-polymer composite implants of urothelial and muscle cells, retrieved at extended times (50 days), showed extensive formation of multi-layered sheet-like structures and well-defined muscle layers. Polymers seeded with cells and manipulated into a tubular configuration showed layers of muscle cells lining the multilayered epithelial sheets. Cellular debris appeared reproducibly in the luminal spaces, suggesting that epithelial cells lining the lu-

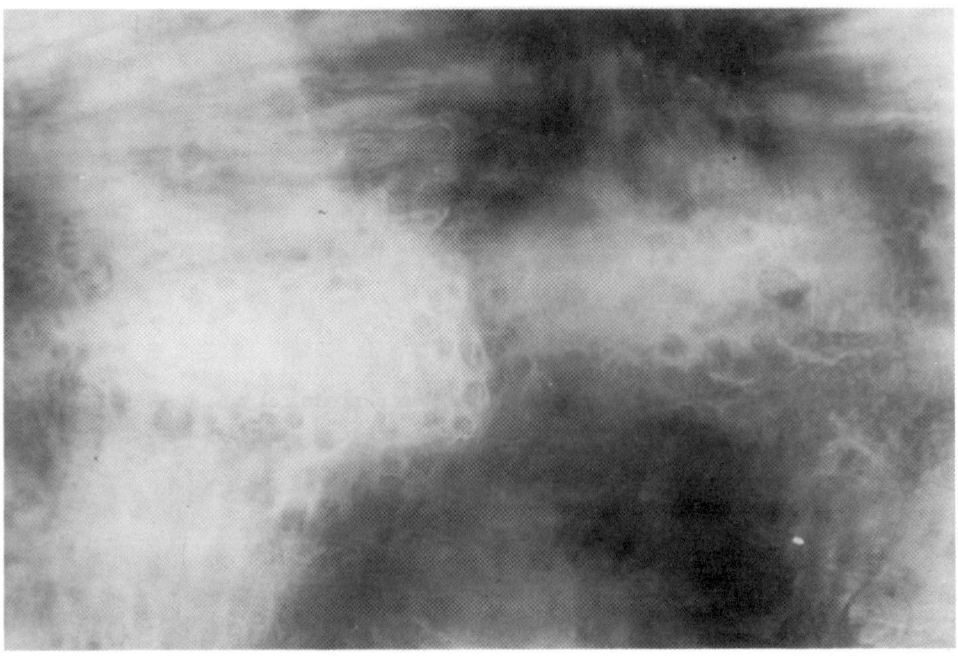

Figure 4. Urothelial cells over polymer matrix. Urothelial cells grow in layers over the polymer scaffold in vitro.

mina are sloughed into the luminal space. Cell polymers implanted with human bladder muscle cells alone showed almost complete replacement of the polymer with sheets of smooth muscle at 50 days. This experiment demonstrated, for the first time, that composite tissue engineered structures could be created de novo. Prior to this study, only single cell type tissue engineered structures had been created.

This approach was expanded to engineer new functional urologic structures. In one study conducted in dogs, urothelial and smooth muscle cells were harvested, expanded in vitro and seeded onto biodegradable polymer scaffolds. These structures were tubularized and used to replace ureteral segments in each animal.[16] The results suggested that the creation of artificial ureters may be achieved in vivo using biodegradable polymers as transplanted cell delivery and native cell expansion vehicles. The malleability of the synthetic polymer allowed for the creation of cell-polymer implants manipulated into pre-formed tubular configurations. The combination of both smooth muscle and urothelial cell-polymer scaffolds is able to provide a template wherein a functional ureter may be created de novo. The same strategy has been used in trying to achieve urethral reconstruction.[17]

In other sets of experiments, bladders were augmented using tissue engineering techniques.[18] Beagle bladder tissue specimens were micro dissected and the mucosal and muscular layers separated. Both urothelial and smooth muscle cells were harvested and expanded separately. The cells were seeded onto 5×5 and 10×10 cm sized sheets of polyglycolic acid polymers in vitro. Cell-polymer scaffolds were created consisting of urothelial and smooth muscle, urothelial, and smooth muscle cell populations. Polymers without cells served as controls. Beagles underwent cruciate cystotomies on the bladder dome and the polymer scaffolds were used to augment the bladder in each animal. Omentum was wrapped over the augmented bladder in order to enhance angiogenesis to the polymer-cell complex. The average increase in bladder capacity was approximately 40%. Furthermore, the average compliance of the augmented bladder increased an average of 30%. Polymer, polymer-urothelial cell, bladder muscle and urothelial/bladder muscle implants were found to contain a normal cellular organization consisting of a urothelial lined lumen surrounded by overlying submucosal tissue and smooth muscle. An angiogenic response and polymer fiber degradation were evident in all animals. Similar results were achieved by using a bladder submucosal collagen based matrix as a cell delivery vehicle.

In all of the studies performed at our laboratory, there was a major difference evident between matrices used with autologous cells and those used only as a regenerating scaffold. Matrix-cell composites retained most of their implanted diameter, as opposed to matrix only, wherein graft contraction and shrinkage occurred.

The results of all our prior studies showed that the creation of artificial functional and anatomical bladders may be achieved in vivo, however, much work remained to be done in terms of the functional parameters of these implants. In order to address the functional parameters of tissue engineered bladders, an animal model was designed which required a subtotal cystectomy with subsequent replacement with a tissue engineered organ.[20]

A total of 14 beagle dogs underwent a trigone-sparing cystectomy. The animals were randomly assigned to one of three groups. Group A (n=2) underwent closure of the trigone without a reconstructive procedure. Group B (n=6) underwent reconstruction with a cell-free bladder shaped biodegradable polymer. Group C (n=6) underwent reconstruction using a bladder shaped biodegradable polymer that delivered autologous urothelial cells and smooth muscle cells. The cell populations had been separately expanded from a previously harvested autologous bladder biopsy. Preoperative and postoperative urodynamic and radiographic studies were performed serially. Animals were sacrificed at 1, 2, 3, 4, 6, and 11 months postoperatively. Gross, histological and immunocytochemical analyses were performed.[20]

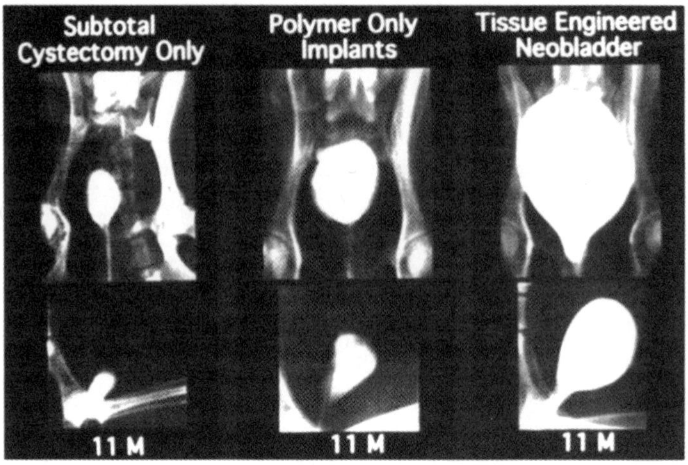

Figure 5. Radiographic cystograms in beagles 11 months after subtotal cystectomy without reconstruction (left); with reconstruction using a polymer without cells (center); and reconstruction with a polymer and cell seeded tissue engineered organ (right) . Trigone-sparing cystectomy organs retained a small sized reservoir. Tissue engineered neo-bladders showed a normal configuration and a larger capacity than the trigones grafted with polymer only.

The cystectomy only controls and polymer only grafts maintained average capacities of 22% and 46% of preoperative values, respectively. An average bladder capacity of 95% of the original pre-cystectomy volume was achieved in the tissue-engineered bladder replacements (Figure 5). The subtotal cystectomy reservoirs which were not reconstructed and polymer only reconstructed bladders showed a marked decrease in bladder compliance (10% and 42%). The compliance of the tissue engineered bladders showed almost no difference from preoperative values that were measured when the native bladder was present (106%). Histologically, the polymer only bladders presented a pattern of normal urothelial cells with a thickened fibrotic submucosa and a thin layer of muscle fibers. The retrieved tissue engineered bladders showed a normal cellular organization, consisting of a tri-layer of urothelium, submucosa and muscle (Figure 6). Immunocytochemical analyses for desmin, alpha actin, cytokeratin 7, pancytokeratins AE1/AE3 and uroplakin III confirmed the muscle and urothelial phenotype. S-100 staining indicated the presence of neural structures. The results from this study showed that it is possible to tissue engineer bladders which are anatomically and functionally normal.[20]

6. FETAL TISSUE ENGINEERING

The prenatal diagnosis of patients with bladder disease is now more prevalent. Prenatal ultrasonography allows for a thorough survey of fetal anatomy. The absence of bladder filling, a mass of echogenic tissue on the lower abdominal wall or a low set umbilicus during prenatal sonographic examination may suggest the diagnosis of bladder exstrophy. These findings and the presence of intraluminal intestinal calcifications suggest the presence of a cloacal malformation.

The natural consequence of the evolution in prenatal diagnosis led to the use of intervention before birth to reverse potentially life threatening processes. However, the con-

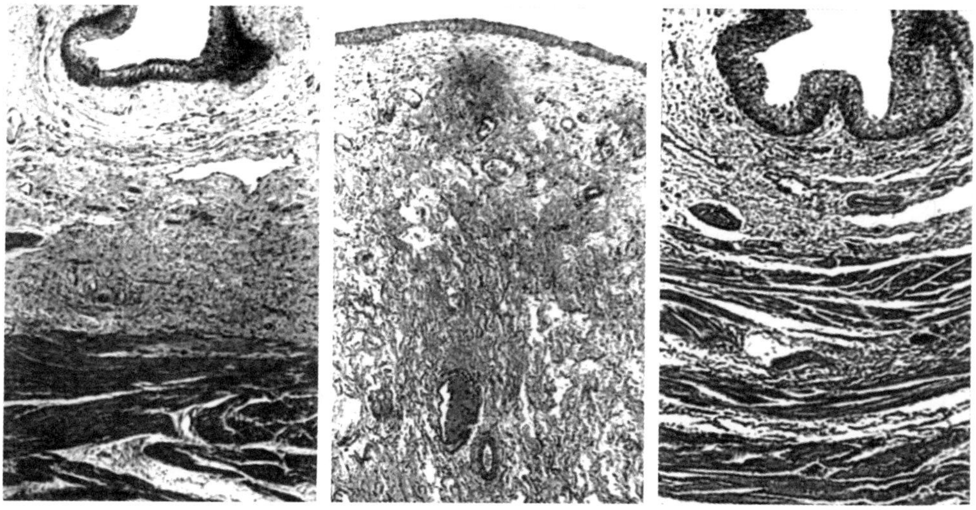

Figure 6. Hematoxylin and Eosin histological results six months after surgery (original magnification: × 250). Left: Normal canine bladder. Center: The bladder dome of the cell-free polymer reconstructed bladder consists of normal urothelium over a thickened layer of collagen and fibrotic tissue. Only scarce muscle fibers are apparent. Right: The tissue engineered neo-organ shows a histo-morphologically normal appearance. A trilayered architecture consisting of urothelium, submucosa and smooth muscle is evident.

cept of prenatal intervention itself is not limited to this narrow group of indications. A prenatal , rather than a postnatal diagnosis of exstrophy, may be beneficial under certain circumstances. There is now a renewed interest in performing a single-stage reconstruction in some patients with bladder exstrophy. Limiting factors for following a single or multiple-stage approach may include the findings of a small, fibrotic bladder patch without either elasticity or contractility, or a hypoplastic bladder.

There are several strategies which may be pursued, using today's technological and scientific advances, which may facilitate the future prenatal management of patients with bladder disease. Having a ready supply of urologic associated tissue for surgical reconstruction at birth may be advantageous. Theoretically, once the diagnosis of bladder exstrophy is confirmed prenatally, a small bladder and skin biopsy could be obtained via ultrasound guidance. These biopsy materials could then be processed and the different cell types expanded in vitro. Using tissue engineering techniques developed at our center and described previously, reconstituted bladder and skin structures in vitro could then be readily available at the time of birth for a one stage reconstruction, allowing for an adequate anatomic and functional closure (Figure 7).

Toward this end, we conducted a series of experiments using fetal lambs.[21,22] Bladder exstrophy was created surgically in ten 90–95 day gestation fetal lambs. The lambs were randomly divided into two groups of five. In group I, a small fetal bladder specimen was harvested via fetoscopy. The bladder specimen was separated and muscle and urothelial cells were harvested and expanded separately under sterile conditions in a humidified 5% CO_2 chamber, as previously described. Seven to ten days prior to delivery, the expanded bladder muscle cells were seeded on one side and the urothelial cells on the opposite side of a 20 cm^2 biodegradable polyglycolic acid polymer scaffold. After delivery, all lambs in group I had surgical closure of their bladder using the tissue engineered bladder tissue. No fetal bladder harvest was performed in the group II lambs, and bladder

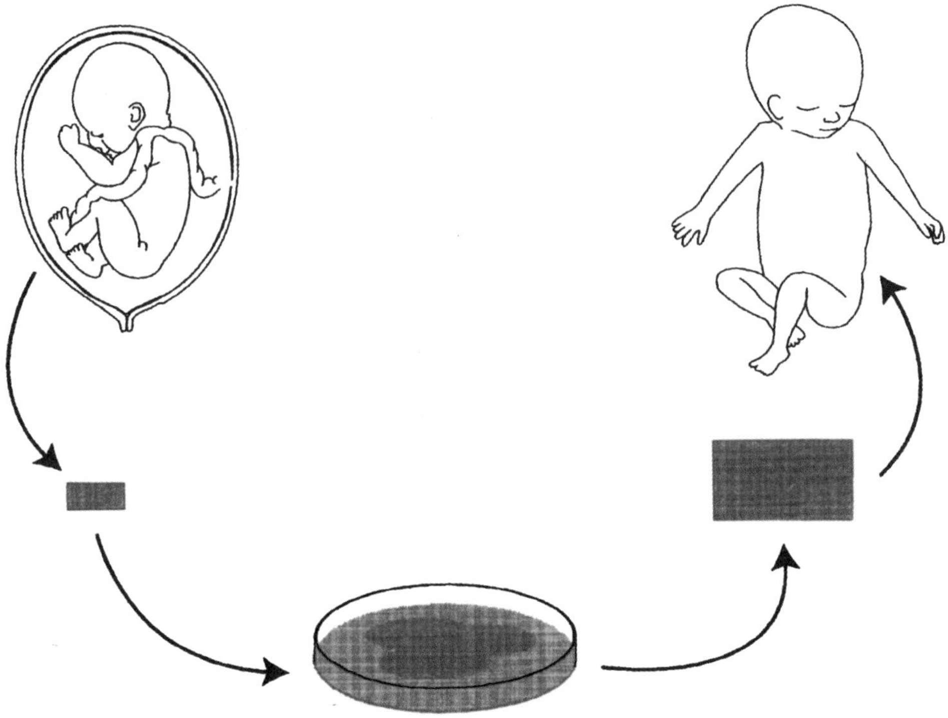

Figure 7. In-utero tissue engineering strategy. Certain conditions, such as bladder exstrophy can be diagnosed prenatally. Bladder, cartilage, skin and other tissues can be retrieved via fetoscopy or percutaneously using ultrasound guidance. Cells can be harvested, expanded in vitro, and reconstituted into tissue structures. Tissue is readily available at the time of birth for reconstruction.

exstrophy closure was performed using only the native bladder. Cystograms were performed 3 and 8 weeks after surgery. The engineered bladders were more compliant (p=0.01) and had a higher capacity (p=0.02) than the native bladder closure group. Histologic analysis of the engineered tissue showed a normal histological pattern, indistinguishable from native bladder at 2 months.[21] Similar prenatal studies were performed in lambs, engineering skin for reconstruction at birth.[22]

Both of the studies described above show that the potential for replicating this technology in humans is possible. Certainly, other tissues, such as cartilage, corpora cavernosa and bowel can also be harvested and expanded in the same manner. Similar studies addressing these tissues are now in progress in our laboratory.

In addition to being able to manage the bladder exstrophy complex in utero with tissue engineering techniques, one could also manage patients after birth in a similar manner, whenever a prenatal diagnosis is not assured. In these instances, bladder tissue biopsies could be obtained at the time of the initial surgery. Different tissues could be harvested and stored for future reconstruction, if necessary. A tissue bank for exstrophy complex patients could preserve different cell types indefinitely.

In addition to having an exstrophy complex tissue bank serve as a repository of cells for future tissue reconstitution, it could also serve as a resource to elucidate the cellular, molecular and genetic mechanisms required for the development and future prevention of these anomalies.

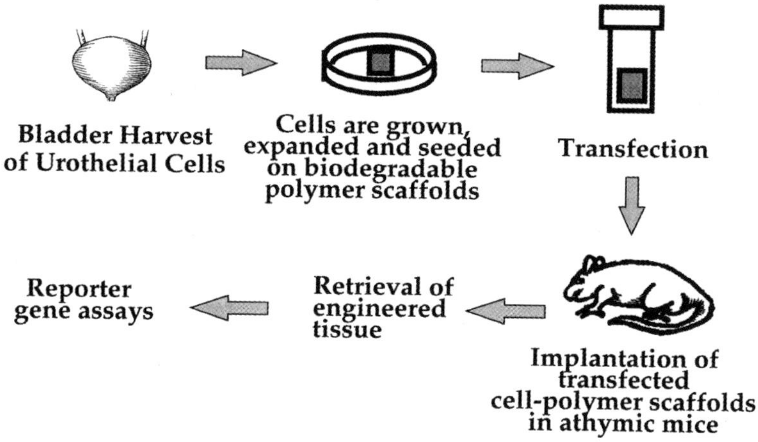

Figure 8. Successful gene transfer can be achieved using the principles of tissue engineering. Biodegradable polymer scaffolds act as bladder cell delivery vehicles. The transfected cell-polymer scaffold can form organ-like structures with functional expression of the transfected genes.

7. GENE THERAPY AND TISSUE ENGINEERING

Based on the feasibility of tissue engineering techniques in which cells seeded on biodegradable polymer scaffolds form tissue when implanted in vivo, the possibility was explored of developing a neo-organ system for in vivo gene therapy. The method of gene delivery at present is most effectively achieved by ex vivo gene transfer, which includes removal of the target tissue, in vitro gene delivery to the target cells, possible selection to enhance the proportion of transfected cells, and reintroduction of the gene-modified cells. Reintroduction of the transformed cells in vivo has been a challenging task.

Results indicated that successful gene transfer could be achieved using biodegradable polymer scaffolds as a bladder urothelial cell delivery vehicle. The transfected cell/polymer scaffold formed organ-like structures with functional expression of the transfected genes (Figure 8).[23]

8. CONCLUSIONS

It is evident that the creation of bladder tissue de novo is possible both in vitro and in vivo. Tissue engineering can be applied not only to bladder tissue or organ replacement, but also as a model system for the study of bladder disease. Gene therapy and fetal intervention are just a few of the associated sciences which can be successfully combined with this technology. Tissue engineering, although in its infancy, has already been applied clinically in other areas of urology.[24–27] Although challenges remain, bladder tissue regeneration for human application is not far in the horizon.

REFERENCES

1. Atala A., Retik A. : Pediatric Urology - future perspectives. In: Clinical Urology. Krane R.J., Siroky M.B., Fitzpatrick J.M. (ed) J.B. Lippincott, Philadelphia., 507, 1994.
2. Atala A. : Tissue engineering in the urinary tract. Dialogues in Pediatric Urology., **18**: 1, 1995.

3. Atala A., Vacanti J.P., Peters C.A., Mandell J., Retik A.B., Freeman M.R. : Formation of urothelial structures in vivo from dissociated cells attached to biodegradable polymer scaffolds in vitro. J. Urol., **148**: 658, 1992.
4. Atala A., Freeman M.R., Vacanti J.P., Shepard J., Retik A.B. : Implantation in vivo and retrieval of artificial structures consisting of rabbit and human urothelium and human bladder muscle. J. Urol., **150**: 608, 1993.
5. Cilento B.J., Freeman M.R., Schneck F.X., Retik A.B., Atala A. : Phenotypic and cytogenetic characterization of human bladder urothelia expanded in vitro. J. Urol., **152:** 665, 1994.
6. Tobin M.S., Freeman M.R., Atala A. : Maturational response of normal human urothelial cells in culture is dependent on extracellular matrix and serum additives. Surgical Forum., **45**: 786, 1994.
7. Freeman M.R., Yoo J.J., Raab G., Soker S., Adam R.M., Schneck F.X., Renshaw A.A., Klagsbrun M. and Atala A. : Heparin-Binding EGF-like growth factor is an autocrine growth factor for human urothelial cells and is synthesized by epithelial and smooth muscle cells in the human bladder. Journal of Clinical Investigation., **99**: 1028 1997.
8. Prigent S.A. : The type 1 (EGFR-related) family of growth factor receptors and their ligands. Prog. Growth Factor Res., **4**: 1, 1992.
9. Atala A. : Replacement of urologic associated mucosa. J. Urol., **156**: 338, 1996.
10. suji I., Ishida H., Fujieda J. : Experimental cystoplasty using preserved bladder graft. J. Urol., **85**: 42, 1961.
11. Sutherland R.S., Baskin L.S., Hayward S.W., Cunha G.R. : Regeneration of bladder urothelium, smooth muscle, blood vessels and nerves into an acellular matrix. J.Urol., **156:** 571, 1996.
12. Kropp B.P., Rippy M.K., Badylak S.F., Adams M.C., Keating M.A., Rink R.C., Thor K.B. : Regenerative urinary bladder augmentation using small intestinal submucosa: urodynamic and histopathologic assesment in long term canine bladder augmentation. J. Urol., **155:** 2098, 1996.
13. Machluff M. and Atala A. : Progress in tissue engineering: Emerging concepts for tissue and organ transplantation. Graft, in press, 1998.
14. Wesolowski S.A., Fries C.C., Karlson K.E., et al. : Porosity: primary determinant of ultimate fate of synthetic vascular grafts. Surg., **50:** 91, 1961.
15. Mikos A.G., Sarakinos G., Lyman M.D., et al. : Prevascularization of porous biodegradable polymers. Biotechnol. Bioeng., **42:** 716, 1993.
16. Yoo J., Satar N., Retik A.B., Atala A. : Ureteral replacement using biodegradable polymer scaffolds seeded with urothelial and smooth muscle cells. J Urol., **153:**4 (suppl), 1995.
17. Cilento B.G., Retik A.B., Atala A. : Urethral reconstruction using a polymer scaffolds seeded with urothelial and smooth muscle cells. J. Urol., **155:**5 (suppl), 1996.
18. Yoo J.J, Satar N., Atala A. : Bladder augmentation using biodegradable polymer scaffolds seeded with urothelial and smooth muscle cells. J. Urol., **155:** 5 (supp), 1996.
19. Yoo J., Oberpenning F., Atala A. : Bladder augmentation using a new biomaterial composed of allogenic bladder submucosa. Urology, in press, 1998.
20. Oberpenning F., Meng J., Yoo J., Atala A. : Bladder Replacement with Tissue-Engineered Neo-Organs. Pediatrics., **100S:** 10, 1997.
21. Fauza D.O., Fishman S., Mehegan K., Atala A. : Videofetoscopically assisted fetal tissue engineering: bladder augmentation. J. Ped. Surg., in press, 1998.
22. Fauza D.O., Fishman S., Mehegan K., Atala A. : Videofetoscopically assisted fetal tissue engineering: skin replacement. J. Ped. Surg., in press, 1998.
23. Yoo J. and Atala A. : A novel gene delivery system using urothelial tissue engineered neo-organs. J.Urol.,**158:**1066, 1997.
24. Atala A., Cima L.G., Kim W., Paige K.T., Vacanti J.P., Retik A.B., Vacanti C.A. : Injectable alginate seeded with chondrocytes as a potential treatment for vesicoureteral reflux. J. Urol., **150:** 745, 1993.
25. A tala A., Kim W., Paige K.T., Vacanti C.A., Retik A.B. : Endoscopic treatment of vesicoureteral reflux with chondrocyte-alginate suspension. J. Urol., **152:** 641, 1994.
26. Atala, A. : Use of autologous substances in VUR and Incontinence treatment. Dialogues in Pediatric Urology., **18:**1, 1995.
27. Joyner B. and Atala A. Endoscopic substances for the management of vesicoureteral reflux, Urology**; 50:** 489, 1997.

4

RECONSTRUCTION OF THE URINARY BLADDER BY AUTO-AUGMENTATION, ENTEROCYSTOPLASTY, AND COMPOSITE ENTEROCYSTOPLASTY

Mary T. Comer,[1] David F. M. Thomas,[2] Ludwik K. Trejdosiewicz,[1] and Jennifer Southgate[1]

[1]Imperial Cancer Research Fund Cancer Medicine Research Unit
[2]Department of Pediatric Urology
St James's University Hospital
Leeds, LS9 7TF, United Kingdom

1. INTRODUCTION

Urinary incontinence resulting from congenital abnormalities of the urinary bladder and its nerve supply is a serious disability with a major impact on the quality of life of affected children and adolescents. Current clinical management centres on enterocystoplasty, the reconstruction or enlargement of the bladder with segments of bowel.[1,2] The use of bowel gives rise to significant complications, including mucus production, formation of bladder stones, metabolic disturbances, growth retardation and probably an increased long-term risk of malignancy.[3] The use of stomach overcomes some of the problems associated with small and large intestine[4] but carries the risk of complications such as haematuria/dysuria syndrome and metabolic alkalosis.[5,6] The complications and undesirable consequences of enterocystoplasty can be attributed to major differences in the properties of the epithelial linings of the intestine and bladder respectively. Intestinal epithelium is involved in mucus secretion, digestion and absorption of nutrients, electrolytes and water. It is therefore poorly adapted to replace the normal urothelial lining of the bladder which serves as a flexible but watertight impermeable barrier to urine.

2. AUTO-AUGMENTATION

Auto-augmentation has been used to increase the capacity of small, poorly compliant bladders by removing sections of bladder detrusor muscle, leaving the mucosa intact.

Advances in Bladder Research, edited by Baskin and Hayward.
Kluwer Academic / Plenum Publishers, New York, 1999.

During bladder filling the mucosa, unsupported by overlying detrusor, acts as a large diverticulum to impart additional capacity and compliance.[7,8,9] Combining auto-augmentation with a seromuscular enterocystoplasty provides a more substantial muscular backing to the mucosa and has been shown to prevent long term fibrosis in mini pigs[4] and sheep.[10,11] The encouraging results obtained in the normal bladder of experimental animals may not be mirrored in clinical practice. Dissection of the submucosal plane may prove very difficult in the trabeculated neuropathic bladder. Furthermore it is unlikely that small contracted neuropathic bladders would be capable of generating sufficient mucosa to serve as an effective auto augmentation.

3. DE-EPITHELIALIZED BOWEL

Another approach to circumventing the problems of enterocystoplasty has been to remove the intestinal epithelium. The feasibility of excising the mucosa to create de-epithelialized bowel segments for augmentation has been confirmed both experimentally[12,13,14,15] and in a limited clinical setting.[16,17] Regrowth of intestinal epithelium does occur if the epithelium is not completely removed.[14,18] Dewan and colleagues[18] have shown that removal of the muscularis mucosa with a portion of the submucosa will prevent regrowth of the intestinal epithelium.

Fibrosis and shrinkage occur in the de-mucosalised patches.[4,10,11,15] A reduction in shrinkage and fibrosis has been reported when islands of bladder mucosa were placed on the bowel or gastric patches.[4,14,15] Urothelium from the islands of bladder mucosa expands onto the gastrointestinal patches with the least amount of fibrosis being observed with the best cover of urothelium.[4,15,19]

A number of groups are currently working towards ëcomposite enterocystoplasty' where in vitro propagated urothelial cells are grafted onto de-mucosalized bowel segments. Although composite enterocystoplasty is still at the conceptual/developmental stage, an increasing weight of evidence from ourselves[20,21] and others[13,16,17,22] indicates this is a realistic approach. For composite enterocystoplasty to be effective, the cultured urothelial cells would need to retain their differentiation potential and to form a watertight barrier similar to that observed in normal bladder in vivo. The only way to obtain sufficient autologous urothelium would be by growing large numbers of urothelial cells in vitro.

4. UROTHELIAL CELLS PROPAGATED IN MONOLAYER CULTURE

Urothelium is a stratified transitional epithelium of 4 to 5 cells thick consisting of basal, intermediate and superficial cells. The barrier function of the urothelium is dependent on the integrity of cell:cell interactions involving adhesion molecules and junctional complexes which are discussed elsewhere[23,24,25] and in this issue. Superficial cells, which represent the terminally-differentiated phenotype, have a unique and highly specialised apical membrane consisting of thickened plaques of asymmetric unit membrane (AUM). The AUM serves to decrease the permeability of the urothelium to urine.

Urothelial cells have been cultured from a number of species, including pig,[26,27] rabbit,[14] rat[28] and mouse.[29] In addition, human urothelial cells have been cultured from all regions of the human urinary tract.[20,23,31] Normal human urothelial cells, which have a slow turnover rate in vivo, show a rapid and sustained proliferation rate when grown as monocultures in vitro.[20,23,31] Sufficient numbers of cells to be surgically useful can be generated from small specimens of urothelial tissues within approximately 3 weeks.[20] Under optimal proliferative conditions, urothelial cells in vitro grow as a non-stratified monolayer. The cells express the phenotype of basal/intermediate urothelial cells and do not express antigens associated with late or terminal differentiation.[23,31] Cilento et al.[31] showed that cultured human urothelial cells implanted subcutaneously into athymic mice formed multiple layers. Stratification of cultured urothelial cells in vitro can be induced by increasing exogenous calcium. This is associated with increased expression of adhesion molecules such as E-cadherin and integrins, as well as assembly of calcium-dependent structures, such as desmosomes.[23,25]

5. IN VIVO COMPOSITE ENTEROCYSTOPLASTY

In vitro propagated urothelial cells have been grafted onto rabbit and sheep de-mucosalized colon and stomach respectively. Merguerian and colleagues[14] reported that rabbit urothelial cells grown on a collagen mesh and grafted onto an autologous de-mucosalized bowel did not attach or grow. Additionally, an inflammatory response was observed in the mucosa. Schaefer et al.[32] cultured sheep urothelial cells on collagen membranes for sheet grafting. The grafts were transplanted onto demucosalized segments of sheep stomach and colon. Two weeks after transplantation the urothelial cells had stratified and were expressing uroplakin III, a component of the AUM. Shrinkage, fibrosis and inflammation also occurred in the mucosa of the stomach and colon segments. Shrinkage and fibrosis do not occur in conventional enterocystoplasty where the epithelium remains intact.[33] Therefore, minimising the time for complete coverage of the mucosa by the neo-epithelium may reduce this effect. We are currently developing a method for transplanting intact urothelial sheets which may reduce the amount of time that the sub-mucosa is without a complete covering of urothelium.

6. IN VITRO ORGAN CULTURE MODELS

In order to study the effects of transplanted urothelial cells in vitro, an organ culture system for urothelium has been established.[21] Pieces of normal intact human bladder can be maintained in culture for at least 18 weeks without the loss of in vivo histioarchitecture. We have shown that in vitro cultured human urothelial cells can reform a histologically normal urothelium when seeded onto de-epithelialized samples of bladder mucosa that is virtually indistinguishable from organ cultures of intact bladder.[21] We now have preliminary evidence that porcine urothelial cells, after expansion in monolayer tissue culture, also have the ability to reform a full-thickness differentiated urothelium when seeded onto a de-epithelialized stroma. A normal urothelium was also observed when porcine cultured urothelial cells were seeded onto stomach submucosa (Comer et al., in preparation). The ability of cultured cells to reform an intact stratified urothelial layer, and form a continuous basement membrane implies that urothelial cells will not transdifferentiate into enterocytes and provides further support to the concept of composite enterocystoplasty.

7. FUTURE WORK

The growth of bladder urothelium, mucosa and muscle in vitro may lead to the development of tissue engineered bladders. It has already been shown that all the cell types can be maintained in culture in collagen gels.[27] Sutherland et al.[34] have reported the generation of urothelium, smooth muscle, blood vessels and nerves on an acellular extracellular matrix scaffold graft, the full thickness of the bladder. This indicates that an artificial scaffold could be produced which could lead to bladder regeneration. Yoo and Atala[35] have shown that urothelial cells grown in culture can be used as a delivery system for the transfer of genes into the bladder. This novel approach to gene transfer could also be used for the delivery of genes in other organs where tissue engineering is possible and tissue-specific gene transfer is desirable.

8. CONCLUSIONS

The complications associated with enterocystoplasty are due to the different properties of gastrointestinal epithelium and urothelium. De-epithelialization of the bowel by removing the mucosa leads to shrinkage, fibrosis and inflammation. Preliminary studies in vivo in sheep suggests that composite enterocystoplasty where in vitro propagated urothelial cells are combined with de-mucosalized bowel do express differentiation characteristics associated with bladder.

ACKNOWLEDGMENTS

We wish to thank Action Research and Ethicon Limited for financial support.

REFERENCES

1. Fisch M., Wammack R., Muller S.C. and Hohenfeller R.: The use of intestine in paediatric urology. Eur Urol Update Series., **2:** 42, 1993.
2. Payne S.R. and Clarke N.W.: Management of urinary incontinence (2): treatment. Hosp Update., **19:** 353, 1993.
3. Nurse D.E. and Munday A.R.: Assessment of the malignant potential of cystoplasty. Br J Urol., **64**: 489, 1989.
4. Frey P., Lutz N. and Leuba A-L.: Augmentation cystoplasty using pedicled and de-epithelialized gastric patches in the mini-pig model. J Urol., **156:** 608, 1996.
5. Reinberg Y., Manivel J.C., Froemming C. and Gonzalez R.: Perforation of the gastric segment of an augmented bladder secondary to peptic ulcer disease. J Urol., **148:** 369, 1992.
6. Nguyen D.H., Bain M.A., Salmonson K.L., Ganesan G.S., Burns M.W. and Mitchell M.E.: The syndrome of dysuria and hematuria in pediatric urinary reconstruction with stomach. J Urol., **150:** 707, 1993
7. Cartwright P.C. and Snow B.W.: Bladder autoaugmentation: Early clinical experience. J Urol., **142:** 505, 1989.
8. Cartwright P.C. and Snow B.W.: Bladder autoaugmentation: Partial detrusor excision to augment the bladder without use of bowel. J Urol., **142:** 1050, 1989.
9. Kennelly M.J., Gormley E.A. and McGuire E.J: Early clinical experience with adult bladder auto-augmentation. J Urol., **152:** 303, 1994.
10. Dewan P.A. and Byard R.W.: Autoaugmentation gastrocystoplasty in a sheep model. Br J Urol., **72:** 56, 1993.
11. Dewan, P.A. and Stefanek, W.: Autoaugmentation gastrocystoplasty: early clinical results. Br J Urol., **74:** 460, 1994.
12. de Badiola F., Manivel J.C. and Gonzalez R.: Seromuscular enterocystoplasty in rats. J Urol., **146:** 559, 1991.
13. Buson H., Manivel J.C., Dayanc M., Long R. and Gonzalez R.: Seromuscular colocystoplasty lined with urothelium: experimental study. Urology., **44:** 743, 1994.

14. Merguerian P., Chavez D.R. and Hakim S.: Grafting of cultured uroepithelium and bladder mucosa into de-epithelialized segments of colon in rabbits. J Urol., **152:** 671, 1994.
15. Lutz N. and Frey P.: Enterocystoplasty using modified pedicled, detubularized, de-epithelialized sigmoid patches in the mini-pig model. J Urol., **154:** 893, 1995.
16. Gonzalez R., Buson H., Reid C. and Reinberg Y.: Seromuscular colocystoplasty lined with urothelium: experience with 16 patients. Urology, **45:** 124, 1995.
17. Nguyen D.H., Mitchell M.E., Horowitz M., Bagli D.J. and Carr M.C.: Demucosalised augmentation gastrocystoplasty with bladder autoaugmentation in pediatric patients J Urol., **156:** 206, 1996.
18. Dewan P.A., Close C.E., Byard R.W., Ashwood P.J. and Mitchell M.E.: Enteric mucosal regrowth after bladder augmentation using demucosalized gut segments. J Urol., **158:** 1141, 1997.
19. Vates T.S., Denes, E.D., Rabah R., Shapiro E., Suzer O., Freedman A.L. and Gonzalez R: Methods to enhance in vivo urothelial growth on seromuscular colonic segments in the dog. J Urol., **158:** 1081, 1997.
20. Hutton K.A.R., Trejdosiewicz L.K., Thomas D.F. and Southgate J.: Urothelial tissue culture for bladder reconstruction: an experimental study. J. Urol., **150:** 721, 1993.
21. Scriven S.D., Booth C., Thomas D.F.M., Trejdosiewicz L.K., and Southgate J.: Reconstitution of human urothelium from monolayer cultures J Urol., **158:** 1147, 1997.
22. Atala A., Freeman M.R., Vacanti J.P., Shepard J.A. and Reik A.B.: Implantation in vivo and retrieval of artificial structures consisting of rabbit and human urothelium and human bladder muscle. J Urol., **150:** 608, 1993.
23. Southgate J., Hutton K.A.R., Thomas D.F.M., Trejdosiewicz L.K.: Normal human urothelial cell in vitro; proliferation and induction of stratification. Lab Invest., **71:** 583, 1994.
24. Southgate J., Kennedy W., Hutton K.A., Trejdosiewicz L.K.: Expression and in vitro regulation of integrins by normal human urothelial cells. Cell Adhes. Commun., **3:** 231, 1995.
25. Southgate J., Trejdosiewicz L.K., Smith B., Selby P.J.: Patterns of splice variant CD44 expression by normal human urothelium in situ and in vitro and by bladder-carcinoma cell lines. Int J Cancer., **62:** 449, 1995.
26. Guhe C., Follmann W.: Growth and characterization of porcine urinary bladder epithelial cells in vitro. Am J Physiol., **266:** F298, 1994.
27. Fujiyama C., Masaki Z. and Sugihara H.: Reconstruction of the urinary bladder mucosa in three-dimensional collagen gel culture: fibroblast-extracellular matrix interactions on the differentiation of transitional epithelial cells. J Urol., **153**: 2060, 1995.
28. Pauli B.U., Anderson S.N., Memoli V.A., Kuettner K.E.: The isolation and characterization in vitro of normal epithelial cells, endothelial cells and fibroblasts from rat urinary bladder. Tissue Cell., **12**: 419, 1980.
29. de Boer W.I., Rebel J.M., Vermey M., de Jong A.A., Van der Kwast T.H.: Characterization of distinct functions for growth factors in murine transitional epithelial cells in primary organotypic culture. Exp Cell Res., **214:** 510, 1994.
30. Wolf J.S., Jr., Soble J.J., Ratliff T.L., Clayman R.V.: Ureteral cell cultures .1. Characterization and cellular interactions. J Urol., **156:** 1198, 1996.
31. Cilento B.G., Freeman M.R., Schneck .F. X., Retik A.B., Atala A.: Phenotypic and cytogenetic characterization of human bladder urothelia expanded in vitro. J Urol., **152**: 665, 1994.
32. Schaefer B.M., Lorenz C., Back W., Moll R., Sun T-T, Schober C., Waag K-L. and Kramer M.D.: Autologous transplantation of urothelium into demucosalized gastrointestinal segments: Evidence for epithelialization and differentiation of in vitro expanded and transplanted urothelial cells., J Urol., **159:** 281, 1998.
33. Motley R.C., Montgomery B.T., Zollman P.E., Holley H.E., Kramer S.A: Augmentation cystoplasty utilizing de-epithelialized sigmoid colon: a preliminary study. J Urol., **143:** 1257, 1990.
34. Sutherland R.S., Baskin L.S., Hayward S.W. and Cunha G.R.: Regeneration of bladder urothelium, smooth muscle, blood vessels and nerves into an acellular tissue matrix. J Urol., **156**: 571, 1996.
35. Yoo J.J. and Atala A.: A novel gene delivery system using urothelial tissue engineered neo-organs. J Urol., **158:** 1066, 1997.

5

EPITHELIAL-MESENCHYMAL INTERACTIONS IN THE BLADDER

Implications for Bladder Augmentation

Laurence S. Baskin, Simon W. Hayward, Michael S. DiSandro, Ying Wu Li, and Gerald R. Cunha

Department of Urology
Box 0738
UCSF Children's Medical Center
San Francisco, California 94143

1. INTRODUCTION

The urinary bladder has two major functions: 1) storage and 2) socially acceptable efficient emptying.[1] To achieve this end the bladder has evolved as a highly compliant muscular sac with a specialized epithelial lining able to accommodate extensive variations in internal dimensions while maintaining an impervious luminal surface. Mesenchymal-epithelial interactions play an important role in the differentiation of both the urothelium and the smooth muscle of the bladder.[2] These interactions are also important in maintaining homeostasis and in responding to injury and insults in adulthood.[3] There is evidence to support the idea that stromal-epithelial interactions also play important roles in both carcinoma of the urinary bladder[4,5] and in fibrotic responses to urinary obstruction due to a number of causes.[6] Such fibrotic responses lead to the formation of a non-compliant bladder, that is, an organ which is unable to expand sufficiently to maintain the low pressure internal environment required to both store urine and protect the kidneys from damage due to urinary reflux.

In children born with congenital abnormalities of the innervation of the bladder (neurogenic bladders) coordinated voiding is rarely possible.[7,8] Starting from the newborn period these patients can achieve effective bladder emptying with intermittent catheterization.[9] This leaves urinary storage as the major issue confronting this group of patients. Initial treatment involves pharmacologic manipulation typically with anticholinergic medication.[7] When this proves inadequate surgical bladder enlargement is the next step. Bladder augmentation is also indicated in adult patients who have small fibrotic bladders

Advances in Bladder Research, edited by Baskin and Hayward.
Kluwer Academic / Plenum Publishers, New York, 1999.

secondary to diseases such as interstitial cystitis, tuberculosis, and spinal cord injuries. Complete bladder replacement is indicated in patients with invasive transitional cell carcinoma. Extensive efforts have gone into creating compliant urinary reservoirs.[10] These low pressure, adequate volume reservoirs must have sufficient outlet resistance to assure urinary continence.

2. PRESENT AUGMENTATION STRATEGIES

As noted above the urothelial lining of the urinary bladder is highly specialized and is able to respond to progressive distension in bladder volume while maintaining an impervious barrier. For the bladder to be a safe and effective storage chamber its epithelial lining should neither absorb urine or secrete electrolytes. Current augmentation strategies are based on the use of sheets of tissue derived from the intestinal tract. However, epithelial cells from this source are not optimal for this purpose since they are not adapted to the continuous mechanical stresses of the bladder and they either secrete or absorb electrolytes.[11,12] This can lead to chronic electrolyte abnormalities such as metabolic acidosis which occurs with ileal augmentation cystoplasties. In the case of gastric augmentations the stomach segment continues to secrete acid which may predispose these patients to metabolic abnormalities such as hypochloremic hypokalemic metabolic alkalosis.[13,14] Chronic electrolyte imbalance may also lead to abnormalities in calcium homeostasis resulting in bone demineralization. The risk of tumors in bladder augmentation seems to be increased based on sporadic case reports and experimental studies.[15,16] Intestinal segments also secrete mucous changing the quality of urine in these patients. The use of bowel in the urinary tract also increases the chance of urinary stones through unknown lithogenic properties.[17]

To avoid complications due to the use of bowel in the urinary tract a number of alternative strategies have been devised. Autoaugmentation as described by Cartwright and Snow has had success in a select group of patients whose compliance is poor based on abnormal pressures.[18] This procedure is technically difficult in patients who have poor volumes secondary to small contracted bladders. The concept of creating a large bladder diverticulum composed of urothelium and lamina propria has been extended by attempting to support the inherently weak autoaugment with deepithelized bowel segments from the stomach or the sigmoid colon.[19] The well vascularized de-epithelized bowel segments can theoretically provide a vascularized stromal support to the urothelial diverticulum. Inherent in this design, however, is cellular contact between the urothelial cells and the intestinal stroma. What type of cellular interactions occur at this foreign interface between bladder urothelial cells and intestinal stroma? Before we can answer this question we first need to define what we mean by "normal" urothelium, stroma and extracellular matrix.

3. NORMAL UROTHELIUM

The phenotype of urothelium can be defined by morphology, protein expression, cellular turnover and function. Histologically, bladder urothelium is a stratified transitional epithelium consisting of basal, intermediate and superficial cells. Functionally, these cells must accommodate gradual changes during filling and rapid changes during voiding.[20,21] During bladder cycling the urothelial cells function as an impermeable barrier to most solutes. The superficial urothelial cells have a highly specialized apical membrane consisting

Table 1. Immunofluorescence on normal human urothelial tissue sections, cultured urothelial cells and intact organ cultures adapted from Hutton and Southgate

		Normal bladder				Organ culture		
Antibody	Type/specificity	Basal	Inter	Sup	T.C. Urothelium	Basal	Inter	Sup
LdS68	Monoclonal keratin 7	++	++	++	++			
LdS103	Monoclonal keratin 7,8,18	++	++	++	++			
LP1K	Monoclonal keratin 7	++	++	++	++			
E3	Monoclonal keratin 17	++	++	++	++/+			
BA16	Monoclonal keratin 19	++	++	++	++			
LP2K	Monoclonal keratin 19	++	++	++	+++/+	++	++	+++
LE41	Monoclonal keratin 8	+	+	++	+	+	+	++
LE61	Monoclonal keratin 18	+	+	++	++	+	+	++
CAM5.2	Monoclonal keratin 18,19	+	+	++	+++/++			
2D7	Monoclonal keratin 13⁺	++	++	–	++			
IC7	Monoclonal keratin 13⁺	++	++	–	++			
KS13.1	Monoclonal keratin 13	++	++	–	++			
CK8.6	Monoclonal keratin 1,10,11	–	–	–	–			
KB37	Monoclonal basal, squamous	–	–	–	–			
6B10	Monoclonal keratin 4	–	–	–	–			
LL001	Monoclonal keratin 14	–	–	–	++			
AUM	Rabbit urothelial membrane antigen	+	+	+++	–	+	+	+/-
ASU	Rabbit asymmetric unit membrane	–	–	++	–			
AE31	Monoclonal uroplakin 1	–	–	++	–			
V9	Monoclonal vimentin	–	–	–	++			
L9	Rabbit fibronectin	–	–	–	++			
F4	Monoclonal fibronectin	–	–	–	++			
Lam	Rabbit laminin	–	–	–	+			

The immunofluorescence reaction is scored subjectively from negative (–) to strongly positive (+++).
T.C. = Tissue Culture.
Inter = Intermediate.
Sup = Superficial.
*Human keratin polypeptide numbers used as assigned by Moll et al.
⁺Positive on only 2 to 5% of cultured cells.
Adapted from Hutton, Trejdosiewicz et al. 1993; Southgate, Hutton et al. 1994; Baskin, Hayward et al. 1996.

of thickened plaques of asymmetric unit membrane (AUM) which can be internalized to decrease the luminal surface area of the bladder during emptying.[22] The AUM is composed of urothelial specific proteins known as uroplakins.[23]

Urothelial cells can be cultured in vitro.[20,24,25,26–31] Depending on the in vitro environment the urothelial cells can form a monolayer or be induced to stratify and mimic the in vivo phenotype. Southgate et al. have been able to show that long-term organ cultures of urothelium retain the same basic phenotype as normal urinary tract tissue is situ. The "normal" phenotype is dependent on the addition of the exogenous calcium to the in vitro cultures as well as presence of the proper stroma.[27,30]

It is therefore possible to develop a table of antigen expression for urothelium in situ and to compare the normal urothelial phenotype with urothelium grown in culture and urothelium subjected to heterotypic stromas as a result of bladder reconstruction (Table 1). The expression of cytokeratins, uroplakins, integrins and E-cadherin have been well described.[23,28,32]

In the normal adult, urothelial cells have an extremely low mitotic activity, remaining quiescent until the need arises for proliferation.[20] In fact normal human urothelial turnover is estimated to occur at a rate of once every year, one of the lowest rate of cellular

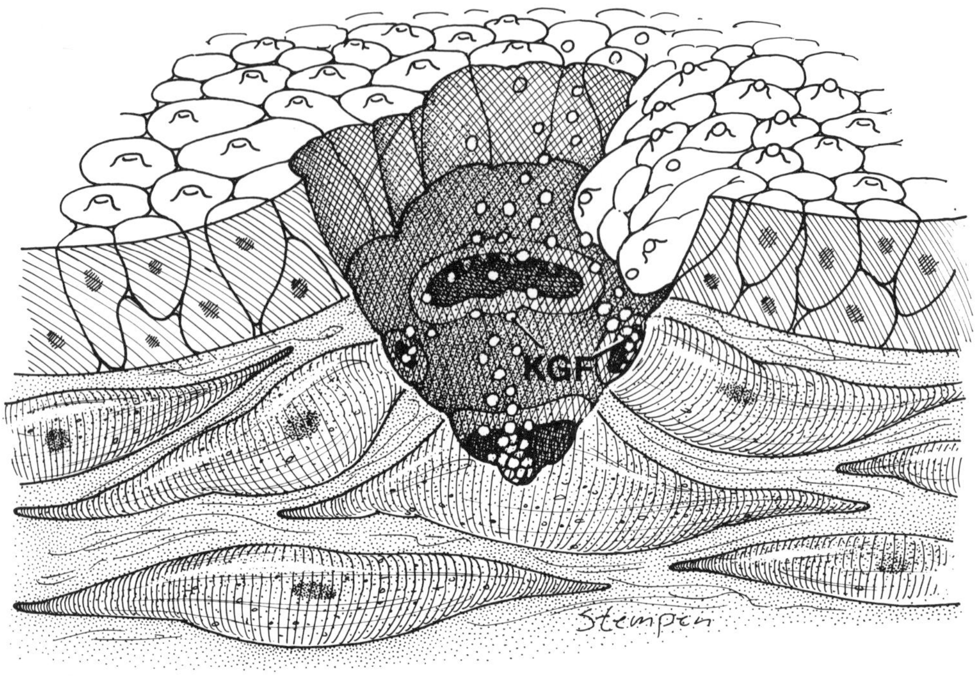

Figure 1. Keratinocyte Growth Factor is released from bladder stroma cells during bladder injury and directly acts through its urothelial located receptor to cause proliferation and regrowth of urothelial cells mending the bladder injury.[84] (Used with Permission.)

proliferation for mammalian epithelia.[33] The potential for urothelial cell proliferation has been well documented. Wong and Martin in a model of bladder injury (guinea pig) showed that mitotic activity occurs within 24 hours in the basal layer of the epithelium adjacent to the wound site.[34] At the injury margin, the epithelium forms a rolled edge, from which a two-layered sheet of flat cells grows over the edematous injury site. Healing is effected within about one week, as mitotic activity declines. Confirmatory studies with scanning electron microscopy showed that new epithelium spreads from a rolled epithelial edge.[34] Sutherland et al in a model of bladder regeneration following transplantation of an acellular matrix also showed that the urothelium from the native bladder grows across an acellular matrix surface completely relining the matrix patch within four days.[35]

What is the mechanism behind this urothelial proliferation? During bladder injury the transcript for keratinocyte growth factor (KGF- also known as fibroblastic growth factor-7) is temporally expressed at the time of urothelial injury and remains transiently elevated as urothelial proliferation repairs the damaged bladder surface.[3] Once healing is completed the message for KGF returns to basal levels. These results are consistent with previously published data on growth factors in bladder as well as changes in growth factor expression in the skin.[6,36] Werner et al showed an approximate 100 fold increase in mRNA encoding KGF one day after surgical skin injury.[36] Furthermore, the increase in KGF message was unique in that the other members of the fibroblast growth factor family (FGF), acidic FGF (FGF-1), basic FGF (FGF-2), and FGF-5 were only slightly elevated during wound healing and FGF-3, FGF-4 and FGF-6 were unchanged. In contrast to many other FGF peptides, KGF acts in a strict paracrine fashion with the receptor located exclusively on the epithelium and KGF secreted from the mesenchyme.[37,38] This is consistent with the stromal (fibroblast or smooth muscle) bladder cells producing the KGF which in turn acts on the KGF receptor located in the urothelium causing proliferation to repair the injury [6] (Figure 1).

Further evidence that KGF is important in urothelial growth has been documented by the systemic administration of KGF into experimental animals which causes direct urothelial proliferation.[3,39]

4. STROMAL CELLS

The stromal compartment of the urinary bladder is complex. It is composed of a thin muscular layer, the muscularis mucosa located immediately adjacent to the epithelial basement membrane. This is surrounded by a loose fibroblastic matrix known as the lamina propria. The lamina propria is, in turn, surrounded by the complex interwoven smooth muscle bundles which comprise the detrusor muscle. It is this detrusor muscle which gradually expands to give the bladder its compliance and contracts in the expulsion of urine.

The stroma of the bladder has a well defined ontogeny with coordinated development of its component parts. For example, the ontogeny of smooth muscle development in the rat bladder follows an orderly sequence of expression.[32] The first mesenchymal marker to be detected in the rat bladder was vimentin at 15 days of gestation (smooth muscle is first detected at day 16). With continued development vimentin expression became localized exclusively to the fibroblasts of the lamina propria and connective tissue between the smooth muscle bundles. The first smooth muscle marker to be expressed is α-actin, followed by the expression of the smooth muscle proteins vinculin, myosin, desmin, and laminin. A similar pattern of expression has been reported in the developing stroma of the rat prostate and seminal vesicle where vimentin is initially widely expressed in the mesenchyme, but with time localizes to the interductal connective tissue and lamina propria.[40] The smooth muscle differentiation markers α-actin, vinculin, myosin, desmin, and laminin also follow the same sequential pattern in the prostate as occurs in the bladder.[40]

Other investigators have shown that smooth muscle cell differentiation, as defined by the temporal expression of smooth muscle markers follows an orderly pattern. For example, Lessard and co workers showed that there is a tissue-specific expression of the isoactin multigene family in the developing rat.[41] Specifically, they were able to show that both α and γ smooth muscle actin are first expressed at ~16 days gestation in the bladder of the fetal rat consistent with our own immunohistochemical studies.[32] In the rabbit bladder, Chiavegato et al have shown that vimentin and an isoform of myosin are expressed in the embryonic period. However, with normal aging these smooth muscle cell markers cease to be expressed but can reappear in the pathologic state of experimental obstruction.[42] Other smooth muscle markers such as desmin, laminin and vinculin have also been shown to be developmentally and pathologically regulated.[43,44]

5. EXTRACELLULAR MATRIX AND ADHESION MOLECULES

Bladder extracellular matrix (ECM) changes as a function of development and pathology. For example, collagen type I and III are the major interstitial collagens that comprise the bladder wall. The amount and ratio of these structural collagens are regulated during development and in the poorly compliant bladder.[45–47] Aging also effects the extracellular matrix with specific changes in the ultrastructure of the submucosa.[48,49] Proteins that make up the elastic fiber such as elastin, fibrillin 15 and Matirx Associated Glycoproteins (MAGP) are also regulated as a function of development and location.[50]

Cell-ECM adhesion molecules expressed on bladder cells also play an important role in the required distensibility of the urinary bladder.[51] Normal human urothelium expresses integrins α 3, α V, β1, and β 4 but does not express β 3. The urothelial basement membrane contains collagen type IV and laminin. Fibronectin and the integrins α 3 and β 4 are found in or near the urothelial basement membrane area, with types I and III collagen and tenascin abutting the area. The patterns of collagen, laminin, tenascin, vitronectin, fibronectin in conjunction with the varied pattern of integrin expression in the lamina propria, vessels, nerves, and smooth-muscle layers of the bladder may facilitate the cellular interactions that occur between the urothelial and stromal cells.

6. MESENCHYMAL AND EPITHELIAL INTERACTIONS

An extensive literature describes the key role of mesenchymal-epithelial interactions in the development of many organs.[52–54] In the bladder the both the urothelial and stromal components depend upon reciprocal cellular signaling for development.[2] Such signalling is presumably also required to maintain the adult organ. These interactions may be disrupted by the use of heterotypic grafts into the bladder wall. For example, bladder augmentation with intestinal tissues may lead to abnormal cellular interactions at the anastomotic border between the urothelium and the foreign stroma. Over time the phenotype of the urothelium may change and in the worst case lead to dysplasia or even cancer. The incidence of primary adenocarcinoma of the small bowel is 0.1% and the incidence of malignancy (adenocarcinoma) in augmented bladders is higher at approximately 1.5%.[55–57] The site of the tumors in both experimental animals and in reported cases of patients with augmentations is almost always at the enterovesical junction.[16,57]

The cellular interactions that occur at the anastomotic site between bladder stroma and urothelium and foreign intestinal stroma and epithelium have not been elucidated. During normal bladder development signaling from the epithelium appears to be required for smooth muscle development[2] (Figure 2). Epithelium is also necessary for the differentiation of mesenchyme into smooth muscle in other organs such as the uterus and intestine.[58,59] Recent studies have shown that the inducing signal for smooth muscle differentiation is not specific to the age of the urothelium or in fact dependent on the origin of the epithelium.[60] For example, embryonic, newborn and adult urothelium all have the capability to induce bladder mesenchyme to form smooth muscle. Furthermore, heterotypic epithelia such ureteral, gastric, uterine, colonic, corneal and ileal will induce the same effect, although the amount of smooth muscle induction varies according to the type of epithelia.[60] In general, the ability of heterotypic epithelia to induce smooth muscle differentiation in bladder mesenchyme was linked to the amount of smooth muscle with which the epithelium is normally associated. This was the case for gastric, ureteral, myometrial and bladder epithelium which all had the ability to induce extensive smooth muscle development in tissue recombination experiments. In contrast, epithelium not typically associated with smooth muscle such as corneal epithelium and epidermis showed little inducing capability.

There exists a large body of scientific evidence to support the concept that abnormal cellular signaling can lead to cellular changes. When embryonic or even more surprising

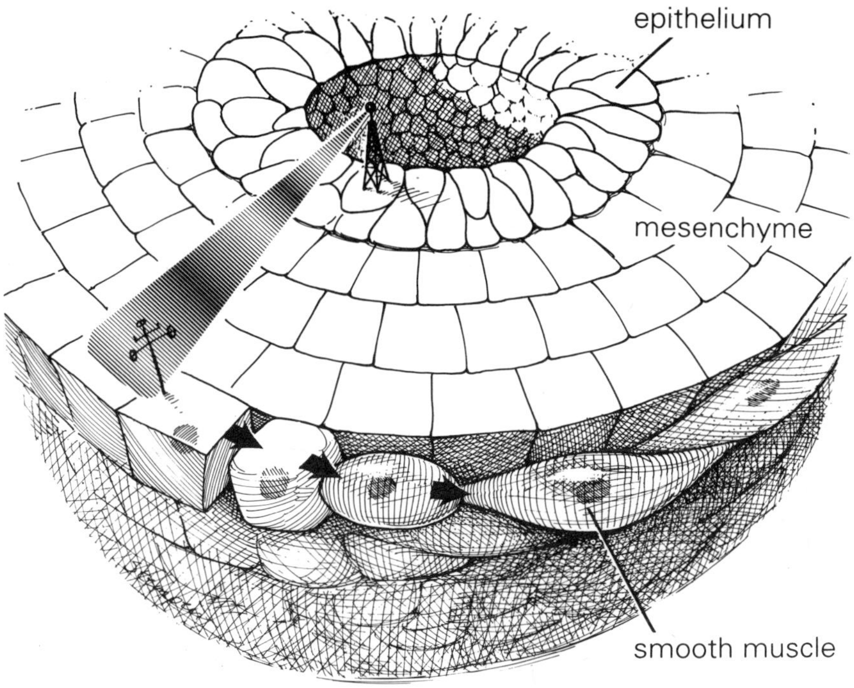

Figure 2. Epithelium is required for bladder mesenchyme to differentiate into bladder smooth muscle. Signaling occurs via an unknown mechanism possibly diffusable growth factors.[84] (Used with Permission.)

adult bladder epithelium is placed in direct contact with urogenital sinus mesenchyme, the bladder epithelium is induced to differentiate into glandular epithelium that resembles prostatic epithelium histologically and ultrastructurally, expresses androgen receptors, and secretes prostate specific proteins.[61–64] For example, both colon and prostate adenocarcinomas are also thought to arise from perturbed epithelial-stromal cell-cell interactions.[65,66 67]

Changes in the stromal environment can profoundly influence adult epithelial morphology and function. This possibility has been tested by transplanting neonatal rat seminal vesicle mesenchyme in situ to the cut end of the mouse adult ureter.[68] The adult urothelium was induced by the seminal vesicle mesenchyme to undergo seminal vesicle morphogenesis, to express seminal vesicle cytodifferentiation, and to produce the complete spectrum of major seminal vesicle proteins characteristic of the mouse. The induced seminal vesicle epithelium also expressed androgen receptors which are not seen in urothelial tissue. Staining with Hoechst dye 33258, which can distinguish cells of mouse and rat origin, further demonstrated that the induced seminal vesicle epithelium was indeed of mouse origin and not a contaminant of the inducing rat seminal vesicle mesenchyme. In addition, neonatal prostate inducing mesenchyme grafted in situ beneath the bladder mucosa of adult male mice induced prostate-like acini, indicating that the above reprogramming of adult organs in situ is not an isolated occurrence. Thus, it is clear that reciprocal epithelial-stromal interactions are important in both organogenesis and in adulthood in the homeostatic maintenance of normal tissue architecture and function and that adult epithelia retain a capacity to respond to stromal signaling. It has been postulated that alterations of these interactions may play a role in the genesis of diseases such as benign prostatic hyperplasia and prostate cancer.[66,69] It is also recognized that stromal changes occur in the genesis of many tumors.[69,70] We have recently hypothesized that the reciprocal communication between incipient prostatic cancer cells and their surrounding stroma can promote of prostatic carcinogenesis.[66,71] A similar scenario may also occur in the promotion of bladder tumors towards an invasive phenotype.

Epithelial-mesenchymal interactions during development continue as epithelia stromal interactions after birth. In numerous experimental systems we have shown that the stroma regulates the epithelium and the interactions are likely to be reciprocal.[72] For example, it has been suggested that both normal prostatic development and prostate carcinogenesis are mediated by stromal-epithelial interactions.[66,71] Recent evidence supporting this model comes from experiments in which the stroma surrounding prostatic tumors was analyzed by immunohistochemistry and shown to be dedifferentiated when compared to normal prostatic stroma.[73] In another experiment epithelial cells from the rat prostatic Dunning tumor were placed in contact with inductive mesenchymal environments (specifically with urogenital sinus and seminal vesicle mesenchyma). This resulted in a dramatic reduction in growth rate of the Dunning tumor cells and a loss of tumorigenicity.[74,75,76,77] These findings demonstrate that the connective tissue environment can have profound regulative effects on neoplastic epithelial cells.

As previously noted tumors in patients with bladder augmentations are primarily adenocarcinomas that originate at the anastomotic junction between the bladder and the intestine.[15] It is known that nitrosamines are elevated in the urine of patients who have undergone enterocystoplasty.[78] The nitrosamines in turn can cause mutagenic changes to the DNA. This hypothesis that nitrosamines are acting as carcinogens is supported by the increase incidence of cancer at the anastomotic site of patients who have had ureterosigmoidostomies.[79]

We propose an alternative hypothesis that abnormal stromal-epithelial interactions at the site of bladder intestinal anastomosis lead to perturbations in reciprocal cell-cell signaling between the intestinal stroma and the urothelium. In vitro studies on bladder cancer cell lines support this hypothesis. Gordon et al showed that altered extracellular matrices influence human bladder cancer cell lines possibly through regulation of the nuclear matrix.[80] The authors conclude that the extracellular matrices derived from transformed stroma producing cells may influence the proliferation, genetic regulation, and maintenance of the overlying urothelial tumor cells.

7. FUTURE DIRECTIONS

7.1. Regeneration of Bladder Tissue via an Acellular Tissue Matrix

Clearly, a better solution to bladder augmentation needs to be found that would avoid the complications of intestinal augmentation and the theoretical concerns of heterotypic stromal-epithelia interactions. A model of bladder wall regeneration using an acellular tissue matrix has allowed the process of cellular migration and signaling to be studied.[35] This model consists of replacing the dome of the rat bladder with a patch of acellular tissue matrix and analyzing the regeneration of bladder cells (smooth muscle, urothelium, vessels, and nerves) within this extracellular matrix scaffold (Figure 3).[35] The acellular matrix tissue graft (~1cm^2) is derived as a full thickness segment from syngeneic rat gastric or bladder specimens. The acellular tissue matrix is prepared by removing the cellular elements with distilled water and mild detergent. Absence of stromal and epi-

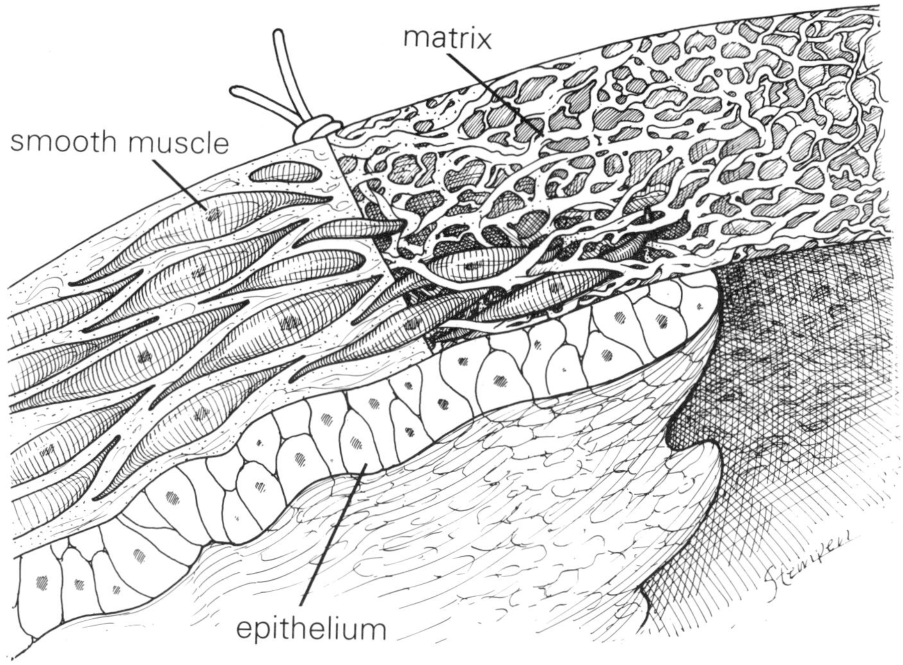

Figure 3. A model of bladder regeneration with an acellular tissue matrix shows that urothelium, smooth muscle and nerves can grow into the tissue matrix most likely under the influence of stromal-epithelial cellular communication.[84] (Used with Permission.)

thelial cellular elements were confirmed histologically. A different type of acellular tissue matrix derived from porcine small intestinal submucosa (SIS) was developed by Badylak and used to reconstruct the urinary tract reconstruction by Kroop.[81–83] While significant differences exist between SIS and the ECM scaffold, both types of matrix grafts have been incorporated successfully into the bladder, becoming populated by both smooth muscle and urothelial cells. Thus, acellular tissue matrices derived from mammalian intestine or bladder can provide an excellent scaffold that facilitates cellular regeneration of the bladder.

In both these model, extracellular matrix appears to organize the regenerative capacity of host bladder cells after bladder augmentation. The assumption in this model is that an appropriate extracellular matrix environment will allow urothelial cells from the native bladder to migrate into and line the acellular matrix patch. In turn, presumably under the influence of the urothelium, stromal cells (smooth muscle or fibroblasts) from the native bladder invade and populate the acellular matrix. Along with the cellular infiltrates composed of urothelium, fibroblasts, and smooth muscle, a new vasculature develops that supplies nutrients to the regenerating bladder tissue.

Epithelization of the acellular tissue matrix occurred by 4 days, accompanied by granulocytic infiltration.[35] Smooth muscle regenerated by 2 weeks post-grafting in juxtaposition to epithelial surfaces and matured into normal sized bundles by 26 weeks. Neovascularity was seen by 2 weeks post-surgery. Neural elements formed around developing smooth muscle bundles as early as 4 weeks. The source of regenerative smooth muscle cells is most likely the host detrusor or muscularis mucosa, as witnessed by the apparent continuity with host tissues. It appears that existing host smooth muscle cells proliferate in response to local stimuli and migrate into the adjacent graft. It is plausible that existing host smooth muscle cells undergo partial de-differentiation to a myofibroblastic phenotype, and then migrate. The spatial orientation of regenerating cellular elements suggest the presence of stromal-epithelial interactions taking place during phenotypic regeneration of the bladder. The fact that the smooth muscle is first noted adjacent to the urothelium supports the hypothesis that the urothelium plays a regulatory role in the formation of smooth muscle, which corresponds to our experimental studies of smooth muscle induction by urothelium.[2] This in vivo model provides a suitable means for the continued study of the stromal-epithelial interactions occurring during bladder cell regeneration.

ACKNOWLEDGMENTS

Supported by Taxpayers of the United States of America via funding from the National Institute of Health Grants K08 DK02397-03 and RO1 DK51397-02, DK51101

REFERENCES

1. Wein, A.J., Levin, R.M. and Barrett, D.M., Voiding Function and Dysfunction. In: Adult and Pediatric Urology. Edited by Gillenwater, J. St. Louis: Mosby, chapt.1, **933**: 1991.
2. Baskin, L.S., Hayward, S., Young, P. and Cunha, G., Role of mesenchymal-epithelial interactions in bladder development. J. Urol., **156:** 1820,1996.
3. Baskin, L., Sutherland, R., Thomson, A., Nguyen, H., Morgan, D., Hayward, S., Hom, Y., DiSandro, M. and Cunha, G., Growth Factors in Bladder Wound Healing. J. Urol., **157:** 2388,1997.
4. Hodges, G.M., Hicks, R.M. and Spacey, G.D., Epithelial-stromal interactions in normal and chemical carcinogen-treated adult bladder. Cancer Res., **37:** 3720,1977.
5. Momose, H., Uchida, K., Kakinuma, H., Yura, Y. and Oyasu, R., Involvement of urine in epithelial-stromal interactions in urinary bladder carcinogenesis. Cancer Lett., **53:** 91,1990.
6. Baskin, L.S., Sutherland, R.S., A.Thomson, A., Hayward, S.W. and Gerald R. Cunha, P.D., Growth Factors and Receptors in Bladder Development and Obstruction. Lab Investigation., **75:** 157,1996.
7. Sutherland, R.S., Mevorach, R.A., Baskin, L.S. and Kogan, B.A., Spinal dysraphism in children: an overview and an approach to prevent complications. Urology, **46:** 294,1995.
8. Bauer, S.B., Hallett, M., Khoshbin, S., Lebowitz, R.L., Winston, K.R., Gibson, S., Colodny, A.H. and Retik, A.B., Predictive value of urodynamic evaluation in newborns with myelodysplasia. JAMA., **252:** 650,1984.
9. Baskin, L.S., Kogan, B.A. and Benard, F., Treatment of infants with neurogenic bladder dysfunction using anticholinergic drugs and intermittent catheterisation. Br J Urol., **66:** 532,1990.
10. Hinman, F.J., Selection of intestinal segments for bladder substitution., physical and physiological characteristics. J Urol., **139:** 519,1988.
11. Koch, M.O., McDougal, W.S., Hall, M.C., Hill, D.E., Braren, H.V. and Donofrio, M.N., Long-term metabolic effects of urinary diversion: a comparison of myelomeningocele patients managed by clean intermittent catheterization and urinary diversion. J Urol., **147:** 1343,1992.
12. Mundy, A.R. and Nurse, D.E., Calcium balance, growth and skeletal mineralisation in patients with cystoplasties. Br J Urol., **69:** 257,1992.
13. Galloway, N.T., Mekras, J.A., Helms, M. and Webster, G.D., An objective score to predict upper tract deterioration in myelodysplasia. J Urol., **145:** 535,1991.
14. Gosalbez, R.J., Woodard, J.R., Broecker, B.H. and Warshaw, B., Metabolic complications of the use of stomach for urinary reconstruction. J Urol., **150:** 710,1993.
15. Barrington, J., Fulford, S., Griffiths, D. and Stephenson, T., Tumors in bladder remnants after augmentation enterocystoplasty. J. Urol., **157:** 482,1997.
16. Buson, H., Diaz, D.C., Manivel, J.C., Jessurun, J., Dayanc, M. and Gonzalez, R., The development of tumors in experimental gastroenterocystoplasty. J Urol., **150:** 730,1993.

17. Blyth, B., Ewalt, D.H., Duckett, J.W. and Snyder, H.3., Lithogenic properties of enterocystoplasty. J Urol.., **148:** 575, 1992.
18. Cartwright, P.C. and Snow, B.W., Bladder autoaugmentation: partial detrusor excision to augment the bladder without use of bowel. J Urol., **142:** 1050,1989.
19. Nguyen, D.H., Mitchell, M.E., Horowitz, M., Bagli, D.J. and Carr, M.C., Demucosalized augmentation gastrocystoplasty with bladder autoaugmentation in pediatric patients. J Urol., **156:** 206,1996.
20. Hicks, R.M., The mammalian urinary bladder: an accommodating organ. Biol Rev Camb Philos Soc., **50:** 215,1975.
21. Baskin, L., Meaney, D., Landsman, A., Zderic, S. and Macarak, E., Bladder Compliance Increases with Normal Fetal Development. J. Urol., **152:** 692,1994.
22. Wu, X.R., Manabe, M., Yu, J. and Sun, T.T., Large scale purification and immunolocalization of bovine uroplakins I, II, and III. Molecular markers of urothelial differentiation. J Biol Chem., **265:** 19170,1990.
23. Wu, X.R., Lin, J.H., Walz, T., Haner, M., Yu, J., Aebi, U. and Sun, T.T., Mammalian uroplakins. A group of highly conserved urothelial differentiation-related membrane proteins. J Biol Chem., **269:** 13716,1994.
24. Howlett, A.R., Hodges, G.M. and Rowlatt, C., Epithelial-stromal interactions in the adult bladder: urothelial growth, differentiation, and maturation on culture facsimiles of bladder stroma. Dev Biol., **118:** 403,1986.
25. Hodges, G.M., Rowlatt, C., Howlett, A.R. and Trejdosiewicz, L.K., An ultrastructural analysis of the physical organization of collagenous (type I) matrices: one determinant of urothelium maintenance in vitro. Scan Electron Microsc.., **15:** 215,1985.
26. de Boer, W.I., Vermeij, M., Diez de Medina, S.G., Bindels, E., Radvanyi, F., van der Kwast, T. and Chopin, D., Functions of fibroblast and transforming growth factors in primary organoid-like cultures of normal human urothelium. Lab Invest., **75:** 147,1996.
27. Fujiyama, C., Masaki, Z. and Sugihara, H., Reconstruction of the urinary bladder mucosa in three-dimensional collagen gel culture: fibroblast-extracellular matrix interactions on the differentiation of transitional epithelial cells. J Urol., **153:** 2060,1995.
28. Hutton, K.A., Trejdosiewicz, L.K., Thomas, D.F. and Southgate, J., Urothelial tissue culture for bladder reconstruction: an experimental study. J Urol.., **150:** 721,1993.
29. Southgate, J., Kennedy, W., Hutton, K.A. and Trejdosiewicz, L.K., Expression and in vitro regulation of integrins by normal human urothelial cells. Cell Adhes Commun., **3:** 231,1995.
30. Southgate, J., Hutton, K.A., Thomas, D.F. and Trejdosiewicz, L.K., Normal human urothelial cells in vitro: proliferation and induction of stratification. Lab Invest., **71:** 583,1994.
31. Baskin, L., Howard, P.S. and Macarak, E., Effect of physical forces on bladder smooth muscle and urothelium. J Urol.., **150:** 601,1993.
32. Baskin, L.S., Hayward, S.W., Young, P. and Cunha, G.R., Ontogeny of the rat bladder: Smooth muscle and epithelial differentiation. Acta Anatomica., **155:** 163,1996.
33. Hainau, B. and Dombernowsky, P., Histology and cell proliferation in human bladder tumors. An autoradiographic study. Cancer., 33: 115,1974.
34. Wong, Y.C. and Martin, B.F., A study of light and scanning electron microscopy of the lining epithelium of the guinea pig bladder following artificial ulceration. Am J Anat., **150:** 219,1977.
35. Sutherland, R.S., Baskin, L.B., Hayward, S.W. and Cunha, G.R., Regeneration of Bladder Urothelium, Smooth Muscle, Blood Vessels and Nerves Into an Acellular Tissue Matrix. Journal Urology., **156:** 571,1996.
36. Werner, S., Smola, H., Liao, X., Longaker, M., Krieg, T., Hofschneider, P. and Williams, L.T., The function of KGF in morphogenesis of epithelium and reepithelialization of wounds. Science., **266:** 819,1994.
37. Finch, P.W., Cunha, G.R., Rubin, J.S., Wong, J. and Ron, D., Pattern of keratinocyte growth factor and keratinocyte growth factor receptor expression during mouse fetal development suggests a role in mediating morphogenetic mesenchymal-epithelial interactions. Dev Dyn., **203:** 223,1995.
38. Rubin, J.S., Osada, H., Finch, P.W., Taylor, W.G., Rudikoff, S. and Aaronson, S.A., Purification and characterization of a newly identified growth factor specific for epithelial cells. Proc Natl Acad Sci U S A., **86:** 802,1989.
39. Yi, E.S., Shabaik, A.S., Lacey, D.L., Bedoya, A.A., Yin, S., Housley, R.M., Danilenko, D.M., Benson, W., Cohen, A.M., Pierce, G.F. and et al., Keratinocyte growth factor causes proliferation of urothelium in vivo. J Urol., **154:** 1566,1995.
40. Hayward, S.W., Baskin, L.B., Haughney, P.C., Foster, B.A., Dahiya, R. and Cunha, G.R., Stromal development in the ventral prostate, anterior prostate and seminal vesicle of the rat. Acta Anat., **155:** 1996.
41. McHugh, K.M., Crawford, K. and Lessard, J.L., A comprehensive analysis of the developmental and tissue-specific expression of the isoactin multigene family in the rat. Dev Biol., **148:** 442,1991.

42. Chiavegato, A., Scatena, M., Roelofs, M., Ferrarese, P., Pauletto, P., Passerini, G.G., Pagano, F. and Sartore, S., Cytoskeletal and cytocontractile protein composition of smooth muscle cells in developing and obstructed rabbit bladder. Exp Cell Res., **207:** 310,1993.
43. Allen, K.M. and Haworth, S.G., Cytoskeletal features of immature pulmonary vascular smooth muscle cells: the influence of pulmonary hypertension on normal development. J. Pathol.., **158:** 311,1989.
44. Glukhova, M., Koteliansky, V., Fondacci, C., Marotte, F. and Rappaport, L., Laminin variants and integrin laminin receptors in developing and adult human smooth muscle. Develop. Biol., **157:** 437,1993.
45. Baskin, L.S., Constantinescu, S., Duckett, J.W., Snyder, H.M. and Macarak, E., Type III collagen decreases in normal fetal bovine bladder development. J Urol., **152:** 688,1994.
46. Ewalt, D.H., Howard, P.S., Blyth, B., Snyder, H.3., Duckett, J.W., Levin, R.M. and Macarak, E.J., Is lamina propria matrix responsible for normal bladder compliance? J Urol., **148:** 544,1992.
47. Howard, P.S., Ewalt, D.H., Duckett, J.W., Snyder, H.M. and Macarak, E.J., Alterations in extracellular matrix gene expression in normal versus non-compliant human bladders. Adv Exp Med Biol., **385:** 215,1995.
48. Levy, B.J. and Wight, T.N., Structural changes in the aging submucosa: new morphologic criteria for the evaluation of the unstable human bladder. J Urol., **144:** 1044,1990.
49. Levy, B.J. and Wight, T.N., The role of proteoglycans in bladder structure and function. Adv Exp Med Biol., **385:** 191,1995.
50. Rosenbloom, J., Koo, H., Howard, P.S., Mecham, R. and Macarak, E.J., Elastic fibers and their role in bladder extracellular matrix. Adv Exp Med Biol., **385:** 161,1995.
51. Wilson, C.B., Leopard, J., Cheresh, D.A. and Nakamura, R.M., Extracellular matrix and integrin composition of the normal bladder wall. World J Urol., **14:** S30,1996.
52. Kratochwil, K., Tissue combination and organ culture studies in the development of the embryonic mammary gland. In: Developmental Biology: A Comprehensive Synthesis. Edited by Gwatkin, R.B.L. New York: Plenum Press., chapt. 4,, 315, 1987.
53. Cunha, G.R., Epithelial-stromal interactions in development of the urogenital tract. Int. Rev. Cytol., **47:** 137,1976.
54. Cunha, G.R., Alarid, E.T., Turner, T., Donjacour, A.A., Boutin, E.L. and Foster, B.A., Normal and abnormal development of the male urogenital tract. Role of androgens, mesenchymal-epithelial interactions, and growth factors. J Androl., **13:** 465,1992.
55. Smith, R.B., van Cangh, P., Skinner, D.G., Kaufman, J.J. and Goodwin, W.E., Augmentation enterocystoplasty: a critical review. J Urol., **118**: 35,1977.
56. Golomb, J., Klutke, C.G., Lewin, K.J., Goodwin, W.E., deKernion, J.B. and Raz, S., Bladder neoplasms associated with augmentation cystoplasty: report of 2 cases and literature review. J Urol., **142:** 377,1989.
57. Shokeir, A.A., Shamaa, M., el-Mekresh, M.M., el-Baz, M. and Ghoneim, M.A., Late malignancy in bowel segments exposed to urine without fecal stream. Urology., **46:** 657,1995.
58. Cunha, G.R., Young, P. and Brody, J.R., Role of uterine epithelium in the development of myometrial smooth muscle cells. Biol Reprod., **40:** 861,1989.
59. Haffen, K., Kedinger, M. and Simon-Assmann, P., Mesenchyme-dependent differentiation of epithelial progenitor cells in the gut. J. Pediatr. Gastroenterol. Nutr., **6:** 14,1987.
60. DiSandro, M., Li, Y., Baskin, L., Hayward, S. and Cunha, G., Mesenchymal-epithelial interactions in bladder smooth muscle development: epithelial specificity.J. Urol., In press.
61. Donjacour, A.A. and Cunha, G.R., Assessment of prostatic protein secretion in tissue recombinants made of urogenital sinus mesenchyme and urothelium from normal or androgen-insensitive mice. Endocrinology., **131:** 2342,1993.
62. Donjacour, A.A. and Cunha, G.R., Seminal vesicle mesenchyme induces prostatic morphology and secretion in urinary bladder epithelium. J. Cell Biol., **107:** 609,1988.
63. Cunha, G.R., Sekkingstad, M. and Meloy, B.A., Heterospecific induction of prostatic development in tissue recombinants prepared with mouse, rat, rabbit and human tissues. Differentiation., **24:** 174,1983.
64. Neubauer, B.L., Chung, L.W., McCormick, K.A., Taguchi, O., Thompson, T.C. and Cunha, G.R., Epithelial-mesenchymal interactions in prostatic development. II. Biochemical observations of prostatic induction by urogenital sinus mesenchyme in epithelium of the adult rodent urinary bladder. J Cell Biol., **96:** 1671,1983.
65. Bosland, M.C., Han, X., Ofner, P., Leav, I. and Liehr, J.G., Induction of DNA adducts and adenocarcinomas in the dorsolateral prostate of NBL rats by combined treatment with estradiol-17B and testosterone. Cancer Res., **33:** 277,1992.
66. Hayward, S.W., Cunha, G.R. and Dahiya, R., Normal Development and Carcinogenesis of the Prostate: A Unifying Hypothesis. Ann N Y Acad Sci., **784:** 50,1996.
67. Kong, W. and Sporn, M., Recent advances in chemoprevention of cancer. Science., **278:** 1073,1997.

68. Lipschutz, J., Young, P., Taguchi, O. and Cunha, G., Urothelial transformation into functional glandular tissue in situ by instructive mesenchymal induction. Kidney International., **49(1):** 59,1996.
69. McNeal, J.E., Morphology and biology of benign prostatic hyperplasia. In: Regulation of androgen action. Edited by Bruchovsky, N., Chapdellaine, A. and Neumann, F. Berlin: Congressdruck R. Bruckner, chapt., 191, 1985.
70. Hayashi, N. and Cunha, G.R., Mesenchyme-induced changes in neoplastic characteristics of the Dunning prostatic adenocarcinoma. Cancer Res., **51:** 4924,1991.
71. Cunha, G.R., Hayward, S.W., Dahiya, R. and Foster, B.A., Smooth muscle-epithelial interactions in normal and neoplastic prostatic development. Acta Anatomica., **155:** 63,1996.
72. Cunha, G.R., Bigsby, R.M., Cooke, P.S. and Sugimura, Y., Stromal-epithelial interactions in adult organs. Cell Diff., **17:** 137,1985.
73. Hayward, S.W., Olumi, A.F., Haughney, P.C., Dahiya, R. and Cunha, G.R., Prostate adenocarcinoma causes dedifferentiation of its surrounding smooth muscle. Proc. Am. Urol. Assn., **155:** 605,1996.
74. Wong, Y.C., Cunha, G.R. and Hayashi, N., Effects of mesenchyme of embryonic urogenital sinus and neonatal seminal vesicle on the cytodifferentiation of the Dunning tumor: Ultrastructural study. Acta Anat., **143:** 139,1992.
75. Hackler, R.H., Hall, M.K. and Zampieri, T.A., Bladder hypocompliance in the spinal cord injury population. J Urol., **141:** 1390,1989.
76. Hayashi, N., Cunha, G.R. and Wong, Y.C., Influence of male genital tract mesenchymes on differentiation of Dunning prostatic adenocarcinoma. Cancer Res., **50:** 4747,1990.
77. Cunha, G.R., Hayashi, N. and Wong, Y.C., Regulation of differentiation and growth of normal adult and neoplastic epithelia by inductive mesenchyme. Cancer Surv., **11:** 73,1991.
78. Nurse, D.E. and Mundy, A.R., Assessment of the malignant potential of cystoplasty. Br J Urol., 64: 489,1989.
79. Crissey, M.M., Steele, G.D. and Gittes, R.F., Rat model for carcinogenesis in ureterosigmoidostomy. Science., **207:** 1079,1980.
80. Gordon, J.N., Shu, W.P., Schlussel, R.N., Droller, M.J. and Liu, B.C., Altered extracellular matrices influence cellular processes and nuclear matrix organizations of overlying human bladder urothelial cells. Cancer Res., **53:** 4971,1993.
81. Badylak, S., Lantz, G., Coffey, A. and Geddes, L., Small intestinal submucosa as a large diameter vascular graft in the dog. J. Surg. Res., **47:** 74,1989.
82. Kropp, B.P., Rippy, M.K., Badylak, S.F., Adams, M.C., Keating, M.A., Rink, R.C. and Thor, K.B., Regenerative urinary bladder augmentation using small intestinal submucosa: urodynamic and histopathologic assessment in long-term canine bladder augmentations. J Urol., **155:** 2098,1996.
83. Kropp, B.P., Badylak, S. and Thor, K.B., Regenerative bladder augmentation: a review of the initial preclinical studies with porcine small intestinal submucosa. Adv Exp Med Biol., **385:** 229,1995.
84. Baskin, L., The fate of the bladder in patients with congenital infravesical obstruction. Current Opinion in Urology., **7:** 325,1997.

6

SEROSAL THICKENING, SMOOTH MUSCLE CELL GROWTH, AND PHENOTYPIC CHANGES IN THE RABBIT BLADDER WALL DURING OUTFLOW OBSTRUCTION AND REGENERATION

Saverio Sartore,[1,2] Marleen Roelofs,[1] Angela Chiavegato,[1] Luigi Faggian,[1] and Rafaella Franch[1]

[1]Department of Biomedical Sciences
University of Padua
[2]CNR Unit for Muscle Biology and Physiopathology
I-35121 Padua, Italy

1. ABSTRACT

In the last five years we have been studying the phenotypic changes that the rabbit bladder wall undergoes at the level of serosal compartment in experimental models known to induce an increase of organ mass. Gene expression of submesothelial mesenchymal cells (SMMC) were monitored by a panel of monoclonal antibodies specific for smooth muscle (SM) and non-muscle (NM) differentiation markers. Both the phenotypic profile and growth pattern of these cells were compared in developing, obstructed and regenerating bladder. In the obstruction model, the whole organ hypertrophy is accompanied by a hypertrophy/hyperploidy of SM cells and a marked serosal thickening due to the proliferation of SMMC. Concomitant with changes in the growth pattern of detrusor muscle, there is a SM cell phenotype switching to that of fetal type. The hyperplastic response of serosal compartment was paralleled by a spatiotemporal modification of the differentiation status of these cells, that was related to the shift of the submesothelial elastic membrane towards the underlying detrusor muscle. As a result, the thickened serosa was divided into two tissues: the 'extrinsic' and the 'intrinsic' region with respect to this membrane. The inherent differentiation plasticity of SMMC was evidenced by the two-step cellular innervation-independent conversion process from resident fibroblast to myofibroblast (or keratin-containing myofibroblast) to fetal-type SM cell which took place in the obstructed bladder. In the regenerating bladder wall this process was faster and small bundles of SM cells could be observed at both sides of the elastic

Advances in Bladder Research, edited by Baskin and Hayward.
Kluwer Academic / Plenum Publishers, New York, 1999.

membrane in the thickened serosa. Similarly, fibroblast-like interstitial cells participated in the detrusor muscle regeneration by a recruitment and subsequent differentiation within the regenerating SM cell fasicles. In chemically denervated bladder, however, interstitial cells could complete their SM-type differentiation before being incorporated in newly formed SM cell bundles, suggesting a repressive role of innervation on differentiation of these cells. Infusion experiments delivering both growth factors and cytokine to the bladder surface indicated that TGFβ1 was able to reproduce serosal thickening, transition from SMMC to myofibroblast and formation of small bundles of SM cells near the elastic membrane.

Taken together, these data indicate that SMMC and interstitial cells possess an intrinsic differentiation potential to become real SM cells. While the differentiative response of SM cells in the growth-altered detrusor muscle recapitulates the bladder ontogeny, the pattern of differentiating "serosal" SM cells is not entirely compatible with the developmental SM cell pathway. A model is proposed to explain the centripetal (from the serosa to the mucosa) gradient of SM cell differentiation during bladder ontogeny and obstruction with the well known urothelium-mesenchymal cell interactions which induce the formation of SM cell in the opposite direction (from the mucosa to the serosa) during development and regeneration.

2. INTRODUCTION

Like other hollow organs, such as the blood vessels, the bladder possesses two peripheral structural-functional interfaces with respect to the intramural mass: the mucosa and serosa (Figure 1). Both of these cell layers, along with the underlying tissues, are

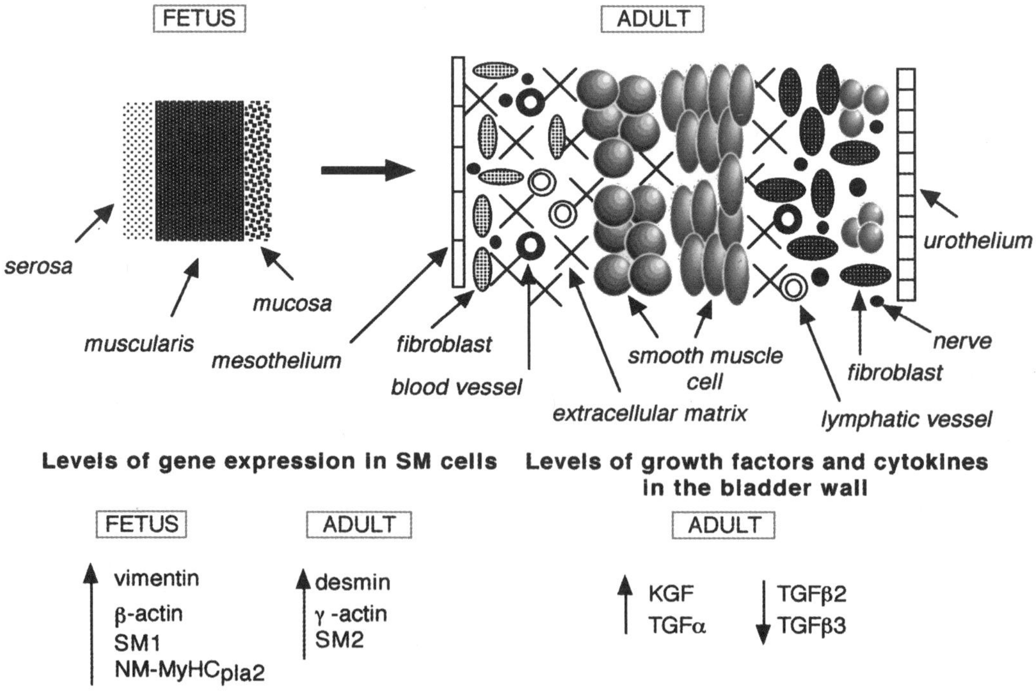

Figure 1. Schematic representation of structural changes and variations in gene expression during bladder development (physiological remodeling). The expression level of some cytoskeletal/cytocontractile proteins and growth factors/cytokines is shown.

likely to be spatially vulnerable to modifications in external environmental conditions and, hence, potentially able to influence the growth and differentiation pattern of the SM tissue and the surrounding cells. In reality, the final outcome of the homeostatic perturbation (adaptation) or the result of the application of a pathological stimulus (disease) is dependent on the fine interplay between the external action and the intrinsic (genetic) ability to confront this alteration.[1,2] The bladder wall response to changes in the basic state comprises modifications in the growth pattern (hypertrophy, hyperplasia, hyperploidy) and gene expression profile of SM cells of detrusor muscle. As accompanying phenomenon and to varying degrees, there are alterations in the extracellular matrix (ECM) deposition, serosal thickening, alterations of the urothelium, lamina propria and innervation.[1]

The hallmark of bladder wall stress or injury is represented by the increase of its mass, which is particularly evident in models where the intravesical pressure is increased. In the largely diffused procedures of inducing partial outflow obstruction in various species, there are modifications in the growth pattern and contractility of detrusor muscle, ECM content, hyperplasia of some cellular components and alteration of innervation. The concept has emerged that organ hypertrophy of bladder is not irreversible but it can be partially retrieved with the removal of obstruction.[1] On this basis, it has been proposed that the bladder wall consists of cellular elements which possess a relevant growth plasticity and the ability to activate the apoptotic process in order to expand or reduce the wall size. Among the various cells involved, two of them have drawn our attention in the last five years: the SM cells and the cells of the serosa.

Basically, we have explored the differentiation level attained by the muscularis and the serosa, monitoring the expression of some cytoskeletal and cytocontractile protein markers that in growth-altered arterial SM cells had furnished useful information. In particular, we have focused on the myosin isoform expression which in all three muscle systems is considered to be a reliable tool to study the differentiation process.[3,4] Myosin, in association with the other contractile and regulatory proteins, plays a crucial role in the physiology and pharmacology of bladder SM cells and several structural variants of SM and NM type have so far been described. We have asked whether changes of growth pattern in the detrusor muscle were accompanied by a recapitulation of an "immature" SM cell phenotype which might potentially adapt better than the "adult" counterpart to the new functional demand imposed on the bladder system by obstruction.

3. GROWTH AND DIFFERENTIATION IN THE NORMAL BLADDER WALL

It is well known that the morphological, biochemical and functional properties of the bladder change markedly during development essentially because the increasing volume of urine produced during growth requires a more compliant bladder.[5] This event is paralleled by an increase in wall thickness due to the increase in the interstitial connective tissue and SM tissue hypertrophy (physiological remodeling). As with the vascular SM tissue, the various phases of development (fetal, postnatal and adult) are characterized by an assortment of cytoskeletal/cytocontractile proteins (see Figure 1). Vimentin, β–actin and type 2 platelet-related myosin heavy chain (NM-$MyHC_{pla2}$) become down regulated around birth or after birth, whereas desmin, SM γ-actin and the SM-type MyHC named SM2 are up regulated.[6–8] Some information accumulated in the vascular system indicates that NM-$MyHC_{pla2}$ up regulation is associated with proliferation (and migration) of SM cells.[2,3] Thus, its down regulation during bladder development might be related to the marked decrease of the intrinsic capacity of these cells to proliferate. As for the SM-type MyHC isoforms, it is known that they are the product of distinct alternative splicing processes at the C- or N-terminus transcript from a single gene, whereas the NM variants are encoded by different genes, though additional subvariants can be produced by alternative splicing at the N-terminus (reviewed in Sartore et al.[2]). It is noteworthy that no firm correlation exists between SM1/SM2 ratio and maximum velocity of shortening when different muscles are compared.[9] This is quite suprising if one considers that visceral and vascular variants of SM-MyHC display distinct velocity of both actin filament movement in vitro and actin-activated Mg^{2+}-ATPase activity.[10] Thus, no direct functional consequence can be drawn from the shift of myosin isoform regulation during development.

Interestingly, the number of S-phase SM cells in detrusor muscle declines dramatically in pre- vs. postnatal rabbit, and in adult about 2% of SM cells are able to incorporate bromo-deoxyuridine (BrdU).[6] Baskin et al.[11] have estimated the level of some growth factors and cytokines during development in the rat bladder and their data about the time course of keratinocyte growth factor (KGF) and transforming growth factor (TGF)-α, -β2 and -β3 fit closely with our own about the proliferation index of rabbit SM cells (Ref.6, see also Figure 1). Thus a direct relationship might exist between changes of gene expression for some differentiation markers and the switching on of growth factors/cytokines in the bladder wall.

In detrusor muscle SM cells, the distribution of some cytoskeletal and cytocontractile proteins appear to be homogeneous,[6] whereas in vitro experiments performed on cultures from neonatal and adult bladder SM tissue have furnished a slightly different picture.[12] While SM myosin and SM-type α-actin are equally expressed in two morphologically distinguishable SM cell populations from adult bladder grown in vitro, α-actin is absent in "short" cells from neonatal rabbits but expressed in the adult counterpart. Like arterial SM cells,[13] distinct SM cell lineages, with different growth responsiveness and differentiative potentials, might be present in the bladder SM tissue.

4. THE SMOOTH MUSCLE AND SEROSA RESPONSE TO OUTFLOW OBSTRUCTION

Depending upon the severity of lesion and the specific propensity of damaged tissue/cell to respond to perturbation or injury, the exposure of the bladder wall to mechanical stress (stretching and increased transmural pressure) is likely to induce dramatic changes in the proliferative/differentiative/apoptotic characteristics of SM and NM cells.[14] When the basic homeostatic range or the "pathological" threshold are overcome, a complex series of molecular and cellular events are set in motion. Remodeling and/or repair of the bladder wall are activated and the increased load induces a compensatory phase followed eventually by a decompensation phase and, subsequently, organ failure as occurs with other hollow organs containing a circulating fluid and subjected to similar conditions.[15]

4.1. The Smooth Muscle

The SM response of the bladder from adult animal to obstruction is temporally well defined. The first "wave" of molecular response is represented by the transient increased activation of some protoncogenes, bFGF (basic fibroblastic growth factor) and stress proteins, whereas TGFβ1 decreases shortly after obstruction (see Figure 2).[11,16–20] DNA syn-

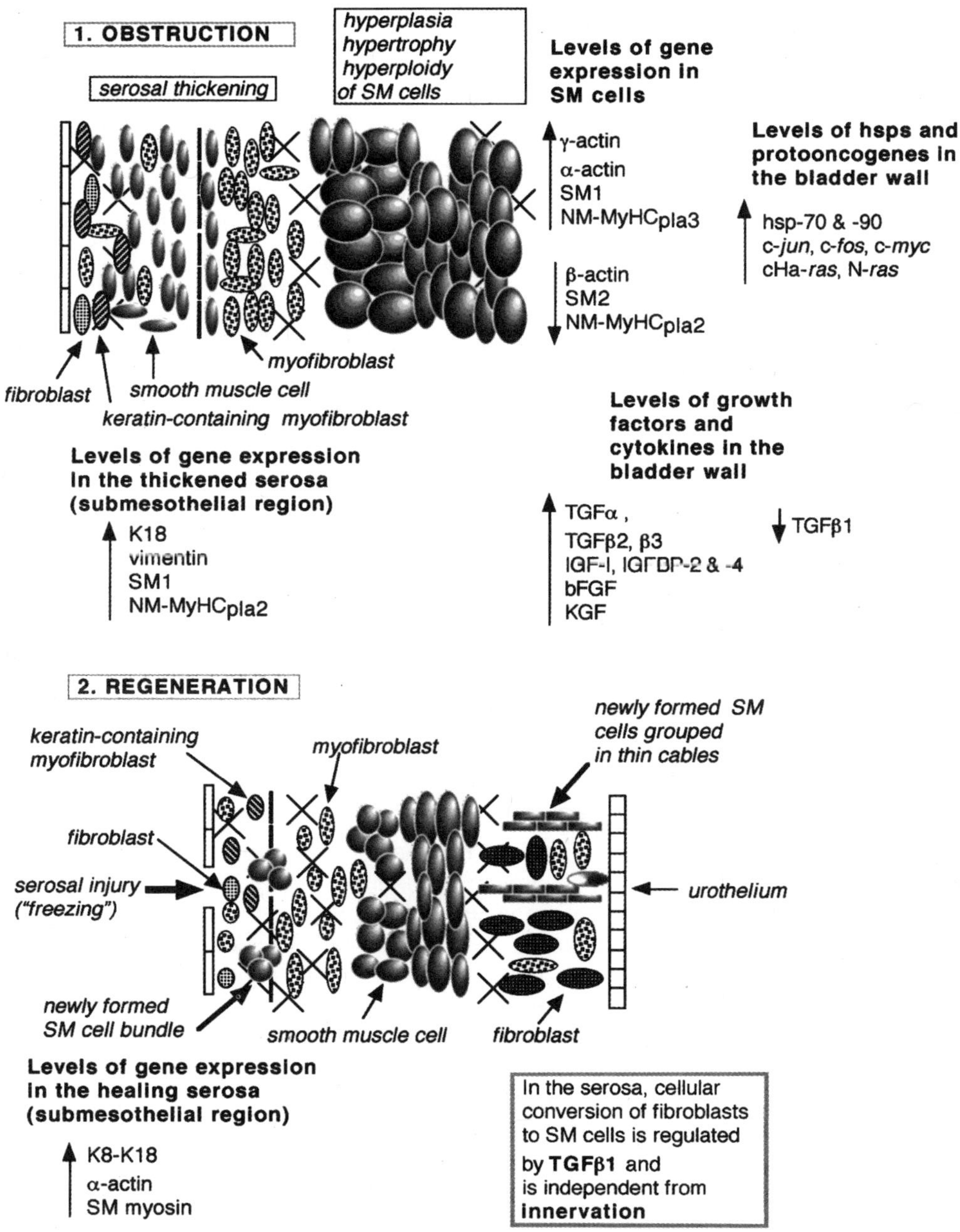

Figure 2. Schematic representation of cell phenotypic changes and variations in gene expression for some cytoskeletal/cytocontractile proteins and growth factors/cytokines during bladder obstruction and regeneration.

thesis, visualized by ^{3}[H]-thymidine incorporation, follows this initial burst of "immediate early" gene activation, but is essentially localized to NM cells.[21] In concordance with Brent and Stephens,[22] we found that BrdU incorporation in SM cells occurs at later times (30 days). Hypertrophy and/or hyperplasia (the former possibly accompanied by hyperploidy[6]) develop in detrusor muscle in the second "wave" of response to the injury1.[17,23,24] Activation of normally quiescent SM cells is accompanied by marked changes in the cytoskeletal and cytocontractile protein expression. As a general rule, there seems to be the reactivation of a "fetal" program of gene expression in the obstructed bladder, i.e., up regulation of vimentin, NM-MyHCpla,[26] and γ-actin[8,9] and down regulation of SM2[6,9,25,26], and β-cytoplasmic actin.[8] In the rabbit model, hypertrophy of SM cell is paralled by the de-novo expression of NM-MyHC$_{pla3}$.[6] Upon removal of obstruction, the up-regulated vimentin and NM-MyHC$_{pla3}$ expression[6] as well as the SM1/SM2 ratio return to nearly normal values,[25] concomitantly with the regression of bladder hypertrophy.

Changes in the growth and differentiation patterns of SM cells are paralleled by morphological[27] and physiological adaptations.[1,9,28] Though myofilaments are ultrastructurally unchanged compared to the normotrophic SM cell, in hypertrophic SM cell there is an alteration in cytoskeletal and thick filament content. There is a decrease in the force generation for cross-sectional area in the face of unchanged tension cost associated with this event. Since a more direct relationship exists between change in the maximun velocity of shortening and the 17-kDa myosin light chain expression,[29] it remains to be established just what the relevance of the qualitative changes in MyHC isoform composition, the quantitative decrease of SM-type MyHC[29] and the variation in contractility are in the hypertrophic SM cell.

4.2. The Serosa

The serosal compartment of the bladder undergoes profound morphological changes which were originally interpreted as evidence of healing from the obstructive stress inasmuch as this tissue contained abundant ECM and an apparent histologically homogeneous cell population of "fibroblasts."[22,30–33] This external "cushion" could be responsible for or associated with the functional modifications of bladder physiology, such as decreased compliance.[1] The use of monoclonal antibodies to specific differentiation markers allows for the identification of the phenotypic properties of these cells named as SMMC.[24,34,35] In combination with molecular biology, image analysis and physiological investigations, the differentiation, function and growth profile of these cells have been defined in detail.

It is important to underline that the serosal thickening which develops as a consequence of obstruction can be substantially reversed with obstruction removal.[20] This finding is indicative of regression phenomenon similar to that of a wound repair where there is proliferation, differentiation, migration and apoptosis of various cell types (endothelial cells, fibroblasts, etc.) within the wound environment.[36] Since all the measurements concerning the transient or permanent activation of stress proteins and "early immediate" genes have been performed on the whole bladder preparation (see Figure 2), it is not clear from which cell or cells these factors are released.[11,16–19,37] Similarly, the cellular source of growth factors and cytokines putatively involved in the hyperplastic process of SMMC which takes places within 3–4 days in the serosa remains unknown.[16] It must also be taken into account that fibroblasts (including, probably, SMMC) act as "sentinel" cells inasmuch as they are able to intitiate inflammation and are key sites of chemokine synthesis.[38] It is highly probable that these factors act locally by autocrine/paracrine mechanisms and their ultimate goals might be exerted on (in combined or separate manner) the urothelium, the

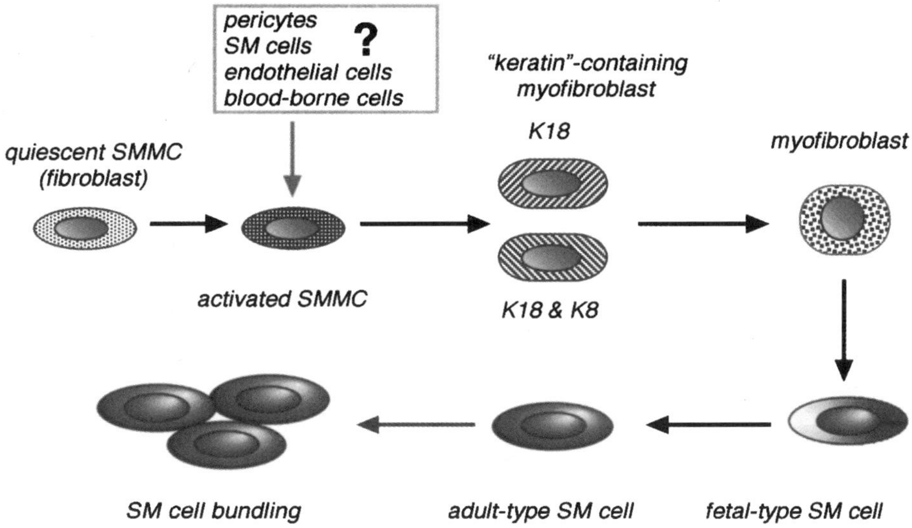

Figure 3. Schematic representation of the time-dependent multistep "serosal" SM cell pathway from SMMC to differentiated SM cells.

resident fibroblasts of serosa and lamina propria, the SM cells of detrusor muscle, the neural components and the blood vessel network. The impact of these transiently or permanently activated factors on the possible functional interdependence between different cellular components within the bladder wall remains to be established.

Both in young[24] and in adult[34] rabbits the obstruction induces an extraordinary thickening of the serosal tissue characterized by a temporal- and spatial-dependent phenotypic changes of SMMC. A few days after obstruction, these cells are converted to myofibroblasts, i.e. cells showing a mixed fibroblast-SM cell structural-functional repertoire.[39–41] This event occurs concomitantly with an increased level of BrdU incorporation by SMMC.[24,35] There is a marked spatial difference in the distribution predominantly localized to the so-called 'extrinsic' region, i.e., in a tissue which is separated by an elastic membrane[31] from the underlying 'intrinsic' region. Since the elastic membrane is in unobstructed bladder present just beneath the mesothelium,[24] it is reasonable to assume that the cells accumulated in the 'extrinsic' region are the result of migration and proliferation of SMMC. Interestingly enough, the differentiation profile is markedly different in cells from the 'extrinsic' vs. 'intrinsic' region:[1] in the 'extrinsic' region myofibroblasts express keratin 18, whereas in the 'intrinsic' region this intermediate filament protein is lacking (see also Figure 2). Also noteworthy is that coexpression of vimentin and simple keratins (8 and 18) in mesenchymal cells has been associated with increased migratory and invasive ability of tumor cells[42] as well as with cell proliferation and differentiation in the regenerating limbs of urodele amphibians.[43] In this latter circumstance, it has been hypothesized that keratin expression in mesenchymal cells is a pre-requisite activating step for starting division and keeping cells in undifferentiated state.[43]

At later stages of obstruction (about day 30), the myofibroblasts undergo a further cell transition to SM cell of fetal type[44] (see Figure 3). The final maturational step reached by SMMC, i.e., the acquisition of a SM cell phenotype, has been identified on the basis of expression of SM-type myosin which is thought to be a reliable marker of SM cell differentiation.[45] As keratin expression is strictly related to the onset of SMMC proliferation, the differentiation of SMMC into SM cells might necessarily require a direct contact of fibroblast with myofibroblast. In fact, when performing in vitro experiments on cultured chicken gizzard SM cells a layer of fibroblasts is necessary for these cells to complete their phenotypic changes.[46]

To confirm that the phenotypic change is also accompanied by a functional competence, we isolated the thickened serosa and analyzed the contractile behaviour of the tissue. While at early stages after obstruction (when NM myosin and SM-type α-actin are present) no contractile response is detectable, at day 30 (when SM myosin is expressed) KCl-stimulated contractility appears.[44] At day 60, when NM-MyHC expression disappears and that of desmin and of phosphoglucomutase (PGM) - related protein appear (adult-type SM cell phenotype), contractility stimulated by bethanechol or KCl is evidenced. Under no circumstances can a field stimulation (via release of neurotrasmitters) develop.[44] Interestingly, the complete differentiation in adult-type SM cells seems to be related to cell grouping, i.e., the tendency of SM cells to group in small bundles. These data clearly demonstrate that activated SMMC can be converted into real SM cells.

While the derivation of myofibroblasts from SMMC has been convincingly proved, that of proliferating SMMC is still under debate. Some authors have postulated SM cells or pericytes as alternative sources.[39] While the transition from SM cells to pericytes has been found in vitro,[47] the opposite might be hypothesized to occur only during morphogenesis of blood vessels.[48] We think that these two hypotheses can be discarded for the following reasons. First, in the 'extrinsic' region of the thickened serosa,[31] there is no trace of blood vessels or SM cells. Second, the activated SMMC possess a cytoskeletal protein which is not present in the pericytes: the simple keratins 8 and/or 18.[35] Third, the zones in villous stroma between central vessels and superficial trophoblast of the mature human placenta are distributed in a spatio-temporal manner similar to that found in the bladder wall, i.e., a gradient ranging from fibroblast (in the periphery) to myofibroblasts (in proximity of the center).[49] Fourth, fibroblasts of porcine carotid artery (upon injury) become myofibroblasts and are incorporated into the tunica media.[50] The final step, from myofibroblast to SM cell, has also been demonstrated to occur during development in the placental blood vessels[49] and in some human pathological settings, such as the hypertrophic scars and the stromal cells of mammary tumors.[51]

5. THE SMOOTH MUSCLE AND SEROSA RESPONSE TO REGENERATION

Regeneration of urinary bladder in the treatment of fibrosis or invasive bladder tumors are approached by bladder augmentation or substitution using intestinal segments or alloplastic prosthetic materials.[52–54] In light of the complications which may follow the reconstitution of this organ, the evaluation of the regenerating ability of the bladder wall major tissue is of paramount importance. Experimentally-induced lesions in bladder with loss of tissue are indeed thought to take place by true regeneration, i.e., outgrowth of SM cells from the edge of remnants into the defect.[55–60] Sutherland et al.[61] have studied in detail the regenerative process using hemi-cystectomy and transplantation of an acellular bladder matrix and have come to the conclusion that the SM tissue regenerates in 2 weeks and is preceded by that of urothelium. On the other hand, healing of wounds and subsequently scar formation without histological evidence for newly formed SMC has also been postulated to occur in various animal species and models.[62,63] From these contradictory results and on the basis of what is reported in Section 3 one might pose the following questions: (1) can the SMMC play a role in lesions to the bladder wall characterized by loss of tissue? and (2) when a localized lesion involving a loss of tissue is produced in the bladder wall, is the differentiation pattern of SMMC the same as the one found with a generalized insult, such as obstruction?

To answer these queries we produced a mild necrotizing lesion by "freezing" a limited region of rabbit bladder (following essentially the protocol of Shirai et al.[59]) and compared the cell phenotypic response of SM tissue and injured serosa with those previously obtained with partial outlet obstruction (Figure 2). Though the BrdU-incorporation level in the surviving SM cell bundles is scarce, it cannot be excluded that part of the newly formed SM tissue derives from pre-existing but de-differentiated and migratory-competent SM cells. In the regeneration of SM cells, we have assigned an important role to the interstitial cells sorrounding the SM cell fascicles and expressing in untouched controls vimentin and NMplα2.[64] In fact, there is incorporation of these cells in regenerating SM fasicles which become at later stages fully differentiated SM cells.[64] During extensive remodeling of lung non-muscular microvessels in the course of hyperoxic pulmonary hypertension a similar phenomenon takes place: interstitial and myofibroblast-like cells become SM cells after being incorporated into the vascular wall.[65] In our model, the cell conversion process of interstitial cells from fibroblast to myofibroblast is accelerated by partial chemical denervation.[64] In addition, alteration of reinnervation during the regenerating process induces a complete differentiation of interstitial myofibroblasts to differentiated SM cells before being incorporated into the SM fasicle, particularly in the lamina propria.[64] These data indicate that: (1) interstitial cells, like SMMC, have the potential to be converted in SM cells, (2) a differentiation gradient exists between the outer vs. the inner layer of bladder wall, suggesting some interactions with the urothelium or growth factors/cytokines released from this tissue, and (3) the lack of innervation in the regenerating bladder stimulates the achievement of the final step in the conversion pathway from interstitial cells to SM cells

The serosa responds initially to injury by producing an intimal thickening (Figure 4) and later by the formation of small bundles of SM cells at both sides of the elastic membrane (Figure 5). Like the thickened serosa in obstructed bladder, there is an elastic membrane which divides the 'extrinsic' from the 'intrinsic' region (Roelofs et al., in preparation). The SMMC accumulated in the 'extrinsic' region display the characteristics of myofibroblasts containing NM-MyHCplα2, vimentin, SM-type α-actin and keratin 8–18 (Figure 4, Roelofs et al., in preparation). The grouping of SM myosin- and SM22-containing SM cells (Figure 5, the expression of SM22 protein is indicative of fully differentiated cells[2]) suggests that the elastic membrane might act as recruiting tissue for SM cell differentiation. Thus, the comparative analysis of serosal thickening in the obstruction vs. regeneration model indicates that in the latter the process of SM cell conversion is faster and fully differentiated SM cells can bundle in a manner resembling the normal SM fasicles.

6. SOME FACTORS INVOLVED IN SM CELL DIFFERENTIATION DURING SEROSAL THICKENING

We reasoned that the serosal response found in the "freezing" model of bladder wall regeneration might involve some growth factor/cytokine typically released as part of the flogistic response which precedes SMMC accumulation.[66] We therefore undertook a series of infusion experiments using TGFβ1, PDGF, EGF, G-CSF or GM-CSF separately deliv-

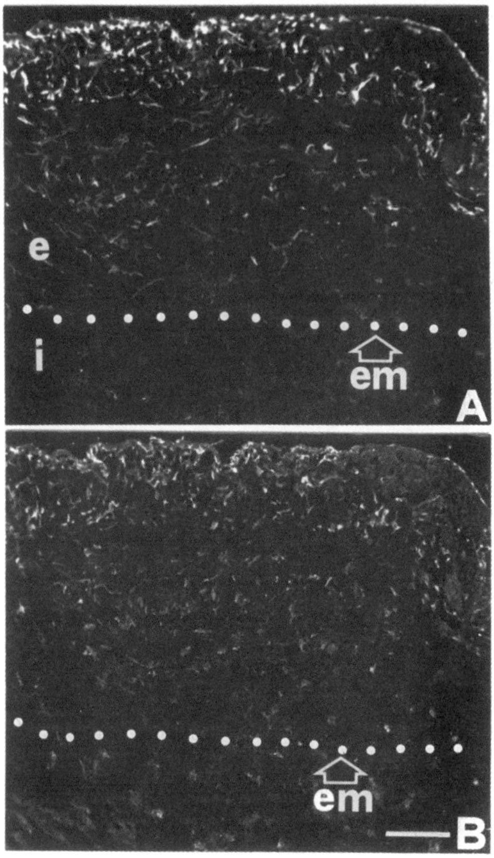

Figure 4. Micrographs showing the immunofluorescence pattern of serosa from injured adult bladder wall (7 days after "freezing") with monoclonal anti-keratin 18 (A) and keratin 8 (B) antibody. Note the accumulation of keratin-positive cells in the submesothelial region (the outermost area of 'extrinsic' region). em, elastic membrane; e and i, 'extrinsic' and 'intrinsic' region of serosal thickening. Bar, 70 mm.

ered onto the serosal surface of the intact bladder via osmotic minipumps.[67] It turned out that the TGFβ1 on its own is able to mimic exactly the pattern of serosal thickening formation as described for the regenerating bladder. In particular, this cytokine induces the formation of myofibroblasts and small bundles of fully differentiated SM cells at both sides of the elastic membrane which divides the 'extrinsic' from the 'intrinsic' region.[67]

The role of TGFβ1 as inducer of fibroblast to myofibroblast transition is well known[41] and some information is available about a similar activity in cultured dedifferentiated human SM cells which can redifferentiate upon addition of this cytokine.[68]

Since other investigators have analyzed the bladder wall response to obstruction and found that TGFβ1 is decreased soon after surgery,[20] the effect played by this cytokine in our model seem to be in contradiction with their results. A possible explanation for this discrepancy might be that (1) these authors have analyzed the whole bladder and not the thickened serosa, and (2) the obstruction and regeneration involve different sets of growth factors/cytokines which act differently on their specific target cells. Baskin et al.[69] in their model of regeneration using an acellularized bladder transplant found that KGF and TGFα, but not TGFβ2 and β3, are consistently more elevated in the regenerating bladder. This finding is interesting because, as reported above, TGFβ2 and β3 are increased in the obstructed bladder. This is a confirmation that even in this regeneration model of the bladder wall there might be a peculiar activation of growth factors/cytokines. A detailed study aimed at the evaluation of the combined growth pattern, differentiation profile and growth

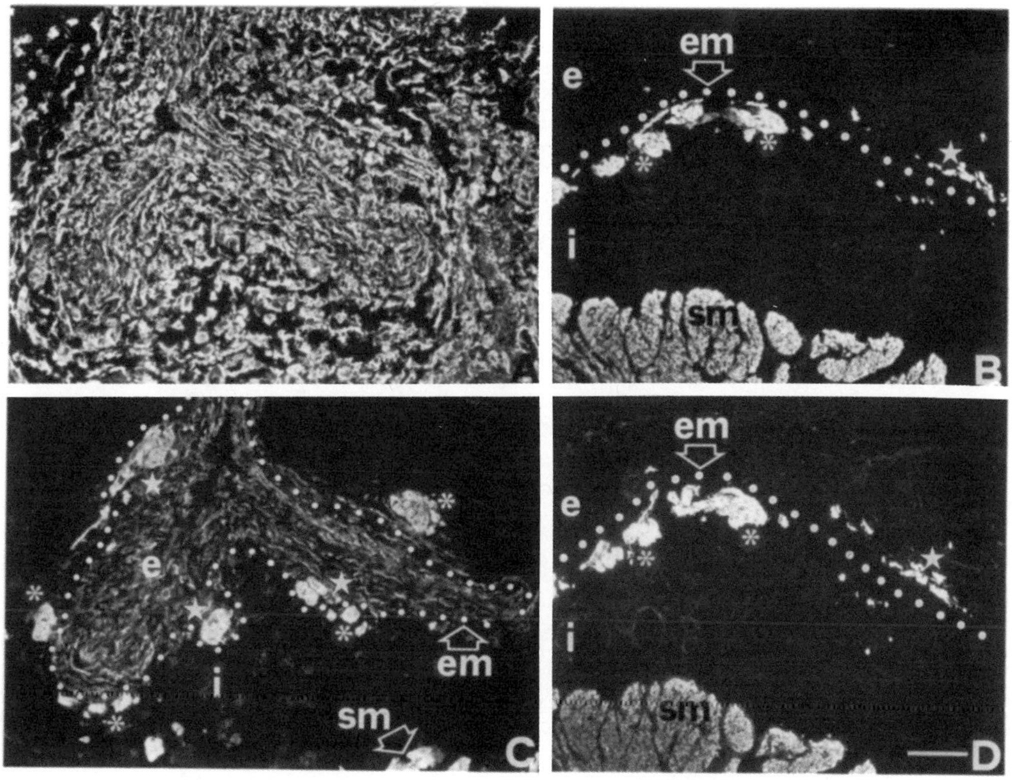

Figure 5. Micrographs showing the immunofluorescence pattern of serosa from injured adult bladder wall (15 days after "freezing") with monoclonal anti-cytoskeletal/cytocontractile protein antibodies. A, anti-vimentin; B, SM-E7 anti-SM myosin; C, anti-SM α-actin; D, E-11 anti-SM22. em, elastic membrane; e and i, 'extrinsic' and 'intrinsic' region of serosal thickening; sm, smooth muscle. Note that small bundles of SM cells are visible in the 'extrinsic' (stars) or 'intrinsic' (asterisks) region, very close to the elastic membrane. A large number of vimetin/actin-positive myofibroblasts is detectable in the 'extrinsic' region in panel A and C. Bar =70 mm.

factor/cytokine expression in each tissue of the bladder wall in obstruction vs. regeneration model would be particularly helpful in elucidating the molecular and cellular responses of the bladder in the two models.

7. A PUTATIVE MODEL OF SMMC ACTIVATION

We have shown that SMMC have the differentiative potential of becoming real SM cells and of being assembled in small SM cell bundles. This morphological aspect of the biology of "serosal" SM cells in conjunction with the development of SM-specific physiological properties provides the unquestionable evidence that the differentiation pathway has achieved its final step. The second conclusion that could be drawn from our studies is that in the obstructed bladder there are two opposite phenotypic processes at work, namely the induction of SM cell differentiation in SMMC and a dedifferentiation process in the detrusor muscle towards a fetal phenotype. At a given time from obstruction, it might seem that the whole bladder wall is in a differentiative condition resembling the early stages of development. If one considers the fetal bladder (see Figure 6 and Ref.[70]), the

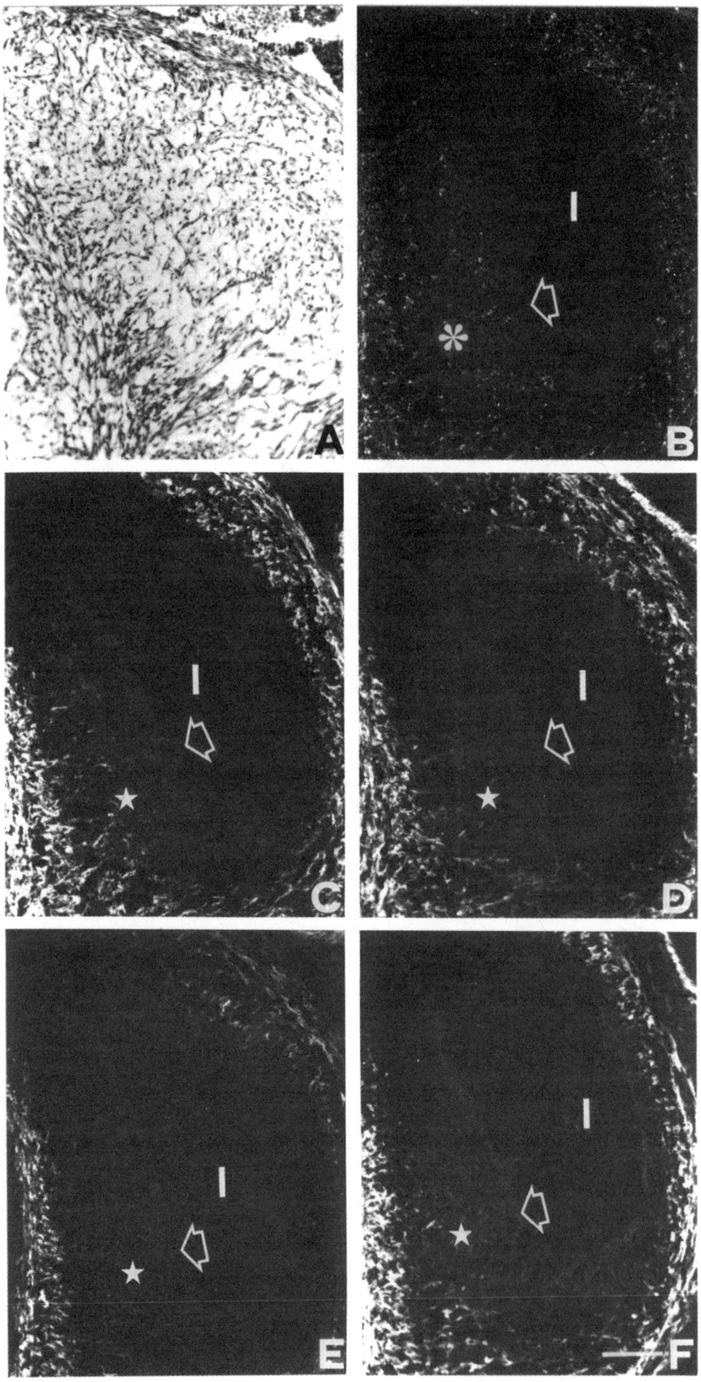

Figure 6. Micrographs showing the hematoxylin-eosin staining (A) and the immunofluorescence pattern (B-F) of fetal bladder (16 days of gestation) with monoclonal anti-cytoskeletal, cytocontractile and regulatory protein antibodies. B, anti-vimentin, C, anti-SM α-actin; D, anti-calponin; E, SM-E7 anti-SM myosin; F, E-11 anti-SM22. Note that the region near the urothelium (arrowhead) is vimentin positive (asterisk in B), but negative or very weakly positive with the other antibodies (filled stars). All antibodies label the peripheral region. l, Lumen. Bar, 70 mm.

process of SM cell differentiation begins at the serosal level where the specific markers (such as SM myosin, SM α-actin, calponin and SM22) are found. On this basis, one might argue that the "serosal dislocation" of SM cell differentiation is reminiscent of the bladder ontogeny. As a consequence of this assumption, it is quite reasonable to expect that the SMMC possess the ability to furnish new SM cell on demand. If the bladder SMMC are a sort of "fetal remnant", it would be expected that keratin 8/18 are also present in the bladder wall of immature bladder. This is not the case given that in the rabbit fetus at day 16 of gestation there is no trace of keratin immunoreactivity except for the urothelium.[35] In light of these results, we can say that activated SMMC during obstruction of adult rabbit are distinguishable from developing SMMC. We have hypothesized that SMMC have a dual differentiative potential and, hence, two options: the epithelial or mesenchymal pathway.[35,67] The existence of the epithelial (mesothelial) option is based on the finding that pleural lesion is able to activate pulmonary SMMC which show the ability to regenerate the injured mesothelium.[71] These activated SMMC, like those found in the injured bladder, contain keratins.

8. "SEROSAL" SMOOTH MUSCLE CELLS AND UROTHELIUM-MESENCHYMAL INTERACTIONS

Taking into account the spatial origin of myogenesis in obstructed and regenerating adult bladder and the histology of the bladder wall, there are two distinct tissue epithelial-mesenchymal interfaces which might be involved in the process of SM cell differentiation: the urothelium/mesenchymal cells and the mesothelium/mesenchymal cells (Figure 7). In the former, the lack of basement membrane allows direct interaction of urothelium with the subjacent mesenchymal cells, whereas in the latter there is the shift of the elastic membrane towards the underlying mesenchymal cells that might be the crucial event. It might be that in the experimental models of bladder wall injury examined above, the initial stimuli for the respective molecular and cellular response are represented by the lack of integrity or positional control of these two membranes with respect to the epithelium.

According to some authors,[72,73] epithelial-mesenchymal interactions are of fundamental importance during morphogenesis of visceral organs, in a manner much similar to the key role played by endothelium-mesenchymal cell interactions in vasculogenesis and angiogenesis.[48] Likewise, in the regenerating bladder from adult rats there is a convincing demonstration that urothelium and the underlying mesenchymal cells interact,[61] possibly via release of KGF and other growth factors,[69] and that SM cells are produced in the vicinity of regenerating urothelium. As shown by the same authors,[74] SM cell differentiation initiates from the "serosal' side in the developing rat and rabbit (Figure 5) bladders. These findings seem to be in contrast with preliminary experiments in which SM cell differentiation is not induced when undifferentiated bladder mesenchyme and serosa are mixed together and grafted in a "neutral" environment.[75] Conversely, when bladder undifferentiated mesenchyme is mixed with bladder urothelium SM cell differentiation develops.[75]

If one considers what is known about the endothelium-mesenchymal cell interactions (Ref.[76] and Figure 7), it is quite difficult to accept that "recruiting" factor(s) is operating during morphogenesis of the bladder wall. In fact, while in the developing arterial vessel there is a gradient of differentiating cells which is high in proximity of the endothelium,[76] in the bladder the opposite occurs. Thus, a different mechanism is likely to be involved. In Figure 7 we put forward two hypotheses to explain the gradient of

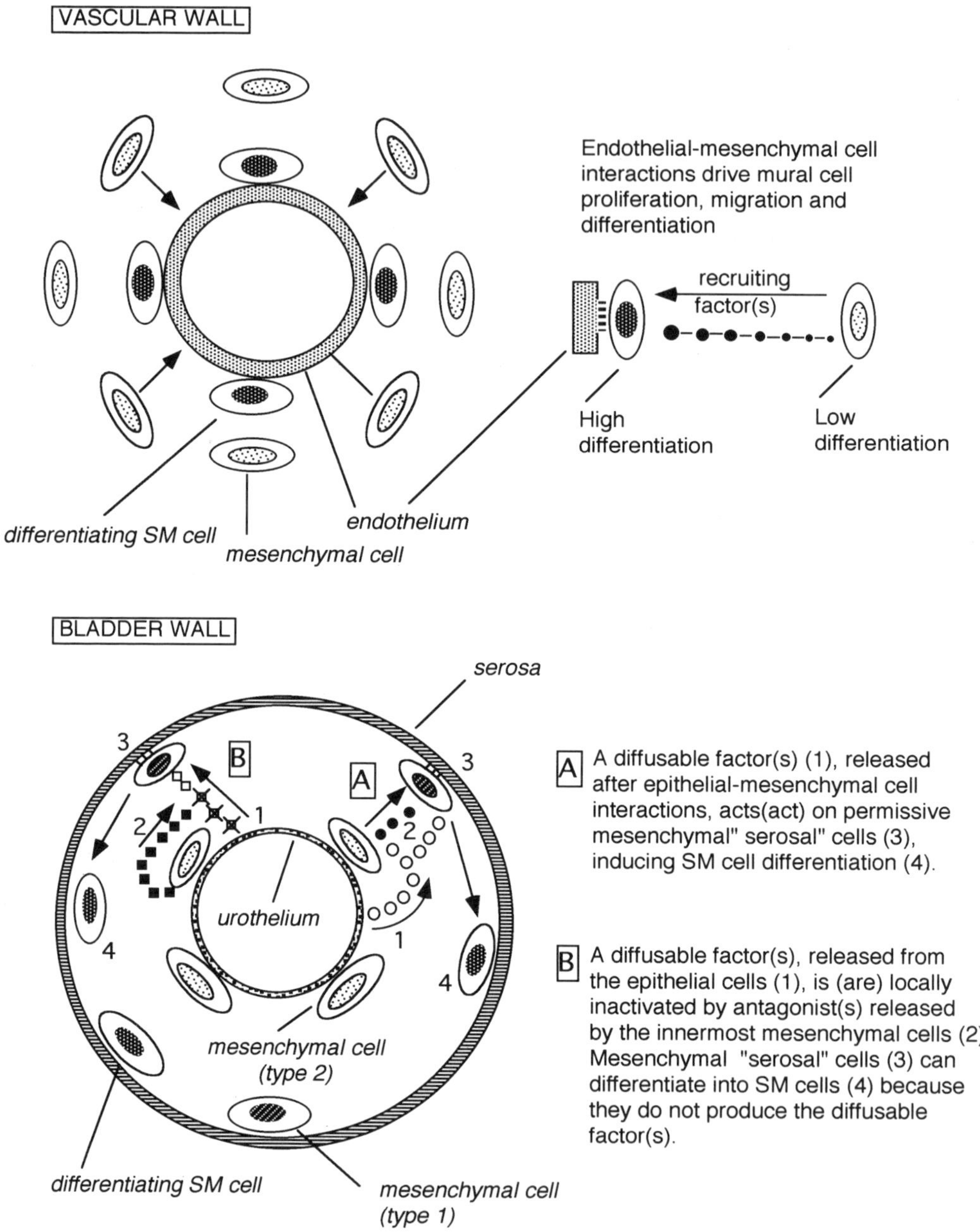

Figure 7. Upper: Model of vasculogenesis in which locally recruited mesenchymal cells become incorporated in the developing arterial vessel according to a gradient, from high to low, of differentiation with respect to the endothelium.[2,76] Lower: Proposed model of SM cell differentiation in the bladder wall based on an opposite gradient of differentiation compared to that of vasculogenesis.

differentiation observed in vivo and the apparent "direct" influence exerted by the urothelium in the graft experiments.[70] Hypothesis "A" suggests that upon interaction with the epithelium, mesenchymal cells release a diffusable factor(s) which acts on "permissive" mesenchymal "serosal" cells. In hypothesis "B", we suggest that the diffusable factor(s), constitutively released by the epithelium, is blocked by a antagonist(s) released in the innermost part of developing bladder wall. Mesenchymal "serosal" cells do not produce this antagonist(s) and, hence, are prone to SM cell differentiation. Both hypotheses focus on the existence of two distinct mesenchymal cell populations and it is clear that in the graft experiments the two cell types might come into direct contact with the urothelium because of the disruption of the in vivo distinct spatial organization.

9. CONCLUSIONS AND FUTURE STUDIES

1. In the hypertrophic detrusor muscle during obstruction there is a cytoskeletal and cytocontractile protein shift towards an "immature" SM cell which might be metabolically and functionally more suited to face with the modified physiology of bladder. The newly formed SM cells in bladder serosa during obstruction and regeneration are likely to be produced from resident SMMC. In the "freezing" model of bladder regeneration, interstitial cells contribute to the formation of new SM tissue through a conversion similar to what happens in the serosa: from fibroblast to myofibroblast to SM cell. While TGFβ1 (in the serosa) acts as inducer of SM cell differentiation, the innervation (in the muscularis) seems to play a repressive role in the late phase of this differentiative process. One might speculate that interstitial cells possess a "default" SM cell-like gene program whose activation is hampered, directly or indirectly, by the innervation input.
2. Analysis of bladder wall response in the experimental models examined is complicated by the fact that the bladder is a contractile organ and, hence, it might be that the proliferation, differentiation and apoptosis processes occurring in the bladder wall tissues are influenced to a various extent by this intrinsic property.[2]
3. The fibroblasts are not a homogeneous population. Therefore, the existence and the extent of fibroblast heterogeneity in the bladder wall must be examined in detail to establish the spatio-temporal SM cell conversion potential. In addition, mesenchymal cell heterogeneity hypothesized in the early stages of bladder ontogeny (Figure 7) and the relevance to the observed dislocation of differentiating SM cells in regenerating vs. developing bladder should be evaluated.
4. There is some evidence that cell heterogeneity is also present in SM bladder tissue.[6,12] Like the vascular SM cells[2,76], it might be that distinct SM cell populations or lineages are segregated in the bladder and are uniquely involved in the hypertrophic, hyperplastic, hyperploidy, apoptotic and migration response of SM cells in the obstruction and regeneration process.
5. One must also keep in mind that, besides the classical blood-borne inflammatory cells, in the "battlefield" of regenerating bladder a new source of "mesenchymal" cells might be involved as putative progenitors of myofibroblasts and, hence, of SM cells: the peripheral blood, bone marrow-derived pluripotent cells of non-hematopoietic potential.[77] It is interesting in this respect that keratin 8/18 are found in cells of hematopoietic derivation[78] and, thus, it might be inferred that some contribution is potentially possible into sites of active angiogenesis as occurs in the regenerating bladder.

6. If epithelial-mesenchymal cell interactions or contacts are specifically required for inducing interstitial cell conversion to myofibroblasts or SM cells, the role of basement membrane laminin $\alpha 1$, platelet-derived growth factor (PDGF) -A or -B and PDGF receptor must all be assessed. In fact, the lack of availability of these two factors have a profound impact on SM cell differentiation in cocultures of epithelial-mesenchymal cells or during organogenesis.[79,80] The generation of mice deficient for specific proteins, growth factors or cytokine is of paramount importance in the study of the molecular and cellular response of the bladder wall to obstruction and regeneration.

ACKNOWLEDGMENTS

We wish to thank Mrs Marissa Galliani Sanavio for her excellent editorial assistance. This work was supported in part by M.U.R.S.T., C.N.R. and Biomedical Association for Vascular Research.

REFERENCES

1. Levin R.M., Monson F.C., Haugaard N., Buttyan R., Roelofs M., Sartore S., Wein A.J.: Genetic and cellular characteristics of bladder outlet obstruction. Urol Clin North America., **22:** 263–284, 1995.
2. Sartore S., Franch R., Roelofs M., Chiavegato A.: Molecular and cellular phenotypes and their regulation in smooth muscle. Rev. Physiol. Biochem. Pharmacol., in press, 1998.
3. Sartore S., Scatena M., Chiavegato A., Faggin E., Giuriato L., Pauletto P.: Myosin isoform expression in smooth muscle cells during physiological and pathological vascular remodeling. J. Vas. Res., **31:** 61, 1994.
4. Sartore S., Chiavegato A., Franch R., Faggin E., Pauletto P.: Myosin gene expression and cell phenotypes in vascular smooth muscle during development, in experimental models, and in vascular disease. Arterioscler. Thromb. Vasc. Biol., **17:**1210, 1997.
5. Coplen D.E., Macarak E.J., Levin R.M.: Developmental changes in normal fetal bovine whole bladder phsyiology. J. Urol., **151:** 1391,1994.
6. Chiavegato A., Scatena M., Roelofs M., Ferrarese P., Pauletto P., Passerini-Glazel G., Pagano F., and Sartore S.: Cytoskeletal and cytocontractile protein composition of smooth muscle cells in developing and obstructed rabbit bladder. Exp. Cell Res., **207:** 310, 1993.
7. Eddinger T.J., Murphy R.A.: Developmental changes in actin and myosin heavy chain isoform expression in smooth muscle. Arch. Biochem. Biophys., **284:** 232, 1991.
8. Kim Y.S., Wang Z., Levin R.M., Chacko S.: Alterations in the expression of the β-cytoplasmic and the γ-smooth muscle actin in hypertrophied urinary bladder smooth muscle. Mol. Cell Biochem., **131:** 115, 1994.
9. Malmqvist U., Arner A., Uvelius B.: Contractile and cytoskeletal proteins in smooth muscle during hypertropphy and its reversal. Am. J. Physiol., **260:** (Cell Physiol 29), C1085, 1991
10. Kelley C.A., Adelstein R.S.: Characterization of isoform diversity in smooth muscle myosin heavy chains. Can. J. Physiol. Pharmacol., **72:**1351, 1994.
11. Baskin L.S., Sutherland R.S., Thomson A.A., Hayward S.W., Cunha G.R.: Growth factors and receptors in bladder development and obstruction. Lab. Invest., **75:** 157, 1996.
12. Lau C.L., Chacko S.: Identification of two types of smooth muscle cells from rabbit urinary bladder. Tissue & Cell, **28:** 339, 1996.
13. Schwartz S.M., DeBlois D., O'Brien E.R.M.: The intima. Soil for atherosclerosis and restenosis. Circ. Res., **77:** 445, 1995.
14. McNeil P.L.: Cellular and molecular adaptations to injurious mechanical stress. Trends in Cell Biol., **3:** 302, 1993.
15. Owens G.K.: Control of hypertrophic versus hyperplastic growth of vascular smooth muscle cells. Am. J. Physiol., **257:** (Heart Circ Physiol 26) H1755, 1989.
16. Buttyan R., Jacobs B.Z., Blaivas J.G., Levin R.M.: Early molecular response to rabbit bladder outlet obstruction. Neurourol. Urodyn., **11:** 225, 1992.

17. Karim O.M.A., Seki N., Mostwin J.L.: Detrusor hyperplasia and expression of "immediate early" genes with onset of abnormal urodynamic parameters. Am. J. Physiol., **263:** (Regulatory Integrative Comp Physiol 32), R1284, 1992.
18. Zhao Y., Chacko S., Levin R.M.: Expression of stress proteins (HSP-70 and HSP-90) in the rabbit urinary bladder subjected to partial outlet obstruction. Mol. Cell Biochem., **130:** 49, 1994.
19. Chen M-W., Krasnapolsky L., Levin R.M., Buttyan R.: An early molecular response induced by acute overdistension of the rabbit urinary bladder. Mol. Cell Biochem., **132:**39, 1994.
20. Santarosa R., Colombel M.C., Kaplan S., Monson F., Levin R.M., Buttyan R.: Hyperplasia and apoptosis. Opposing cellular processes that regulate the response of the rabbit bladder to transient outlet obstruction. Lab. Invest., **70:** 503, 1994.
21. Monson F.C., Wein A.J., Eika B., Murphy M., Levin R.M.: Stimulation of DNA synthesis in rabbit wall after partial outlet obstruction and acute overdistension. Neurourol. Urodyn., **13:** 51, 1994.
22. Brent L., Stephens F.D.: The response of smooth muscle cells in the rabbit urinary bladder to outflow obstruction. Invest. Urol., **12:** 494, 1975.
23. Uvelius B., Persson L., Mattiasson A.: Smooth muscle cell hypertrophy and hyperplasia in the rat detrusor after short-time infravesical outflow obstruction. J. Urol., **131:** 173, 1984..
24. Pampinella F., Roelofs M., Castellucci E., Passerini-Glazel G., Pagano F., and Sartore S.: Time-dependent remodeling of the bladder wall in growing rabbits after partial outlet obstruction. J. Urol., **157:** 677, 1996.
25. Wang Z., Gopalakurup S.K., Levin R.M., Chacko S.: Expression of smooth muscle myosin isoforms in urinary bladder smooth muscle during hypertrophy and regression. Lab. Invest., **73:** 244, 1995.
26. Cher M.L., Abernathy B.B., McConnell J.D., Zimmern P.E., Lin V.K.: Smooth-muscle myosin heavy-chain isoform expression in bladder-outlet obstruction. World J. Urol., **14:** 295, 1996.
27. Gabella G., Uvelius B.: Urinary bladder of rat: fine structure of normal and hypertrophic musculature. Cell Tissue Res., **262:** 67, 1990.
28. Arner A., Malmqvist U., Uvelius B.: Metabolism and force in hypertrophic smooth muscle from rat urinary bladder. Am. J. Physiol., **258:** (Cell Physiol 27), C923, 1990.
29. Sjuve R., Haase H., Morano I., Uvelius B., Arner A.: Contraction kinetics and myosin isoform composition in smooth muscle from hypertophied rat urinary bladder. J. Cell Biochem., **63:** 86, 1996.
30. Kato K., Monson F.C., Longhurst P.A., Wein A.J., Haugaard N., Levin, R.M.: The functional effects of long-term outlet obstruction on the rabbit urinary bladder. J. Urol., **143:** 600, 1990.
31. Monson F.C., Goldschmidt M.H., Zdzric S.A., Ruggieri M.R., Levin R.M., Wein A. Use of a previously undescribed elastic lamina of the serosa to characterize connective tissue hypertrophy of the rabbit bladder wall following partial outlet obstruction. Neurol. Urodyn., **7:** 385, 1988.
32. Ghoniem G.M., Regnier C.H., Biancani P., Johnson L., Susset J.G.: Effect of vesical outlet obstruction on detrusor contractility and passive properties in rabbits. J. Urol., **135:** 1284, 1986.
33. Kato K., Wein A.J., Kitada S., Haugaard N., Levin R.M.: The functional effect of mild outlet obstruction on the rabbit urinary bladder. J. Urol., **140:** 880, 1988.
34. Buoro S., Ferrarese P., Chiavegato A., Roelofs M., Scatena M., Pauletto P., Passerini-Glazel G., Pagano F., and Sartore, S.: Myofibroblast-derived smooth muscle cells during remodeling of rabbit urinary bladder wall induced by partial outflow obstruction. Lab. Invest., **69:** 589, 1993.
35. Pampinella F., Roelofs M., Castellucci E., Chiavegato A., Guidolin D., Passerini- Glazel G., Pagano F., and Sartore S.: Proliferation of submesothelial mesenchymal cells during early phase of serosal thickening in the rabbit bladder is accompanied by transient keratin 18 expression. Exp. Cell Res., **223:** 327, 1996.
36. Schaffer C.J., Nanney L.B.: Cell biology of wound healing. Int. Rev. Cytol., **169:** 151, 1996.
37. Chen Y., Arner A., Bornfeldt K.E., Uvelius B., Arnqvist H.J.: Development of smooth muscle hypertrophy is closely associated with increased gene expression of insulin-like growth factor binding protein-2 and -4. Growth Regul., **5:** 45, 1995.
38. Smith R.S., Smith T.J., Blieden T.M., Phipps R.P.: Fibroblasts as sentinel cells. Synthesis of chemokines and regulation of inflammation. Am. J. Pathol., **151:** 317, 1997.
39. Sappino A.P., Sch,rch W., and Gabbiani G.: Differentiation repertoire of fibroblastic cells: expression of cytoskeletal proteins as marker of phenotypic modulations. Lab. Invest., **63:** 144, 1990.
40. Rönnov-Jessen L., Petersen O.W., Bissel M.J.: Cellular changes involved in conversion of normal to malignant breast: importance of the stromal reaction. Physiol. Rev., **76:** 69, 1996.
41. Schmitt-Gräff A., Desmouliére A., Gabbiani G.: Heterogeneity of myofibroblast phenotypic features: an example of fibroblastic plasticity. Virchows Archiv., **425:** 3, 1994.
42. Chu Y.W., Runyan R.B., Oshima R.G., Hendrix M.J.C.: Expression of complete keratin filaments in mouse L cells augments cell migration and invasion. Proc. Natl. Acad. Sci. USA., **90:** 4261, 1993.
43. Corcoran J.P., Ferretti P.: Keratin 8 and 18 expression in mesenchymal progenitor cells of regenerating limbs is associated with cell proliferation and differentiation. Dev. Dyn., **210:** 355, 1997.

44. Roelofs M., Wein A.J., Monson F.C., Passerini-Glazel G., Koteliansky V., Sartore S., Levin R.M. Contractility and phenotype transitions in serosal thickening of obstructed rabbit bladder. J. Applied Physiol., **78:** 1432, 1995.
45. Miano J.M., Cserjesi P., Ligon K.L., Periasamy M., Olson E.N.: Smooth muscle myosin heavy chain exclusively marks the smooth muscle lineage during mouse embryogenesis. Circ. Res., **75:** 803, 1994.
46. Ko J-A., Murahashi S., Arata T., Inoue A.: Differentiation of smooth muscle cells from undifferentiated cells of chicken gizzard occurs on the layer of fibroblast-like cells. Cell Tissue Res., **285:** 395, 1996.
47. Nicosia R.F., Villaschi S.: Rat aortic smooth muscle cells become pericytes during angiogenesis in vitro. Lab. Invest., **73:** 658, 1995.
48. Beck L. Jr., D'Amore P.A.: Vascular development: cellular and molecular regulation. F.A.S.E.B. J., **11:** 365, 1997.
49. Kohnen G., Kertschanska S., Demir R., Kaufmann P.: Placental villous stroma as a model system for myoblast differentiation. Histochem. Cell Biol., **105:** 415, 1996.
50. Zalewski A., Shi Y.: Vascular myofibroblasts. Lessons from coronary repair and remodeling. Arterioscler. Thromb. Vasc. Biol., **17:** 417, 1997.
51. Chiavegato A., Bochaton-Piallat M-L., D'Amore E., Sartore S., Gabbiani G.: Expression of myosin heavy chain isoforms in mammary epithelial cells and in myofibroblasts from different fibrotic settings during neoplasia. Virchows Archiv., **426:** 77, 1995.
52. Gleeson M.J., Griffith D.P.: The use of alloplastic biomaterials in bladder substitution. J. Urol., **148:** 1377, 1992.
53. Knapp P.M., Lingeman J.E., Siegel Y.I., Badylak S.F., Demeter R.J.: Biocompatability of small-intestinal submucosa in urinary tract as augmentation cytoplasty graft and injectable suspension. J. Endourol., **8:** 125, 1994.
54. Kropp B.P., Eppley B.L., Prevel C.D., Rippy M.K., Harruff R.C., Badylak S.F., Adams M.C., Rink R.C., Keating M.A.: Experimental assessement of small intestine submucosa as a bladder wall substitute. Urology, **46:** 396, 1995.
55. Tsuji I., Kuroda K., Fujreda J., Shiraishi Y., Kassai T., Ishida H.: A clinical and experimental study on cytoplasty not using the intestine. J. Urol., **89:** 214, 1963.
56. Rasmussen F.: Healing of urinary bladder wounds, morphological and biochemical studies. Proc. Soc. Exp. Biol. Med., **123:** 470, 1966.
57. Kiviat M.D., Ross R., Ansell J.S.: Smooth muscle regeneration in the ureter. Am. J. Pathol., **72:** 403, 1973.
58. Liang D.S., Liang M.D.: Bladder muscle regeneration. Invest. Urol., **12:** 5, 1974.
59. Shirai T., Cohen S.M., Fukushima S., Hananouchi M., Nobuyuki I.: Reversible papillary hyperplasia of the rat urinary bladder. Am. J. Pathol., **91:** 33, 1978.
60. Gorham S.D., French D.A., Shivas A.A., Scott R.: Some observations on the regeneration of smooth muscle in the repaired urinary bladder of the rabbit. Eur. Urol., **16:** 440, 1989.
61. Sutherland R.S., Baskin L.S., Hayward S.W., Cunha G.R.: Regeneration of bladder urothelium, smooth muscle, blood vessels and nerves into acellular tissue matrix. J. Urol., **156:** 571, 1996.
62. Ross G. jr, Thompson I.M., Keown K.K., Judy B.B., Gammel GE.: Further observations on the role of smooth muscle regeneration. J. Urol., **102:** 49, 1969.
63. Hastings J.C., Van Winkle W. jr, Barker E., Hines D., Nichols W. The effect of suture materials on healing wounds of the bladder. Surg. Gynec. Obstet., **140:** 933, 1975.
64. Faggian L., Pampinella F., Roelofs M., Paulon T., Franch R., Chiavegato A., Sartore S.: Phenotypic changes in the regenerating rabbit bladder muscle. Role of interstitial cells and innervation on smooth muscle cell differentiation. Histochem. Cell Biol., **109:** 25, 1998.
65. Jones R.C.: Role of interstitial fibroblasts and intermediate cells in microvascular wall remodelling in pulmonary hypertension. Eu.r Resp. Rev., **3:** 569, 1993.
66. Kovacs E.J., DiPietro L.A.: Fibrogenic cytokines and connective tissue production. F.A.S.E.B. J., **8:** 854, 1994.
67. Roelofs M., Faggian, Pampinella F., Paulon T., Franch R., Chiavegato A., Sartore S.: TGFβ1 involvement in the conversion of fibroblast to smooth muscle cell in the rabbit bladder serosa. Histochem. J, **30:** 393, 1998.
68. Bjökerud S.: Effects of trasforming growth factor-beta 1 on human arterial smooth muscle cells in vitro. Arterioscler. Thromb., **11:** 892, 1991.
69. Baskin L., Sutherland R.S., Thomson A.A., Nguyen H-T., Morgan D.M., Hayward S.W., Hom YK., DiSandro M., Cunha G.R.: Growth factors in bladder wound healing. J. Urol., **157:** 2388, 1997.
70. Baskin L.S., Hayward S.W., Young P., Cunha G.R.: Role of mesenchymal-epithelial interactions in normal bladder development. J. Urol., **156:** 1820, 1996.

71. Davila R.M., Crouch E.C.: Role of mesothelial and submesothelial stromal cells in matrix remodeling following pleural injury. Am. J. Pathol., **142:** 547, 1993.
72. Cunha G.R., Battle E., Young P., Brody J., Donjacour A., Hayashi N., Kinbara H.: Role of epithelial-mesenchymal interactions in the differentiation and spatial organization of visceral smooth muscle. Epithelial Cell Biol., **1:** 76, 1992.
73. Birchmeier C., Birchmeier W.: Molecular aspects of mesenchymal-epithelial interactions. Annu. Rev. Cell Biol., **9:** 511, 1993.
74. Baskin L.S., Hayward S.W., Young P.F., Cunha G.R.: Ontogeney of the rat bladder: smooth muscle and epithelial differentiation. Acta Anat., **155:** 163, 1996.
75. Baskin L.S., Hayward S.W., Young P.F., Cunha GR.: Role of mesenchymal-epithelial interactions in normal bladder development. J. Urol., **156:** 1820, 1996.
76. Owens G.K.: Regulation of differentiation of vascular smooth muscle cells. Physiol. Rev., **75:** 487, 1995.
77. Prockop D.J.: Marrow stromal cells as stem cells for nonhematopoietic tissues. Science, **276**: 71, 1997.
78. Traweek S.T., Liu J., Battifora H.: Keratin gene expresion in non-epithelial tissues. Am. J. Pathol., **142:** 1111, 1993.
79. Schuger L., Skubitz A.P.N., Zhang J., Sorokin L., He L.: Laminin α1 chain synthesis in the mouse development lung: requirements for epithelial-mesenchymal contact and possible role in bronchial smooth muscle development. J. Cell Biol., **139:** 553, 1997.
80. Boström H., Willetts K., Pekny M., Levéen P., Lindhal P., Hedstrand H., Pekna M., Hellström M., Gebremedhin S., Schalling M., Nilsson M., Kurland S., Törnell J., Heath J.K., Betsholtz C.: PDGF-A signaling is a critical event in lung alveolar myofibroblast development and alveogenesis. Cell, **85:** 863, 1996.

7

REPLICATIVE SENESCENCE IN HUMAN UROEPITHELIAL CELLS

Jairaj A. Puthenveettil, Melissa S. Burger, and Catherine A. Reznikoff*

University of Wisconsin Comprehensive Cancer Center
and Environmental Toxicology Center
Madison, Wisconsin 53792

ABSTRACT

Purpose

Normal human uroepithelial cells (HUCs) proliferate rapidly in culture during early passage and then spontaneously undergo replicative senescence. We previously reported that the cyclin D1-CDK4/6 inhibitor, p16^{INK4a}, is elevated at senescence in HUCs. Hence, we proposed that p16^{INK4a} may play a critical role in mediating senescence in this cell type. In the current study, we further characterized the senescent state in HUCs. We also tested the possible roles of changes in other cell cycle proteins, including p53, p21^{WAF1}, pRb, and cyclin D1 in HUC senescence.

Methods

Normal HUCs cultured from explants of ureteral mucosa were used for these studies. Senescence associated-β-galactosidase activity (SA-β-gal) was used to identify cells in senescence. Flow cytometric analysis was used to determine changes in cell cycle distribution at senescence. Response of cells to serum stimulation was determined by Northern analysis of c-fos. Western analysis was used to assess changes in p53, p21^{WAF}, p16^{INK4a}, cyclin D1 and plasminogen activator inhibitor-1 (PAI-1) levels at senescence.

* Requests for reprints: University of Wisconsin Comprehensive Cancer Center, K6/548 Clinical Science Center, 600 Highland Avenue, Madison, Wisconsin 53792.

Advances in Bladder Research, edited by Baskin and Hayward.
Kluwer Academic / Plenum Publishers, New York, 1999.

Results

β–gal-positive HUCs were blocked at G1/S in senescence and failed to show c-fos induction in response to serum stimulation. As previously reported, senescent HUCs also showed elevated p16^{INK4a}. However, unlike human fibroblasts, neither p53 nor p21^{WAF1} elevation accompanied HUCs senescence. PAI-1 levels were also not elevated in HUC senescence.

Conclusion

These findings support a model in which elevation of p16^{INK4a}, but not p53 or p21^{WAF1} plays a critical role in HUC replicative senescence. These findings elucidate the tumor suppressor mechanism of p16^{INK4a} and the frequent loss of either p16^{INK4a} or pRb in invasive human bladder tumors.

INTRODUCTION

Hayflick first reported in 1961 that normal diploid cells show limited replicative doublings when cultured in vitro.[1] This limited lifespan has been termed cellular senescence. Certain rodent species, notably mice, show spontaneous escape from senescence and formation of essentially immortal cell lines at a rate (estimated at 1 in 10^{-4} to 10^{-6}) that is readily detectable.[2] In contrast, there are virtually no examples of spontaneous escape from senescence and immortalization of normal diploid human cells in culture. This limitation on cell doublings in normal cells has been thought to reflect the inevitable aging at the cellular level. More recent studies suggest that cellular senescence also plays a role as a powerful tumor suppressor mechanism.[3,4] This latter hypothesis is based on the observation that malignant transformation requires multiple mutations that are unlikely to accumulate in the limited lifespan of human cells.[5]

Many laboratories have studied the characteristics of senescence in human cells in an attempt to determine what physiological changes underlie this phenomenon. For the most part, such studies have been done using human fibroblasts. Results show that senescent cells are viable[5] and metabolically active,[5] but fail to enter the DNA synthesis phase of cell cycle after stimulation with serum.[5] Thus, proliferation is thought to be irreversibly blocked in senescent cells.[5] Increased levels of plasminogen activator inhibitor-1 (PAI-1), a regulator of pericellular proteolytic activity, have been associated with senescence in fibroblast cultures.[3] An interesting feature of senescent fibroblasts is the increased activity of proteolytic enzymes, including collagenase and stromelysin.[4] A distinct form of β-galactosidase

(β-gal) has also been associated with senescence,[5] and has been used as a marker for senescent cells. Results of somatic cell hybridization studies suggest that senescence is dominant in nature.[5] But the identification of such senescence-associated characteristics has not clarified the mechanism(s) underlying cellular senescence.

Recent studies have focused on the role of increased levels at senescence of proteins involved in cell cycle regulation. Elevation of p53, and its downstream target p21^{WAF1} (an inhibitor of cyclin dependent kinases (cdks) 1, 2 and 4) have been observed in human fibroblasts at senescence.[6,7] Notably, neither p53 nor p21^{WAF1} elevation has been reported for senescent human epithelial cells. In contrast to p21^{WAF1}, elevation of p16^{INK4a}, an inhibitor of cyclin D1-cdk4/6 dependent phosphorylation of pRb, has been observed in both human fibroblasts[8] and human epithelial cells,[9] including HUCs.[10] Elevation of p16^{INK4a} would explain replicative arrest at G1. We and others have hypothesized that elevated p16^{INK4a} may mediate replicative senescence[9] by virtue of its ability to block pRb phosphorylation and subsequent E2F release.[9,10] Indeed, pRb has been shown to be hypophosphorylated at senescence in human fibroblasts.[11] In support of this latter hypothesis, restoration of either p16^{INK4a} 12 or pRb[13] to immortal cell lines lacking these proteins induces a senescence-like state. Finally, it has recently been proposed that a second tumor suppressor protein, p19^{ARF} transcribed from an alternate reading frame of INK4a/p16 plays a role in rodent cell transformation.[14] At present, data to support a role for p19^{ARF} in human cell senescence is lacking.

Finally, it has been proposed that telomere shortening represents a cellular ìclockî leading to senescence.[2] Such telomere shortening, attributed to the end replication problem of DNA synthesis and the hypothesized lack of telomerase in normal cells[2] has been demonstrated in human fibroblasts.[15] Telomere shortening is thought to lead to chromosome alterations.[4] It has been proposed that such shortened telomeres and altered chromosomes may be interpreted by normal cells as DNA damage.[16] This would lead to elevated p53 and p21^{WAF1}, two features that have been associated with senescent human fibroblasts.[6,7] However, our laboratory recently reported that proliferating normal human bladder, ureteral, mammary and prostate epithelial cells do contain telomerase activity.[17] Furthermore, we did not observe telomere shortening in HUCs at senescence. Therefore, the model that senescence is associated with telomere shortening in telomerase negative normal somatic cells does not fit all human epithelial cell types.

In the present study, we examine HUCs for markers that have been associated with human fibroblast senescence, but not described for human epithelial cell senescence. Our results again demonstrate that p16^{INK4a} is elevated at senescence in HUCs. However, we report here for the first time that neither elevation of p21^{WAF1} nor p53 accompany HUC senescence.

MATERIALS AND METHODS

Cells and Culture Methods

Cultures of HUC were established from the mucosal epithelium of residual pieces of ureter left over from kidney transplantation surgery, as described.[18] Briefly, 1mm^2 explants were plated onto 100mm petri dishes pre-coated with Type I rat tail collagen and pre-equilibrated with HUC growth medium. The growth medium, F12+1%FBS, consisted of a supplemented Ham's F12 (GIBCO, Grand Island, NY) with 1% fetal bovine serum (FBS; GIBCO). After 7–10 days of cell proliferation, explants were removed, HUCs were dispersed using 0.1% EDTA, and split at a ratio of one to three. Cells were passed weekly until proliferation stopped. The number of cells was determined using a hemocytometer. Viability was determined by dye exclusion using 0.1% trypan blue stain.

SA-β-gal activity was used as a marker for senescence.[5] For this assay, cells were fixed with formalin/glutaraldehyde, rinsed, and incubated in the SA-β-galactose reaction mixture for 18 hrs, as previously described.[22] The blue nuclear and cytoplasmic staining characteristic of senescent cells was visualized using a light microscope. Controls were early passage proliferating HUCs.

Passage and expansion of HUCs for all experiments reported herein were done as follows. HUCs were pooled at primary passage (P0) and reseeded for P1 using 5×10^5 viable cells per 100mm dish. After seven days, when cells were still in log phase of proliferative growth, cells were dispersed, counted, and reseeded as above for P2. This process was repeated for 3–5 passages (depending on the culture) until HUCs senesced as determined by positive SA-β gal activity. For the c-fos induction studies, the following changes were made. Forty-eight hours after seeding, HUCs at passages 1, 3, and 5 were placed in basal F12 medium without FBS. After 96 hours, F12+10%FBS was added for 30 minutes. Then, cells were lysed and RNA was extracted for northern analysis of c-fos mRNA levels, as described below.

Northern Blot Analysis

Northern blot analysis was performed according to a standard protocol (Current Protocols in Molecular Biology, 1997). Total RNA was extracted from HUCs using the RNeasy kit (Qiagen, Inc., Santa Clarita, CA). The probes for c-fos and GAPDH were made from HUC mRNA by RT-PCR using the Gene Amp Kit (Perkin Elmer, Branchburg, NJ). These probes were radio labeled (specific activity, >10^8 dpm/μg) by random priming using the RediPrime Kit (Amersham Life Science, England). The unincorporated nucleotides were removed by G-50 Sephadex spin columns (Quick Spin Columns, Boehringer Mannheim, Indianapolis, IN). 1 μg of RNA was resolved in 1% agarose-formaldehyde gel and was transferred by pressure blotting (Stratagene, San Diego, CA) to a nylon membrane (Hybond-N, Amersham) followed by UV-crosslinking. The membrane was prehybridized for 3 hrs at 42°C and hybridized with the radio-labeled probe at 42°C overnight. The radioactive signals were detected by PhosphorImager (Molecular Dynamics, Sunnyvale, CA) and quantified (ImageQuant, Amersham). 28S (4.7 kb) and 18S (1.9 kb) ribosomal RNAs were used as size markers. GAPDH mRNA was used as the loading control.

Western Analyses

Western blot analysis was done as previously described.[18] Briefly, cells were lysed using a detergent lysis buffer in the presence of proteinase inhibitors. 25 μg of proteins were resolved using 15% SDS-polyacrylamide gel, and transferred to a polyvinyl difluoride membrane (Immobilon P, Millipore, Bedford, MA). This membrane was hybridized with primary mouse monoclonal antibodies against p53, $p21^{WAF1}$, $p16^{INK4a}$, PAI-1 (Oncogene Science, Cambridge, MA), pRb (Pharminigen, San Diego, CA) and cyclin D1 (Santa Cruz Biotechnology Inc., Santa Cruz, CA). After the secondary hybridization with a horseradish peroxidase-conjugated secondary antibody against mouse IgG (Pierce, Rockford, IL), the protein bands were detected using chemiluminescence (Pierce). Equal loading of proteins in different lanes was verified by hybridization using a mouse monoclonal antibody against α-tubulin (Oncogene Science). All blots were performed at least twice.

Flow Cytometric Analysis

Cell cycle analysis was done using a standard technique.[18] Cells were incubated for 2 hrs with 20μM bromodeoxyuridine (BrdU; Sigma), and fixed in 70% ethanol. These cells were permeablilized, and incubated with a primary antibody against BrdU (Caltag, Burlingame, CA) followed by another incubation with a FITC-conjugated secondary antibody against mouse IgG (Sigma, St.Louis, MO). After labeling with 50 μg/ml propidium iodide, these cells were analyzed on a FACScan (Becton Dickinson, San Jose, CA). Cell cycle distribution was determined using CellQuest software package (Becton Dickinson).

RESULTS

Senescent HUCs Are Viable, but Growth Arrested

Cells were determined to be in senescence when proliferation stopped and cultures were strongly positive for SA-β-gal activity (Figure 1). Senescent HUCs were examined for elevation of PAI-1, which has been observed in senescent human fibroblasts. As shown in Figure 2, PAI-1 was not elevated at senescence in HUCs. Cell viability at senescence was compared with viability of proliferating early passage HUCs. Senescent HUCs showed no significant loss of cell viability (78% +/- 1.4 at passage 1 (P1) compared to 81% +/- 1.39 at senescence, average of three experiments using three different HUC cul-

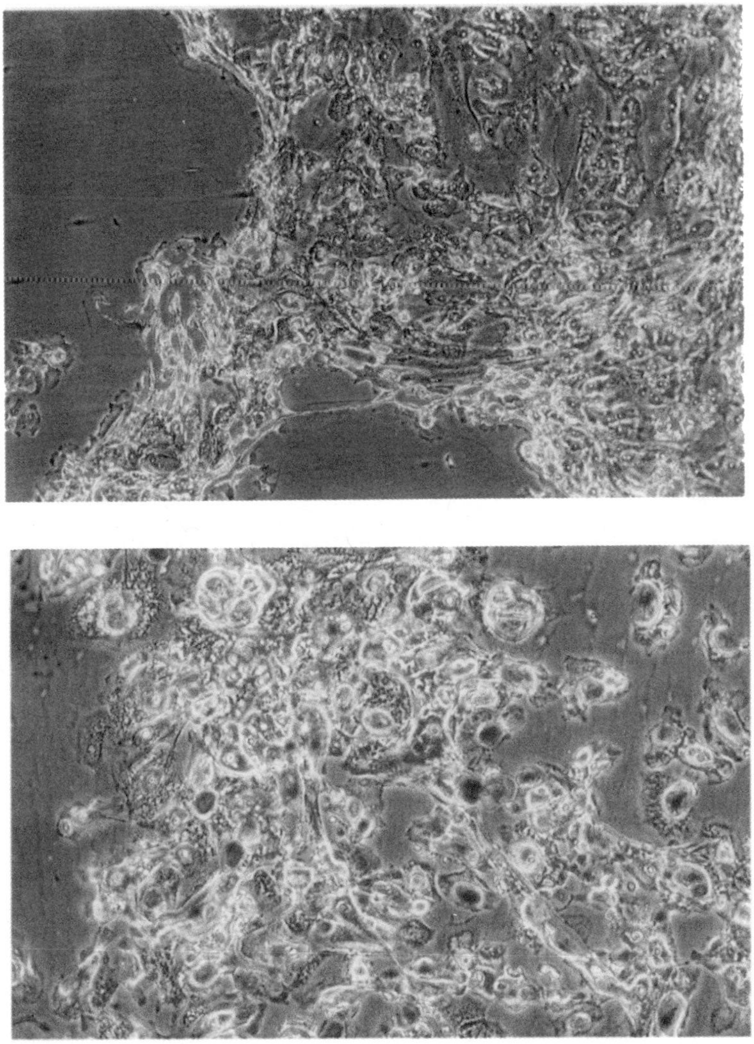

Figure 1. Photomicrographs of P1 HUCs and P5 HUCs after staining for senescence associated β-galactosidase activity (SA-β-gal). The late passage senescent cells (bottom panel) exhibit SA-β-gal activity, while the early passage (top panel) do not.

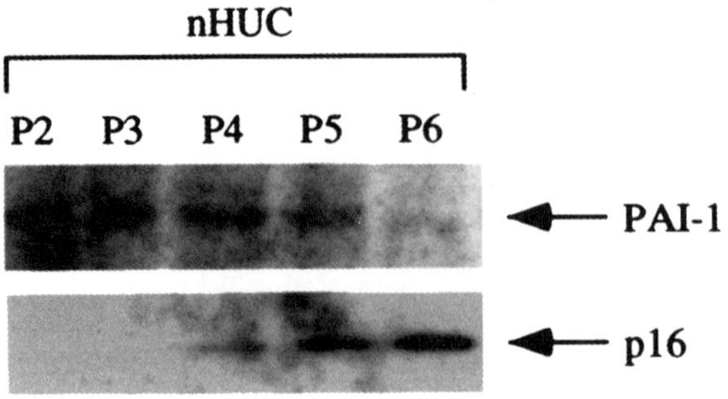

Figure 2. PAI-1 levels are not elevated by western analysis in senescent HUCs. Protein extracts from HUCs at passages two to six (P2 to P6) were analyzed by western blotting. Expression of p16 was also determined on the same blot to verify the onset of senescence.

tures). Senescent cultures were then examined using flow cytometry for alterations in cell cycle distribution. Results showed a reduction of S phase cells, from 15.9% at P1 to 0.16% at senescence consistent with results reported from other groups (Figure 3). We also observed a corresponding increase in cells at G1, 58.3% at P1 to 71.4% at senescence.

Early Passage Proliferating HUCs, but Not Senescent HUCs Respond to Serum Stimulation by c-fos Induction

To further examine replicative senescence, senescence HUCs were tested for their ability to respond to serum stimulation by inducing the immediate early gene c-fos. Induction of c-fos is critical for cells to enter s-phase.[4] Results showed that serum starved HUCs at P1 respond to serum stimulation within 30 minutes with elevated c-fos transcription, while senescence cells fail to respond (Figure 4).

HUCs Show Elevated p16^{INK4a}, but Not Elevated p53, p21^{WAF1}, pRb or Cyclin D1 at Senescence

As previously reported, elevation of p16^{INK4a} is reproducibly seen in association with HUC senescence. Here, we for the first time examine levels of p53 and p21^{WAF1}, two cell

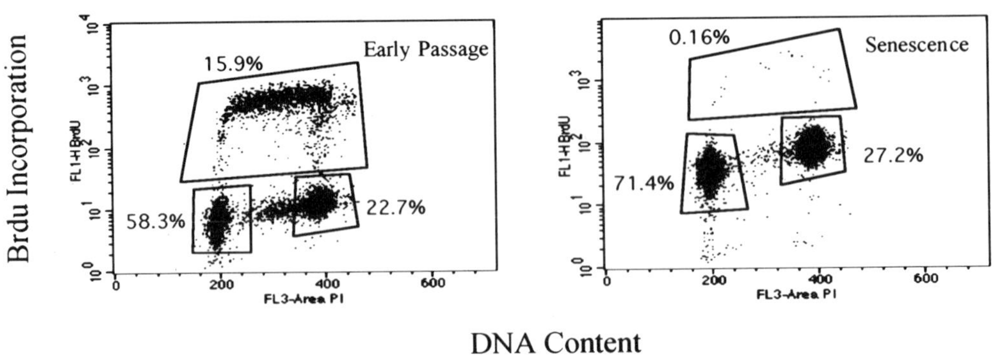

Figure 3. Reduction of S phase cells at senescence in HUCs by flow cytometry. Early passage and senescent HUCs were labeled with bromodeoxyuridine and were assayed for incorporation. BrdU incorporation is indicated on the y-axis with the total DNA content on the x-axis. The percentage of cell in G1, S and G2 phases of cell cycle are indicated next to each compartment.

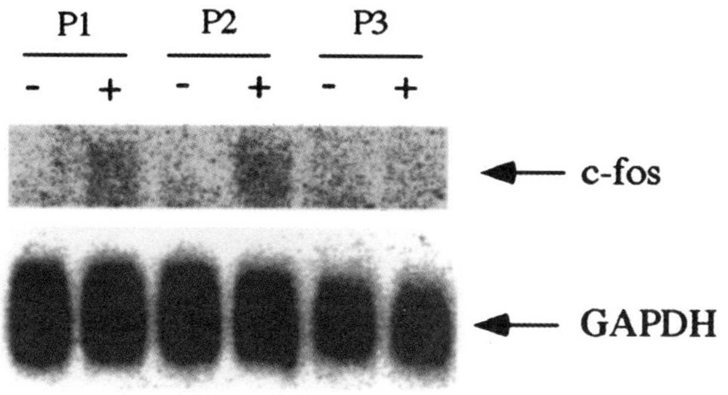

Figure 4. Failure of serum to induce c-fos in senescent HUC. Serum-starved HUCs (-) at passages one, two and three (P1, P2 and P3) were stimulated with 10%FBS (+) and their RNA levels after 30 minutes were analyzed by northern blotting. GAPDH mRNA was used as loading control.

cycle inhibitors elevated at senescence in normal human fibroblasts. Our results reproducibly showed in three separate experiments that neither p53 nor p21^{WAF1} levels increase at senescence (Figure 5). In addition, as shown in Figure 5, expression of cyclin D1 and pRb also did not change at senescence.

DISCUSSION

We studied the characteristics of human uroepithelial cells at senescence. It has been demonstrated that human and rodent fibroblasts are arrested in G1 at senescence.[2] We report here that HUCs, a representative human epithelial cell type, also arrest at the G1/S boundary at senescence. We previously reported that p16^{INK4a} is elevated in HUCs at senescence and hypothesized that this cell cycle inhibitor may play a critical role in HUC

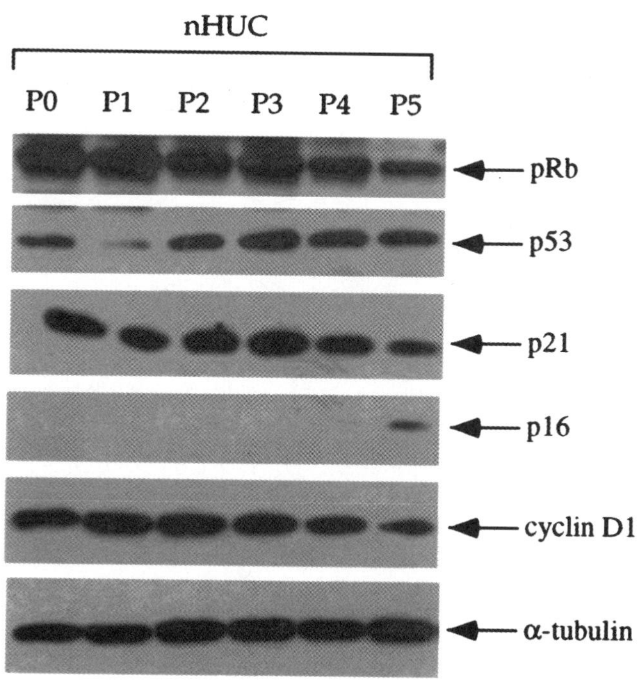

Figure 5. Elevation of p16^{INK4a}, but not p21^{WAF1}, p53, pRb or Cyclin D1 at senescence by western. Protein extracts from HUCs at different passages (P0 to P5) were analyzed by western blotting. The α-tubulin was used as loading control.

senescence.[10] Others had reported elevation of both p21^{WAF1} and p53, as well as p16^{INK4a}, at senescence in rodent and human fibroblasts.[6,7] Here we report for the first time that neither p53 nor p21^{WAF1} are elevated at senescence in HUCs. Thus, the mechanisms underlying replication senescence in epithelial cells may be different from those in fibroblasts. We also did not observe any significant change in the level of cyclin D1, a key regulator of the transition from G1 to S phase of cell cycle, at senescence. Hence, these new results support our earlier model in which p16^{INK4a} plays a critical role in mediating replicative senescence in HUCs.

In the current study we again used SA-β-gal activity as a senescence marker for senescence in HUC.[5] Other markers, such as elevated levels of plasminogen activator inhibitor-1 (PAI-1) have been described for senescent fibroblasts.[3] Thus, we tested whether elevated PAI-1 is also a useful marker of senescence for HUC. Results failed to show elevated PAI at HUC senescence. PAI controls the activation of proteolytic enzymes such as uroplasminogen. It is also a gene that is upregulated by transforming growth factor-β.[19,20]

Elevation of c-fos protein and consequent AP-1 activity is critical for the initiation of S-phase.[4] The immediate early genes including c-fos promote S-phase entry through pathways distinct from those regulated by pRb. We have shown that senescent HUCs do not show induction of c-fos in response to serum stimulation. This implies that pathways other than p16^{INK4a}/pRb pathway, may also contribute to senescence in HUCs. This is consistent with the observation that abrogation of the pRb pathway using a high risk Human Papillomavirus16 E7 is insufficient for immortalization, and other genetic alterations are required.[21]

In our studies, we have further defined important parameters of senescence in a representative human epithelial cell type, HUC. Studies of senescence in epithelial cells are particularly relevant as most solid cancers are epithelial in origin. Recent data from our laboratory supports a model in which bypassing senescence is important in tumorigenesis.[22] Consistent with this model, many human carcinomas show inactivation of p16^{INK4a} or pRb. Some show alteration of p21^{WAF1}, cyclin D1 overexpression or amplification, or mutations in CDKs. Finally, our data suggest that fundamental differences in control of replicative senescence in epithelial cells and fibroblasts make it inappropriate to extrapolate all results obtained in fibroblast systems to epithelial cells.

REFERENCES

1. Hayflick L, Moorhead PS. The serial cultivation of human diploid cell strains. Experimental Cell Research, 25:585–621, 1961.
2. Campisi J. Replicative senescence: An old lives' tale? Cell, 84: 497 500,1996
3. Serrano M, Lin AW, McCurrach ME, Beach D, Lowe SW. Oncogenic ras provokes premature Cell senescence associated with accumulation of p53 and p16INK4a. Cell, 88:593–602, 1997.
4. Smith JR and Pereira-Smith OM. Replicative Senescence: Implications for in vivo aging and tumor suppression. Science, 273:63–67, 1996.
5. Dimri GP, Campisi J. Molecular and cell biology of replicative senescence. Cold Spring Harbor Symposia on Quantitative Biology, Vol. LIX, Cold Spring Harbor Laboratory Press, pp 67–73, 1994.
6. Kulju KS, Lehman JM. Increased p53 protein associated with aging in human diploid fibroblasts. Experimental Cell Research 217(2):336–45, 1995 Apr.
7. Sherr CJ, Roberts JM. Inhibitors of mammalian G1 cyclin-dependent kinases. Genes & Development, 9:1149–1163, 1995.
8. Hara E, Smith R, Parry D, Tahara H, Stone S, Peters G. Regulation of p16CDKN2 expression and Its implication for cell immortalization and senescence. Molecular and Cellular Biology, 16:859–867, 1996.

9. Loughran O, Malliri A, Owens D, Gallimore PH, Stanley MA, Ozanne B, Frame MC, Parkinson EK. Association of CDKN2A/p16INK4A with head and neck keratinocyte replicative senescence: relationship of dysfunction to immortality and neoplasia. Oncogene, 13:561–568, 1996.
10. Reznikoff CA, Yeager TR, Belair CD, Savelieva E, Puthenveettil JA, Stadler W M. Elevate p16 at senescence and loss of p16 at immortalization in Human Papillomavirus 16 E6, but not E7, transformed human uroepithelial cells. Cancer Research 56:2886–2890, 1996.
11. Barrett JC, Annab LA, Alcorta D, Preston G, Vojta P, Yin Y. Cellular Senescence and Cancer. Cold Spring Harbor Symposia on Quantitative Biology, Vol. LIX, Cold Spring Harbor Laboratory Press, pp 411–417, 1994.
12. Uhrbom L, Nister M, Westermark B. Induction of senescence in human malignant glioma cells by p16INK4A. Oncogene 15(5):505–14, 1997 July 31.
13. Xu H-J, Zhou Y, Ji W, Perng G-S, Kruzelock R, Kong C-T, Bast RC, Mills GB, Li J, Hu S-X. Reexpression of the retinoblastoma protein in tumor cells induces senescence and telomerase inhibition. Oncogene 15:2589–2596, 1997.
14. Kamijo T, Zindy F, Roussel MF, Quelle DE, Downing JR, Ashmun RA, Grosveld G, Sherr CJ. Tumor suppression at the mouse INK4A locus mediated by the alternative reading frame product p19ARF. Cell, 91:649–659, 1997.
15. Allsop RC, Vaziri H, Patterson C, Goldstein S, Younglais, EV, Futcher AV, Greider CW and Harley CB. Telomere length predicts replicative capacity of human fibroblasts. Proc. Natl. Acad. Sc., 89:10114–10118, 1992.
16. Harley CB, Sherwood SW. Telomerase, checkpoints and cancer. Cancer Surveys 29:263–84, 1997.
17. Belair CD, Yeager TR, Lopez PM, Reznikoff CA. Telomerase activity: A biomarker of cell proliferation, not malignant transformation. Proc Natl Acad Sci USA 94:13677–13682, 1997.
18. Puthenveettil JA, Frederickson SM and Reznikoff CA. Apoptosis in human papillomavirus16 E7 - but not E6-immortalized human uropeithelial cells. Oncogene, 13:1123–1131, 1996.
19. Yingling JM, Datto MB, Wong C, Frederick JP, Liberati NT and Wang XF. Tumor suppressor Smad4 is transforming growth factor beta-inducible DNA binding protein. Molecular and Cellular Biology, 17:7019–7028, 1997.
20. Mu XC and Higgins PJ. Differential growth state-dependent regulation of plasminogen activator inhibitor type-1 expression in senescent IMR-90 human diploid fibroblasts. Journal of Cellular Physiology 165(3):647–57, 1995.
21. Reznikoff CA, Belair C, Savelieva E, Zhai Y, Pfieffer K, Yeager T, Thompson KJ, DeVries S, Bindley C, Newton MA, Sekhon G and Waldman F. Long term genome stability and minimal genotypic and phenotypic alterations in HPV16 E7-, but not E6-, immortalized human uroepithelial cells. Genes & Development, 8:2227–2240, 1994.
22. Yeager TR, DeVries S, Jarrard DF, Kao C, Nakada SY, Moon TD, Bruskewitz R, Stadler WM, Meisner LF, Gilchrist KW, Newton MA, Waldman FM, Reznikoff CA. Overcoming cellular senescence in human cancer pathogenesis. Genes & Development 12:163–174, 1998.

8

ULTRASTRUCTURAL SMOOTH MUSCLE ONTOGENY OF THE RAT BLADDER

Hsi-Yang Wu, Laurence S. Baskin, Carrie Blakey, Joseph Goodman, and Gerald R. Cunha

Departments of Urology, Pediatrics, and Anatomy
University of California School of Medicine
San Francisco, California 94143

1. ABSTRACT

The transmission electron microscope characteristics of developing rat bladder smooth muscle are described at 14 and 18 days of gestation, birth, and adulthood. Caveolae, microfilaments, and dense bodies increase during smooth muscle development. Collagen content in the extracellular matrix also increases. These changes may reflect increased bladder emptying in the rat, and also allow for comparison of smooth muscle cells in studies of bladder wound healing and tissue substitutes.

2. INTRODUCTION

We have previously shown that smooth muscle development in the rat bladder follows an orderly sequential expression of smooth muscle markers.[1] As we continued our investigation into the use of acellular matrix as a bladder substitute material,[2] we wished to further characterize the cells that infiltrated the matrix and turned into smooth muscle bundles. Unfortunately, identification of smooth muscle markers and intermediate filaments alone is insufficient to determine the origin of a given cell, as myofibroblasts share many of the same characteristics as smooth muscle cells.[3] We therefore sought to better identify the ultrastructural characteristics of normal rat bladder smooth muscle cells in development.

The smooth muscle of the bladder is derived from mesenchymal cells, in contrast to the endodermally derived epithelial cells. At 14 days gestation, the bladder is lined by an inner epithelium, which is surrounded by mesenchyme. By 16 days gestation, we were first able to detect smooth muscle cells by α-actin staining, which were located in the sub-

Advances in Bladder Research, edited by Baskin and Hayward.
Kluwer Academic / Plenum Publishers, New York, 1999.

serosal position. At 18 days, smooth muscle cells were also located in an intermediate position between the serosa and the mucosa. By 20 days, well-differentiated smooth muscle fascicles of the muscularis mucosa were noted immediately deep to the urothelium. Additional investigations have shown that the bladder epithelium is necessary for the development of bladder smooth muscle, as bladder mesenchyme grafted alone under the kidney capsule failed to develop into bladder smooth muscle, whereas smooth muscle differentiation occurred in homotypic tissue recombinants of bladder mesenchyme and bladder epithelium.[4]

Epithelial development in the rat bladder has been well characterized, due to its usefulness as a model for induction and invasion of transitional cell carcinoma. Starting at 11 days of gestation, a single layer of small round epithelial cells are seen, which differentiate into polygonal epithelial cells by day 15. At 17 days of gestation, there are 2 cell layers of polygonal cells, which differentiate into a multilayer urothelium by day 20. This structure is maintained into adulthood, when separate superficial, intermediate, and basal layers can be demonstrated.[5] Basal urothelial cells are undifferentiated, with poorly developed endoplasmic reticulum, a large, deeply indented nucleus, and large numbers of ribosomes. Intermediate urothelial cells are about twice as large as basal cells, and have a smaller nucleus/cytoplasm ratio and fewer ribosomes than basal cells. The rough endoplasmic reticulum and Golgi apparatus are more clearly seen. Intermediate cells also contain large numbers of vesicles adjacent to the superficial cells. Superficial urothelial cells have a nucleus that only occupies 10% of the cell volume, larger vesicles than intermediate cells, and filaments that run from nucleus to cell borders.[6] In human fetal specimens, a 2 layer epithelium is found at 10 weeks, progressing to 3 layers by 13 weeks. Specific features, such as microvilli, binucleated umbrella cells, vesicles, and lysosomes are present at 19 weeks.[7] The development of the epithelium precedes that of the smooth muscle by one week in the rat and two weeks in the human.

Smooth muscle has certain characteristics when viewed with the electron microscope. It has a high surface to volume ratio, peripheral caveolae (round invaginations of the cell membrane), microfilaments of actin and myosin, and dense bands, which are attachments on the cell membrane for the microfilaments.[8] Human adult bladder smooth muscle cells are spindle shaped, have an elongated nucleus at the center of the cell, an electron dense basal lamina, multiple caveolae, extensive apposition with other smooth muscle cells, and organelles packed at the end of the cells.[9] Human fetal bladder smooth muscle at 12 weeks has been described as arranged in thin bundles with abundant myofibrils in a immature connective tissue. Collagen was apparent at 13 weeks and extended to superficial muscle bundles by 15 weeks. From 21 weeks to term, the muscles are organized into interwoven bundles separated by loose connective tissue.[7]

While immunohistochemistry is useful for distinguishing smooth muscle cells from most other cell types, this technique is unreliable for distinguishing smooth muscle cells from myofibroblasts, which also express smooth muscle actin and other smooth muscle markers (vimentin, actin, desmin) during wound repair. Electron microscopy is useful for distinguishing between smooth muscle cells, myofibroblasts, and fibroblasts. Smooth muscle cells and myofibroblasts may be distinguished from fibroblasts by the presence of microfilaments, attachment plaques, dense bands, basal lamina, caveolae, intercellular junctions, and cell-stroma attachments, all of which are absent in fibroblasts. A fibroblast has a smooth or stellate border, a smooth nucleus, well developed Golgi and dilated rough endoplasmic reticulum, and makes few attachments to other fibroblasts. A myofibroblast may be distinguished from a smooth muscle cell by a basal lamina that does not completely surround the cell, a more irregular shape with long cytoplasmic extensions, a

Table 1. Ultrastructure characteristics of bladder smooth muscle, myofibroblasts, and fibroblast

	Smooth muscle	Myofibroblast	Fibroblast
Shape	Spindle	Irregular	Smooth
Nucleus	Elongated	Deeply indented	Smooth
Basal lamina	Present	Partial	Absent
Apposition	Present	Intermediate	Absent
Caveolae	Present	Present	Absent
Golgi	Poor	Well developed	Well developed
Rough endoplasmic reticulum	Poor	Well developed	Dilated
Dense bands	Present	Present	Absent
Microfilaments	Present	Present	Absent
Attachment plaques	Present	Present	Absent
Intercellular junction	Intermediate	Intermediate and gap	Absent
Cell-stroma attachment	Present	Well developed	Absent

deeply indented nucleus, the presence of gap intercellular junctions, a more well developed Golgi and rough endoplasmic reticulum, and the presence of well developed cell-stroma attachments called the fibronexus[10] (See Table 1).

During normal wound healing, myofibroblast cell numbers were found to peak at 15 days post-wounding, and diminish as they de-differentiate to fibroblasts or die by 30 days post-wounding. Apoptotic figures were seen as the expression of smooth muscle actin decreased with time. Based on the time course of microfilament expression and the location of fibroblasts and blood vessels, it was felt that myofibroblasts more likely derive from local fibroblasts, rather than dedifferentiated pericytes or smooth muscle.[3] However, myofibroblasts are persistent in pathologic processes such as Dupuytren's disease and are responsible for the hard, sclerotic tissues associated with breast and colon carcinomas.[10] Therefore, the persistence of myofibroblasts in a healing wound suggests the eventual formation of scar tissue.

Research into ultrastructural changes of bladder smooth muscles due to obstruction has shown that fibroblastic cells from the sub-serosal layer can be induced to transform into myofibroblasts and subsequently into fetal smooth muscle cells.[11] In this situation, myofibroblasts can be distinguished from fibroblasts by the presence of a basal lamina. Myofilaments are still present in the myofibroblast, but they are obscured because they are tend to be pushed to the periphery. Reversal of the obstruction leads to a decrease in the size of the smooth muscle cells without evidence of apoptosis. In the rat, both the weight of the bladder and cystometric function return to normal following relief of obstruction.[12] However, electron microscopy reveals that smooth muscle cells from bladders which have undergone obstruction and subsequent release of obstruction maintain a significantly larger profile areas, more irregular outlines, and increased intracellular collagen compared to smooth muscle cells from control bladders.[13]

In human fetal specimens, obstruction of the bladder by posterior urethral valves leads to an increase in bladder mass that is proportional to smooth muscle and collagen.[14] Extensive work has been done linking different types of human geriatric voiding dysfunction with specific smooth muscle changes. It has been suggested that the structural basis of detrusor hyperactivity or impaired contractility is due to altered cell junctions, collagen deposition, hypertrophied smooth muscle cells, or enlarged dense bands.[15–18] Likewise, the long-term outcome of the use of acellular bladder matrix may depend on correct cell-cell signaling during the repair phase.

Just underneath the epithelium, a dense set of capillaries lies closely associated with the urothelial basement membrane and the lamina propria. The density of vessels increases as animals mature. While the main purpose of this bed may be to exchange metabolic products, to us, the importance of this structure is that fibroblasts are located nearby this capillary bed, slightly deeper in the lamina propria, and may be the source of myofibroblasts which would play a role in bladder remodeling after obstruction or bladder regeneration in the use of acellular bladder matrix.[19, 20]

3. MATERIALS AND METHODS

Urinary bladders were dissected from Fisher 344 rats at day 14 and 18 of gestation, birth, and 60 days post-natal. Four rats were used at each time point. Following dissection, the bladders were fixed in 1% glutaraldehyde at pH 7.2 overnight at 4 degrees Celsius. They were then post-fixed in 1% OsO_4 in PBS, sequentially dehydrated in graded alcohols, then embedded in LR White (medium grade) plastic. Thin sections were cut. Transmission electron microscopy was carried at 3000 magnification, beginning at the epithelial border, and proceeding peripherally until smooth muscle bundles were encountered. Contact prints were made, and scanned into Adobe Photoshop using an HP scanner on a PowerMac. Representative sections of smooth muscle were used to reconstruct a montage of the bladder wall.

4. RESULTS

At 14 days, smooth muscle is barely recognizable, as it forms very thin layers of single cells in an extracellular matrix containing sparse collagen fibers. At this stage, smooth muscle cells cannot be identified by α-actin staining. These undifferentiated mesenchymal cells are distributed from the serosal to mucosal borders, and are very similar in morphol-

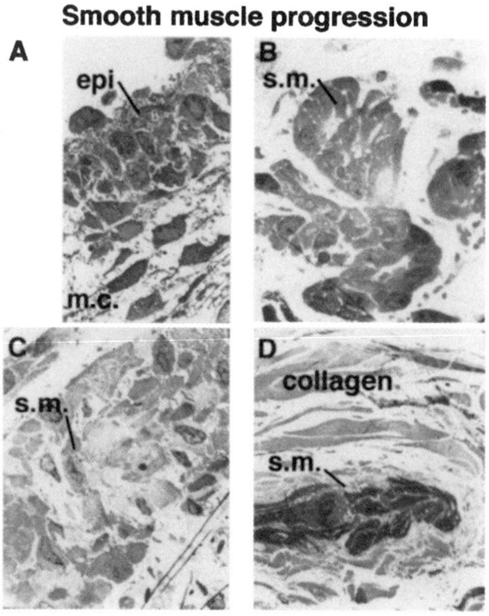

Figure 1A. Smooth muscle progression, 3000 magnification. A) 14 day gestation. Epi = epithelial cell, M.C. = mesenchymal cell; B) 18 day gestation. S.M. = smooth muscle; C) Newborn. S.M. = smooth muscle; D) Adult. S.M. = smooth muscle.

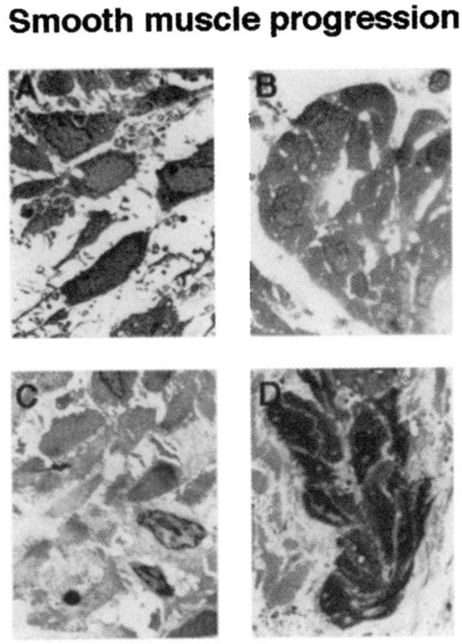

Figure 1B. Smooth muscle progression, 6000 magnification. A) 14 day gestation;B) 18 day gestation; C) Newborn; D) Adult.

ogy to fibroblasts. They have an elongated shape, with a round central nucleus, minimal basal lamina, and are opposed only at their long cell processes. No caveolae were seen, and organelles were centrally located. Microfilaments and dense bands were not seen, and few collagen fibers were seen in the extracellular matrix. (Figures 1A and 1B)

The smooth muscle cells at 18 days were organized into fascicles in a position intermediate between the mucosa and the serosa, as had been seen under light microscopy. The cells themselves were less elongated than those at birth, or in the adult. The nucleus assumed a rounder shape, and the basal lamina surrounding individual cells was present but poorly developed. While apposition was present, the cells were much more loosely connected as compared to the newborn and adult smooth muscle cells. Caveolae were absent, and organelles were mainly centrally located. Both microfilaments and dense bands were visible, but they were not prominent in the cytoplasm. Collagen fibers again were sparse in the bladder wall (Figures 1A, 1B, and 2).

At birth, the smooth muscle cells were more elongated, and also had an oblong nucleus, like adult smooth muscle cells. The basal lamina was thickened around individual smooth muscle cells, and attachment plaques between cells were frequently observed. The smooth muscle cells themselves were located in a luminal position. However, they maintained characteristics of immature cells: caveolae were few, organelles were still centrally located, collagen fibers were more abundant, and both microfilaments and dense bands were in an intermediate stage of development (Figure 1A, 1B and 3).

Finally, in the adult bladder, the usual picture of spindle shaped cells with elongated nuclei, thick basal lamina, extensive apposition, caveolae, microfilaments, dense bands, and peripheral organelles was seen (Figures 1A, 1B, 4, and 5).

Fibroblasts were seen at all stages of development, just below the lamina propria. They tended to lie close to the small blood vessels which course over the lamina propria and also above the collagen layer that usually divides the bladder epithelium and smooth muscle. The orderly progression of epithelium from single layer of round cells to a 3-layers of polygonal cells, which was previously described, was confirmed in this study (See Table 2).

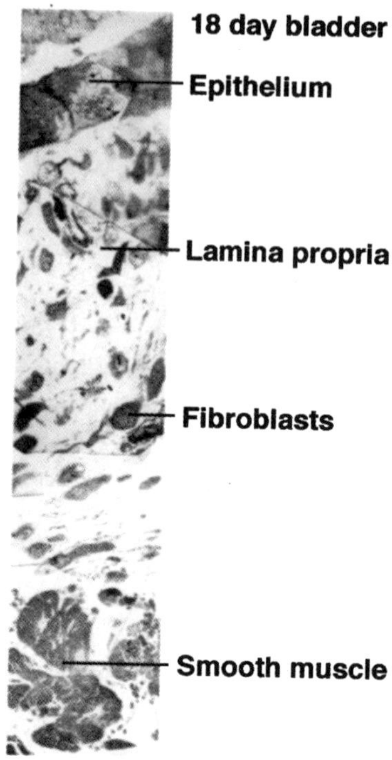

Figure 2. 18 day gestation bladder, 2000 magnification.

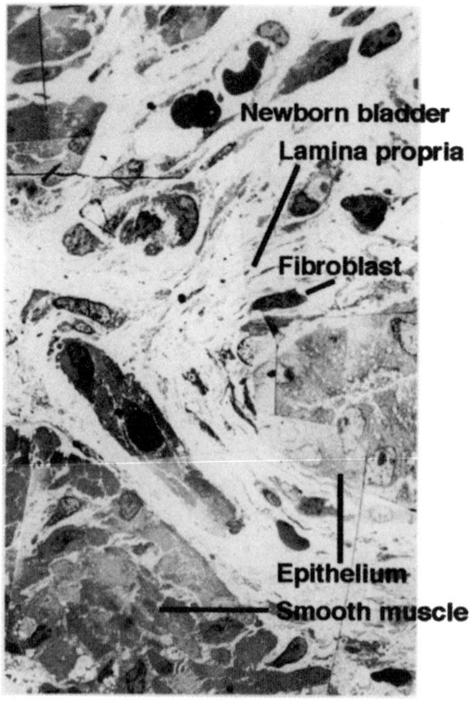

Figure 3. Newborn bladder, 4000 magnification.

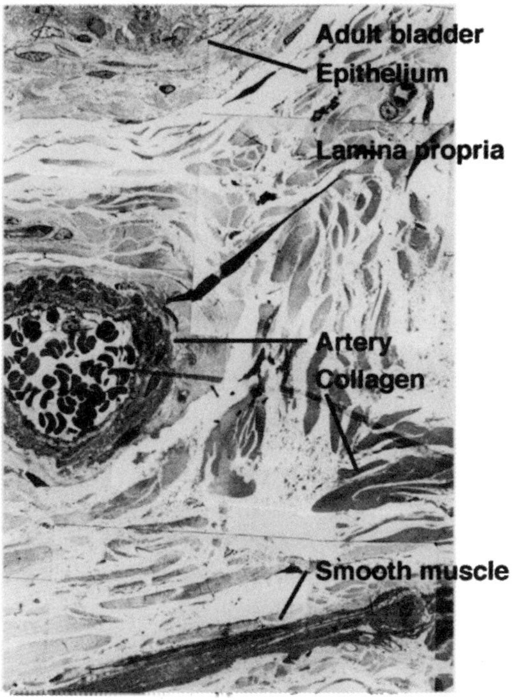

Figure 4. Adult bladder, 2500 magnification.

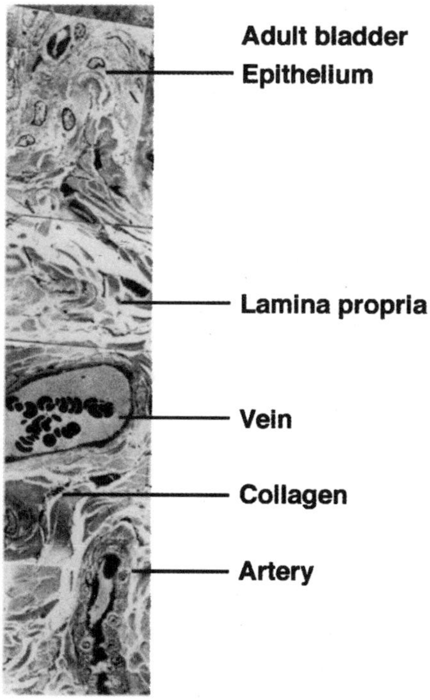

Figure 5. Adult bladder, emphasizing structures near lamina propria. 2500 magnification.

Table 2. Ultrastructure characteristics of bladder smooth muscle cells from embryonic 14 and 18 days, newborn, and adult rat bladders

Smooth muscle	14 day gestation	18 day gestation	Newborn	Adult
Shape	Spindle	Rectangular	Elongated	Spindle
Nucleus	Round	Round	Elongated	Elongated
Basal lamina	0	1+	2+	3+
Apposition	1+	2+	3+	3+
Caveolae	0	0	1+	3+
Organelles	Central	Central	Central	Peripheral
Dense bands	0	1+	2+	3+
Microfilaments	0	1+	2+	3+
Location	Serosal	Intermediate	Lamina	Lamina
Collagen	0	0	1+	3+
Epithelium	Single layer, round cells	Double layer, polygonal	Triple layer, polygonal	Triple layer, polygonal

5. DISCUSSION

Bladder smooth muscle cells acquire ultrastructural features during development that eventually set them apart from fibroblasts and myofibroblasts. Initially at 14 days, the mesenchymal cells which differentiate into smooth muscle cells do not express smooth muscle markers and resemble fibroblasts. At 18 days, these differentiating smooth muscle cells express α-actin, vimentin, laminin, vinculin, myosin, and desmin, and continue to grow in size.[1] The smooth muscle cells may be distinguished from fibroblasts by the presence of microfilaments. The migration inward from the serosal edge of smooth muscle cells, in contrast to the luminal origin of fibroblasts, argues that in normal development, smooth muscle cells are derived from a different subset of mesenchymal cells than fibroblasts. As the smooth muscle cells are only beginning to show a basal lamina, they must be distinguished from myofibroblasts by their shape and less developed Golgi, rough endoplasmic reticulum, and cell-stroma attachments. At the time of birth, the elongated nucleus and cytoplasm, as well as thickening basal lamina allow a distinction to be made between smooth muscle cells and myofibroblasts.

Teleologically, the increase in dense filaments and microfilaments in normal development suggests that as urinary output is increased after birth, increased muscle contractility and cell stabilization are required, therefore the cell increases actin-myosin units and their corresponding attachments. Fetal urine output in humans gradually increases with gestation, reaching 50 cc/hr at delivery.[21] This is followed post-natally by a rapid rise in glomerular filtration rate (GFR) and bladder volume.[22] Increased coupling of cells may also serve to better coordinate a voiding contraction.

In studying the ontogeny of bladder smooth muscle in other species, it has been noted that sheep show a switch between myosin heavy chain isoforms from fetal to neonatal growth.[23] Extensive work has also been done on the development of collagen in bovine fetal bladders. Compliance has shown to increase with development, paralleling a drop in collagen III/I ratio.[24] The synthesis of both collagen III and I is higher in the epithelial-lamina propria component than the detrusor portion of the bladder.[25] In this study, the most marked increase in collagen was also seen at the epithelial-lamina propria level, suggesting an increase synthesis of extracellular matrix, possibly driven by the abundant local blood supply.

While rat bladder smooth muscle cells remain readily identifiable after 16 days of gestation, changes in ultrastructure occur which parallel functional changes brought on by delivery. Our further investigations will focus on determining whether the a-actin positive cells that repopulate acellular matrix grafts in the bladder are maturing smooth muscle cells or myofibroblasts, as long-term function of these grafts will depend on normal smooth muscle development.

REFERENCES

1. Baskin, L.S., Hayward, S.W., Young, P.F. and Cunha, G.R.: Ontogeny of the rat bladder: smooth muscle and epithelial differentiation. Acta. Anat. (Basel)., **155:** 163,1996.
2. Sutherland, R.S., Baskin, L.S., Hayward, S.W. and Cunha, G.R.: Regeneration of bladder urothelium, smooth muscle, blood vessels and nerves into an acellular tissue matrix. J. Urol., **156:** 571,1996.
3. Darby, I., Skalli, O. and Gabbiani, G.: Alpha-smooth muscle actin is transiently expressed by myofibroblasts during experimental wound healing. Lab. Invest., **63:** 21,1990.
4. Baskin, L.S., Hayward, S.W., Young, P. and Cunha, G.R.: Role of mesenchymal-epithelial interactions in normal bladder development. J. Urol., **156:** 1820,1996.
5. Cohen, S.M., Cano, M., Sakata, T. and Johansson, S.L.: Ultrastructural characteristics of the fetal and neonatal rat urinary bladder. Scanning Microsc., **2:** 2091,1988.
6. Walton, J., Yoshiyama, J.M. and Vanderlaan, M.: Ultrastructure of the rat urothelium in en face section. J. Submicrosc. Cytol., **14:** 1,1982.
7. Newman, J. and Antonakopoulos, G.N.: The fine structure of the human fetal urinary bladder. Development and maturation. A light, transmission and scanning electron microscopic study. J. Anat., **166:** 135,1989.
8. Gabella, G.: General aspects of the fine structure of smooth muscles. In: Ultrastructure of Smooth Muscle. Edited by Motta, P. Boston: Kluwer, chapt. 1, 1990.
9. Dixon, J.S. and Gosling, J.A.: Ultrastructure of smooth muscle cells in the urinary system. In: Ultrastructure of Smooth Muscle. Edited by Motta, P. Boston: Kluwer, chapt. 8, 1990.
10. Schurch, W., Seemayer, T. and Gabbiani, G.: Myofibroblast. In: Histology for Pathologists. Edited by Sternberg, S. New York: Raven Press, chapt. 5, 109, 1992.
11. Buoro, S., Ferrarese, P., Chiavegato, A., Roelofs, M., Scatena, M., Pauletto, P., Passerini-Glazel, G., Pagano, F. and Sartore, S.: Myofibroblast-derived smooth muscle cells during remodelling of rabbit urinary bladder wall induced by partial outflow obstruction. Lab. Invest., **69:** 589,1993.
12. Gabella, G. and Uvelius, B.: Urinary bladder of rat: fine structure of normal and hypertrophic musculature. Cell Tissue Res., **262:** 67,1990.
13. Gabella, G. and Uvelius, B.: Reversal of muscle hypertrophy in the rat urinary bladder after removal of urethral obstruction. Cell Tissue Res., **277:** 333,1994.
14. Freedman, A.L., Qureshi, F., Shapiro, E., Lepor, H., Jacques, S.M., Evans, M.I., Smith, C.A., Gonzalez, R. and Johnson, M.P.: Smooth muscle development in the obstructed fetal bladder. Urology, **49:** 104,1997.
15. Elbadawi, A., Yalla, S.V. and Resnick, N.M.: Structural basis of geriatric voiding dysfunction. I. Methods of a prospective ultrastructural/urodynamic study and an overview of the findings. J. Urol., **150:** 1650,1993.
16. Elbadawi, A., Yalla, S.V. and Resnick, N.M.: Structural basis of geriatric voiding dysfunction. II. Aging detrusor: normal versus impaired contractility. J. Urol., **150:** 1657,1993.
17. Elbadawi, A., Yalla, S.V. and Resnick, N.M.: Structural basis of geriatric voiding dysfunction. III. Detrusor overactivity. J. Urol., **150:** 1668,1993.
18. Elbadawi, A., Yalla, S.V. and Resnick, N.M.: Structural basis of geriatric voiding dysfunction. IV. Bladder outlet obstruction. J. Urol., **150:** 1681,1993.
19. Inoue, T. and Gabella, G.: A vascular network closely linked to the epithelium of the urinary bladder of the rat. Cell Tissue Res., **263:** 137,1991.
20. Inoue, T. and Gabella, G.: The interface between epithelium and lamina propria in the rat urinary bladder. Arch. Histol. Cytol., **55 Suppl:** 157,1992.
21. Rabinowitz, R., Peters, M.T., Vyas, S., Campbell, S. and Nicolaides, K.H.: Measurement of fetal urine production in normal pregnancy by real-time ultrasonography. Am. J. Obstet. Gynecol., **161:** 1264,1989.
22. Chevalier, R. and Mandell, J.: Renal function in the fetus and neonate. In: Campbell's Urology. Edited by Walsh, P., Retik, A., Stamey, T. and Vaughan, J., ED. Philadelphia: Saunders, chapt. 33, 1344, 1992.
23. Chern, J., Kamm, K.E. and Rosenfeld, C.R.: Smooth muscle myosin heavy chain isoforms are developmentally regulated in male fetal and neonatal sheep. Pediatr. Res., **38:** 697,1995.

24. Baskin, L.S., Constantinescu, S., Duckett, J.W., Snyder, H.M. and Macarak, E.: Type III collagen decreases in normal fetal bovine bladder development. J. Urol., **152:** 688,1994.
25. Koo, H.P., Howard, P.S., Chang, S.L., Snyder, H.M., Duckett, J.W. and Macarak, E.J.: Developmental expression of interstitial collagen genes in fetal bladders. J. Urol., **158:** 954,1997.

DISCUSSION

Epithelial-Mesenchymal Interactions in the Bladder

Q & A with Tung-Tien Sun, Ph.D.

Edward Macarak, Philadelphia.

Q: What kinds of proteins do you propose would exist in the hinge region?

A: We know relatively little about that. We have previously shown using a monoclonal antibody, which is how we first got into the uroplakin area. We made the antibody AE-32, which selectively stains the hinge area, so that was really the first indication, that the hinge is not as innocent as it may seem. We always say that these hinges are just an Interplak area, but we think that biochemically there is also something very peculiar about the hinge. That antibody defines a molecule which we call a neurohingen because of its hinge association. This is about 85–86Kd in molecular weight, and heavily glycosylated. We have not yet been able to clone that; we are presently in the process. Today I showed you one picture where we can experimentally create a hinge by bending the urothelial apical surface. And so, in that particular case, to our surprise, you are not just bending all the particles neatly, you know, move to both sides, and then creating a hinge. We actually start to see the deformation of the uroplakin particles, and they becoming shallower, become shorter, and so on. So eventually, then you see this area that is particle free. That makes you wonder whether or not this hinge area may actually have some deformed uroplakin molecules, and that this is something that is quite interesting because actually, when we isolate these urothelial plaques by using very stringent detergent wash, the hinge is insoluble. So something is very, very peculiar about the hinge.

Gerald Cunha, UCSF.

Q: I have a question Henry, in regard to the embryonic development or primary differentiation of the urothelium; do the uroplakins come on sequentially? What can you tell us about the initial differentiation of urothelium?

A: That's an interesting question., I think it is about 14 days of mouse embryonic development that we start to be able to detect uroplakin. We used several antibodies to stain that, and as far as we can tell, they pretty much express simultaneously. Xue-Ru Wu has done some chemical cross-linking experiments showing that these four uroplakins actually form two pairs, so uroplakin 1a is cross-linked to UP2, 1b is cross-linked to UP3, so whether one pair is expressed earlier than the other is something that we are very interested in finding out.

Barry Kogan, Albany, New York.

Q: Did I hear you say that there's not a mucopolysaccharide layer on top of these uroplakins?

A: No, I said that morphologically we cannot see a striking glycocalyx forest on top of the AUM plaques. We have actually done control experiments using, for example, the frog bladder epithelium which is of course, a different animal so to speak. But that is covered by beautiful, delicate, long, forests of the glycocalyx, and we can see that using the quick-freeze deep edge technique. Meaning that this technique is mild enough to preserve a real extensive glycocalyx structure, but we did not see that. We cannot rule out the possibility there is a small amount of protein or urinary protein or urothelial protein secreted that then comes back to coat the surface.

Barry Kogan, Albany, New York.

Q: What do you propose is the actually function of the uroplakins?

A: There are three theories that have been proposed over the years. One of the previous diagrams shows that maybe the cytoplasmic phase of the AUM is anchored into the cytoskeleton, and we have some preliminary data in support of that. And so this hypothesis states that it's a mechanical integration, dispersion of the mechanical stress when the bladder is expanded, so that to prevent the apical surface from fracturing, so that's number one. Number two, is that AUM proteins or plaques contribute to the remarkable permeability barrier, function of the bladder epithelium, and we have some evidence to suggest that. And then finally, the most striking theory is that these AUM vesicles can fuse with the apical surface, open up to increase the surface area during bladder distention, and then when the bladder contracts some of these plaques can re-form and retrieve back into the cytoplasm. There are some rather strong physiological measurements to support that. I don't know what is the answer, and I don't think these three functions are mutually exclusive, but we are in the process of knocking out some of the uroplakin molecules, for example, UP3, just to see what would be the resultant pathology or physiological abnormality in order to answer that question.

Steve Zderic, Philadelphia.

Q: The question I wanted to ask is related to your coupling of the growth hormone to the UP promoter. In those transgenic mice, were there any differences in their bladder weights compared to their litter mates? Have you thought at all about coupling another growth hormone, say IGF, and in targeting it to that area to see if you could produce a difference in bladder mass? Recently, Fagan has published a paper showing that if you couple IGF to promote smooth muscle specifically, you actually get a larger bladder mass. Were these bladder masses different?

A: The bladder mass is normal. The animals become very fatty because of the human growth hormone. There is some, you know, secreted backwards, so therefore, so you can see that the hormone is biologically active. But, yes, two points. One is that it would be very interesting to see whether or not one would be able to induce some physiologic changes in the underlying tissues or overall function, that would be very interesting. Another question from a bioreactive point of view, we are in the process of trying to put other model hormones, because of the inherent dificulty with human growth hormone It's an endocrine hormone which is normally secreted backwards anyway. So for us to see a low level of secretion into the urine is already better than we expected, but it is not a good model from a bioreactive point of view.

Simon Hayward, UCSF.

Q: Are uroplakins expressed in other urothelial cells, for example, in the ureter and the urethra? Because if they are, then those areas are not expanding and contracting in the way that the bladder is, but would still have to be involved in keeping urine out of the general circulation, and maybe that would feed into a mechanistic role for these cells.

A: No, no, absolutely. As Jerry mentioned, the ureteral urothelium and the bladder urothelium are embryologically distinct. Functionally, you know, as far as one can reason, they could also have different functional requirements. That is why maybe uroplakins are designed in such a small way, that you really can accommodate all these different functions, and that is why I said that I feel these three functions may not be mutually exclusive.

Q: But the proteins are expressed all the way through the urinary tract?

A: Yes.

Sam Chacko, Philadelphia.

Q: You mentioned the uroplakins 2 and 3 bind to cytoskeletal protein. Are the cytoskeletal proteins, using the uroplakin as attachment plaques? And which cytoskeletal component binds to uroplakins.

A: I only meant to say that about uroplakin 3, which is the only uroplakin which has a significant cytoplasmic tail. UP2, for example, does not have any cytoplasmic tail to speak of. But the UP3 cytoplasmic tail has 50 amino acids and multiple potential phosphorylation sites, and so that actually raises an interesting question, as to how this membrane - cytoskeleton interaction may be regulated. We don't know the detail as to how this is binding occurs. So far, we have some preliminary data using conventional biochemical techniques.

Q & A with Jennifer Southgate, Ph.D.

Laurence Baskin, UCSF.

Q: Have you looked at urothelial differentiation in any of the TGF-beta knock-out mice? There is beta-1, beta-2 knock-out, is there anything unusual about these animals ?

A: We haven't really done very much in rodent systems. We find that the culture systems for rodent cells are very different from human, and have found it very difficult to correlate the two systems. Our feeling, being based in a Medical School, is that we are more interested in the human state. So, yes, there are certain things we can't do in the human model, but I think our next stage will be to begin to manipulate the normal human urothelial cells genetically and to see what happens when we study them in the different in vitro models.

Monica Liebert, M.D. Anderson Cancer Center, Texas.

Q: Can you grow the urothelium with the stroma, but have them physically separated?

A: We have considered doing that but haven't yet tried it. I agree that it's important to try to identify whether soluble factors are involved or whether cell-cell contact is required.

Q: What cell types are present in the stroma, and may they contribute to the urothelial phenotype?

A: We haven't done much to characterise the stromal cells. They may well be important, although the work from Mary Comer, which is only in the porcine system at the moment, suggests that the tissue-specificity of stromal cells may not be too relevant in terms of inducing expression of urothelial differentiation antigens. However, I would say that we're probably not achieving full maturation of the urothelium in vitro.

Henry Sun, New York University.

Q: I'm very impressed by the anti-AUM staining when you grow the cells on top of Matrigel. Can you start with urothelial cells in low calcium medium where the cells are uniformly AUM negative and put them on the Matrigel where they turn on like that? What are the growth properties of the cells on Matrigel? Can they form a confluent layer?

A: The urothelial cells form a stratified epithelium, and again, the proliferation rate is reduced. We find it very difficult to process these tissues on Matrigel, and we have to disaggregate the cells to look at them. But we see some induction in the majority of cells. At the moment, we are still trying to characterize exactly what is being expressed in terms of, for example, uroplakins. We don't yet have very clear answers as to whether all differentiation-associated proteins are being expressed, or whether there is a selective expression of some of them.

Q: Also, you mentioned that neutralizing antibody to TGF-beta can block differentiation. Can you add TGF-beta to induce?

A: No. Addition of TGF-beta to a monoculture doesn't induce anything. So there are some things, about either the three dimensionality of placing cells onto the Matrigel, or other factors that we haven't yet identified. Just to add to that, I think that the different signaling networks are very highly inter-related, and it's very difficult to sort out one pathway in the absence of another.

Gerald Cunha, UCSF.

Q: I was struck by the orderly arrangement of, I think, the RK tumor cell line. You had three cell lines, and they varied in their degree of anaplasia.

A: The RT4 cell line is very well differentiated.

Q: Have you done any experiments where you grow those various tumor cell lines in association with, if you will, tumor fibroblasts, that is the stroma that's normally associated with the carcinoma?

A: Just in two-dimensional culture? No we haven't, but one of the things we have done in conjunction with the normal human urothelial cells has been to look at what the tumor cell lines express in terms of differentiation antigens. I think that the tumor cells can ectopically express some of the uroplakins when the normal cells don't in similar microenvironments, so there appears to be some dysregulation of differentiation following neoplastic transformation.

Q & A with Avraham Raz, Ph.D.

John Park, Boston.

Q: Is there any known rule or association for autocrine motility factor receptor in the development? Because there is quite a bit of cell migration and motility. For example, in the renal development, the uteric bud branches out and moves quite a bit, and I wonder whether there is anything known about this molecule and the receptor for that environment.

A: The only thing that we have done is in Xenopus development, and we think that it is involved in the motility of the normal cells. Unfortunately, we don't have the expertise, and we have not took it further.

M'Liss Hudson, St. Louis, MO.

Q: I too am interested in these different markers to predict which of the superficial tumors are going to be the bad actors, if you will. Have you substratified those groups, because I think many people now accept that there is a biological difference between TA, TI, and TIS tumors, and does that hold up when you do the subgroup analysis.

A: Yes, it does.

Patricia Harden, St. James Hospital, Leeds, U.K.

Q: In your model systems, have you actually looked at the metastases to see whether they express this marker or not, and whether that is related to outcome? Because, we now know that cell motilities are only a very small part of the development of metastases. Cells circulate very readily in the vasculature, and change position very elegantly, that in fact they circulate very easily and travasate and extravasate very easily, and that it's really the postextravasation phenomena that determines whether you get clinically significant metastases or not. In particular, whether they proliferate or not is a significant determinant. So, does this work as a growth-factor related system in the postextravasation system?

A:That's a very good question. People don't realize the majority of the cells that enter the circulation will not end up as clinical metastases. 99.9% of the cells that do enter the circulation perish or are unable to proliferate at the new site. When I'm saying we believe, you have to know that it means that we don't know, we can only speculate. We do believe that it does play a role, I don't want you to go out with the impression this is the first and only thing that controls metastasis. This is only a small part of the cumbersome phenomenon of metastasis. We find out that the cell does proliferate in new sites, and secretes enormous amounts of this AMF, so we believe that AMF does assist in some way in the proliferation at the newer site.

Q: I was wondering whether you specifically looked at what cells expressed AMF, circulated, extravasated, and then either stopped expressing it, or didn't stop expressing it.

A: They don't stop expressing. We believe that is one of the reasons why the phenomenon of metastasis from metastasis do work here.

Bob Hurst, University of Oklahoma, Oklahoma City.

Q: How do your GP-78 and E-cadherin data, associate with aneuploidy? Because George Hemstreet and I did a study a number of years ago, and were able to predict the bad actors among G-1 tumors simply by finding aneuploid cells. Presumably, these should be associated. Did you look? In other words, the tumors that you predict by AMF and E cadherin to be bad actors, are they also aneuploid?

A: We have no idea.

Jenny Southgate, St. James Hospital, Leeds, U.K.

Q: I was just wondering whether it really refers back to the first question you had, whether this had a role in normal cell migration and wound healing, and whether it would be looking either in a normal human urothelial cell culture system, or in a wound-healing model to see whether this was expressed as part of the normal response repertoire to tissue wounding.

A: Yes, it does. And this is a topic for a completely different action. We found out that each is secreted in the player role, at least in experimental systems, and the promoter activity is highly sensitive to cell-to-cell contact and a normal consideration. And whereas the wound healing, as the cells start moving, and at least the promoter of GP-78 became active the moment the cell reached confluency, there is a shut off of a relation of expression.

Q: Would this be a problem when you're trying to use this as a tumor marker, having confusion between a normal or benign state where you've got tissue injury and expression?

A: I don't believe so, but I don't know.

Q:Have you looked to normal groups of tissue in which benign and inflammatory conditions have been included?

A:Yes, we look in many pathological bladder diseases which are noncancerous, and we never find this relationship to hold.

Q & A with Catherine A. Reznikoff, Ph.D.

Aria Olumi, UCSF.

Q: You showed some primary tumors, I believe they were the muscle-invasive tumor types, you had six examples, three of which had P16 expression, and three of which did not have P16 expression. Then you went on to show that three of these samples were methylated. Were the three samples that were methylated, the same samples that did not expressed P16 on your Western blots?

A: Yes.

Q: So, they were the samples exactly.

A: Exactly the same. So there were six samples, three had no P16 and those were methylated or mutated.

Q: The second question is, it may be a bit premature, have you considered putting back, recently there was a publication by the Geron Corporation where they put portion of the telomerase gene back into the cells and expanded the life span. So have you considered putting that same gene back into these primary cells to see whether you may halt the expression of P16 gene for these cells.

A: Yes, that's a wonderful idea, yeah, thanks. And we are doing gene manipulation. We are putting over expressed Cyclin B in to see if we can bypass it, so I think this is a wonderful idea.

Bob Hurst, University of Oklahoma, Oklahoma City.

Q: That was a splendid paper, by the way. The two invasive cell lines that had the wild-type P-53, did you by chance check to see whether they expressed microsatellite instability?

A: Oh, thank you. This is a good question, and our students are doing that right now. Because I think that that is completely related to what I suggested, that there are other mechanisms for genome instability.

Q: One would predict that they probably would, or that there's another mechanism for genomic instability, I don't know.

A: Yeah, one would expect that they would have lots of mutations.

Q: I think they have to keep shuffling the deck.

A: Yes, thanks.

Q & A with Laurence S. Baskin, M.D.

Aria Olumi, UCSF.

Q: You demonstrated nicely that when you inject KGF into the nude mice, you get proliferation of urothelium. Have you looked to see in your wound-healing model, when you subcutaneously or systemically inject KGF, does it improve your healing process.

A: Actually, when you inject the KGF subcutaneously, not only do you get proliferation of bladder epithelium, but you also get proliferation of epithelium throughout the body, for example, the uterus and other organs. How could this potentially be useful, you don't want the rest of the body to proliferate. For example, if you do bladder surgery maybe you can paint it right into the wound allowing the bladder, to seal things off even quicker. The answer to your question is that we haven't done the experiment, but it seems reasonable.

Bob Weiss, Yale, Connecticut.

Q: Larry, very nice talk. Have you tried to inhibit the KGF and see what occurs as far as the bladder healing at that point in time, and it may have to do with other types of healing processes.

A: There are neutralizing monoclonal antibodies available to KGF which we've used in a prostatic system, but those have not been applied to the bladder system as yet. What's interesting, and what I didn't touch on, is all the other growth factors, except TGF-alpha seems to respond very similarly to KGF. We know TGF-alpha is important, for example, for gastric mucosa repopulation, so you wonder if KGF is so clean and so nice, but you wonder if there are alternative pathways that might be involved, but that's a good plan for an experiment.

Sam Chacko, University of Pennsylvania, Philadelphia.

Q: Have you measured the force produced by these newly formed muscle strips in the regenerated bladder using acellular matrix.

A: I have not, but some people in this room have. Brad Kroop, I know in his SIS model has done that and showed that they do contract.

Q: Are there any differences in the contracted characteristics?

A: Brad Kroop University of Oklahoma, responding. We have previously shown that both in the rat and dog, the regenerated muscle in the SIS model, that the carbachol response is approximately 50%. The amount of muscle regeneration overall is about 50%. It also responds to adrenergic and the purinergic stimulation, as well as electrical field stimulation.

Q: One more question. Do you see an increased proliferative activity at the area where the smooth muscles from the normal region migrate to the matrix.

A: Laurence Baskin responding. We haven't studied it carefully enough.

David Thomas, Leeds, U.K.

Q: I hesitate to ask this question because it's rather peripheral, I had hoped to ask it earlier. But this is about what's happening with interactions at the other end of the urinary tract. We've said this morning that it's remarkable that urothelium produced by endoderm and urothelium produced by the mesoderm, the ureteric bud, are so similar. But I wonder at what point in the urinary tract that ceases to happen. Embryologically, we're taught that in nephrogenesis, the collecting ducts are derived from the ureteric bud, and that the kidney are derived from the metanephros, and I just wondered if anyone, Dr. Cunha, or Dr. Sun know at what point these asymmetric, forces cease to be present in the urinary tract, because the waterproofing qualities that we look for lower down the urinary tract are positively unwelcome within the kidney itself and in the tubules. So at what point does this transition occur? Can you identify it clearly within the kidney, or is there a gradual transition and loss of this asymmetric unit membrane?

A: So it sounds like the ureteral epithelium and the bladder epithelium, even though they're different embryologically, stain for uroplakins all the way up north in the renal pelvis to all the way south.

Q: Into the calyces and up into the collecting ducts in the renal parenchyma? So it goes as far as the calyces?

A: Pelvis. To the renal pelvis.

Q: Have any of the factors which are known to be involved in reciprocal interactions between the ureteric bud and the metanephric mesenchyme, been tested in your model, looking at the signaling between the urothelium and the pelvic mesenchyme?

A: No, that hasn't been done. There's quite a large amount of data on those signaling molecules.

Q: As Larry's saying, we don't know what those signaling molecules are. It might be interesting to start with, just to look at those, and perhaps if nothing else, rule them out.

Penny Longhurst, Albany, New York.

Q: Larry, do you catheterize these rats after surgery? For how long? And to back up from Sam's question, do you seek functional changes?

A: We can only do females because we can't catheterize the males, and about 10 to 15% of our animals do not do well secondary to acute obstruction after we do the surgery. They just look punkish, and so we euthanize them.

Q: Well, if they're punkish, are the bladders distended?

A: Yes, but even the ones that do well have huge bladders, so I don't know whether they're in overflow incontinence, or how they're surviving. I just know that from my experience, rats are tough.

Q: So maybe the distention is stimulating the proliferation.

A: Possibly.

Patricia Hardern, Leeds, U.K.

Q: I was very struck by one of your last pictures where you're showing that the smooth muscle when it's growing is actually very much closer to the urothelium than you would normally seen in an in situ bladder which hasn't been traumatized in that way. In fact, if you look at normal bladder in the trigone, the smooth muscle is actually very much closer to the urothelium than in the sort of body detrusor which distends. So, this shows my ignorance of embryology, but initially, when smooth muscle is developing, is it close to the epithelium, and then does the sort of what we call the lamina propria develop later when the bladder becomes more functional? And is that why you're seeing, because they're not actually fully functional, the smooth muscle near the urothelium? Because that would explain why there isn't this gap in your model.

A: It's a good question. In the very first pictures I've shown, we've focused more the top of the bladder, or what Bob Levin really has defined as the body of the bladder. The **base** of the bladder is more confusing, and we know there's different receptors. But just in the body of the bladder, the muscle forms in the periphery, and I've no idea why. It's a great puzzle, but very confusing.

Gerald Cunha, UCSF.

A: You know, I suppose another point that should be made is it is in the region of the trigone where the interface presumably exists between the mesodermally derived urothelium and the endodermally-derived urothelium. And there are suggestions that as the ureters take up their final position, and that involves tremendous rearrangement from their original position coming off of the Wolffian duct, there is a suggestion that mesodermally-derived Wolffian duct epithelium gets incorporated into the posterior wall of the urethra and into the trigonal region. So it may be that the trigone actually develops in association with a urothelium from an entirely different dermal area, namely the mesoderm. And that may be involved in structuring the smooth muscle in that area differently from the rest of the bladder. Something that's testable.

Avraham Raz, Detroit.

Q: Two questions. First of all, how do you know it's smooth muscles? Second, this is very similar to what happened during lung development. And their differentiation to smooth muscle is mediated by laminin. Have you tried to block the differentiation with antibody against laminin?

A: Your first question is kind of a killer question, because we're actually trying to answer the question, how do you define smooth muscle? We're defining it by electron microscopy now. Yang Wu's in the audience. We have beautiful pictures, and it's actually documented all types of different parameters. So that's one way to define it. We've defined it by an EM, using smooth muscle a-actin gold particle staining, as well as using a whole series of smooth muscle markers: smooth muscle alpha-actin, laminin, myosin, desmin, and it's not perfect, but I think it's a pretty good way. Finally, I've just given my definition, and people can argue about it. So, that's how we're defining it. It's not perfect, but I think it's reasonable. The electron microscopy is very helpful. Your second question is a good idea. We haven't done it.

John Park, Boston.

Q: Larry, I know your original question was, "how does smooth muscle form?", and the data you presented, both in the tissue recombination and acellular matrix indicate that it is really the epithelium that has sort of a dominant effect in the differentiation of the bladder. Do you have any information or an speculation as to what role undifferentiated mesenchyme plays in bladder development?

A: Well, smooth muscle won't form unless you have the epithelium, and that's the only thing I can offer that we see biologically. But I think there's reciprocal communication.

Q: In the 14-day bladder that you show that histologically, it is not really differentiated transitional cell epithelium. So, what is the fate of those cells without the mesenchyme, the mesenchyme from the developing bladder?

A: The fate of the epithelial cells? They don't survive well without mesenchyme. We would speculate, and this is something that we plan on testing in the very near future, that the undifferentiated mesenchyme of that embryonic bladder is involved in instructively specifying that the epithelium in that area will in fact undergo urothelial differentiation. That's the hypothesis, and that's something that we can test.

Q: I did some work in Josie Brigg's laboratory the last two years. She did some work with organ culture, where she was doing a recombination experiment with the ureteric bud and the mesnonephric mesenchyme, but as a urologist, a plumber who is concerned about what happens downstream, it sort of made me wonder, like Dr. Thomas pointed out, why does a certain portion of the ureteric bud become ureter, and then certain part of the ureteric bud become the collecting duct and the renal epithelium? Probably the answer is because of the influence of the surrounding mesenchyme, and whether there is some sort of default program that differentiates the mesenchyme surrounding that epithelium has an affect on it.

James Roberts, Tulane.

Q: It's interesting Jerry that you brought up the mesodermal origin of the trigone. It's very important that we remember that because the trigone does respond to hormones, and it doesn't matter if you treat a female or a male with estrogen, the trigone becomes squamous tissue, and that is a different epithelium, certainly, embryologically.

Brad Kroop, Oklahoma.

Q: Larry, you've really pioneered this area in terms of trying to figure out where it's going, and we've talked a lot about it. A couple comments. You've noticed in your model that you only had smooth muscle differentiation underneath the epithelium. John Pope published a paper that recently showed muscle both under the mucosa, as well as out in the periphery in the small intestinal submucosa model. I think there is some signaling going on, but I'm not sure, and we commented in a paper that we're just not sure about that. Second comment. About the bladder distention. We know that in the dog model, and we don't catheterize our rats either, but in the dog model, if we leave a catheter in for more than seven to ten days, essentially we get no smooth muscle regeneration. It's just pure fibroblast. And I think there is something with the mechanical stretch that you had previously done six years ago. The final comment was on in terms of how to define what a smooth muscle is. I think you can do it with immunostaining, but it's very difficult to get the right myosins, to get the right actins. We're also trying to do it with functional, you know, does a myofibroblast from a fibroblast respond the same as a myofiblast from a regenerating smooth muscle.

Q & A with Anthony Atala, M.D.

David Thomas, Leeds, U.K.

Q: Can I just focus on two aspects of the bioengineered bladder, just practical questions? What's the duration, the interval between the time you seat the cells into your biodegradable scaffold and the time you implant that into the experimental animal?

A: From the time we take the biopsy, it's about four to six weeks to the time that we seat it. So, we allow four to six weeks to expand, and then we seat the cells, and seating the cells is only a three-day process We seat the muscle cells first. We then flip and seat the urothelial cells second. Once we do that, we allow the cells to attach for an extra three days. Once a whole attachment series is done, and then it's ready for implantation.

Q: Do you divert the urine from the bladder for any period of time once you've put that onto the bladder.

A: Yes, it depends on what we do. We're using the polymer-only or the acellular matrix only. We only divert for a few days for about 48–72 hours. If we're using matrix with cells for these experiments, we divert them for about four weeks.

Q: The biodegradable material—have you any idea how long it retains its physical characteristics.

A: Yes. We can actually control the properties. The nice thing about the biodegradable polymer is that we can control a lot of its characteristics, including the degradation curve. We have designed these to degrade at about three months, fully degrade, you can't see any more polymer after three months. It's important because what you want to do is you want to be sure that the polymer retains its structural characteristics until the tissue that you put in there has been able to totally regenerate.

Q: What switches the muscle off? What are the signals that tell the muscle to stop? Because you can imagine, it's a balance between a distending force, and the inherent contractility of the bladder, and you can imagine that if these muscle cells were exposed to distending pressures too early, that you'd get the equivalent of a diverticulum. Equally, if the muscle went on multiplying and growing, you'd have a contracted bladder. So what is it that strikes the balance.

A: Yes, well I think that the fact that you're putting it back into the organ which is doing what it's supposed to do is what's making it go. I'm a firm believer that, as we know, stretch is equal to growth. As a matter of fact, John Park, this afternoon is going to talk a little bit about stretch in the muscle. But I think it's very important that you're putting this back in the condition that it's known for, you're putting the bladder muscle back into a bladder environment where you do have this cycling going on for its development.

Barry Kogan, Albany, New York.

Q: My question's a really simplistic one. We know that bladders that are denervated don't function normally. Now you've shown us the epithelium looks pretty good, the muscle looks pretty good, although I don't know about EMs or whatever, but it looks pretty good, and you've shown us nerves, but clearly these nerves are not centrally innervated. So how does this bladder function so normally if it doesn't have appropriate innervation? And I guess, related to that is you didn't show us contractile characteristics of this, but, how does this bladder function so normally without any neural input?

A: It's an excellent question, and I think that we have really underestimated the value of the trigone. I really do. And I think that the trigone is really providing a lot of what you see in bladder function. And I think that a lot of bladder disease that we see may be, once you get to the trigone level, and you see the pathology there, that's what may be happening. Certainly, if this were a totally de novo system, I don't think you would see contraction like you did now. And I'm not looking for contraction, because again, like Larry mentioned, all we want is storage. So we were very impressed. We were not expecting the results that we got. But certainly, as the goal is, is just to not use some other type of tissue such as intestine. The goal is really to use something that's totally compatible. So from that perspective, I think we're okay. It remains to be seen what'll happen in the clinical setting.

Earl Chang, Chicago.

Q: I'm curious as to what your feelings are about what your polymers would like from cells that have been cultured from abnormal bladders? As you know, clinically, the majority of kids that we deal with in pediatric urology are going to have an abnormal bladder. And most of the stuff that you're showing from the laboratory are cells cultured from normal bladders. So, what's your feeling as to what the cells would look like if they were taken from abnormal bladders? Do you feel that they would de-differentiate in culture and then eventually grow as normal cells, or would they act differently?

A: That's a very good question Earl. Basically, it depends what you start with. That you're talking about a bladder which is diseased because of the abnormal components of collagen and fibrosis, etc. Once you take the urothelial cells out of its environment, and the muscle cells out of their environment, they act perfectly fine, they act perfectly normal. We've looked at cells from patients with myotubular myopathy, which has a genetic component, a genetic disease where the cells have eccentric nuclei, and the cells just don't fully differentiate. Those cells remain abnormal. So it depends what you're starting with. If you're starting with a genetic defect which is inherent within the cell, then there's nothing you can do about it. But if you're starting with a cell type which is just in the wrong milieu, and you put it in the right environment, then it's gonna do okay.

Section II

MUSCLE AND EXTRACELLULAR MATRIX IN THE BLADDER

OVERVIEW OF MUSCLE AND EXTRACELLULAR MATRIX IN THE BLADDER

Edward J. Macarak

Department of Anatomy and Histology
University of Pennsylvania
Philadelphia, Pennsylvania 19104-6003

The bladder wall is a complex structure whose physiologic function is dependent upon the interaction between the active or force generating components and the passive or force distributing components. The smooth muscle cells of the detrusor actively participate in the process of tension generation while the constituents of the extracellular matrix such as collagen, elastin, and proteoglycans passively distribute or store force during the micturition cycle. Fundamental to an overall understanding of bladder function is the way in which active and passive components interact with one another to store urine and at low pressure and expel it at the proper time and with the proper force. In their chapter, Nguyen and Kogan review our current understanding of fetal bladder physiology. They point out the importance of functional evaluation of bladder tissue using whole organ or muscle strip studies. Such studies provide the baseline data with which data from studies of pathological tissue can be compared. It is important to consider the interactions and potential influences of the various components of the bladder that must be integrated to achieve normal function. Interactions between muscle, matrix and nerves become more difficult to understand during fetal development when additional hormonal and maternal factors influence cell behavior. Nevertheless, a number of investigators have been successful in studying bladder development.

Evidence for the complexity of these interactions is provided in the chapter by Kropp in which he reviews exciting studies based upon the use of small intestinal submucosa (SIS) as a graft material for bladder reconstruction. Previous studies from his laboratory have shown that SIS is a collagen-rich material that promotes regeneration of a variety of host tissues including the urinary bladder. Over 1000 cross-species transplants using this material have been performed with excellent results. These extensive studies show that SIS is an viable option for genitourinary construction. As well, the model is suitable for studies to investigate the mechanism by which this acellular matrix controls the de novo regeneration of the bladder. Nevertheless, despite these elegant results, it is not clear why this material, which is largely collagen, works better than other collagen

Advances in Bladder Research, edited by Baskin and Hayward.
Kluwer Academic / Plenum Publishers, New York, 1999.

based materials. This underscores the necessity for additional studies to elucidate why this materials composition and architecture is so successful in promoting bladder regeneration.

Chacko et al. have investigated the active properties of bladder smooth muscle. In obstructed bladders, they have characterized the expression of myosin isoforms in the bladder smooth muscle to determine if the bladder wall remodeling is associated with an alteration in their composition. During the compensatory phase of the bladder's response to partial obstruction, the SM1 isoform increases relative to that of SM2 resulting in an altered SM2:SM1 ratio. Associated with this change is altered contractile function. This altered pattern of isoform expression is a direct consequence of obstruction since it returns to normal when the obstruction is removed. This important study provides insight into the potential mechanism responsible for altered active properties of bladder smooth muscle.

Zderic et al. have emphasized that key to the active function of the bladder is the role of calcium. Both the storage and release of calcium is tightly controlled in detrusor smooth muscle. The sarcoplasmic reticulum (SR) is the actual intracellular reservoir into which calcium ions may be actively transported, stored and released via the action of an ion transporting ATPase and the sarcoplasmic reticulum $Ca^{2+}MG^{2+}$ATPase (SERCA) is located on the membrane surface of the SR. Since the SR increases in amount during development, the pattern of SERCA expression was found to increase with gestational age. Increased levels of mRNA encoding for SERCA were also found. This suggested that functional requirements for bladder active processes were dependent upon the presence of adequate amounts of SERCA. Currently unknown is the level of SERCA activity at the protein and mRNA levels during and as a consequence of obstruction. These studies on the molecular control mechanisms of detrusor contractility will be key to the understanding, at the molecular level, of the excitation - contraction coupling in the bladder.

To further underscore the importance of active forces in the bladder, Park and his colleagues describe their studies on the characterization of cyclooxygenase (COX) isoforms in the bladder. Cyclooxygenase is crucial to prostaglandin synthesis and may play an important biological role during fetal development and after acute bladder obstruction. Although COX-1 was not detectable during development, COX-2 was differentially expressed in the murine embryos. In addition, COX-2 was rapidly induced in response to obstruction. This isoform was localized to the stromal layer after outlet obstruction suggesting that its source was the smooth muscle cells. Using an elegant in vivo model of regional bladder wall distension, they demonstrated that stretch and not pressure was the primary stimulus for changes in COX-2 gene expression.

The cellular processes that control the response to obstruction are currently not well understood. To examine these cellular mechanisms more extensively, the role of angiotensin II is discussed in the chapter by Cheng et al. Because of the well-described actions of this octapeptide on the systemic vasculature, it is likely that it may play a role in the response of the bladder to obstruction. To investigate its effects on bladder wall cells, the investigators used pharmacologic agents that inhibit angiotensin II production and/or block angiotensin II receptors. Captopril, an angiotensin converting enzyme inhibitor, was administered to obstructed neonatal rabbits. Animals receiving the drug exhibited a marked reduction in the physical changes associated with obstruction: increased bladder weight, detrusor hypertrophy and transmural deposition of collagen. Similar studies were carried out in a rat model of obstruction. However, these studies did not show any effects of either captopril or lorsatan (angiotensin subtype AT-1 receptor antagonist). Nevertheless, in vitro studies have shown strong effects on rabbit stromal cell division suggesting that some responses are species specific. These potentially important and exciting studies are being extended to humans although the investigators caution about the use of these drugs in the pediatric population because of their potent effects on the renin-angiotensin system.

In her studies, Ellen Shapiro emphasized the importance of an intact nervous system in the development of normal smooth muscle cells. Her previous elegant studies on children with myelomeningocele (MMC) emphasized that structural changes associated with affected individuals occurred in the developing fetus. Detailed histologic and immuno-histochemical studies in human fetal samples suggest that a global defect in the development of smooth muscle occurs in children with myelomeningocele. Since such tissues also exhibit diminished peripheral nerve density, these data suggest that an intact nervous system is of fundamental importance to the normal development of the detrusor layer.

Howard et al. present a study of several patients with bladder diverticuli, who had Ehlers-Danlos syndrome (EDS), a group of inherited abnormalities of connective tissue. Bladder tissue samples were studied histologically, immunohistochemically and biochemically. This is the comprehensive study of stromal composition and arrangement in bladder tissue from EDS patients. These bladders displayed an abnormal wall architecture with disorganized muscle bundle structure. Since EDS tissue was also more compliant than normal tissue, these data suggest that the disorganization of the detrusor results in reduced active "tone" in the bladder wall rendering the organ more compliant.

The chapter by Cheng et al. has focused on the passive elements of the bladder, the extracellular matrix. Using laser confocal microscopy and antibodies to type III collagen, these investigators have examined bovine bladders at several stages of the filling cycle. They note pronounced changes in the morphologies of type III collagen fibers that correlate with the degree of tension being sustained by the bladder wall. The tension-bearing collagen fibers transfer tension from the active muscle components throughout the bladder wall. The authors suggest that the resiliency of the bladder likely results from the structural accommodation of coiled collagen fibers that occurs during the micturition cycle.

These contributions in this section are testimony to the wealth of new and important information that is derived from basic research. It is truly exciting to see the innovative ideas and new directions being taken to help understand the molecular mechanisms that control bladder development and fuction. The new information will hopefully help provide more comprehensive and useful treatments for individuals with lower urinary tract dysfunction.

FETAL BLADDER PHYSIOLOGY

Hiep T. Nguyen[1] and Barry A. Kogan[2]

[1]Department of Urology
University of California
San Francisco, California 94143
[2]Division of Urology
Albany Medical College
Albany, New York 12208-3479

1. INTRODUCTION

The function of the bladder is to store urine at a low pressure and then to periodically expel the urine via coordinated and sustained contractions. Failure of the bladder to perform these functions will result in abnormally high pressure transmitted to the kidneys resulting in renal damage. The process by which the bladder acquires these important functions remains to be fully understood. Much research has been dedicated to understanding the physiology of the fetal lower urinary tract. Currently, the methods of studying the fetal lower urinary tract are those in which the human fetus is observed by ultrasound, histologic and physiologic measurements are made in whole organ preparation or muscle strip studies, and physiologic measurements are made in an animal preparation. In this chapter, we will summarize our current understanding of fetal bladder physiology based on studies in the human fetus and animal preparations.

2. ULTRASONOGRAPHIC STUDIES OF THE HUMAN FETAL BLADDER

As early as the 10th week of gestation, the fetal bladder can be observed on maternal ultrasonography as a rectangular shaped, anechoic structure in the fetal pelvis. The fetal bladder has a very thin wall that acts as a specular reflector on ultrasound. Consequently, only a few cc of urine in the bladder are required for visualization. However, expert sonography is necessary as there are pitfalls. For example, the internal iliac arteries course along the lateral wall of the bladder and may sometimes lead to the false interpretation of the bladder wall as thickened.[1]

Advances in Bladder Research, edited by Baskin and Hayward.
Kluwer Academic / Plenum Publishers, New York, 1999.

The fetal bladder can be seen emptying approximately every 30 minutes at 28 weeks and every 60 minutes at 40 weeks gestation.[2] Based upon measuring the fetal bladder at sequential intervals, urine production is estimated to increase from 5 ml/hr at 20 weeks gestation to 51 ml/hr at 40 weeks gestation. The maximal bladder volume increases from 1 cc (0.5–1.0 ml/kg) at 20 weeks gestation and 36–54ml (6 ml/kg) at 40 weeks of gestation.[3] On the average, voiding last for 9.5 seconds[3] and occurs almost exclusively during periods of fetal wakefulness.[4] In a study of 101 normal fetuses at 20 weeks' gestation, complete bladder emptying was observed in only 28% of the fetuses studied. 54% of the fetuses had a post-void residual of less than 25% of estimated capacity, but 18% had a post-residual of 26–65% of estimated capacity.[3] In contrast, by 40 weeks, all had complete bladder emptying. From these observations, it can be concluded that the human fetal bladder is able to store urine and empty as early as 20 weeks gestation. However, with increasing fetal age, these functions mature with increasing bladder capacity and more complete emptying.

3. WHOLE ORGAN AND MUSCLE STRIP STUDIES IN THE DEVELOPING BLADDER

To further our understanding of bladder functions, many have turned to in vitro studies of the developing bladder in different animal preparations. Keating et al.[5] observed that in the newborn rabbit, bladder capacity increased proportionally with body and bladder weight until the eight weeks of age. Thereafter, bladder capacity increased in greater magnitude than body and bladder weight. The observed increase in capacity reflects the ability of the developing bladder to accommodate progressively greater amount of urine while maintaining low intravesical pressure; this is termed compliance.

In whole organ studies, compliance is measured by determining the change in intravesical volume per change in intravesical pressure. Coplen et al.[6] and Koo et al.[7] found that in the fetal calf, bladder compliance increased as the fetus matured, up to a 3 fold difference between the second trimester and term. In addition to using whole organs, compliance can also be measured in bladder strips. Baskin et al.[8] developed a novel testing system in which circularly clamped bladder tissues were repetitively pressurized causing the samples to deflect upward in the shape of the spherical cap. When the centerline deflection was greater than the tissue thickness, the elastic modulus could be determined by linear regression techniques. Bladder compliance, the inverse of elastic modulus, was determined in the fetal bovine bladder as a function of age. The authors found that during normal development, there was a progressive change from a rather stiff noncompliant bladder characterized by a high elastic modulus to a compliant bladder with a lower elastic modulus.

Compliance may be divided into two components: passive (i.e., tension) and active (i.e. tone). Passive compliance results from the inherent tension in the bladder wall, provided in part by collagen and other components of the extracellular matrix. In contrast, active compliance results from the tone of the bladder wall, provided by smooth muscles. In previous studies,[9–12] many investigators have noted an increase in collagen content in the noncompliant bladder affected by conditions such as myelomeningocele and outlet obstruction. To assess whether similar changes in collagen can account for the observed increased compliance in the fetal bladder, Baskin et al.[8] determined total collagen per unit of bladder tissue in the fetal calf by assaying for the total hydroxyproline content. This amino acid is a marker for collagen since it is not found in significant amount in other proteins. The authors found that total collagen content decreased with gestation in the fetal bovine bladder. In analyzing the subtypes of collagen, they found that the ratio of type III/I collagen and total type III collagen decreased during fetal development. These findings in the fetal bovine bladder are similar to those observed in the fetal human bladder.[13] To further these observations, Koo et al.[14] measured collagen gene expression in the fetal bovine bladder at different stages of gestation and found that type I collagen gene expression peaked during the late second to early third trimester, while type III collagen gene expression progressively decreased throughout fetal development. With immunohistochemical methods, they further found that as expected, the collagen expression was localized to the urothelium-lamina propria of the bladder rather than the detrussor layer. Based upon these studies, it is postulated that the reduction in type III collagen is responsible for the observed increase in compliance as the fetal bladder develops.

However, collagen is only one determinate of passive compliance and the overall compliance is determined by both bladder tension and tone. Coplen et al.[6] (in whole bladder preparations) and Dean et al.[15] (using muscle strips) evaluated the relative contributions of the passive and active components to the overall compliance of the fetal bovine bladder. Passive compliance was measured by incubating and studying the samples in a calcium-free medium; this prevented smooth muscle function, eliminating the contributions of active compliance to overall compliance. In contrast, when the samples were studied in a calcium-containing buffer, the compliance characteristics reflected both the passive and active components. These studies demonstrated that passive compliance was greatest in the youngest bladder and decreased gradually with increasing fetal age. The opposite pattern was seen with overall compliance, in which the younger bladder had the least active compliance and the older bladder the greatest. These observations suggest that in the immature bladder, there is high active smooth muscle tone, resulting in relatively poor compliance. As the fetal bladder matures, smooth muscle tone decreases, allowing for storage of increasing amount of urine. It appears that changes in bladder active compliance may be of more significance than changes in its passive component.

In addition to storage, the other major function of the bladder is to expel the urine. Levin et al.[16] observed that newborn rabbit bladder strips contracted poorly in response to methoxamine, an alpha-adrenergic agonist and isoproterenol, a beta-adrenergic agonist. In contrast, these bladder strips had excellent contractile response to bethanechol, a muscarinic cholinergic agonist and ATP, a purinergic agonist. Using histochemical techniques, the authors found that at birth the density of adrenergic fibers in the rabbit bladder was sparse while that of cholinergic fibers abundant. Over the next 6 weeks of development, there was a rapid increase in the density of adrenergic fibers while there was no significant change in cholinergic innervation. Using radio-ligand binding assays, they also found that the density of adrenergic alpha and beta receptors was low while that of muscarinic receptors was high in the newborn bladder. With increasing age, the density of alpha and beta receptors increased rapidly while that of muscarinic receptors remained unchanged.

As with bladder compliance, there appears to be a gradual development in the fetal bladder contractile function. Using whole bladder preparations, Keating et al.[5] demonstrated that in response to cholinergic stimulation, greater bladder contractions were seen in newborn rabbits compared to the adults. Zderic et al.[17] similarly found that nonadrenergic, noncholinergic agonists such as serotonin, histamine and substance P induced a higher degree of contractile response in the newborn rabbit bladder than in the older ones.

Because free calcium is important to smooth muscle contraction, investigators speculated that developmental changes in contractility may be due to changes in the way smooth muscle cells handle cytosolic free calcium. Zderic et al.[18] again demonstrated that in the presence of physiologic concentrations of calcium (2.5 μmol/l) and in response to bethanechol, newborn rabbit bladder strips contracted 98% of their maximal tension compared to only 68% in those from older rabbits. This contractile response in the newborn bladder strips was inhibited by 1 μmol/l of diltiazem, a calcium-channel blocker, whereas the older bladder strips did not demonstrate any inhibition at this concentration. The authors further demonstrated, using spectrofluorometry with FURA-2 methyl ester, that the effect of extracellular calcium correlated with the intracellular free calcium concentration. These findings suggest that the increased contractile response seen in younger rabbit bladder strips is in part due to an increase in maximal intracellular concentration of free calcium.

Zderic et al.[19] evaluated whether the developmental-dependent effect of calcium on bladder contractility was due to an increase in permeability of membrane-bound calcium channels. Using the radioligand, H3 PN-200, the authors demonstrated no differences in either receptor density or dissociation constants between newborn and older rabbit smooth muscle. Similarly with radioimmunoassay techniques, Zderic et al.[19] did not demonstrate any developmental differences in intermediary regulatory proteins, specifically calmodulin, that transduce the calcium signal into force. These observations led the authors to speculate that there must be developmental changes in intracellular storage and release of calcium to account for differences in contractility response to calcium by younger and older rabbit bladder strips. Ryanodine is an alkaloid that binds to ion channels located on the smooth endoplasmic reticular membrane where it inhibits the efflux of intracellular stores. Zderic et al.[20] and Gong et al.[21] demonstrated that the binding of ryanodine was significantly less in newborn rabbit bladder strips compared to older ones (6.6 versus 44 fmol/mg protein) without any associated changes in the dissociation constants. Furthermore, the authors found that the contractile response by the younger rabbit bladder strips was less inhibited by ryanodine than the older strips. From these findings, the authors concluded that there was an increase in the expression of ryanodine-sensitive ion channels with increasing age, suggesting that with development, the bladder smooth muscle cells acquired an increased pool of sequestered intracellular calcium. Similarly, this pattern of development was seen in rabbit myocardium. Thus, it appears that contraction in the newborn rabbit smooth muscle is more dependent upon the influx of extracellular calcium and with increasing age, mechanisms for intracellular calcium storage and release develop.

More recently, Koo et al.[7] characterized the contractile response in the fetal bladder further. In fetal calf whole bladder preparations, they noted that bethanechol induced a sustained contractile response resulting in emptying that is almost complete in mid and late gestational bladder but was only 50% complete in those of early gestation. Lee et al.[22] demonstrated that this developmental pattern was regional. The authors separated the fetal calf bladder into upper and lower bladder segments and found that the maximal contractile response of the upper bladder strips to bethanechol increased significantly from early to mid gestation with no further changes from mid to late gestation. In contrast, strips from the lower bladder segments did not demonstrate any age-related differences in contractility. Currently, it remains unknown what accounts for the more complete emptying observed in the older fetal bladder in these experiments.

4. PHYSIOLOGIC STUDIES OF THE FETAL BLADDER IN ANIMAL PREPARATIONS

Although anatomic and functional information determined from in vitro studies has provided much insight into the function of fetal bladder, in vivo studies are necessary to fully understand the complex functional environment of the fetal bladder, particularly the effect of the central nervous system. Because of their size and similar urogenital anatomy to human, fetal sheep have been used for in vivo urodynamic studies and pharmacological manipulation of the fetal bladder. The surgical preparation consists of placing catheters in the aorta and vena cava and in the fetal bladder through the urachus and urethra. Pressure changes in the bladder (i.e. cystometrogram) can then be measured allowing for investigations of the fetal bladder with the normal interactions among the bladder, sphincter and central nervous system. Using this preparation, Kogan and Iwamoto[23] demonstrated that in vivo, the fetal bladder was compliant and that fetal bladder contractions could be discerned following furosemide infusion. Intravesical pressure at time of contraction varied from 10–24 mmHg with less than 1 cc residual urine.

More importantly, Kogan and Iwamoto[23] demonstrated that fetal bladder functions could be altered pharmacologically. They found that bethanechol decreased bladder capacity while atropine, a muscarinic-receptor blocking drug, significantly inhibited bladder contractions, increased bladder capacity and caused marked residual urine. In contrast, infusion with alpha-adrenergic drugs had little effect on bladder storage or emptying, while infusion with beta-sympathomimetic drugs (specifically ritodrine) inhibited bladder contractions and increased residual urine. These findings suggested that in utero, bladder functions were under both cholinergic and beta-adrenergic control. The observation that fetal bladder function was in part under beta-adrenergic control, was not seen previously with in vitro fetal or in vivo adult bladder studies. The authors speculated that this was most likely due to the sensitization of beta-adrenergic receptors by progesterone, the same mechanism that exists in the uterus. Interestingly, the authors also found that medications commonly used to inhibit preterm uterine contractions such as magnesium sulfate and diltiazem also inhibited fetal bladder contractions. They speculated that this finding may explain partially the high incidence of fetal hydronephrosis that resolves spontaneously after birth and the occasional finding of urinary retention in the neonate in whom preeclampsia was treated with magnesium sulfate.

In the fetal sheep preparation, Mevorach et al.[24] evaluated the role of nitric oxide, a nonadrenergic, noncholinergic neurotransmitter active throughout the central nervous system. Nω-nitro-L-arginine was used to inhibit NO synthesis and L-arginine to stimulate NO activity. The authors found that inhibition of NO synthesis resulted in increased numbers of ineffective contractions at relatively low bladder volumes, a larger overall capacity and poorer emptying. These changes were reversed when L-arginine was infused. When only L-arginine was infused, constant urinary drainage was observed. To determine its effects on the bladder, the bladder neck was ligated. Although Nω-nitro-L-arginine still increased bladder capacity, L-arginine infusion had no effects on bladder function. Thus, the poor emptying without alterations in pressure generated during bladder contraction resulting from Nω-nitro-L-arginine infusion is most likely due to a failure of sphincteric relaxation rather than primary effects on the bladder. These findings suggest that nitric oxide system is important in bladder neck and striated external sphincteric function.

Using a similar fetal sheep preparation, Wlodek et al.[25] evaluated the relationship between voiding and fetal cortical activity during late gestation. They found that voiding coincided with periods of low voltage electrocortical activity in the fetus. This suggested that spontaneous voiding in the fetus was in part regulated by descending information from the brain. Of interest however, continuous catheter drainage of the bladder did not alter the previously noted electrocortical activity, suggesting that the development of sensory neural pathways and connections to the motor pathways was incomplete.

Due to technical complexity, the relatively few in vivo studies of the fetal bladder are limited to late gestation. It is unknown whether the developmental changes in bladder compliance and contractility seen in the in vitro bladder strip studies reflect what is seen in vivo. Much research is still needed until we can understand fully the function of the fetal bladder in vivo.

5. STUDIES OF THE OBSTRUCTED DEVELOPING BLADDER

Based upon the above in vitro and in vivo studies of the developing bladder, we gradually are beginning to comprehend normal fetal bladder physiology; this understanding is necessary when we begin to evaluate the effects of pathologic conditions such as bladder outlet obstruction on fetal bladder function. In humans, investigators have observed that the obstructed fetal bladder has obvious histologic changes. The bladder wall thickness is increased markedly. However, the percent area density of smooth muscle and connective tissue as well as ratio of smooth muscle to connective tissue are similar in the obstructed and normal fetal bladder.[26,27] These findings suggest that bladder outlet obstruction in the fetus results in increased bladder mass with a concomitant increase in both smooth muscle and collagen. However, the spatial distribution of connective tissues is altered with fetal bladder outlet obstruction.[26] In the obstructed bladder, connective tissues are seen infiltrated into smooth muscle bundles.

Several studies have attempted to replicate fetal bladder outlet obstruction in animal preparations in order to evaluate histologic and physiologic changes in the obstructed developing bladder. Peters et al.[28] created complete bladder obstruction in the fetal sheep at mid gestation. The authors found that total bladder weight in the obstructed bladder was significantly higher compared to age-matched controls. Total DNA content was higher in the obstructed fetal bladder, while total DNA to protein ratio decreased with obstruction. These findings suggest that obstruction induces both cellular hyperplasia and hypertrophy. Karim et al.[29] created partial bladder obstruction in the fetal sheep at mid gestation and found that the obstructed bladder mass and total DNA content were increased compared to controls. They also found that there was an increase in total choline acetyltransferase activity, an index of cholinergic innervation, in proportion to the increase in cell number. These findings suggest that normal levels of innervation are maintained with fetal bladder outlet obstruction. In a different study, Peters et al.[30] evaluated collagen synthesis and degradation in the obstructed fetal sheep bladder. The authors found that despite an increase in collagen content in the obstructed fetal bladder, there was no increase procollagen gene expression, the precursor for collagen. This finding suggested that there were other mechanisms, such as a reduction in collagen degradation, that may account for the increased collagen content in the fetal obstructed bladder. To add support to this notion, the authors demonstrated that matrix metalloproteinases-2 (MMP-2) activity, which is important in collagen degradation, was not detected in the obstructed fetal bladder but was present in normal fetal bladder. Furthermore, they found that both the amount of mRNA and protein for tissue inhibitors of metalloproteinases (TIMP) were significantly increased in the obstructed bladder. These findings taken together suggest that as a result of obstruction, accumulation of collagen in the fetal bladder results from inhibition of its degradation.

In a different animal preparation, Rohrmann et al.[31,32] evaluated the storage and emptying functions of the obstructed fetal bladder. Partial bladder outlet obstruction was created in fetal rabbits. The authors found that histologically, there was smooth muscle cell hypertrophy and increased connective tissue in the obstructed bladders. In whole bladder preparations, the obstructed bladder had increased compliance.[32] They found that by removing the calcium from the organ bath and thus, eliminating the active component of compliance, effected no change in the obstructed bladder compliance. This finding suggests that the increased compliance of the obstructed fetal rabbit bladder is due to the passive properties of the bladder wall rather than from smooth muscle tone. Additionally, Rohrmann et al.[31] found that the contractile function of the obstructed fetal bladder was altered. The obstructed bladder strips had greater response to bethanechol compared to normal bladder strips. The authors suggested that this may represent denervation supersensitivity, as a result of obstruction. It should be noted, however, that since the bladders were studied approximately seven days following obstruction, the above findings represent only the early effects of obstruction. It is likely that with prolonged obstruction the fetal bladder will begin to decompensate, losing both its compliance and contractility.

6. CONCLUSIONS

Our understanding of normal and abnormal fetal bladder function remains limited. In considering bladder function, one must keep in mind the complex interactions between smooth muscle, connective tissues and neural elements. In vitro studies using bladder strips and whole organ will help to elucidate these interactions. However, there are also complex interactions between the bladder, the fetus as a whole and the in utero environment. In vivo studies are also needed to elucidate these later interactions. Based upon both in vivo and in vitro studies, it appears that as with any other organ, the fetal bladder undergoes a series of developmental changes, acquiring increasing smooth muscle function and neural inputs. Alterations in this developmental pattern will result in changes in bladder compliance and contractile function. Further research will help to define normal and abnormal development of the fetal bladder in hope of identifying processes that can be manipulated to produce adequate bladder storage at low pressure and complete emptying.

REFERENCES

1. Grannum P.A.: The Genitourinary Tract, in DA Nyberg, Mahony BS and Pretorius DH: Diagnostic Ultrasound of Fetal Anomalies: Text and Altas. Chicago, Year Book Medical Publishers Inc., 433, 1990.
2. Hata T. and Deter R.L.: A review of fetal organ measurements obtained with ultrasound: normal growth. J. Clin. Ultrasound., **20:** 155, 1992.
3. Rabinowitz R., Peters M.T., Vyas S., Campbell S. and Nicolaides K.H.: Measurement of fetal urine production in normal pregnancy by real-time ultrasonography. Am. J. Obstet. Gynecol., **161:** 1264, 1989.
4. Ohel G., Haddad S. and Samueloff A.: Fetal urine production and micturition and fetal behavioral state., Am. J. Perinatol., **12:** 91, 1995.
5. Keating M.A., Duckett J.W., Snyder H.M., Wein A.J., Potter L. and Levin R.M.: Ontogeny of bladder function in the rabbit. J. Urol., **144:** 766, 1990.
6. Coplen D.E, Macarak E.J. and Levin R.M.: Developmental changes in normal fetal bovine whole bladder physiology. J. Urol., **151:** 1391, 1994.
7. Koo H.P., Macarak E.J., Zderic S.A., Duckett J.W., Synder H.Mr. and Levin R.M.: The ontogeny of bladder function in the fetal calf. J. Urol., **154:** 283, 1995.
8. Baskin L., Meaney D., Landsman A., Zderic S.A. and Macarak E.: Bovine bladder compliance increases with normal fetal development. J. Urol., **152:** 692; discussion 696, 1994.
9. Shapiro E., Becich M.J., Perlman E. and Lepor H.: Bladder wall abnormalities in myelodysplastic bladders: a computer assisted morphometric analysis. J. Urol., **145:** 1024, 1991.
10. Mayo M.E. and Hinman F.: Structure and function of the rabbit bladder altered by chronic obstruction or cystitis. Invest Urol., **14:** 6, 1976.
11. Susset J.G., Servot-Viguier D., Lamy F., Madernas P. and Black R.: Collagen in 155 human bladders. Invest. Urol., **16:**204, 1978.

12. Gilpin S.A., Gosling J.A. and Barnard R.J.: Morphological and morphometric studies of the human obstructed, trabeculated urinary bladder. Br. J. Urol., **57:** 525, 1985.
13. Kim K.M., Kogan B.A., Massad C.A. and Huang Y-C.: Collagen and elastin in the normal fetal bladder. J. Urol., **146:** 524, 1991.
14. Koo H.P., Howard P.S., Chang S.L., Snyder H.M., Ducket J.W. and Macarak E.J.: Developmental expression of interstitial collagen genes in fetal bladders. J. Urol., **158:** 954, 1997.
15. Dean G.E., Cargill R.Sr., Macarak E., Snyder H.M., Duckett J.W. and Levin R.: Active and passive compliance of the fetal bovine bladder. J. Urol., **158:** 1094, 1997.
16. Levin R.M., Malkowicz S.B., Jacobowitz D. and Wein A.J.: The ontogeny of the autonomic innervation and contractile response of the rabbit urinary bladder. J. Pharmacol. Exp. Ther., **219:** 250, 1981.
17. Zderic S.A., Duckett J.W., Wein A.J., Snyder H.Md. and Levin R.M.: Development factors in the contractile response of the rabbit bladder to both autonomic and non-autonomic agents. Pharmacology , **41:** 119, 1990.
18. Zderic S.A., Hypolite J., Duckett J.W., Snyder H.Md., Wein A.J. and Levin R.M.: Developmental aspects of bladder contractile function: sensitivity to extracellular calcium. Pharmacology, **43:** 61, 1991.
19. Zderic S.A., Sillen U., Liu G.H., Snyder Hd., Duckett J.W., Wein A.J. and Levin R.M.: Developmental aspects of bladder contractile function: evidence for an intracellular calcium pool. J. Urol., **150:** 623, 1993.
20. Zderic S.A., Sillen U., Liu G.H., Snyder M.Cr., Duckett J.W., Gong C. and Levin R.M.: Developmental aspects of excitation contraction coupling of rabbit bladder smooth muscle. J. Urol., **152:** 679, 1994.
21. Gong C., Zderic S.A. and Levin R.M.: Ontogeny of the ryanodine receptor in rabbit urinary bladder smooth muscle. Mol. Cell Biochem., **137:** 169, 1994.
22. Lee J.G., Coplen D., Macarak E., Wein A.J. and Levin R.M.: Comparative studies on the ontogeny and autonomic responses of the fetal calf bladder at different stages of development: involvement of nitric oxide on field stimulated relaxation. J. Urol., **151:** 1096, 1994.
23. Kogan B.A. and Iwamoto H.S.: Lower urinary tract function in the sheep fetus: studies of autonomic control and pharmacologic responses of the fetal bladder. J. Urol., **141:** 1019, 1989.
24. Mevorach R.A., Bogaert G.A. and Kogan B.A.: Role of nitric oxide in fetal lower urinary tract function. J. Urol., **152:** 510, 1994.
25. Wlodek M.E., Thorburn G.D. and Harding R.: Bladder contractions and micturition in fetal sheep: their relation to behavioral states. Am. J. Physiol., **257:** R1526, 1989.
26. Kim K.M., Kogan B.A., Massad C.A. and Huang Y-C.: Collagen and elastin in the obstructed fetal bladder. J. Urol., **146:** 528, 1991.
27. Freedman A.L., Qureshi F., Shapiro E., Lepor H., Jacques S.M., Evans M.I., Smith C.A., Gonzalez R. and Johnson M.P.: Smooth muscle development in the obstructed fetal bladder. Urology, **49:** 104, 1997.
28. Peters C.A., Vasavada S., Dator D., Carr M., Shapiro E., Lepor H., McConnell J., Retik A.B. and Mandell J.: The effect of obstruction on the developing bladder. J. Urol., **148:** 491, 1992.
29. Karim O.M., Cendron M., Mostwin J.L. and Gearhart J.P.: Developmental alterations in the fetal lamb bladder subjected to partial urethral obstruction in utero. J. Urol., **150:** 1060, 1993.
30. Peters C.A., Freeman M.R., Fernandez C.A., Shepard J., Wiederschain D.G. and Moses M.A.: Dysregulated proteolytic balance as the basis of excess extracellular matrix in fibrotic disease. Am. J. Physiol., **272:** R1960, 1997.
31. Rohrmann D., Monson F.C., Damaser M.S., Levin R.M., Duckett J.W., Jr. and Zderic S.A.: Partial bladder outlet obstruction in the fetal rabbit. J. Urol., **158:**1071, 1997.
32. Rohrmann D., Zderic S.A., Duckett J.W., Jr., Levin R.M. and Damaser M.S.: Compliance of the obstructed fetal rabbit bladder. Neurourol. Urodyn., **16:**179, 1997.

DEVELOPMENTAL ASPECTS OF THE CONTRACTILE SMOOTH MUSCLE COMPONENT IN SMALL INTESTINAL SUBMUCOSA REGENERATED URINARY BLADDER

Bradley P. Kropp

Children's Hospital of Oklahoma
University of Oklahoma Health Science Center
Oklahoma City, Oklahoma 73104

1. INTRODUCTION

Urinary bladder reconstruction through bladder regeneration requires the use of a material that is readily available, increases bladder volume and compliance, does not violate the gastrointestinal tract, and is non-immunogenic. The major obstacle has been developing a biomaterial, either permanent or biodegradable, that will function as a suitable scaffold to allow the natural process of regeneration to occur.

For the past 5 years, my laboratory has been researching small intestinal submucosa (SIS), a biodegradable, acellular, xenogeneic, collagen-based tissue matrix graft, that has been shown to promote regeneration of a variety of host tissues including the urinary bladder.[1,2,3,4] Small intestinal submucosa (SIS) represents a biodegradable material for clinical use in bladder augmentation. This article will highlight the preclinical studies pertaining to the use of SIS for bladder augmentation as well as review the use of SIS as a model to study the developmental aspects of smooth muscle contractions in regenerating urinary bladder.

SIS is a xenogeneic membrane harvested from pig small intestine in which the tunica mucosa is mechanically removed from the inner surface, and the serosa and tunica muscularis are mechanically removed from the outer surface. This produces a thin, translucent graft (0.1 mm wall thickness) composed mainly of the submucosal layer of the intestinal wall. Production of SIS is similar to the manufacturing of sausage casing. The submucosal layer of animal intestine has an established background in surgery as gut suture. SIS is a collagen-rich membrane has been previously shown to function well as an arterial or venous

Advances in Bladder Research, edited by Baskin and Hayward.
Kluwer Academic / Plenum Publishers, New York, 1999.

graft with rapid replacement by native tissues.[5,6,7] It has also been shown to have excellent host compatibility and remodeling when submucosal bladder injection of minced SIS were performed in pigs.[8] To date, SIS has been shown to be non-immunogenic with over 1000 cross-species transplants and direct challenge testing elucidating no response.[9,10]

2. PREPARATION OF SIS GRAFTS

Sections of porcine jejunum were harvested from 400–600lb. sows within 10 minutes of euthanasia and prepared according to the methods of Badylak et al.[5] A brief summary of this preparation is included here for review. Mesenteric tissue was removed and the intestine everted. The mucosal epithelium and lamina propria were removed by gentle abrasion. The segment was then everted to normal orientation, and the tunica serosa and tunica muscularis externa also were mechanically removed, leaving a thin (0.1 mm) white tube of SIS consisting of tunica submucosa with attached stratum compactum and muscularis mucosa of the tunica mucosa. Subsequently, the graft was sterilized with 0.1% peracetic acid and stored in sterile water for up to 3 months.

2.1. Surgical Procedure

The surgical procedures for the rat and dog studies were similar. The bladders were exposed through a midline abdominal incision and the thin attachments of the perivesical fat were sharply freed from the bladder exposing the bladder neck. The entire ventral surface of the bladder (35–45 % of the detrusor) was resected, weighed, and discarded. A single layer SIS graft, slightly larger than the resected bladder and oriented with the stratum compactum surface (side where previous mucosa was present) facing the bladder lumen, was sewn into place with mild chromic suture in a water tight fashion. The average SIS patch size was 5cm × 4cm (dogs) or 1cm × 1cm (rats). Four permanent marking sutures were placed at the 3, 6, 9, and 12 o'clock positions around the SIS augmentation graft for future reference. Adjacent perivesical fat or a small piece of omentum was used to cover the graft, and the abdomen was then closed in layers with absorbable suture. All dogs were diverted with a transurethral catheter for 3 days and then allowed to void spontaneously. Rats did not undergo urinary diversion.

2.2. In Vitro Contractility and Compliance Studies

To better characterize the regenerated smooth muscle cells, a series of similar in vitro contractility and compliance experiments were designed for both the rat and canine models. SIS regenerated muscle strips were obtained from within the area outlined by the permanent marking sutures in 11 month and 15 month post-augmentation rats and dogs, respectively. For the contractility studies, full thickness longitudinal strips were then suspended from a platinum/iridium electrode in a 10ml organ bath of modified Krebs solution and attached to a Grass FT .03 force transducer. Changes in isometric tension were recorded and analyzed with a M5000 Signal Processing Center using XYZ Realtime Software (Modular Instruments, Inc., Malvern, PA).

For the compliance studies, full thickness longitudinal strips were suspended between special tissue supports in the Vitrodyne V-200 organ bath (Liveco, Inc., Burlington, VT.) in modified Kreb's solution. The average slope of each curve obtained during the stress/strain compliance protocol were calculated by a computer line fit program included in the Vitrodyne software package.

3. INITIAL AUGMENTATION RESULTS IN RATS

The initial augmentation study was performed in 22 Sprague-Dawley rats which underwent the augmentation technique described above.[11] Urinary diversion or drains were not used in the immediate postoperative period for the rats nor did any of the rats receive antibiotics during the study period. The rats were sacrificed at periods ranging between 2 weeks and 11 months post-augmentation, and their bladders were studied grossly and with routine histology.

Macroscopically, the inflated bladder specimens were uniformly dilated without evidence of diverticular formation in the region of the graft. Grafts were barely discernible as a slightly whitened area compared to the surrounding native bladder. There were minimal adhesions to the surrounding perivesical fat or other intraabdominal structures. The permanent marking sutures left at the time of augmentation were easily identifiable, and grossly the SIS augmentation graft had not contracted. There was no macroscopic evidence of hydroureteronephrosis in any of the animals.

At two weeks, histologically, the SIS grafts were completely covered by apparently normal transitional epithelium. Grafts showed a prominent vascular proliferation at their edges and infiltration of their entire surface by fibroblasts and new capillaries. There was a moderately intense mononuclear inflammatory cell response in the periphery of the graft. Centrally, inflammation was less intense, and fibroblasts and small capillaries appeared to infiltrate the collagen fibers of the SIS.

By four weeks, the grafts were completely infiltrated by small well-formed blood vessels and fibroblasts. In the periphery there were scattered fibers with morphologic and staining characteristics consistent with smooth muscle. Mononuclear inflammatory cell infiltrates were minimal at the edges but moderately intense in the central region.

At three months, inflammatory activity throughout the graft had largely subsided, and small but well-formed blood vessels persisted. Collagen was organized into a linear array, and scattered fibroblasts were interspersed among the individual collagen fibers. Small fibers with morphologic and staining properties of smooth muscle were scattered throughout the graft. In addition, distinct bundles of smooth muscle now were evident in small portions of the grafts.

In the six and eleven month animals, all three layers of the normal bladder (urothelium, smooth muscle and serosa) were present and grossly and microscopically indistinguishable from the normal rat bladder. The marking sutures were the only guide to where the SIS graft had been surgically implanted.

From the above results it was evident that complete bladder regeneration was occurring. Histologically, large bundles of smooth muscle were present. To determine if the SIS regenerated muscle cells were 1) contractile, 2) express receptors for neurotransmitters, and 3) functionally innervated, a series of in vitro contractility experiments on SIS regenerated muscle strips were performed.[12]

In the rat, the results of the in vitro contractility show that both the SIS regenerated and control bladder strips demonstrated contractile responses to a muscarinic agonist (carbachol 1μm) and a purinergic agonist (a,β-methylene ATP 1μm) and relaxation to α-adrenergic agonist (isoprenaline 1 μm). In addition, both tissues contracted in response to electrical field stimulation (EFS) (100V, 1 msec duration, 5 sec. trains) in a frequency-dependent manner. These contractions were antagonized in both control and SIS strips by atropine (1μm) and showed desensitization to α β-methylene ATP (100 μm). These results were consistent with what has been previous published on in vitro contractility in the bladder.[13,14,15,16] The only functional difference between the SIS regenerated bladder and control bladder was that the maximum contractile force generated by the SIS bladder was only 50% of control bladder contractile force.

3.1. Long Term Augmentation Results in Dogs

The encouraging results obtained from the pilot study in rat led to studying SIS bladder augmentations in a long term canine model.[17] Augmentation was performed in eight adult male beagles as described in detail above. All dogs had extensive pre-operative and periodic post-operative evaluations with complete blood counts (CBC), serum chemistries, urine cultures, cystograms, intravenous urograms (IVU) and cystometrograms (CMG).

All eight dogs recovered from the initial surgery and survived the entire study period (15 months) without complication. All dogs were observed to void spontaneously without difficulties or evidence of retention. All eight dogs had normal blood work, negative urine cultures, normal IVU's and cystograms throughout the entire study period. Preoperative CMG's revealed a mean bladder capacity of 51mls (range 39–66mls) with a mean maximal voiding pressure of 52cm H_20 (range 39–69cm H20). CMG's remained stable and similar throughout the study period and at 14 months post-augmentation the bladders remained compliant with a flat tonus limb, no evidence of unstable contractions, a mean bladder capacity of 55mls (range 42–70mls) with a mean maximal voiding pressure of 45cm H_20 (range 28–52cm H20).

All eight dogs were sacrificed at 15 months post-augmentation and grossly the SIS regenerated bladder was remarkably similar to the remaining normal bladder. There were minimal adhesions and no evidence of bladder calculi in any of the animals. There was no significant change in the measured area of the permanent marking suture left around the original graft material, thereby excluding graft shrinkage. Histologically, the SIS regenerated bladder resembled normal canine bladder (e.g. all three layers; transitional mucosa, smooth muscle, and serosa were present and the thickness of the transitional cell layer was indistinguishable from the normal bladder). The only notable difference between SIS regenerated bladder and normal bladder was the quantity and organization of the smooth muscle fibers. Specifically, the SIS regenerated bladder had a larger amount of collagen interspersed between what appeared to be a smaller number of smooth muscle bundles.

To further define the difference between SIS-regenerated bladder and normal bladder computer assisted image analysis was perform and demonstrated a decreased muscle to collagen ratio with a loss of the normal architecture in the SIS-regenerated bladders.

In vitro contractility studies were also performed on the SIS regenerated dog bladder,[18] as described in the rat studies above, which confirmed 1) contractile activity of the smooth muscle cells, 2) expression of muscarinic, purinergic, and α-adrenergic receptors, and 3) functional cholinergic and purinergic innervation that is similar to the normal canine bladder muscle. As in the rat there was a difference between the SIS regenerated dog bladder and the control dog bladder with the maximal contractile response being decreased by 50%.

3.2. Cellular Events of Bladder Regeneration

In addition to the potential clinical use offered by SIS for bladder augmentation, it has provided an opportunity to study the early cellular factors involved in bladder development and function.[19] Twenty-two dogs underwent partial cystectomy with immediate augmentation with a SIS patch graft. The SIS graft/native bladder interface was marked with permanent marking sutures for future reference. The SIS-regenerated bladder were harvested at 2, 3, 4, 6, 8, and 10 weeks post-augmentation. The tissue was then studied with routine histology and immunohistochemistry using Factor VIII, smooth muscle specific actin (1A4), and neurofilament staining.

Epithelialization of the SIS grafts occurred between 2 and 4 weeks after augmentation from the cut edges of the native bladder. In the early time periods neovascularization was prominent throughout the entire graft, as shown by factor VIII staining. Later more mature vessels were noted. Smooth muscle formation appeared to begin at 2 weeks from dense parallel sheets of elongated smooth muscle actin positive spindle cells emanating from the incised native smooth muscle margins and running parallel to the mucosal surface. This spindle cell proliferation completely traversed the graft by 4 weeks. Trichrome stained sections of the 4-week-old grafts demonstrated no evidence of muscle differentiation and the spindle cells appeared to be fibroblasts. Interestingly, these cells stained positive for smooth muscle specific actin (1A4), indicating myogenic potential. Over the next 2 weeks, the spindle cells became more haphazardly arranged and separated by loose interstitium. Between eight and ten weeks, distinct smooth muscle bundles formed with these areas of proliferating myocytes. Innervation of the graft, as documented by neurofilament positive cells, appeared to coincide with the ingrowth of the muscle forming spindle cells.

3.3. Contractile Activity in Early SIS-Regenerated Smooth Muscle Cells

In order to investigate the contractile properties of early (4 weeks) regenerating smooth muscle cells, 9 rats underwent partial cystectomy with immediate SIS augmentation. Four weeks post-augmentation, in vitro smooth muscle contractility and immunostaining were assessed in longitudinal muscle strips from both SIS-regenerated and normal rat bladders.

The results demonstrated that the absolute magnitude of contractions generated by KCL (3.33mM), Carbachol (10–6M), Angiotensin II (10–6M), Lysophosphatidic acid (10–6M) and Endothelin I (10–7M) were less in the SIS-regenerated muscle strips that in normal bladder strips. However, when standardized to KCL, SIS-regenerated bladder strips obtained significantly greater ($p<0.05$) responses to Angiotensin II, LPA and Endothelin I than normal bladder strips, and decreased responsiveness to carbachol.

Immunohistochemical staining with Masson's trichrome and α-smooth muscle actin demonstrated smooth muscle fiber/bundles within the submucosa of the regenerated bladder and at the margins between native and SIS-regenerated bladder. In addition, α-smooth muscle positive spindle shaped cells (myofiberblasts) were noted throughout the regenerating bladder wall.

This study indicates that newly regenerating smooth muscle cells have contractile activiity. In addition, binding sites for Angiotensin, LPA and Endothelin are present on normal and SIS-regenerated detrusor smooth muscle cells. The relative role of non-cholinergic contractile factors appears to be significantly more important in the regenerated smooth muscle cells during the early stages of development.

4. INTRINSIC PROPERTIES OF SIS

The above research has demonstrated that a functional bladder augmentation using SIS as a scaffold for bladder regeneration is a potential viable option for genitourinary reconstruction. However, it is difficult to explain why the results with SIS are seemingly better than with other collagen based materials. Is there something special about the 3-dimensional architecture of SIS? Are the improved regenerative results related to intrinsic growth factors present within the SIS graft material? Although, the exact composition of SIS is still under investigation. A combination of cytokines, structural proteins, glycoproteins and proteoglycans have been found to exist within SIS in a 3-dimensional arrangement. It is possible that it is these factors that provide the necessary environment to promote cell migration, cell to extracellular matrix interaction, as well as cell growth and differentiation.[20] Additional studies by Brightman et al. have demonstrated that TGFβ could be extracted from SIS in both its latent as well as its active forms.[21] Therefore it may be the presence of TGFβ and/or other unidentified growth factors that are contributing to the improved regenerative results. To better characterize the importance of the natural occurring components of SIS, Cheng et. al. investigated in vitro human bladder smooth muscle cell and human urothelial cell culture growth on SIS sheets and in a SIS suspension.[22] This group has been able to demonstrate rapid multilayer cellular proliferation in both native SIS as well as a suspension of SIS. Therefore, from this preliminary data, it is tempting to speculate that both the 3-dimensional architecture as well as the natural occurring growth factors present within the SIS material are important in the improved regenerative process. In addition, preliminary data from my laboratory suggests that SIS promotes and maintains smooth muscle cells in their differentiated forms.

In conclusion, native bladder has the propensity to proliferate and regenerate given the proper environment. SIS provides a biodegradable, collagen based scaffold with natural occurring growth factors that seem to give improved regenerative results. In addition, SIS provides a model to study the regenerative process from a cellular level in order to better define the cellular events that are occurring during bladder regeneration and development.

REFERENCES

1. Aiken S.W., Badylak S.F., Toombs J.P., Shelbourne K.D., Hiles M.C., Lantz G.C., Van Sickle D.: Small intestinal Submucosa as an intra-articular ligamentous repair material: A pilot study in dogs., Vet. Comp. Orthopedics Traumatology., **7:** 124, 1994.
2. Badylak S.F., Voytik S.L., Kokini K., Shelbourne K.D., Klootwyk T., Kraine M.R., Tullius R., Simmons C.: The use of xenogeneic small intestinal submucosa as a biomaterial for Achilles' tendon repair in a dog model. J. Biomed. Mat. Res., **29:** 977, 1995.
3. Lantz G.C., Badylak S.F., Hiles M.C., Coffey A.C., Geddes L.A., Kokini K., Sandusky G.E., Morff R.J..: Small intestinal submucosa as a vascular graft: a review. J. Invest. Surg., **6:** 297, 1993.
4. Prevel C.D., Eppley B.L., Smmerlin D.J., Jackson J.R.: Small intestinal submucosa (SIS): Utilization as a wound dressing in full-thickness rodent wounds. Annals of Plastic Surgery, **35:** 381, 1995.
5. Badylak S.F., Lantz G.C., Coffey A., Geddes L.A.: Small intestinal submucosa as a large diameter vascular graft in the dog. J. Surg. Res., **47:** 74, 1989.
6. Lantz G.C., Badylak S.F., Coffey A.C., Geddes L.A., Blevins W.E.: Small intestinal submucosa as a small-diameter arterial graft in the dog. J. Invest. Surg., **217,** 1990.
7. Lantz G.C., Badylak S.F., Coffey A.C., Geddes L.A., Sandusky G.E.: Small intestinal submucosa as a superior vena cava graft in the dog. J. Surg. Res., **53:** 175, 1992.
8. Knapp P.M., Lingeman J.E., Siegel Y.I., Badylak S.F., Demeter R.J.: Biocompatability of Small-intestinal Submucosa in Urinary Tract as Augmentation Cystoplasty Graft and Injectable Suspension. J. Endourol., **8:** 125, 1994.
9. Badylak S.F.:The Immunogenic Response of SIS.Personal communication, 1994.
10. Metzger D.W., Moyad T.F., McPherson T., Badylak S.F.: Cytokine and antibody responses to xenogeneic SIS transplants First SIS Symposium Orlando, FL, 1996.

11. Kropp B.P., Eppley B.L., Prevel C.D., Harruff R.C., Badylak S.F., Adams M.C., Rink R.C., Keating M.A.: Experimental Assessment of Small Intestine Submucosa as a Bladder Wall Substitute. Urology., **46:** 396, 1995.
12. Vaught J.D., Kropp B.P., Sawyer B.D., Shannon H.E., Thor K.B.: Detrusor Smooth Muscle Regeneration in the Rat Using Procine Small Intestinal Submucosal Grafts: Functional Innervation and Receptor Expression. J. Urol., **155:** 374, 1996.
13. Brading A.F., Williams J.H.: Contractile responses of smooth muscle strips from rat and guinea-pig urinary bladder to transmural stimulation: effects of atropine and α, β-methylene ATP. Br. J. Pharmacol., **99:** 493, 1990.
14. Burnstock G., Dumsday B., Smythe A.: Atropine-resistant excitation of the urinary bladder: the possibility of transmission via nerves releasing a purine nucleotide. Br. J. Pharmacol., **44:** 451, 1972.
15. Kasakov L., Burnstock G.: The use of the slowly degradable analog, α, β-methylene ATP, to produce desensitization of the P2- purinoceptor: effect on non-adrenergic, non-cholinergic responses of the guinea-pig urinary bladder. Eur. J. Pharmacol., **86:** 291, 1982.
16. Maggi C.A., Meli A.: Modulation by beta-adrenoreceptors of spontaneous contractions of rat urinary bladder. J. Auton. Pharmacol., **2:** 255, 1982.
17. Kropp B.P., Rippy M.K., Badylak S.F., Adams M.C., Keating M.A., Rink R.C., Thor K.B.: Small Intestinal Submucosa: Urodynamic and Histopathologic Evaluation in Long Term Canine Bladder Augmentations. J. Urol ,**155:** 2098, 1996.
18. Kropp B.P., Sawyer B.D., Shannon H.E., Rippy M.K., Badylak S.F., Adams M.C., Keating M.A., Rink R.C., Thor K.B.: Characterization of Small Intestinal Submucosa-Regenerated Canine Detrusor: Assessment of Re-innervation, In Vitro Compliance and Contractility. J. Urol., **156:** 599, 1996.
19. Pope IV J.C., Davis M.M., Smith Jr. E.R., Walsh M.J., Ellison P.K., Rink R.C., Kropp B.P.: The ontogeny of canine SIS-regenerated urinary bladder. J. Urol., **158:** 1105, 1997.
20. Badylak S.F.: Speculation (with a little evidence) for the roles of cell proliferation, differentiation, neovascularization, and environmental stressors in SIS-induced remodeling, presented at the First SIS Symposium Orlando Florida, 1996.
21. Brightman A.O., Waisner B., Klich D.L., Voytik-Harbin S.L.: Extraction of a biologically inactive form of transforming growth factor (from small intestinal submucosa. First SIS Symposium Orlando, Florida, 1996.
22. Kolligian M.E., Furness III P.D., Voytik-Harbin S., Waisner B., Kaplan W.E., Firlit C.F., Cheng E.Y.: Div. of Urol, Childrens Mem. Hosp., Chicago, IL and Purdue Univ, West Lafayette, IN, Small Intestinal Submucosa (SIS) Promotes 3-Dimensional Growth and Differentiation of Human Bladder Cells In Vitro, presented at the American Academy of Pediatrics, 1997.

11

CONTRACTILE PROTEIN CHANGES IN URINARY BLADDER SMOOTH MUSCLE FOLLOWING OUTLET OBSTRUCTION

Samuel Chacko,[1,2] Michael DiSanto,[2] Chandrakala Menon,[2] Yongmu Zheng,[2] Joseph Hypolite,[2] and Alan J. Wein[2]

[1]Department of Pathobiology
[2]Division of Urology
University of Pennsylvania
Philadelphia, Pennsylvania 19104-4274

1. INTRODUCTION

Force production by the bladder body smooth muscle and relaxation of the bladder outlet, required for bladder emptying, are regulated by Ca^{2+} *via* myosin light chain (MLC) phosphorylation by a Ca^{2+}-calmodulin-dependent MLC kinase (for review, [1–3]). Besides this mode of regulation, evidence from most laboratories, including ours, supports the existence of a thin-filament-mediated regulation due to protein-protein interaction by actin-binding proteins acting in concert with tropomyosin and a calcium-binding protein, most likely, calmodulin. Caldesmon (CaD), a thin-filament-associated protein, inhibits the actin-myosin interaction and actomyosin ATPase, and this inhibition is reversed by calmodulin in the presence of Ca^{2+}.[3–5] Urinary bladder outlet obstruction interferes with the ability of the bladder to empty its contents, thereby inducing changes in the bladder wall smooth muscle cells which enable the bladder to produce the increased contractile force required to expel urine through the obstructed urethra. In the initial phases of outlet obstruction, there is transient decompensation of the bladder smooth muscle, which initiates the molecular events that lead to hypertrophy of the detrusor smooth muscle cells enabling it to generate and maintain force. However, continuation of the outlet obstruction induces molecular, cellular, and structural alterations in the muscle cells of the bladder wall, leading ultimately to decreased compliance and impaired emptying. The ability of smooth muscles to compensate for increased functional demand is associated with alterations in the expression of proteins in the thick- and thin-filaments. Changes in the amounts and the isotype of the proteins in the contractile apparatus are likely to affect the ability of smooth muscle cells in the bladder wall to generate and maintain force, two events that are impor-

Advances in Bladder Research, edited by Baskin and Hayward.
Kluwer Academic / Plenum Publishers, New York, 1999.

tant for proper emptying of the bladder. In this paper, we briefly review the molecular mechanisms for muscle contraction and describe the obstruction-induced changes in the isoforms of (a) myosin, the molecular motor for muscle contraction, (b) actin, a major component of the thin-filament, and (c) caldesmon, an actin-associated calmodulin-binding protein in response to partial outlet obstruction.

2. MOLECULAR MECHANISM FOR MUSCLE CONTRACTION

As in striated muscle, smooth muscle contraction is regulated by the concentration of free calcium in the cytosol.[1,6] The ability of smooth muscle to generate tone is dependent on the interaction between the major contractile proteins, actin and myosin. A sliding filament model, described in detail for striated muscle contraction (for review see [1,7]), also applies to smooth muscle, although the contractile apparatus in smooth muscle is not highly ordered as in striated muscle. During muscle contraction, the heads of the myosin molecules that project out of the thick myosin filaments go through a cycle of high- and low- affinity binding to actin, driven by ATP hydrolysis.[1] A change in the angle of the myosin heads bound to actin from 90° to 45° with respect to the long axis of the thin filament and the thick filament causes the sliding of the thick filament past the thin filament.[8] The direction of force depends on the polarity of the filaments in the contractile apparatus. Filaments of opposing polarity must extend from opposing ends of the structure from which the filaments anchor in order for the smooth muscle cells to shorten or develop isometric tension. The characteristics of the contractile proteins, which are components of the thick and thin filaments, involved in cross-bridge cycling and contraction in smooth muscles in the bladder wall, are altered in smooth muscle hypertrophy induced by partial outlet obstruction.

Although tropomyosin, a component of the regulatory system in striated muscle, is present in smooth muscle thin filaments at the same location and stoichiometry;[9,10] the troponin complex is absent in smooth muscle.[2,11] Calcium regulation of smooth muscle contraction is accomplished by Ca^{2+}/calmodulin-dependent myosin light chain phosphorylation, since myosin phosphorylation is a prerequisite for actin activation of myosin MgATPase (see below).

3. REGULATION OF ACTIN-MYOSIN INTERACTION AND CONTRACTION IN SMOOTH MUSCLE

Contraction of the urinary bladder smooth muscle, as in smooth muscle from other sources, is initiated by an increase in the free Ca^{2+} concentration in the cytoplasm of smooth muscle cells.[12,13] Studies using intact muscle and isolated muscle proteins from a variety of smooth muscle sources showed that free Ca^{2+} binds to calmodulin, a calcium-binding protein, and that Ca^{2+}-calmodulin activates the MLCK.[1,2] The MLCK catalyzes the phosphorylation of the 20 kDa LC at a specific (position 19) serine residue.[14] Phosphorylation of the myosin LC, which is present in the junction between the head region of the myosin molecule and the rod, causes conformational changes in myosin and favors assembly into thick filaments.[15–17] Phosphorylation of myosin is prerequisite for the actin activation of the Mg-ATPase activity of myosin, an *in vitro* correlate of muscle contraction.[18–21] Development of force by smooth muscle is also correlated with light chain phosphorylation.[2,22–24] This is shown schematically in Figure 1.

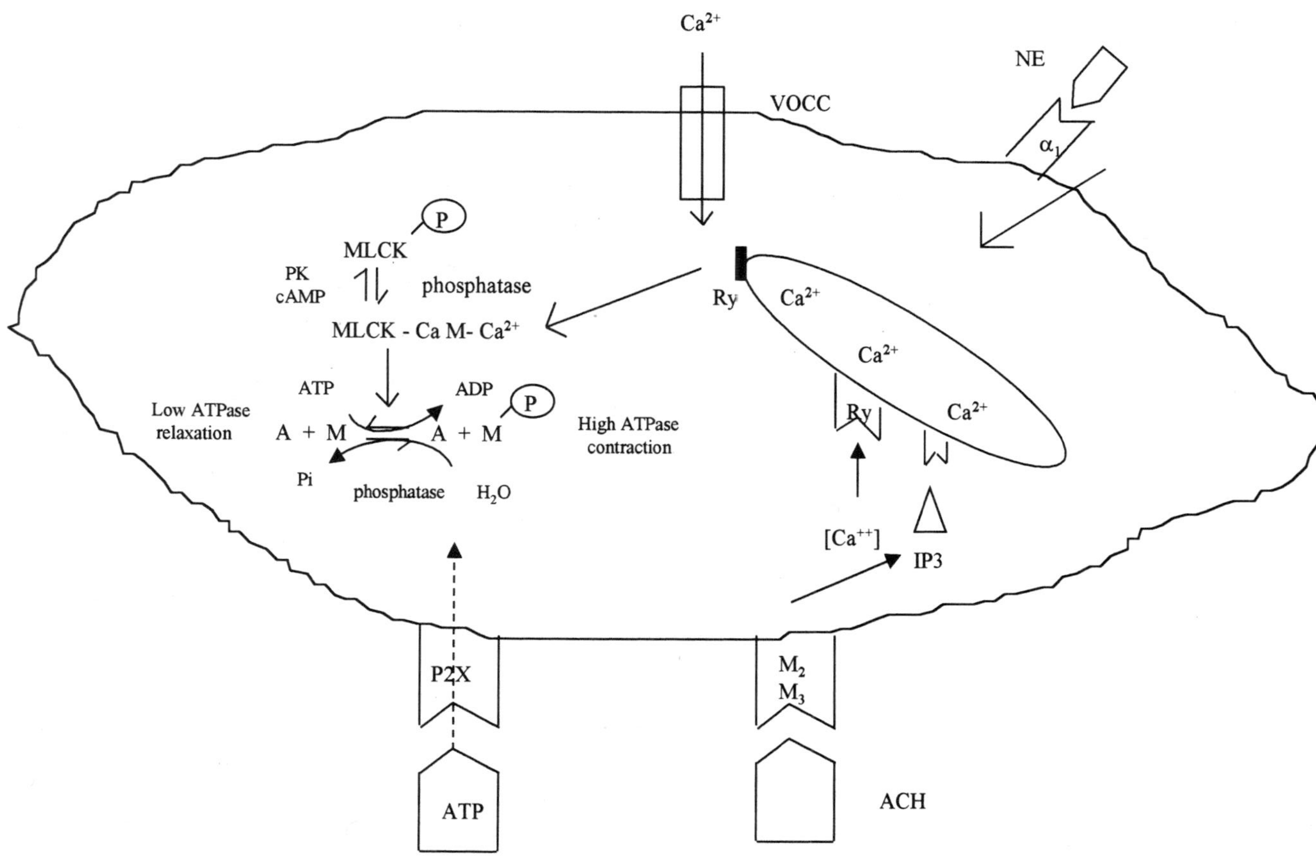

Figure 1. Schematic drawing of the signal transduction pathway leading to contraction of a smooth muscle cell. Upon stimulation of the smooth muscle cells, the cytosolic Ca^{2+} rises. Myosin light chain kinase (MLCK) is phosphorylated by cAMP-dependent protein kinase (PK) and dephosphorylated by phosphatase. The free Ca^{2+} binds to calmodulin (CaM), and the Ca^{2+}-calmodulin binds to dephosphorylated MLCK. The activated MLCK phosphorylates the 20 kDa myosin light chain. The actin-activated ATPase activity of phosphorylated myosin is higher than that of unphosphorylated myosin. The myosin light chain phosphorylation and the high actin-activated ATPase activity are correlated with increased contraction. Relaxation results when Ca^{2+} concentration diminishes and the phosphatase dephosphorylates the myosin.

Force maintenance in smooth muscle, whether induced by the action of excitatory mediators or spontaneously activated by myogenic mechanisms, does not require the phosphorylated state of myosin.[24,25] Utilization of ATP during force maintenance in all types of smooth muscles is very low and myosin cross-bridges cycle slowly, if at all.[24,26] This state has been termed "latch";[24] the biochemical basis of the latch phenomenon is not understood. It may involve some variation in the Ca^{2+}-dependent phosphorylation so that force can be maintained at very low levels of ATP utilization and slow myosin cross bridge cycling. It is also likely that the thin-filament associated proteins such as caldesmon and calponin play a role in the maintenance of force at a low level of phosphorylation and low cross-bridge cycling.[27–30] In vitro motility assay using myosin phosphorylated at various levels revealed that the velocity of actin filaments containing tropomyosin and caldesmon slowed upon reaching a phosphorylation level below 50%.[28] A similar result was obtained when myosin rods, instead of unphosphorylated myosin, were used. These data indicate that the decreased filament velocity in the presence of caldesmon reflects the mechanical load produced by the "tethering" of actin filament to myosin through the interaction of the caldesmon N-terminal domain and the myosin S-2 region.[31] Deletion of CaD residues 24–53 nearly abolished the ability of CaD to inhibit actin filament velocity in the in vitro motility experiments, but had no effect on the CaD-induced inhibition of actin-activated ATPase activity. These data provide direct evidence for the inhibition of actin filament velocity in the in vitro motility assay caused by the tethering of myosin to actin through binding of both the CaD N-terminal region to myosin and the C-terminal region to actin.[32] The "tethering" is believed to play a role in force maintenance with very low ATP hydrolysis seen in the latch state. It has been suggested that when CaD is phosphorylated at a site near the myosin-binding region, the binding of CaD to myosin is low,[33] and the "tethering" is decreased when CaD is phosphorylated. However, the data for supporting this notion is lacking

Other data to support a role for CaD in the regulation of contraction comes from studies on the conformational changes induced by CaD on the actin filaments in "ghost fibers" which do not contain any muscle proteins except actin filaments, which are retained in their native organization.[27] Malmqvist et al. [34] showed that the removal of CaD from the chemically "skinned" smooth muscle increases the force production and the relaxation time, indicating that CaD suppresses the force and decreases relaxation (a condition that would favor the "latch" state).

4. ROLE OF MYOSIN ISOFORMS IN THE CONTRACTILE CHARACTERISTICS OF SMOOTH MUSCLES FROM DIFFERENT SOURCES

Myosin, the major component of the thick (15 nm diameter) filaments in all types of muscle, is a long asymmetrical molecule containing a globular head and a long tail.[1,3,35] A myosin molecule contains two long polypeptide chains of around 200 kDa, called heavy chains. A major portion of the C-terminal region of the heavy chains is α-helical, and the two chains are wound around each other (Figure 2A). This region, which forms the rod portion of the myosin molecule, can self-assemble under physiological conditions to form the thick-filaments.[36–39] The N-terminal regions of both heavy chains are folded into globular structures to form the head. Therefore, each myosin molecule has two heads and a rod portion (Schematic drawing, Figure 2A). The ATP binding sites, enzymatic activity, and the actin binding site reside in the head region of the molecule. Each head also has

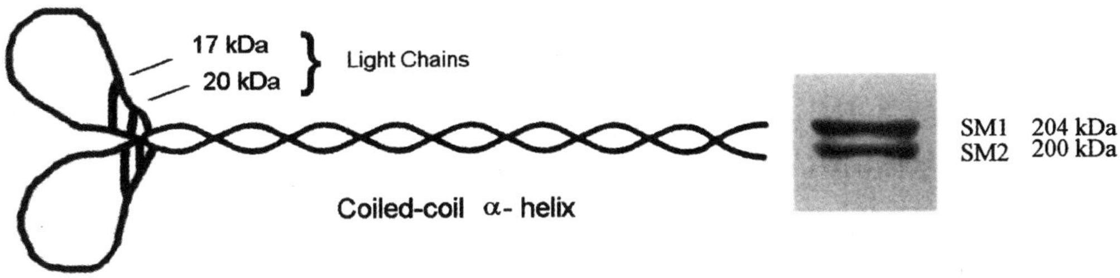

Figure 2A. Schematic representation of a smooth muscle myosin molecule. The SM1 and SM2 myosin heavy chains (204 kDa and 200 kDa) separated by electrophoresis on a 3.5% SDS-polyacrylamide gel are shown on the right side of the panel. These two heavy chains, formed by alternative splicing of the pre-mRNA at the 3' end, are different at the C-terminal rod region of the molecule. The N-terminal regions of both heavy chains are folded into globular structures to form the head (left side). The location of the two light chains (20 kDa and 17 kDa) on each head is shown at the base of the head.

two smaller polypeptides (light chains, LC) of 20 kDa (phosphorylatable) and 17 kDa (SH-LC). The 20 kDa LC is called the regulatory light chain, since the contraction and relaxation of smooth muscle are regulated by phosphorylation and dephosphorylation of this LC, respectively.[40] The 17 kDa LC contains SH-groups; thus they are called the SH-light chain. Removal of this light chain from the myosin molecule is associated with the loss of ATPase activity of myosin.[41] Hence, this light chain is termed the essential light chain. The crystalline structure of the myosin molecule reveals a head region that contains the sites that are important for cross bridge formation and an adjacent neck region, where the light chains are located.[42]

The mRNAs which encode vertebrate smooth muscle myosin heavy chains (MHC) originate from a single smooth muscle MHC gene through an alternative splicing mechanism at the 3' end of the coding region.[43] The types of MHC isoforms present in the muscle may determine its biomechanical behavior. Two myosin heavy chain (MHC) isoforms, SM1 (204 kDa) and SM2 (200 kDa), present in smooth muscle cells [44] are depicted in Figure 2A and B. Their ratios vary in smooth muscles from different sources.[45–47] These two isoforms differ in the amino acids at the C-terminus, which is in the rod portion of the myosin molecule.[48]

Tissue-specific MHC isoforms that differ in the amino terminal head region have been reported.[49] MHC mRNAs from uterine and vascular smooth muscles are different from intestinal and bladder smooth muscle MHC mRNA in the 5'-end coding region near the 25 kDa-50 kDa junction that is close to the region encoding the ATP-binding region of myosin.[49–52] The position of this insert is shown schematically in Figure 2B. Two myosin II isoforms, SM-A and SM-B, produced by alternative splicing of the pre-mRNA at the 5'-ends, differ in the presence of a 7-amino acid insert near the ATP-binding region in the myosin head of the latter.[51] The SM-A isoform has low actin-activated ATPase activity and shows lower velocity of actin filament movement in the *in vitro* motility assay compared to SM-B.[49] We have shown that the maximum velocity of shortening (V_{max}) of smooth muscles that contain the SM-B (for example, phasic muscles such as bladder) is ~ 2-fold higher than those containing SM-A isoform (for example, aortic muscle which exhibits tonic characteristics). The results from a recent study from our laboratory showed that the level of MHC mRNA with the 21-nucleotide insert in arterial smooth muscle increases as the abdominal aorta branches into smaller distributing arteries.[53] The MHC mRNA from rabbit femoral artery and stomach muscle also contains the 21-nucleotide insert at the 5'-end coding region (Figure 2B). An antibody raised against the 7-amino acid peptide en-

PRIMERS

P1:	5'TACAGGAGCATGCTG3'
P2:	5'TGGCGGATGGCTCGTGA3'
P3:	5'AGTCGCAGCGCATCA3'
P4:	5'TACGCCCTCCAGACCTTCT3'
P5:	5'GACCGTGGCGAAGAACAA3'
P6:	5'CAGCCATTCAGCACCATCCG3'

MHC

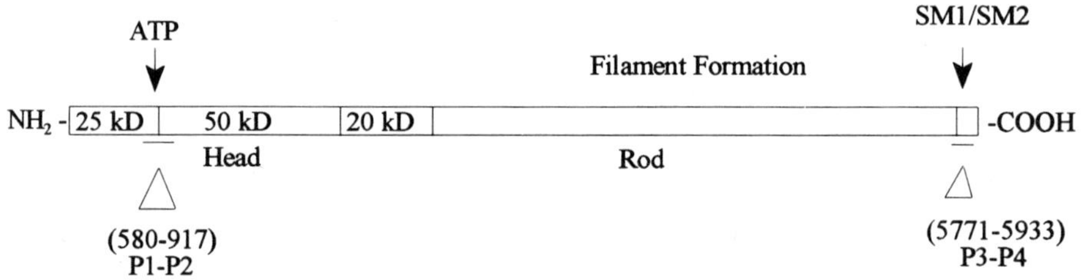

Chicken Gizzard MHC	Q	G	P	S	**F**	**S**	Y
Rabbit Stomach MHC	Q	G	P	S	**L**	**A**	Y
Rat Smooth Muscle MHC	Q	G	P	S	**F**	**A**	Y
Rabbit Femoral Artey MHC	Q	G	P	S	**L**	**A**	Y
Rabbit Saphenous Artery MHC	Q	G	P	S	**L**	**A**	Y

MLC

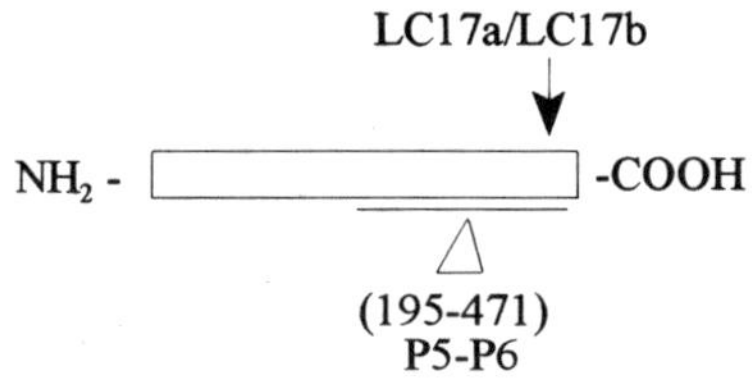

Figure 2B. Alternatively spliced isoforms of smooth muscle myosin. Panel A: Schematic of myosin heavy chain (MHC) and myosin light chain (MLC) pre-mRNA showing positions of the SM-A/SM-B, SM1/SM2 and LC17a/LC17b alternative splice sites. The RT-PCR primers that were designed to detect the relative ratios of these alternatively spliced isoforms for SM-A/SM-B (P1 and P2), SM1/SM2 (P3 and P4) and LC17a/LC17b are shown at the top. These primers were designed according to the rabbit uterus smooth muscle MHC full-length cDNA sequence[51] and the mouse MLC sequence (Halstones and Gunning, Gene Bank, UO4443). The PCR products for SM-B from the saphenous and femoral arteries were sequenced. Shown at the bottom are the deduced amino acid sequences of he SM-B insert in these arteries and also the insert in rabbit stomach,[50] chicken gizzards[92] and rat[93] smooth muscle MHC.

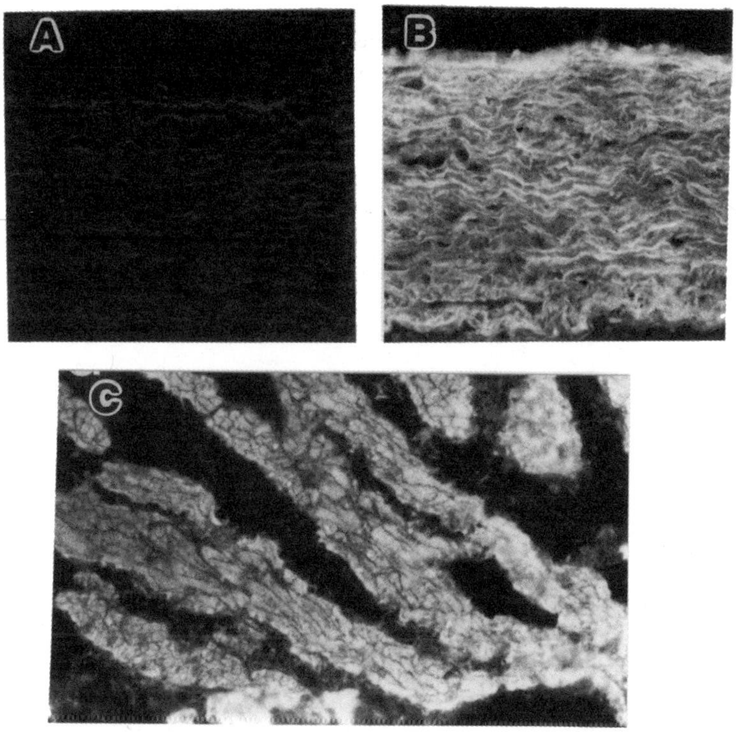

Figure 3. Localization of the SM-B isoform in the urinary bladder smooth muscle. Sections of aorta (A) and bladder (C) were treated with antibody against the 7-amino acid peptide. Immunofluorescence microscopy reveals that the aorta does not react with the antibody against the 7-amino acid insert (see A), whereas the bladder smooth muscle tissue did (see C). The faint fluorescence in panel A is the autofluorescence coming from the connective tissue fibers between the smooth muscle cells. B is a section from the aorta treated with antibody against the whole myosin molecule.

coded by the 21-nt insert does not cross-react with the myosin in the aorta (Figure 3). However, smooth muscle cells in the bladder wall consists entirely of myosin with the 7-amino acid insert (Figure 3, Panel C).

5. ANIMAL MODEL FOR PARTIAL OUTLET OBSTRUCTION

The molecular mechanisms underlying smooth muscle dysfunction in urinary bladder outlet obstruction associated with benign prostate hyperplasia (BPH) are not understood. In the rabbit model for partial outlet obstruction, the detrusor smooth muscle undergoes a remodeling which results in detrusor hypertrophy to compensate for the increased muscle contractility required to overcome the increased urethral resistance during micturition.[54–57] With continued obstruction, the bladder undergoes a stage of decompensation resulting in an over distension of the bladder with its complication (for example, overflow and decreased renal function). The rabbit and other small experimental animal models for partial outlet obstruction, although not analogous to human outlet obstruction associated with BPH, have been used successfully to study the pathophysiology and pharmacology of outlet obstruction in several laboratories.[58–62] An understanding of the cellular and molecular basis of the contractile dysfunction associated with urinary bladder outlet obstruction in BPH is essential for developing therapy to maintain the contractility of the smooth muscle, to prevent or delay decompensation, to rehabilitate decompensated detrusor, and/or to assess the effectiveness of surgical treatment and the reversibility of contractile function of the detrusor after surgery.

6. CHANGES IN THE COMPOSITION OF CONTRACTILE PROTEIN ISOFORMS IN OBSTRUCTION-INDUCED HYPERTROPHY

6.1. Myosin Isoforms in Smooth Muscle Cells in the Bladder Wall following Outlet Obstruction

Partial ligation of the urethra in rabbit causes the bladder wall to undergo an increase in mass,[63] presumably to compensate for the increased functional demand impinged upon the detrusor to expel the urine against the outlet obstruction. We investigated the expression of myosin isoforms in the bladder smooth muscle to determine if the remodeling of the bladder wall in response to the obstruction is associated with an alteration in the composition of the smooth muscle myosin isoforms.

6.1.1. Isoforms of Myosin Heavy Chain. The SM1 and SM2 myosin heavy chains migrate at 204 kDa and 200 kDa, respectively, as shown in Figure 2A. In our studies on the expression of myosin isoforms in the detrusor in response to outlet obstruction, both isotypes were detected in tissue extracts from normal, hypertrophied, 7-day reversal, and 14-day reversal muscle tissues.[62] In response to hypertrophy, the SM1 heavy chain increases compared with the SM2 protein, lowering the SM2:SM1 ratio. Expression patterns of both species return close to normal when the obstruction-induced stimulation is removed. As shown in Figure 4, the SM2:SM1 ratios of 2.8 for normal muscle and 1.03 for hypertrophied muscle are similar to published values.[46] Seven days after removal of the obstruction, the SM2:SM1 ratio increases only very slightly (1.08), whereas after 14 days of the reversal, the ratio reaches 1.78. These data show that the relative proportions of SM1 and SM2 in bladder smooth muscle are associated with the change in contractile function. The high abundance of SM1 and the low abundance of SM2 in hypertrophied smooth muscle may affect contractility of the detrusor muscle. Although the mRNA transcripts have been shown to correlate with the myosin isoforms at the protein level (compare Figures 4 and 5), it is not known if the variation in the composition of the myosin isoforms is due to a change in the turnover of these isoforms, associated with an overexpression of proteases, or due to alterations in the transcription and/or the translation of these proteins.

The mRNA isolated from the detrusor smooth muscle from normal rabbit bladder contains the 21-nt insert, thereby encoding the myosin isoform that contains a 7-amino acid insert near the ATP-binding region of the myosin head. The actin-activated ATPase activity of bladder myosin is 2-fold higher than that of the myosin from large arteries.[64,65] Outlet obstruction induces the expression of myosin without the 7-amino acid insert in the detrusor smooth muscle. The expression of myosin heavy chain mRNAs that differ in the presence of a 21-nt insert at the 5' end is depicted in Figure 6.

6.1.2. Isoforms of Myosin Light Chain. The LC subunits of myosin exist as isoforms, and these isoforms may be different for smooth muscle myosin from different sources. Al-

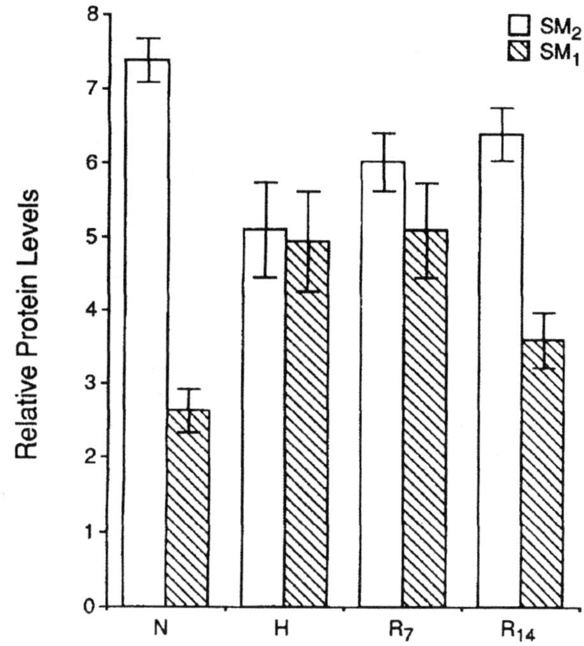

Figure 4. Analysis of the levels of expression of SM2 and SM1 at the protein level. Relative protein levels as determined by densitometric scanning of the SM1 and SM2 heavy chain bands separated by electrophoresis on a 3.5% SDS-polyacrylamide gel. The N, H, R_7, and R_{14} represent normal, hypertrophied, and 7-day reversed and 14-day reversed bladder smooth muscle, respectively. Values are mean ± SE for five determinations from five different bladders. The difference between the relative SM2:SM1 protein ratios of the normal and hypertrophied bladders and the difference between the relative SM2:SM1 protein ratios of the 14-day reversal and hypertrophied bladders are statistically significant (p<0.01). [Reproduced from[62] with permission.]

ternative splicing of the pre-mRNA in the 3' end of the coding region results in two 17 kDa LC mRNAs. These mRNAs encode LCs of 150 amino acids that are identical except in 5 amino acids within the last 9 amino acid residues in the C-terminal region. The two types of light chains are separable as two isoelectric variants (LC 17a & LC 17b). Malmqvist & Arner showed an inverse relationship between maximal shortening velocity (V_{max}) and LC17b in smooth muscles.[66] It has been suggested that phasic smooth muscles contain only LC 17a, whereas tonic muscles contain both LC 17a & LC 17b.[67] A difference in the actin-activated myosin ATPase between tonic and phasic smooth muscles has also been attributed to a difference in the 17 kDa LC of myosin;[68] however, the functional significance of LC17 isoforms in cross-bridge cycling and force development remains uncertain. The mRNA isolated from the detrusor smooth muscle from normal rabbit bladder possesses predominantly of the mRNA which codes for the LC 17a isoform (approximately 97% LC17a and 3% LC17b). However, outlet obstruction induces the expression of the mRNA coding for the LC 17b isoform in the detrusor and switches the composition to 70% LC 17a and 30% LC 17b (data not shown). Removal of the ligature causes the LC_{17} isoform to revert almost completely back to the isoform composition seen in the normal detrusor.

6.2. Composition of Thin Filament-Associated Proteins

A major component of the thin-filament (6–8 nm in diameter) is actin, a protein of around 42 kDa that polymerizes to form a two stranded α-helical filamentous structure. The actin filament constitutes the major component of the cytoskeleton of both smooth and non-muscle cells. In addition to forming part of the contractile apparatus, actin also activates the hydrolysis of ATP by myosin ATPase, an enzymatic activity that provides the energy required for the sliding filament mechanism to operate.[1] Six isoforms of actin, α-skeletal muscle, α-cardiac muscle, α-vascular smooth muscle, γ-visceral smooth muscle,

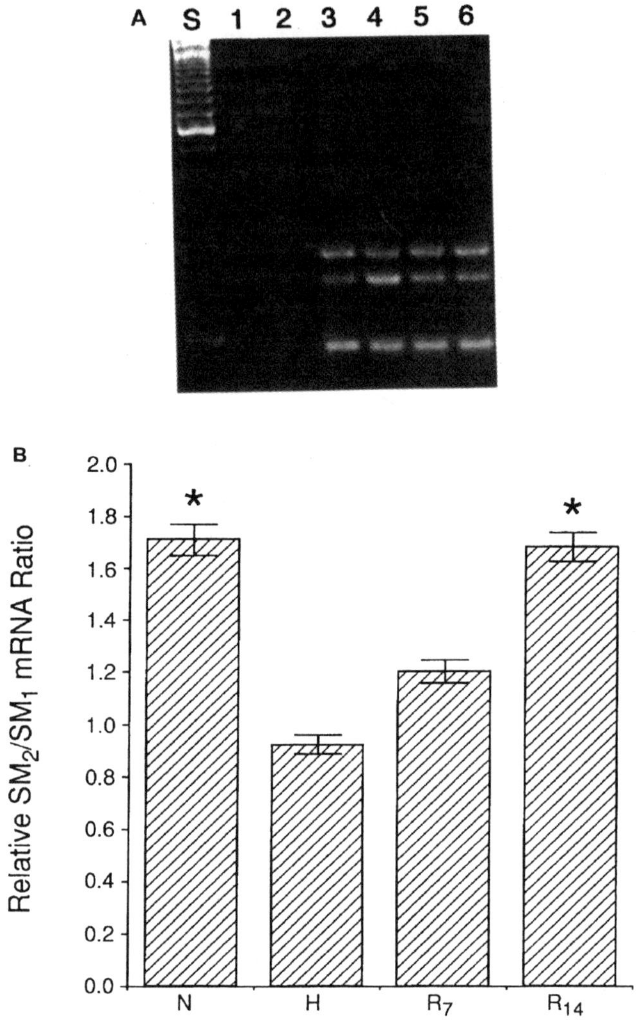

Figure 5. Analysis of SM1 and SM2 mRNA using RT-PCR. The primers (5'AGTCGCAGCGCATCAi3 and 5'TACGCCCTCCAGACCTTCTi3) were used to competitively amplify SM1 and SM2 cDNA fragments. The rabbit uterine smooth muscle α-actin was used as an internal standard. A: Lane S: 1 μg of a 100-bp DNA ladder as molecular size standard. Lanes 1 and 2: PCR products from pBluescript plasmid DNA and total RNA of Sf9 insect cells, respectively. Lanes 3, 4, 5, and 6: PCR products using total RNA from normal, hypertrophied, 7-day reversed, and 14-day reversed rabbit bladder smooth muscle, respectively. The 200-nt and 161-nt bands represent the specific amplification DNA products of SM2 and SM1, respectively, as confirmed by DNA sequence analysis. The lower band is the amplified α-actin DNA. B: Relative SM2:SM1 mRNA ratios as determined by densitometric scanning. N, H, R_7, and R_{14} represent normal, hypertrophied, 7-day reversed, and 14-day reversed bladder smooth muscle, respectively. Values are mean ± SE for six determinations from six different bladders. Significant difference from the hypertrophy is indicated by asterisk (p<0.01). [Reproduced from[62] with permission.]

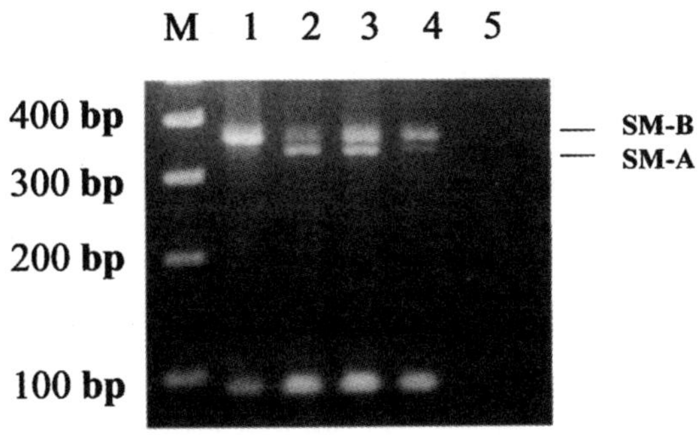

Figure 6. Modulation of the SM-A/SM-B composition by the bladder in response to outlet obstruction. The expression of myosin isoforms with 7-amino acid insert at the N-terminal ATP-binding region was determined using RT-PCR with specific upstream and downstream nucleotide probes designed to amplify a region of the MHC containing this alternative splice site. In this experiment rabbits were obstructed for 7 days and the obstruction was reversed for 7-days and 14-days before sacrifice. The normal bladder smooth muscle contains predominantly inserted myosin as shown from the RT-PCR (lane 1) whereas smooth muscle from obstructed bladder begins to express non-inserted mRNA transcript (lane 2). Upon reversal, the expression of noninserted myosin decreases and inserted myosin increases to more than 2 times that of the noninserted myosin (see lane 3). By 14 days after reversal by removal of the ligature, the bladder again expresses predominantly SM-B.

and γ- and β-cytoplasmic actins have been reported in mammals.[1,69] These actin isoforms originate from a multigene family, but differ in amino acid sequence at the NH2-terminal end, although highly homologous in primary structure. The actin isoforms are highly conserved across species lines, but the proportions of the various actin isoforms are different in smooth muscles from different sources. The actin isoform in the vascular smooth muscle is α-actin, while γ-actin is the predominant form in the visceral smooth muscle. Actin isolated from all types of smooth muscle consists of small amounts of γ- and β- cytoplasmic actins, either from the non-muscle cells present in the smooth muscle layer or from the cytoskeleton of smooth muscle cells.

Tropomyosin, a coiled-coiled fibrous protein that consists of two subunits of around 35 kDa, is located in the thin filaments of smooth muscle at the same location and stoichiometry (1:7 molar ratio of tropomyosin to actin) as in the thin filaments of the striated muscle.[10] Tropomyosin exists as three isoforms in smooth muscles (α-, γ- and β-tropomyosin). The avian gizzard smooth muscle contains γ- and β-subunits of tropomyosin [70], whereas the vascular smooth muscle contain α- and β-subunits.[71] The role played by tropomyosin in the regulation of smooth muscle is different from that in striated muscle, since smooth muscle does not contain the troponin complex. Binding of smooth muscle tropomyosin makes the actin filament cooperative, and this cooperativity is associated with a two to three-fold increase in V_{max} in the actin-activated ATPase activity of smooth muscle myosin.[72]

Smooth muscle thin filaments also contain caldesmon, an actin/calmodulin binding protein with a molecular mass of 89 kDa (h-caldesmon). An isoform of this protein (l-caldesmon) with a smaller molecular mass (69 kDa) is found in the stress fibers (formed predominantly by actin) in nonmuscle cells. These proteins are the alternatively spliced product of a single gene.[73] Both proteins bind to actin and their affinity to actin is decreased in the presence of Ca^{2+}-calmodulin.[74] In addition to its ability to bind actin and calmodulin, caldesmon can also interact with tropomyosin[75–77] and the S2 portion of myosin,[78] the region between the myosin head and the rod portion. The region of caldesmon that binds the myosin S2 portion is a locus near the N-terminus of the molecule.[31,79] Binding of caldesmon to actin inhibits the actin-activated ATP hydrolysis by smooth muscle myosin, and this inhibition is reversed by Ca^{2+}-calmodulin.[4,80–82] Tropomyosin enhances the caldesmon-induced inhibition of the ATPase activity of myosin.[80,81] Kinetic experiments show that inhibition of actin-activated ATPase activity in the absence of tropomyosin is caused by blocking the binding of myosin head to actin; on the other hand, the inhibition in the presence of tropomyosin is caused mainly by an effect on the V_{max}.[80] Binding of the caldesmon N-terminus to myosin and the C-terminus to actin forms a link between myosin and actin as described above. This "non-productive" attachment between myosin and actin has been suggested as a possible mechanism for the maintenance of the "latch state".[31]

Another thin-filament associated protein that may play a regulatory role in smooth muscle is calponin, a 34 kDa protein.[83,84] Calponin also inhibits the actin-activated Mg-ATPase activity of smooth muscle myosin and this inhibition is reversed by Ca^{2+}-calmodulin. The kinetics of the actin-activated ATP hydrolysis in the presence of calponin is influenced by tropomyosin.[85]

6.3. Changes in Actin Isoforms and Actin-Associated Proteins

Using the rabbit model of partial outlet obstruction, we demonstrated that obstruction-induced bladder hypertrophy is associated with increases in both γ-actin and γ-actin specific mRNA, and with decreases in β-actin and β-actin mRNA.[86,87] The α-actin composition is not changed significantly. These changes are depicted in Figure 7. Similar results have been found in the obstructed rat bladder.[88] Based on immunohistochemical studies, using antibodies against muscle proteins, Buoro et al. suggested that the mesenchymal cells in the bladder serosa transform into myofibroblasts which differentiate into fetal type smooth muscle cells during remodeling of the bladder wall following outflow obstruction.[89]

Partial outlet obstruction also causes a decrease in the relative amounts of the actin-associated protein, CaD. The composition of the smooth muscle specific CaD, h-CaD decreases and the non-muscle CaD, l-CaD increases.[90] The expression of the smooth muscle specific h-caldesmon at the protein and mRNA levels is decreased upon obstruction, and upon reversal of the obstruction, the expression of h-caldesmon return to that of the normal detrusor.[91] If caldesmon plays a role in force maintenance, as indicated above, a decrease in the h-caldesmon and/or an increase in l-caldesmon, which is unlikely to "tether" the actin filament to myosin due to its small size, could affect the ability of bladder smooth muscle to maintain force during emptying.

7. SMOOTH MUSCLE CONTRACTILE DYSFUNCTION ASSOCIATED WITH OUTLET OBSTRUCTION

A major effect of benign prostatic hyperplasia (BPH) in men is the gradual obstruction of the urinary bladder outlet by the growing tumor that obliterates the passage of urine during micturition. Although the precise molecular mechanism for the pathophysiology of the bladder dysfunction associated with outlet obstruction is not understood, outlet obstruction induces molecular, cellular, and structural alterations in the smooth muscle cells in the bladder wall, leading to impairments in the ability of bladders to store and empty urine.[54] In the initial phases of outflow obstruction, there is a transient decompensation of the bladder smooth muscle, which initiates the molecular events that lead to hypertrophy of the detrusor smooth muscle and a compensatory increase in detrusor pressure to maintain flow in the face of increased outflow resistance.[56] As obstruction persists, there is a decompensation of the detrusor characterized by increased amounts of extracellular ma-

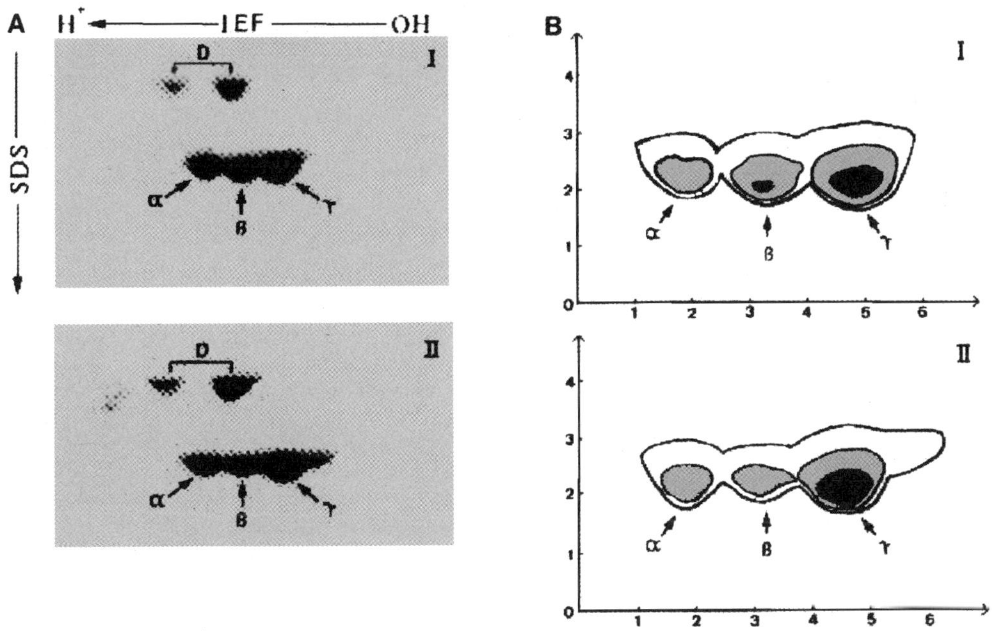

Figure 7. Two-dimensional gel electrophoresis of total protein extracts from control (I) and obstructed (II) rabbit bladders. Panel A: The amount of protein loaded per gel was 30 μg. The α, β, and γ represent α-, β-, and γ-actin isoforms, respectively. The two protein spots shown above actin are identified as desmin isoforms and indicated as D above the bracket. Desmin was increased in the obstructed rabbit bladder. Panel B: Computer images of 2D gels presented in A. The numbers on the X and Y axes indicate the position of the protein spots using arbitrary units. The area of the image with absorbance values between 0.2 and 0.5 (white area) and between 0.5 and 0.8 (shaded area) and 0.8 and 1.1 (dark area) are distinguished. [Reproduced from[87] with permission.]

trixes and alterations in the ability of detrusor smooth muscle cells to generate and maintain force, leading to decreased compliance and impaired emptying.[55] The ability of smooth muscle cells to compensate for increased functional demand is associated with alterations in the expression and function of many proteins involved in the excitation-contraction coupling and active force generation. The alterations in the proteins that form the contractile apparatus may affect the interaction between actin and myosin heads and actin-activated ATP hydrolysis, with subsequent changes in the cross-bridge cycling. Similarly, decompensation of these bladder muscles with persistent outlet obstruction is likely to be caused by a reversal of the changes or breakdown of the structure and function of the proteins that enable the smooth muscle cells to take up, store, and release Ca^{2+}, affecting the calcium-activation of the contractile apparatus.

REFERENCES

1. Adelstein, R.S. and Eisenberg, E.: Regulation and kinetics of the actin-myosin-ATP interaction. [Review]. Annu. Rev. Biochem., **49:** 921, 1980.

2. Kamm, K.E. and Stull, J.T.: The function of myosin and myosin light chain kinase phosphorylation in smooth muscle. [Review]. Ann. Rev. Pharmacol. Toxicol., **25:** 593, 1985.
3. Chacko, S. and Longhurst, P.A.: Regulation of actomyosin and contraction in smooth muscle. [Review]. World J. Urol., **12:** 292, 1994.
4. Marston, S.B. and Redwood, C.S.: The molecular anatomy of caldesmon. [Review] [189 refs]. Biochem. J., **279:** 1, 1991.
5. Sobue, K. and Sellers, J.R.: Caldesmon, a novel regulatory protein in smooth muscle and nonmuscle actomyosin systems. [Review] [89 refs]. J. Biol. Chem., **266:** 12115, 1991.
6. Ebashi, S. The Croonian lecture, 1979: Regulation of muscle contraction. Proc. R. Soc. Lond. - Series B: Biol. Sci., **207:** 259, 1980.
7. Chacko, S. and Longhurst, P.A.: Contractile proteins and their response to bladder outlet obstruction. [Review]. Adv. Exp. Med. Biol., **385:** 55, 1995.
8. Huxley, H.E. and Kress, M.: Crossbridge behaviour during muscle contraction. [Review] [28 refs]. J. Muscle Res. Cell Motil., **6:** 153, 1985.
9. Nonomura, Y. and Ebashi, S.: Calcium regulatory mechanism in vertebrate smooth muscle. Biomed. Res., **1:** 1, 1980.
10. Parry, D.A. and Squire, J.M.: Structural role of tropomyosin in muscle regulation: analysis of the x-ray diffraction patterns from relaxed and contracting muscles. J. Mol. Biol., **75:** 33, 1973.
11. Chacko, S., Rosenfeld, A., and Thomas, G.: Calcium regulation of smooth muscle actomyosin. In: Calcium and contractility. Edited by A.K. Grover and E.E. Daniel, Clifton,NJ: Humana Press, pp. 175–190,1985.
12. Fay, F.S., Shlevin, H.H., Granger, W.C., and Taylot, S.R.: Aequorin luminescence during activation of single isolated smooth muscle cells. Nature, **280:** 506, 1979.
13. Somlyo, A.P. and Somlyo, A.V.: Signal transduction and regulation in smooth muscle [published erratum appears in Nature 1994 Dec 22–29; 372 (6508): 812]. [Review]. Nature, **372:** 231, 1994.
14. Hartshorne, D.J.: Physiology of the gastrointestinal tract, 2nd ed. New York, Raven, 1987.
15. Kendrick-Jones, J., Cande, W.Z., Tooth, P.J., Smith, R.C., and Scholey, J.M.: Studies on the effect of phosphorylation of the 20,000 Mr light chain of vertebrate smooth muscle myosin. J. Mol. Biol., **165:** 139, 1983.
16. Suzuki, H., Onishi, H., Takahashi, K., and Watanabe, S.: Structure and function of chicken gizzard myosin. J. Biochem.(Tokyo), **84:** 1529, 1978.
17. Trybus, K.M. Filamentous smooth muscle myosin is regulated by phosphorylation. J. Cell Biol., **109:** 2887, 1989.
18. Chacko, S., Conti, M.A., and Adelstein, R.S.: Effect of phosphorylation of smooth muscle myosin on actin activation and Ca2+ regulation. Proc. Natl.Acad. Sci. U.S.A., **74:** 129, 1977.
19. Gorecka, A., Aksoy, M.O., and Hartshorne, D.J.: The effect of phosphorylation of gizzard myosin on actin activation. Biochem. Biophys. Res. Commun., **71:** 325, 1976.
20. Sellers, J.R., Pato, M.D., and Adelstein, R.S.: Reversible phosphorylation of smooth muscle myosin, heavy meromyosin, and platelet myosin. J. Biol. Chem., **256:** 13137, 1981.
21. Sobieszek, A. and Small, J.V.: Regulation of the actin-myosin interaction in vertebrate smooth muscle: activation via a myosin light-chain kinase and the effect of tropomyosin. J. Mol. Biol., **112:** 559, 1977.
22. Barany, M. and Barany, K.: Phosphorylation of the myofibrillar proteins. Ann. Rev. Physiol., **42:** 275, 1980.
23. Butler, T.M., Siegman, M.J., and Mooers, S.U.: Chemical energy usage during shortening and work production in mammalian smooth muscle. Am. J. Physiol., **244:** C234, 1983.
24. Dillon, P.F., Aksoy, M.O., Driska, S.P., and Murphy, R.A.: Myosin phosphorylation and the cross-bridge cycle in arterial smooth muscle. Science, **211:** 495, 1981.
25. Gerthoffer, W.T., Murphey, K.A., Mangini, J., Boman, S., and Lattanzio, F.A., Jr.: Myosin phosphorylation and calcium in tonic and phasic contractions of colonic smooth muscle. Am. J. Physiol., **260:** G958, 1991.
26. Butler, T.M. and Siegman, M.J.: Chemical energy usage and myosin light chain phosphorylation in mammalian smooth muscle. Federation Proc., **42:** 57, 1983.
27. Borovikov, Y.S., Khoroshev, M.I., and Chacko, S.: Comparison of the effects of calponin and a 38-kDa caldesmon fragment on formation of the "strong-binding" state in ghost muscle fibers. Biochem. Biophys. Res. Commun., **223:** 240, 1996.
28. Horiuchi, K.Y. and Chacko, S.: Effect of unphosphorylated smooth muscle myosin on caldesmon-mediated regulation of actin filament velocity. J. Muscle Res. Cell Motil., **16:** 11, 1995.
29. Pfitzer, G., Zeugner, C., Troschka, M., and Chalovich, J.M.: Caldesmon and a 20-kDa actin-binding fragment of caldesmon inhibit tension development in skinned gizzard muscle fiber bundles. Proc. Natl. Acad. Sci. U.S.A., **90:** 5904, 1993.

30. Horowitz, A., ClementChomienne, O., Walsh, M.P., Tao, T., Katsuyama, H., and Morgan, K.G.: Effects of calponin on force generation by single smooth muscle cells. Am. J. Physiol., **39:** H1858, 1996.
31. Hemric, M.E. and Chalovich, J.M.: Effect of caldesmon on the ATPase activity and the binding of smooth and skeletal myosin subfragments to actin. J. Biol. Chem., **263:** 1878, 1988.
32. Wang, Z., Jiang, H., Yang, Z.Q., and Chacko, S.: Both N-terminal myosin-binding and C-terminal actin-binding sites on smooth muscle caldesmon are required for caldesmon- mediated inhibition of actin filament velocity. Proc. Natl. Acad. Sci. U.S.A., **94:** 11899, 1997.
33. Sutherland, C. and Walsh, M.P.: Phosphorylation of caldesmon prevents its interaction with smooth muscle myosin. J. Biol. Chem., **264:** 578, 1989.
34. Malmqvist, U., Arner, A., Makuch, R., and Dabrowska, R.: The effects of caldesmon extraction on mechanical properties of skinned smooth muscle fibre preparations. Pflugers Arch., **432:** 241, 1996.
35. Chacko, S. and Longhurst, P.A.: Contractile proteins and their response to bladder outlet obstruction. In: Muscle, matrix, and bladder function. Edited by S.A. Zderic, New York: Plenum Press, pp. 55–63,1995.
36. Craig, R. and Megerman, J.: Assembly of smooth muscle myosin into side-polar filaments. J. Cell Biol., **75:** 990, 1977.
37. Harrington, W.F., Burke, M., and Barton, J.S.: Association of myosin to form contractile systems. Symposia on Quantitative Biology, **XXXVII**: 1972.
38. Somlyo, A.P., Devine, C.E., Somlyo, A.V., and Rice, R.V.: Filament organization in vertebrate smooth muscle. Phil. Trans. R. Soc. Lond. - Series B: Biol. Sci., **265:** 223, 1973.
39. Wachsberger, P.R. and Pepe, F.A.: Purification of uterine myosin and synthetic filament formation. J. Mol. Biol., **88:** 385, 1974.
40. Kamm, K.E. and Stull, J.T.: Myosin phosphorylation, force, and maximal shortening velocity in neurally stimulated tracheal smooth muscle. Am. J. Physiol., **249:** C238, 1985.
41. Dreizen, P. and Gershman, L.C.: Relationship of structure to function in myosin. II. Salt denaturation and recombination experiments. Biochemistry, **9:** 1688, 1970.
42. Rayment, I., Rypniewski, W.R., Schmidt-Base, K., Smith, R., Tomchick, D.R., Benning, M.M., Winkelmann, D.A., Wesenberg, G., and Holden, H.M.: Three-dimensional structure of myosin subfragment-1: a molecular motor [see comments]. Science, **261:** 50, 1993.
43. Babij, P. and Periasamy, M.: Myosin heavy chain isoform diversity in smooth muscle is produced by differential RNA processing. J. Mol. Biol., **210:** 673, 1989.
44. Rovner, A.S., Thompson, M.M., and Murphy, R.A.: Two different heavy chains are found in smooth muscle myosin. Am. J. Physiol., **250:** C861, 1986.
45. Mohammad, M.A. and Sparrow, M.P.: The heavy-chain stoichiometry of smooth muscle myosin is a characteristic of smooth muscle tissues. Aust. J. Biol. Sci., **41:** 409, 1988.
46. Samuel, M., Kim, Y., Horiuchi, K.Y., Levin, R.M., and Chacko, S.: Smooth muscle myosin isoform distribution and myosin ATPase in hypertrophied urinary bladder. Biochem. Int., **26:** 645, 1992.
47. Sartore, S., Demarzo, N., Barrione, A., Zanellato, A., Saggin, L., Fabbri, L., and Schiaffino, S.: Myosin heavy chain isoforms in human smooth muscle. Eur. J. Biochem., **179:** 79, 1989.
48. Eddinger, T.J. and Murphy, R.A.: Two smooth muscle myosin heavy chains differ in their light meromyosin fragment. Biochemistry, **27:** 3807, 1988.
49. Kelley, C.A., Takahashi, M., Yu, J.H., and Adelstein, R.S.: An insert of seven amino acids confers functional differences between smooth muscle myosins from the intestines and vasculature. J. Biol. Chem., **268:** 12848, 1993.
50. Babij, P. Tissue-specific and developmentally regulated alternative splicing of a visceral isoform of smooth muscle myosin heavy chain. Nucleic Acids Res., **21:** 1467, 1993.
51. Babij, P., Kelly, C., and Periasamy, M.: Characterization of a mammalian smooth muscle myosin heavy-chain gene: complete nucleotide and protein coding sequence and analysis of the 5' end of the gene. Proc. Natl. Acad. Sci. U.S.A., **88:** 10676, 1991.
52. DiSanto, M.E., Wang, Z., Menon, C., Zheng, Y., Chacko, T., Hypolite, J., Broderick, G., Wein, A.J., and Chacko, S.: Expression of myosin isoforms in smooth muscle cells in the corpus cavernosum penis. Am. J. Physiol., **275:** C976, 1998.
53. DiSanto, M.E., Cox, R.H., Wang, Z., and Chacko, S.: NH_2-terminal-inserted myosin II heavy chain is expressed in smooth muscle of small muscular arteries. Am. J. Physiol., **272:** C1532, 1997.
54. Coolsaet, B. and van Duyl, W.: Pathophysiology of outlet obstruction. In: The bladder. Edited by J.M. Fitzpatrick and R.J. Krane, New York: Churchill Livingstone. pp. 91–117,1995.
55. Steers, W.D. Physiology of the urinary bladder. In: Urology. Edited by P.C. Walsh, A.B. Retik, T.A. Stamey, and E.D. Vaughan, Jr., Philadelphia: Saunders, pp. 142–176,1992.
56. Wein, A.J. Bladder outlet obstruction- an overview. [Review]. Adv. Exp. Med. Biol., **385:** 3, 1995.

57. Wein, A.J., Levin, R.M., and Barrett, D.M.: Voiding function: Relevant anatomy, physiology, and pharmacology. In: Adult and Pediatric Urology. Edited by J.Y. Gillenwater, J.T. Grayhack, S.S. Howards, and J.D. Duckett, Philadelphia: Mosby Year Book, pp. 933–999,1991.
58. Levin, R.M., Malkowicz, S.B., Wein, A.J., Atta, M.A., and Elbadawi, A.: Recovery from short-term obstruction of the rabbit urinary bladder. J. Urol., **134:** 388, 1985.
59. Lin, V.K. and McConnell, J.D.: Molecular aspects of bladder outlet obstruction. [Review]. Adv. Exp. Med. Biol., **385:** 65, 1995.
60. Mostwin, J.L., Karim, O.M.A., VanKoeveringe, G., and Brooks, E.L.: The guinea pig as a model of gradual urethral obstruction. J. Urol., **145:** 854, 1991.
61. Uvelius, B., Persson, L., and Mattiasson, A.: Smooth muscle cell hypertrophy and hyperplasia in the rat detrusor after short-term intravesical outflow obstruction. J. Urol., **131:** 173, 1984.
62. Wang, Z.E., Gopalakurup, S.K., Levin, R.M., and Chacko, S.: Expression of smooth muscle myosin isoforms in urinary bladder smooth muscle during hypertrophy and regression. Lab. Invest., **73:** 244, 1995.
63. Malkowicz, S.B., Wein, A.J., Elbadawi, A., Van Arsdalen, K., Ruggieri, M.R., Levin, and RM.: Acute biochemical and functional alterations in the partially obstructed rabbit urinary bladder. J. Urol., **136:** 1324, 1986.
64. Chacko, S. and Rosenfeld, A.: Regulation of actin-activated ATP hydrolysis by arterial myosin. Proc. Natl. Acad. Sci. U.S.A., **79:** 292, 1982.
65. Samuel, M., Chowrashi, P.K., Pepe, F.A., and Chacko, S.: Effects of phosphorylation, magnesium, and filament assembly on actin-activated ATPase of pig urinary bladder myosin. Biochemistry, **29:** 7124, 1990.
66. Malmqvist, U. and Arner, A.: Correlation between isoform composition of the 17 kDa myosin light chain and maximal shortening velocity in smooth muscle. Pflugers Archiv., **418:** 523, 1991.
67. Somlyo, A.P. Myosin isoforms in smooth muscle: how may they affect function and structure?. [Review]. J. Muscle. Res. Cell Motil. **14:** 557, 1993.
68. Helper, D.J., Lash, J.A., and Hathaway, D.R.: Distribution of isoelectric variants of the 17,000-dalton myosin light chain in mammalian smooth muscle [published erratum appears in J. Biol. Chem 1989 Feb 5; 264 (4): 2391]. J. Biol. Chem., **263:** 15748, 1988.
69. Vandekerckhove, J. and Weber, K.: At least six different actins are expressed in a higher mammal: an analysis based on the amino acid sequence of the amino- terminal tryptic peptide. J. Mol. Biol., **126:** 783, 1978.
70. Lau, S.Y., Sanders, C., and Smillie, L.B.: Amino acid sequence of chicken gizzard gamma-tropomyosin. J. Biol. Chem., **260:** 7257, 1985.
71. Fatigati, V. and Murphy, R.A.: Actin and tropomyosin variants in smooth muscles. Dependence on tissue type. J. Biol. Chem., **259:** 14383, 1984.
72. Chacko, S. and Eisenberg, E.: Cooperativity of actin-activated ATPase of gizzard heavy meromyosin in the presence of gizzard tropomyosin. J. Biol. Chem., **265:** 2105, 1990.
73. Bryan, J., Imai, M., Lee, R., Moore, P., Cook, R.G., and Lin, W.G.: Cloning and expression of a smooth muscle caldesmon. J. Biol. Chem., **264:** 13873, 1989.
74. Sobue, K., Muramoto, Y., Fujita, M., and Kakiuchi, S.: Purification of a calmodulin-binding protein from chicken gizzard that interacts with F-actin. Proc. Natl. Acad. Sci. U.S.A., **78:** 5652, 1981.
75. Fujii, T., Ozawa, J., Ogoma, Y., and Kondo, Y.: Interaction between chicken gizzard caldesmon and tropomyosin. J. Biochem., **104:** 734, 1988.
76. Graceffa, P. Evidence for interaction between smooth muscle tropomyosin and caldesmon. FEBS Lett., **218:** 139, 1987.
77. Horiuchi, K.Y. and Chacko, S.: Interaction between caldesmon and tropomyosin in the presence and absence of smooth muscle actin. Biochemistry, **27:** 8388, 1988.
78. Ikebe, M. and Reardon, S.: Binding of caldesmon to smooth muscle myosin. J. Biol. Chem., **263:** 3055, 1988.
79. Novy, R.E., Lin, J.L., and Lin, J.J.: Characterization of cDNA clones encoding a human fibroblast caldesmon isoform and analysis of caldesmon expression in normal and transformed cells. J. Biol. Chem., **266:** 16917, 1991.
80. Horiuchi, K.Y., Samuel, M., and Chacko, S.: Mechanism for the inhibition of acto-heavy meromyosin ATPase by the actin/calmodulin binding domain of caldesmon. Biochemistry, **30:** 712, 1991.
81. Sobue, K., Takahashi, K., and Wakabayashi, I.: Caldesmon150 regulates the tropomyosin-enhanced actin-myosin interaction in gizzard smooth muscle. Biochem. Biophys. Res. Commun., **132:** 645, 1985.
82. Zhuang, S., Wang, E., and Wang, C.L.: Identification of the functionally relevant calmodulin binding site in smooth muscle caldesmon. J. Biol. Chem., **270:** 19964, 1995.
83. Lehman, W. Calponin and the composition of smooth muscle thin filaments [news]. J. Muscle Res. Cell Motil., **12:** 221, 1991.

84. Takahashi, K., Hiwada, K., and Kokubu, T.: Vascular smooth muscle calponin. A novel troponin T-like protein. Hypertension, **11:** 620, 1988.
85. Horiuchi, K.Y. and Chacko, S.: The mechanism for the inhibition of actin-activated ATPase of smooth muscle heavy meromyosin by calponin. Biochem. Biophys. Res. Commun., **176:** 1487, 1991.
86. Kim, Y.S., Levin, R.M., and Chacko, S. Alteration of the composition of actin isoforms in urinary bladder hypertrophy. FASEB J., **5:**A1738, 1991.(Abstract)
87. Kim, Y.S., Wang, Z., Levin, R.M., and Chacko, S.: Alterations in the expression of the beta-cytoplasmic and the gamma-smooth muscle actins in hypertrophied urinary bladder smooth muscle. Mol. Cell Biol., **131:** 115, 1994.
88. Malmqvist, U., Arner, A., and Uvelius, B.: Contractile and cytoskeletal proteins in smooth muscle during hypertrophy and its reversal. Am. J. Physiol., **260:** C1085, 1991.
89. Buoro, S., Ferrarese, P., Chiavegato, A., Roelofs, M., Scatena, M., Pauletto, P., Passerini-Glazel, G., Pagano, F., and Sartore, S.: Myofibroblast-derived smooth muscle cells during remodelling of rabbit urinary bladder wall induced by partial outflow obstruction. Lab. Invest., **69:** 589, 1993.
90. Lin, V.K., Lee, I.L., and McConnell, J.D.: Expression of nonmuscle-caldesmon in obstruction-induced detrusor hypertrophy is regulated at mRNA level. J. Urol., **147:** 314A, 1992.
91. Menon, C., Zheng, Y., DiSanto, M.E., Nigro, D., Hypolite, J., Wein, A.J., and Chacko, S. Molecular mechanism for contractile dysfunction in the detrusor following outlet obstruction: expression of caldesmon, a protein that regulates actin-myosin interaction in the smooth muscle. J. Urol. **157:**148, 1997.(Abstract)
92. Yanagisawa, M., Hamada, Y., Katsuragawa, Y., Imamura, M., Mikawa, T., and Masaki, T.: Complete primary structure of vertebrate smooth muscle myosin heavy chain deduced from its complementary DNA sequence. Implications on topography and function of myosin. J. Mol. Biol., **198:** 143, 1987.
93. White, S., Martin, A.F., and Periasamy, M.: Identification of a novel smooth muscle myosin heavy chain cDNA: isoform diversity in the S1 head region. Am. J. Physiol., **264:** C1252, 1993.

CALCIUM ION HOMEOSTASIS IN URINARY BLADDER SMOOTH MUSCLE

Stephen A. Zderic, Chaoliang Gong, Mike Desanto, Joseph Hypolite, Joel Hutcheson, Alan J. Wein, and Samuel Chacko

Urology Research Laboratories
The Children's Hospital of Philadelphia
The University of Pennsylvania School of Medicine
Philadelphia, Pennsylvania 19104

1. INTRODUCTION

The purpose of the urinary bladder is to store urine at low pressures, and allow for effective evacuation at an appropriate time, location and social situation. For this complex set of requirements to be carried out, it is essential that the major components of the lower urinary tract interact in a unified fashion.[1] The basic elements of the bladder which contribute to the active (contractile) and passive properties (resistance to deformation) are listed in Table 1. It is important to note that even the passive properties of the urinary bladder are in part determined by actin myosin interactions that are calcium mediated and energy dependent. Failure of any one (usually several) of these components will contribute to a failure to empty or to store, or some combination thereof. Although each component within this table makes an important contribution to shaping the bladder's biologic properties, the purpose of this review will be to focus on the role of cell calcium regulation in which the sarcoplasmic reticulum (SR) plays a major role.

The two major areas where we have concentrated our laboratory efforts have revolved around the changes that take place in detrusor smooth muscle during normal bladder development in the fetal and neonatal periods, and the response of the detrusor to bladder outlet obstruction. Normal fetal and neonatal development offer the investigator a chance to study the normal hypertorphy of growth. What stimuli allow for normal bladder growth and development as is seen in utero? What are some of the molecular changes taking place within the bladder smooth muscle, which allow it to generate greater contractile force and gain compliance? Cyclic filling and emptying of the bladder is a major growth stimulus that clinicians have long recognized. Bilateral ectopic ureters in females result in a miniscule bladder with poor compliance, and the small bladder resulting from long term

Advances in Bladder Research, edited by Baskin and Hayward.
Kluwer Academic / Plenum Publishers, New York, 1999.

Table 1. The bladder

Smooth muscle	Extracellular matrix	The nervous system
contractile filaments	collagens	peripheral
mitochondria	elastin	central
endoplasmic reticulum	glycosaminoglycans	
sarcoplasmic reticulum		

urinary diversion may be rehabilitated by regular cycling prior to reconstruction.[2] What are some of the molecular changes induced by cycling and mechanical distention that result in growth? In the case of outlet obstruction it is important to try and understand why the clinical manifestations are so variable. Some cases of outlet obstruction produce major bladder wall changes and dramatic urodynamic findings, while others lead to bladders that have adapted to their new workload and whose function is nearly normal. What is the molecular basis for this diversity of responses?

2. IONIC REGULATION

The importance of ionic regulation in normal cellular function cannot be underestimated. Whether one considers the $Na^{+}K^{+}$ATPase that maintains neural membrane potential (and rapidly re-establishes it post depolarization), or the sarcoplasmic endoplasmic reticulum $Ca^{++}Mg^{++}$ATPase of cardiac muscle, the cell expends a tremendous amount of energy maintaining a set of proper ionic gradients. In cardiac muscle, it has been estimated that up to 25% of the ATP expenditure is directed towards maintaining proper calcium gradients.[3] This energy is consumed directly in the form of active transporting pumps (translocating ATPases) and indirectly in terms of the assembly and maintainance of those cellular structures which serve as ionic reservoirs. The cellular ultrastructure is crucial for ionic homeostasis; calcium ion pumps alone would only be capable of driving cytosolic calcium out across the plasma membrane.

Ionic homeostasis within the cell is maintained by the use of energy to develop concentration gradients in conjunction with membranes that are impermeable to the ions of interest. The only spots within the membrane that are permeable correspond to either energy dependent (ATP) pumps or ion channels that when opened in response to a signal, allow for transmembrane ionic flux. The energy required to maintain an ionic gradient is defined by the Gibbs free energy equation : $\Delta G = -RT \ln K'eq$ where K'eq represents the equilibrium constant across the membrane. To maintain calcium concentrations at 1 μM in the cytosol, and 1mM in the SR gives an equilibrium constant of 1000 representing an energy expediture of 17 kJ/mol. This figure gives a rough idea of what kinds of energy demands are made upon a living cell if it is to maintain the ionic gradients that are required for normal homeostasis. It does not include the energy expended by the cell to assemble the SR and its many constituent proteins. This additional energy cost reflects the work a cell must do to overcome entropy and maintain ionic order.

The ion transporting ATPases are located in both the SR and the plasma membrane and have been studied extensively, most commonly in skeletal and cardiac muscle. The sarcoplasmic endoplasmic reticulum $Ca^{++}Mg^{++}$ATPase (SERCA)[4,5,6] plasma membrane $Ca^{++}Mg^{++}$ATPase (PMCA)[7,8] and $Na^{+}K^{+}$ATPase[9,10] have been well characterized especially within striated muscle systems. Furthermore monoclonal antibodies are available against these pumps, and their genes have been cloned. Since these genes tend to be highly conserved, many of these reagents work well in several species. In this manner antibodies and cDNA probes developed for work in human, bovine, or canine muscle have been able to cross react with ion pumps in the rabbit which has been our preferred experimental system.

The presence of the sarcoplasmic reticulum creates an intracellular reservoir into which calcium ions may be actively transported, stored, and released. The pumps such as SERCA are located on the membrane surface of the SR; studies suggest that immediately adacent to these ion pumps are a rich supply of membrane bound glycolytic enzymes.[11] This provides indirect evidence for the importance of ionic regulation; smooth muscle can function for some time under anaerobic conditions, but the ATP produced by the glycolytic pathway is immediately available for ionic regulation by virtue of its proximity. While other energy dependent cellular functions may be deferred upon for short periods of time, cytosolic calcium regulation appears to be a priority in terms of ATP expenditure. Once the calcium has been pumped across the SR it must be maintained within the lumen. In striated muscle this is made possible by calsequestrin, a glyocoprotein rich in sialic acid residues. This protein has a high number of calcium ion binding sites with a low affinity meaning that it can readily allow for calcium ion efflux from the SR. While calsequestrin is expressed in some smooth muscle[12] we have not found evidence for its expression within the urinary bladder, and are considering other alternatives such as calreticulin. Once the calcium ions are pumped into and stored within the SR, it is important that release mechanisms exist allowing for efflux into the cytosol. In rabbit bladder smooth muscle, the ryanodine sensitive ion channel serves as the conduit for cell calcium from the SR back into the cytosol. The existance of this channel in bladder smooth muscle has been documented in both functional[13] and ligand binding studies.[14] In addition to this ion channel, a receptor for inositol tri-phosphate is also expressed on the SR surface[15] though we have not yet demonstrated this in bladder. In striated muscle electron microscopy studies reveal a complex junction between the sarcoplasmic reticulum and the plasma membrane and/or t-tubules referred to as feet (foot processes).[16] Similar ultrastructural findings have not been reported in smooth muscle; sarcoplasmic reticulum definately exists in smooth muscle[17,18] but its 3-dimensional arragement within the cytosol requires further study. Even less is known about how the morphology of the sarcoplasmic reticulum changes with normal development or in response to outlet obstruction.

3. EXCITATION CONTRACTION COUPLING

Figure 1 represents a brief overview of the many steps taking place as smooth muscle generates force in response to stimulation, and then subsequently undergoes relaxation. Contraction of the detrusor is initiated by parasympathetic fibers within the pelvic nerve in which acetylcholine is the primary (but not exclusive) neurotransmitter. The interaction between acetylcholine and its specific muscarinic receptor at the cell membrane of the smooth muscle cell triggers the process of excitation contraction coupling. This receptor ligand interaction is coupled to a G-protein which leads to (1) an influx of extracellular calcium to the cytosol via adjacent plasma membrane bound calcium channels and (2) the production of a second intracellular messenger in the form of inositol triphospate (IP_3). The binding of IP_3 to its receptor on the cytosolic surface of the SR in turn triggers the release of intracellular calcium. This release of intracellular calcium is regulated by the ryanodine sensitive ion channels located in the SR membranes. The calcium induced calcium release (CICR) mechanism refers to a second mechanism of calcium release from

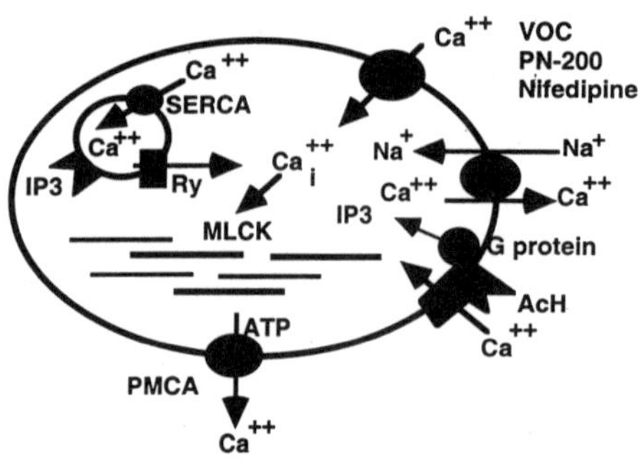

Figure 1. A schematic simplification of excitation contraction coupling. Ry = ryanodine sensitive calcium release channel. VOC = voltage operated calcium channel that is site of action of PN-200 or nifedipine. AcH = acetylcholine which binds to a muscarinic receptor to trigger calcium influx or activates a G protein which in turn leads to the formation of inositol tri-phosphate (IP_3). IP_3 or free cytosolic calcium can in turn trigger calcium release from the sarcoplasmic reticulum. Free cytosolic calcium (Ca_i) activates myosin light chain kinase which is felt to initiate contractile force generation. SERCA - Sarcoplasmic Endoplasmic Reticulum $Ca^{++}Mg^{++}$ ATPase is the ATP consuming ion pump that is responsible for the translocation of free cytosolic calcium out of the cytosol back into the intracellular storage sites. Also shown on the cell surface is the sodium calcium exchanger which does not consume ATP directly (it does in the sense that it is dependent upon the Na^+K^+ATPase) and the plasma membrane $Ca^{++}Mg^{++}$ATPase (PMCA) which is ATP dependent.

the SR. With the initial rise in cytosolic calcium, there is activation of the ryanodine ion channel to allow for greater SR release. This mechanism has been demonstrated in vascular smooth muscle, but has not yet been shown to occur in the urinary bladder.[19] Membrane depolarization in bladder smooth muscle is mediated by influx of the calcium ion[19] (as opposed to nerve where sodium influx dominates) which in turn activates the voltage operated calcium (VOC) channels providing yet another mechanism for the rise in cytosolic calcium needed to trigger contraction.

With a rise in cytosolic calcium from 1 μM to 10 μM, activation of the contractile apparatus occurs, and the force generation is initiated. One (of probably several) step in contractile force production is the phosphorylation and activation of myosin light chain kinase.[20] The work of DiSalvo et al offers evidence for an important role of the tyrosine kinase $pp60^{c\text{-}src}$ in the generation of contractile force in carotid smooth muscle.[21] Cytosolic calcium plays a role in modulating contractility as was shown by Gerthoffer's work in trachealis smooth muscle. His experiments demonstrated that force production was directly related to variations in the calcium concentration in the medium; in contrast the light chain phosphorylation remained constant at 60–70%.[22,23]

Relaxation is brought about as cytosolic calcium is driven back into several storage sites. The two major calcium ion pumps are driven by ATP are located on the SR and the plasma membrane and serve to restore the cytosolic calcium concentration back to normal precontraction levels. ATP is utilized to move calcium ions into either the SR or back out across the plasma membrane into the extracellular space. The Sarcoplasmic Endoplasmic Reticulum Calcium ATPase (SERCA) sits on the surface of the sarcoplasmic reticulum and serves to pump one free calcium ion into the SR per ATP hydolyzed. Since its initial description by MacLennan over 20 years ago, it has been extensively studied at both the protein[4] and nucleic acid level.[6] Three isoforms of SERCA have been identified in fast muscle (skeletal and cardiac), slow muscle (smooth), and non muscle (kidney, brain, liver & other). They are the product of alternative splicing mechanisms.[6] The plasma membrane calcium ATPase (PMCA) has been extensively charaterized by Carafoli[7] who defined it first in erythrocytes. Moncolonal antibodies have been raised against it and the base sequence is known for its smooth muscle isoform.[8] The $Na^{+}Ca^{++}$ exchanger is a "passive" mechanism for calcium homeostasis which is energy dependent due to the $Na^{+}K^{+}$ ATPase of the plasma membrane. With depolarization, this exchanger allows for excess cytosolic calcium to be driven back out of the cell since the natural gradient favors sodium influx. It has been well characterized in cardiac muscle,[24] but its role in bladder smooth muscle remains unclear at this time.

4. ONTOGENY DATA

The process of normal growth and development offers an excellent opportunity to study what happens to a bladder that is required to do increasing amounts of volume work in the absence of any outlet obstruction. Prenatal ultrasound studies have shown bladder filling and emptying in the human [25] as well as fetal sheep.[26] It is well established that fetal urine output increases during gestation with an exponential shape to the curve with major increases in the first part of the third trimester.[25] If one plots fetal bovine bladder mass as a function of crown rump length, then this rise in fetal urine output also corresponds to the time when one sees a dramatic rise in bladder mass.[27] Furthermore, if one measured compliance defined as $\Delta V/\Delta P$, improved bladder compliance was aquired with normal development as bladder mass and fetal urine output increased.[27] Numerous lines of evidence suggest that volume based work and bladder wall cycling are essential for normal bladder development. Clinically this is supported by the urodynamic outcomes seen in patients with bilateral ectopic ureters or bladder exstrophy. At a molecular level what changes take place that can account for all these physiologic findings? As outlined in Table 1, many changes must be taking place in utero that will contribute to the observed physiology. This review will focus on the role of cytosolic calicum regulation and how it changes with normal development in the neonatal rabbit.

The basic premise underlying our study of the ontogeny of bladder function in the rabbit was that with normal maturation, the rabbit bladder smooth muscle aquired increasing amounts of SR. Much electrophysiologic and molecular evidence from the fetal and neonatal heart suggest that with normal maturation, the myocardium aquires increased amounts of SR.[28,29] Our initial physiology studies supported this notion in rabbit bladder smooth muscle as well; neonatal bladder smooth muscle performed poorly in calcium free buffers in contrast to that smooth muscle from mature animals.[30] The response of these neonatal and mature bladder smooth muscle strips to the ryanodine and nifedipine also differed. The enhanced sensitivity of the newborn bladder to nifedipine, and minimal response to ryanodine suggested that in this age range the SR had a lesser role in contractile force generation.[13] In contrast ryanodine which serves to block efflux from the SR (Figure 1), had a more pronounced effect in the mature group, suggesting that with normal maturation, the contribution of the SR to force generation increased.

Using radioligand binding studies, we were able to demonstrate that with normal development, (1) no major shift is observed in the voltage operated calcium

channels associated with the plasma membrane,[30] and (2) a major upregulation is seen in the expression of the ryanodine receptor[14] (Figure 2). This was significant, because expression of the voltage operated calcium channels is restricted to the plasma membrane, whereas the ryanodine sensitive ion channel is a specific marker of the SR. Taken together with our prior physiology studies, this data was consistent with our hypothesis that the amount as well as the functional role of the SR increases with normal development.

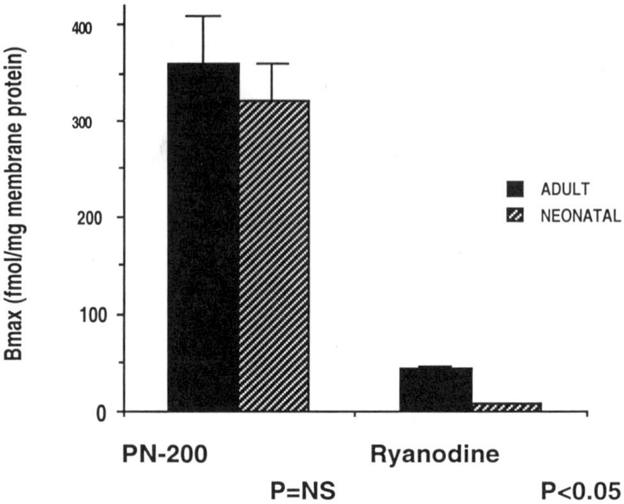

Figure 2. This graph summarizes the maximal number of binding sites as determined by Scatchard analysis for the voltage operated calcium channel detected with PN-200[30] and the SR associated calcium release channel which binds ryanodine.[13,14] The comparison was made between neonatal (3 day old) and mature (8 week old) rabbit bladder smooth muscle. Only the differences for ryanodine are highly signficant.

If this hypothesis were true however, other markers of the SR should also increase in amount during normal development. To prove this point we set out to study the develomental changes in the expression of SERCA in rabbit bladder smooth muscle. To do so we first utilized an inhibition assay based on the sesquiterpene thapsigargin[31,32] which is a specific inhibitor of the SERCA isoforms. The proof of its specificity for the SERCA family (as opposed to PMCA or the Na^+K^+ ATPase) has been established in liposomes into which the pump of 1 gene product has been inserted and ionic transport measured.[32] It is important to point out that thapsigargin does not distinguish between the 3 basic isoforms of SERCA. The assay (is based on the same principles as the oubain inhibition assay for the Na^+K^+ ATPase[33] measures the rate of free PO_4 release in the presence and absence of thapsigargin. The difference between the total ATPase activity and that in the presence of thapsigargin reflect the activity of SERCA, with the data expressed as nmol PO_4/min/mg protein. When this assay was applied to microsomal fractions of 3 day old and 8 week old rabbit bladder smooth muscle, there was a major increase in expression of SERCA (Figure 3). This was true in

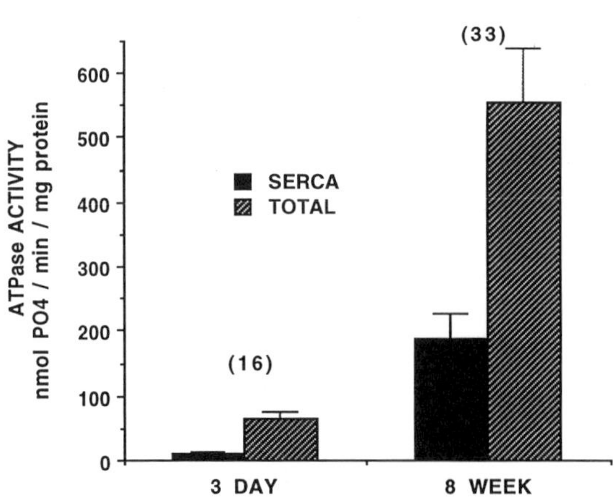

Figure 3. This graph demonstrates the increased activity of SERCA as determined in rabbit bladder smooth muscle during the course of normal development. The comparison was made between neonatal (3 day old) and mature (8 week old) rabbit bladder smooth muscle; the differences are highly significant ($P<0.05$), both in terms of total activity for SERCA and for its activity as a percentage of total ATPase activity.

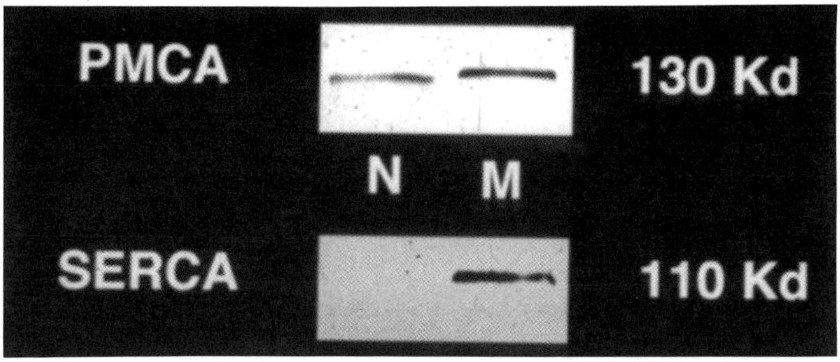

Figure 4. The Western blot analysis of microsomal membranes isolated from neonatal (3 day old) and mature (8 week old) rabbit bladder smooth muscle. The primary antibodies (monoclonal) were raised against SERCA2 and PMCA, and detection was with a second antibody coupled to horseradish peroxidase. SERCA expression increased dramatically, whereas PMCA expression increased, but not as dramatically.

terms of total activity, but also in terms of the percentage of total ATPase activity which increased from 16% to 32%.[34]

With this assay data supporting our hypothesis, we sought to confirm these findings with a second technique. The same membrane fractions used to assay for the biochemical activity of SERCA were then probed with a monocolonal antibody against SERCA2b and PMCA using Western blot techniques. The results shown in Figure 4 are consistent with our hypothesis that with normal development there is an increase in SERCA expression, while plasma membrane bound pumps and channels do not increase to the same degree. Questions of how such data is normalized always arise. These blots were done by loading equal amounts of membrane protein per lane. The yield of μg membrane protein per gram of wet tissue weight proved to actually be higher in the neonatal group (1548 ± 267) versus the mature group (1246 ± 162) though this difference was not significant. The finding of increased expression of SERCA or the ryanodine sensitive ion channel is not an artifact related to changes in the collagen content of the tissues. Furthermore the antibody against SERCA2 is specific for muscle as was shown by dual labeled immunofluorescence studies (Figure 5); this figure shows that SERCA staining outlined with Texas Red is colocalized to those cells that also stain positive for alpha smooth muscle actin as defined by fluorescein.

The increase in SERCA2 expression and activity might be due to transcriptional or translational regulatory mechanisms. In the rabbit myocardium, increased SERCA expression is correlated with increased SERCA gene expression as detected by Northern blot analysis.[29] To answer this question, total RNA was isolated from both groups, and subjected to Northern blot analysis using a cDNA probe raised against the 740 base pairs at the 3' end of the SERCA2b gene from rabbit stomach smooth muscle.[35,36] As shown in Figure 6, the signal corresponding to the 4.2 Kb region where SERCA2 mRNA is found is dramatically increased in the mature groups; this corroborates our hypothesis that the increased SERCA2 expression is a result of increased gene expression. It is interesting to note that a band appears at 9.1 Kb; this band is unlikely to represent an artifact as it persists even after washes at 0.1SSC at 65°C. The probe used in these experiments was cloned from rabbit stomach smooth muscle, and when we compared total RNA from rabbit stomach and bladder together, the 9.1 Kb signal was seen only in the bladder smooth muscle. We speculate that this 9.1 Kb band seen in bladder smooth muscle reflects hybridization to a long RNA segment

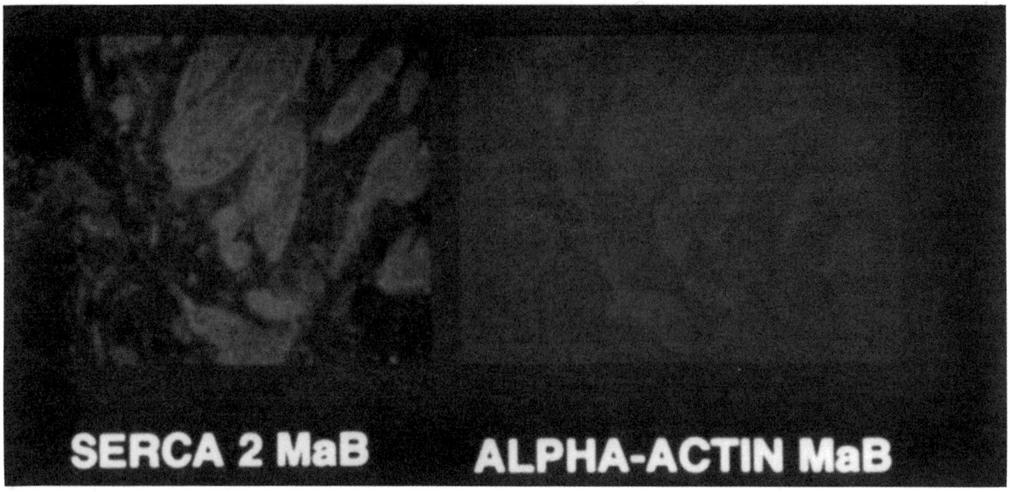

Figure 5. Dual labeled immunofluorescence studies demonstrating that SERCA2 expression is colocalized to those cells that stain positive for smooth muscle alpha actin. The same smooth muscle bundles having undergone dual staining were photographed with the Texas Red filter (to detect SERCA) and the FITC filter (to detect actin).

that is in the stages of alternative splicing. It also suggests that the SERCA RNA from stomach and bladder differ in terms of their base sequences at the 3' end. Further work will be needed to understand the molecular basis for these findings.

5. OBSTRUCTION DATA

The study of bladder outlet obstruction offers an opportunity to understand what molecular events must take place in order for a smooth muscle to adapt to a new workload.

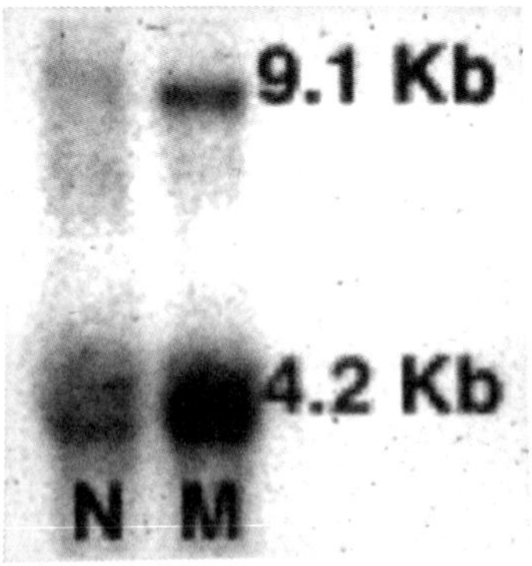

Figure 6. A Northern blot analysis of equal amounts (10 μg) total RNA isolated from neonatal (3 day) and mature (8 week) old rabbit bladder smooth muscle.[34] Detection was with a ^{32}P labeled cDNA probe cloned from the 740 base pairs at the 3' end of the SERCA2 gene form rabbit stomach smooth muscle. Hybridizations with probe were performed at 42°C and the membranes were washed at 65 °C. The 9.1 Kb band persisted despite the multiple washings at 65 °C for all bladder specimens, but was not noted in the stomach preparations (data not shown). The cDNA probe for rabbit stomach SERCA2 was cloned by Dr. Ashok Grover, and inserted into the Bluescript vector. He kindly supplied us with a sample of this vector and probe.

Adaptive hypertrophy or compensation means that the muscle is able to generate enough force that in the aggregate the whole bladder produces sufficient pressure to empty. Ample experimental evidence supports the motion that major changes can occur in each of the bladder components shown in Table 1; however this discussion will focus on the role of cytosolic calcium regulation. It is important to point out that there exist many models for experimental outlet obstruction, and each one is associated with some degree of experimental variability. While complete outlet obstruction avoids this problem of experimental variability and is an excellent way to study early changes in response to bladder wall distension, it is rarely seen clinically and hence is less relevant. However partial obstruction proves difficult to create on a reliable basis. Our own model of outlet obstruction has called for the use of a 3–0 silk ligature to be placed around the urethra of a male rabbit which temporarily contains an 8 French feeding tube to serve as a calibrating rod. While this has worked well for us over many years, it also became apparent that the effects of obstruction were not uniform over time.[37,38] Despite identical periods of outlet obstruction at 14 and 28 days some bladders performed better, and others poorly despite a standardized surgical approach with one very experienced surgeon. Similar heterogeneity was observed in a model of partial bladder outlet obstruction created in the fetal sheep utilizing a 5 mm jeweler's jump ring.[26] While this might seem discouraging, the fact is that clinically, bladder outlet obstruction presents with a wide spectrum of severity and findings whether one is considering older men with BPH or the fetus or neonate with posterior urethral valves.

The human response to outlet obstruction is certainly subject to a great deal of individual variation. Yalla and Sullivan pointed out in a human urodynamics study that the detrusor increased its ability to generate pressure. They defined the detrusor reserve as the difference between the isovolemic voiding pressure, and the maximal voiding pressure.[39] For patients with a minimal residual, the detrusor reserve remained elevated above 40 cm H_20. However as patients developed an increased residual urine, this was correlated with a drop in detrusor reserve. They took this data to imply that the detrusor is able to compensate in response to obstruction, but only to a certain point. In a second human study looking at urodynamics in patients presenting in urinary retention, it was very clear that "detrusor failure" does occur and that it is not always reversible.[40] Urinary retention as a manifestation of BPH is the exception, and not the rule. Still, this points out the widely varying presentations of this syndrome. In an animal model Connor et al.[41] demonstrated adaption to outlet obstruction in the rat bladder as manifested by increased bladder weights, volumes, and enhanced pressure generating abilities. Interestingly enough they reported fairly wide standard errors for the parameters of bladder weight and volume. Again this would imply that even in an excellent model of outlet obstruction, there is bound to be some variability. This is not detrimental, provided that any molecular studies are correlated with the actual physiology of each individual bladder.

6. BLADDER OUTLET OBSTRUCTION/ACTIVE PROPERTIES

The effects of outlet obstruction on contractile function were first studied as a function of time and for each individual bladder, correlated with measures of SERCA activity (Figure 7). In this double y plot we show that by 7 days of outlet obstruction contractility drops to 20% of control values, but recovers to about 50% of control values by 14 days. Could it be that following an initial imposed load, a transient decompensation takes place followed by a partial recovery? We suspect that this is the case, but would emphasize that these are not the same bladders sampled at differing time points. We strongly suspect that

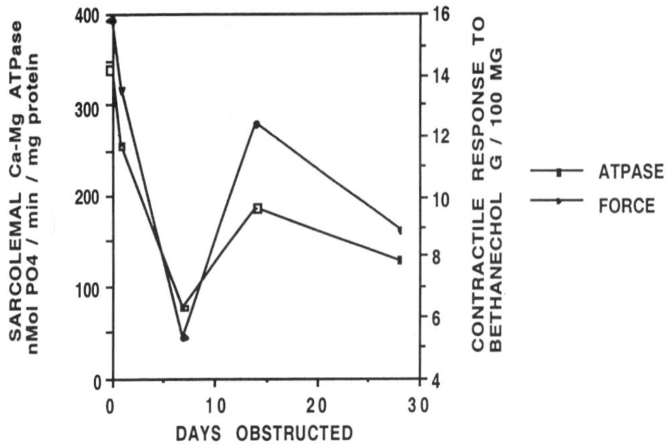

Figure 7. The relationship between the duration of outlet obstruction, contractile force, and SERCA activity. The differences from control values are highly significant (reproduced with permission from Zderic, S.A. et al, J. Urol. 156: 587–592, 1996).

the obstructions produced are heterogenous, with a period of initial decompensation being followed by variable rates and degrees of recovery.

Figure 8 considers this same data plotting contractile function as a function of SERCA activity. The correlation coefficient of 0.7 is remarkable for a biological system and is remarkably similar to data reported in a rat model of aortic stenosis. We feel this data supports our notion that loss of contractile force after outlet obstruction can be correlated with a loss of sarcoplasmic reticulum (of which SERCA serves as but one marker). SERCA activity dropped in all obstructed bladders, with the most pronounced drops in those bladders with the greatest loss of contractile force (this was also proven by a Western blot on the same membrane fractions Figure 9). For bladders with contractile abilities in excess of 50% of control (defined as compensated) total membrane bound ATPase activity was unchanged, despite a statistically significant drop in the SERCA specific fraction. For those bladders we deemed decompensated with contractile force generation below 50% of control total membrane bound ATPase activity was diminished as well as the SERCA specific fraction. Undoubtedly other membrane bound ATPases such as the $Na^{+}K^{+}$ATPase and PMCA are crucial to maintaining cell volume and the proper set of ionic gradients. Bladder smooth muscle adapts to an imposed workload by undergoing a combination of hypertrophy and hyperplasia and maintaining ionic homeostasis. With de-

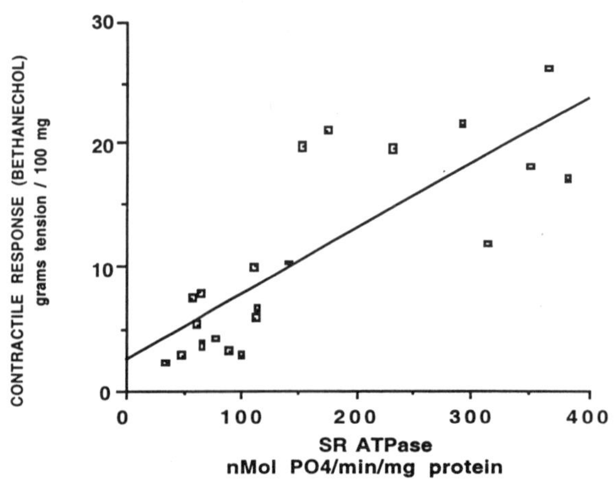

Figure 8. The relationship between contractile force and SERCA activity. The coefficient of regression is 0.7. This data implies that loss of sarcoplasmic reticulum (SERCA is a measure of SR) is associated with a loss of contractile force generation following outlet obstruction. (Reproduced with permission from Zderic, S.A. et al, J. Urol. 156:587–592, 1996).

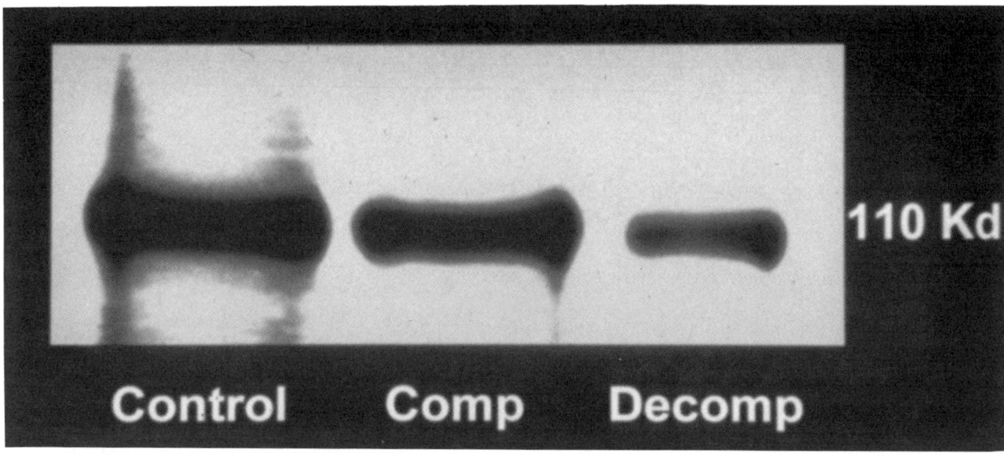

Figure 9. The relationship between the status of the bladder with regard to control, compensated (force > 50% of control) or decompensated (force < 50% of control) is confirmed by Western blotting. (Reproduced with permission from Zderic, S.A. et al., J. Urol. 156:587–592, 1996).

compensation, the bladder loses its ability to keep up with the additional workload. A loss of ionic homeostatic mechanisms is a crucial step as decompensation sets in. This includes a loss of both sarcoplasmic reticulum ion pumping mechanisms as well as those localized to the plasma membrane PMCA, the $Na^{+}K^{+}$ ATPase, and the $Na^{+}Ca^{++}$ exchanger.

With preliminary data such as these demonstrating a loss of SERCA activity both by biochemical assay and by Western Blot, it is essential to ascertain the mechanism(s) for this decline. Diminished protein expression may reflect a downregulation of gene expression or transcriptional regulation, translational regulation, or altered degradative pathways. In the case of SERCA it is well established in the heart that in end-stage congestive heart failure, loss of SERCA expression is highly correlated with a loss of SERCA gene expression in both human and experimental series.[42,43,44] It is important to point out however that in these human studies, the hearts are taken at the time of transplantation from the most severely affected myocardium. In striated muscle, it has been shown that severing the nerve leads to atrophy, and associated with this are changes in the pattern of SERCA gene expression.[45] The same thing has been shown for a model of disuse atrophy.[46] These data would all suggest that the routine use of muscle and cyclic mechanical stretching is an essential component for the normal expression of the SR. It is also important to consider the possibility of SR degradation either by free radicals (which have been shown degrade SERCA in coronary artery smooth muscle[47]), or proteolytic enzymes some of which may be released by the breakdown of lysosomes or through activation of calpain.[48] It may well be possible that a decrease in gene expression and an increase in degradation are in operation, and that that those bladders, which ultimately recover, are capable of generating enough SERCA mRNA that they are ultimately able to compensate and adapt to their new workload. In contrast those bladders that move on to the permanently decompensated states, may be the ones in whom the ability to produce message(s) for the proteins controlling ionic regulation are lost. However this hypothesis must be proven by sampling the same bladder at differing time points and correlating the pattern of gene expression with the ultimate fate of bladder function.

Our preliminary study of SERCA gene expression in the rabbit bladder after outlet obstruction was done using a PCR analysis (Figure 10). It shows that there is an increase

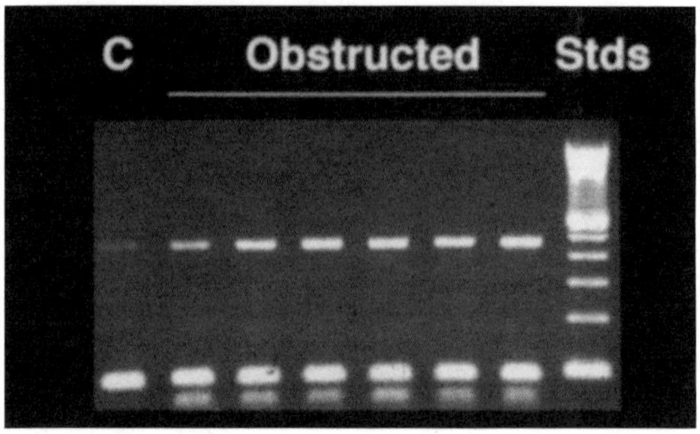

Figure 10. This preliminary PCR analysis for SERCA expression is demonstrated for control versus obstructed specimens. From each specimen, cDNA was produced with the application of reverse transcriptase. The cDNAs were then amplified with upstream and downstream primers for alpha actin and SERCA, and subjected to 30 amplification cycles. Cycle conditions were 0:30 minutes at 94°C, 1:30 minutes at 55°C, and 2:30 minutes at 72°C with a 7 minute final extension after the 30 cycles. The upstream primer corresponded to bases 2289–22312, and the downstream primer to bases 2704–2726 of the SERCA smooth muscle gene.[6] The PCR product was a fragment of 438 base pairs which is the first band at the top of the gel. The actin gene expression remained constant during this PCR analysis, and was utilized to provide evidence for equal amplification. The smaller actin PCR product migrated to the bottom of the gel. The SERCA signal from the control was less intense than that of the obstructed groups implying that SERCA gene expression is increased in response to outlet obstruction.

in the PCR fragments corresponding to SERCA RNA for the obstructed groups versus the control group. This would imply that for this population of rabbit bladders, the loss of SERCA expression would have to be explained by increased degradative activity. However it is also important to note that we are repeating these studies using a number of bladders with known contractile properties to carry out both Western and Northern Blot analysis on the same tissues. Thus we will be able to demonstrate in each individual bladder the protein expression as well as the associated gene expression. If these results are replicated using Northern analysis, it will offer evidence in support of increased degradation of the SR as opposed to transcriptional control.

7. CONCLUDING THOUGHTS AND FUTURE DIRECTIONS

This discussion has centered primarily on the role that the sarcoplasmic reticulum plays in the active properties of the urinary bladder. It is clear however that cell calcium regulation also plays a role in shaping the passive properties of urinary bladder smooth muscle.[49] We are now beginning to investigate in detail the effects of cytosolic calcium on the bladder's resistance to passive strain. The sensitivity of the fetal and adult bovine bladder to extracellular calcium has been established;[50] but the molecular mechanisms for these findings are not understood. It is very clear that bladder compliance is a complex biomechanical phenomenon with numerous components from Table 1 contributing to it. Using a computerized positioning system to apply a constant strain of anywhere from 5% to 20% per minute, we have shown that there is a definite "active" component to these passive properties, and that this is greatly influenced by the presence (or absence) of cell calcium as shown in Figure 11. The role of various ion channels, pumps, and the SR in determining the passive bladder wall properties under conditions of normal fetal development, outlet obstruction, and hypoxia are under active investigation in our laboratory.

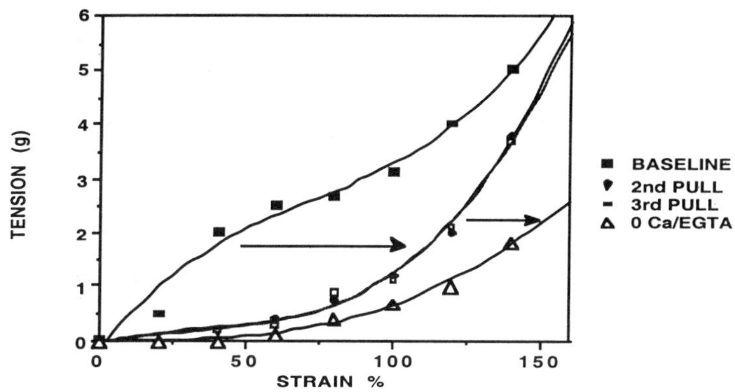

Figure 11. This shows our preliminary data comparing stress strain patterns in the fetal bovine bladder (midgestation) after a baseline stretch to a strain of 140% at 20%/min. The motorized drive unit was then reversed and the strip returned to Lo, at which point the cycle was repeated, in a forward direction. The 2nd and 3rd pulls and all relaxations (not shown) were highly reproducible. The differences between the curves for the forward stretch and the reversal or relaxation reflect the hysteresis seen in smooth muscle. Then the strip was returned to Lo, and subjected to 0 Ca^{++} tyrodes buffer in the presence of 4 mM EGTA. After addition of the calcium free buffer and EGTA the strips were allowed to equilibrate for 30 minutes and field stimulated at 5-minute intervals to deplete intracellular SR stores as best as possible. Then the strip was strained again to 140% in an identical manner at 20%/ min (A similar pattern is seen at strain rates of 5%/min). This data shows that there are substantial tone contributions to compliance that are calcium mediated.

Our ultimate goal is to understand at a molecular level how a bladder responds to outlet obstruction by undergoing compensatory hypertrophy such that it can store urine at low pressure, and maintain a minimal post void residual. Would bladder smooth muscle biopsies from patients deemed by Sullivan and Yalla to have an increased detrusor reserve show increased expression for genes controlling ionic regulation? What molecular factors cause some bladder smooth muscle to lose this struggle and slide into a decompensated state with loss of force generation, a decrease in detrusor reserve, and increased residual urine? In contrast, what patterns of gene expression are seen in the patient with an end stage bladder? Will it prove possible to use the cell's state of ionic regulation in a prognostic manner to predict impending decompensation or recovery after relief of obstruction?

ACKNOWLEDGMENTS

Supported by NIH R-29 DK44144 and NIH K-12 DK02196.

REFERENCES

1. Zderic, S.A., Wein, A.J., Levin, R.M.: Voiding Function: Relevant anatomy, physiology, pharmacology, and molecular aspects. in: Gillenwater, J., Grayhack, J., Howards, S., Duckett, J. (eds). Adult and Pediatric Urology. 3rd Edition, Year Book, 1995, Chicago, pages 1159.
2. Zderic, S.A.,: Bladder smooth muscle: stimuli for growth. in Bladder Development and Function. Cheng, E.Y. (ed), Dialogues in Ped. Urology **20 (9):** 3,1997.
3. Langer, G.: Calcium and the heart: exchange at the tissue, cell, and organelle levels. FASEB Journal, **6:** 893, 1992.

4. Maclennan, D.: Purification and properties of an adenosine triphosphatase from sarcoplasmic reticulum. J. Biol. Chem. **245:** 4508, 1970.
5. Jorgenson, A.D., Arnold, W., Pepper,D.R., Kahl, S.D., Mandel, F., Campbell, K.P.: A monoclonal antibody to the Ca-ATPase of cardiac sarcoplasmic reticulum cross reacts with slow type I but not with fast type II canine skeletal muscle fibers on immunocytochemical and immunochemical study. J. Cell Motil. Cytoskel. **9:** 164, 1988.
6. Lytton, J., Zarian-Herzog, A., Periasamy, M., Maclennan, D.H.: Molecular cloning of the mammalian smooth muscle sarco(endo)plasmic reticulum $Ca^{++}Mg$++ ATPase. J. Biol. Chem. **264:** 7059, 1989.
7. Carafoli, E.: Biogenesis: Plasma membrane calcium ATPase: 15 years of work on the purified enzyme. FASEB J., **8:** 993, 1994.
8. Khan, I., Grover, A.K.: Expression of cyclic nucleotide sensitive and insensitive isoforms of the plasma membrane Ca^{2+} pump in smooth muscle and other tissues. Biochem. Journal **277:** 345,1991.
9. Ishida, Y., Riesinger, I., Wallimann, T., Paul, R.J.: Compartmentation of ATP synthesis and utilization in smooth muscle: roles of aerobic glycolysis and creatine kinase. Molecular and Cellular Biochemistry **133/134:** 39,1994.
10. Book, C.S., Moore, R.L., Semanchik, A., Ng, Y,C.: Cardiac hypertrophy alters expression of Na^+k^+ATPase subunit isoform levels at mRNA and protein levels in rat myocardium. J. Mol. Cell. Cardiol. **26:** 591,1994.
11. Madsen,K, Franch, J., Clausen, T.: Effects of intensified endurance training on the concentration of the Na^+K^+ATPase and Ca++ATPase in human skeletal muscle. Acta Physiol. Scand. **50:** 251,1994.
12. Volpe, P., Martini, A., Furlan, S., Melodesi, J.: Calsequestrin is a component of smooth muscles; the skeletal and cardiac muscle isoforms are both present, although in highly variable amounts and ratios. Biochem. Journal **201:** 465,1994.
13. Zderic, S.A., Gong, C., Sillen, U., Liu, G.H., Duckett, J.W., Snyder, H.M., Levin, R.M.: Developmental aspects of excitation contraction coupling in rabbit bladder smooth muscle. J. Urol. **152:** 679, 1994.
14. Gong, C., Zderic, S.A., Levin, R.M.: Ontogeny of the ryanodine receptor in rabbit bladder smooth muscle. Mol. & Cellular Biochem **137:** 169, 1994.
15. Nixon, G.F., Migerny, G.A., Somlyo, A.V.: Immunogold localization of inositol 1,4,5-triphosphate receptors and characterization of ultrastructural features of the sarcoplasmic reticulum in phasic and tonic smooth muscle. J. Muscle Research and Cell Motility **15:** 682,1994.
16. Franzini-Armostrong, C.: Utrastructural studies on feet/ryanodine receptors. In: Ryanodine Receptors, Sorrentino, V. (ed), CRC Press, Orlando, 1995, pages 1–6.
17. Raymaekers, L., Wuytack, F.: Calcium pumps in smooth muscle cells. J. Muscle Research and Cell Motility, **14:** 141, 1993.
18. Van Breeman, C., Chen, Q., Laher, I.: Superficial buffer barrier function of smooth muscle sarcoplasmic reticulum. Trends in Pharm. Sci. **16(3):** 98,1995.
19. Fry, C.H., Wu, C.: Initiation of contraction in detrusor smooth muscle. Scand. J. Urol. & Nephrol. **184:** 7,1997.
20. Aksoy, M.O., Mras, S., Kamm, K.E., Murphy, R.A.: Calcium, cAMP, and changes in myosin phosphorylation during contraction of smooth muscle. Am. J. Physiol. **245:** C255, 1983.
21. DiSalvo, J., Nelson, S.R., Kaplan, N.: Protein tyrosine phosphorylation in smooth muscle: a potential coupling mechanism between receptor activation and intracellular calcium. Proc. Society Exptl. Med. Biol. **214:** 285,1997.
22. Gerthoffer, W.T.: Dissociation of myosin phosphorylation and active tension during muscarinic stimulation of tracheal smooth muscle. J. Pharm. Exptl Ther. **240:** 8, 1987.
23. Gerthoffer, W.T.: Regulation of the contractile element of airway smooth muscle. Am. J. Physiol. **261:** L15, 1992.
24. Studer, R., Reinecke, H., Bilger, J., Exchenhagen, T., Bohm, M., Hassenfuss, et al: Gene expression of the cardiac Na^+Ca^{++} exchanger in end stage human heart failure. Circ. Res. **75:** 443, 1994.
25. Rabinowitz, R., Peters, M.T., Vyas, S., Campbell, S., Nicolaides, K.H.: Measurement of fetal urine production in normal pregnancy by real time ultrasonography. Am. J. Obs. & Gyn. **161:** 1262, 1989.
26. Kirsch, A.J., Quinn, M.T., Zderic, S.A., et al: Bladder and renal effects of partial urethral obstruction in the fetal lamb. J. Urol.: submitted, 1998.
27. Koo, H.P., Macarak, E.J., Zderic, S.A., Duckett, J.W., Snyder, H.M., Levin, R.M.: The ontogeny of bladder function in the fetal calf. J. Urol. **154:** 283,1995.
28. Klitzner, T.S., Friedman, W.F.: A diminished role of sarcoplasmic reticulum in newborn myocardial contraction: the effects of ryanodine. Ped. Res., **26:** 98, 1989.
29. Fisher, D., Tate,C., Phillips, S.: Developmental regulation of the sarcoplasmic reticulum calcium pump in the rabbit heart. Ped. Res., **31:** 474, 1992.

30. Zderic, S.A, Sillen, U., Liu, G., Snyder, H.M., Duckett, J.W., Wein, A.J., Levin, R.M.: Developmental aspects of bladder contractile function: evidence for an intracellular calcium pool. J. Urol. **150:** 623, 1993.
31. Thastrup, O., Dawson, A.P., Scharff, O., Foder,B, et al: Thapsigargin, a novel molecular probe for studying intracellular calcium release and storage. Agents and Actions **27:** 17, 1989.
32. Lytton, J., Westlin, M., Hanley, M.R.: Thapsigargin inhibits the sarcoplasmic or endoplasmic reticulum Ca-ATPase family of calcium pumps. J. Biol. Chem. **266:** 17067.
33. Graham, E.G., Mishra, O.P., Delivoria-Papadopoulos, M. (1993), Brain cell $Na^{+}K^{+}$ATPase modificiation following hypoxia in the guinea pig fetus. Neuroscience Letters, **153:** 93.
34. Zderic, S.A., Gong, C.: Developmental regulation of the $Ca^{2+}Mg^{2+}$ATPase of the the sarcoplasmic reticulum of rabbit urinary bladder smooth muscle. Mol. & Cell. Biochem.: submitted, 1998.
35. Khan, I., Grover, A.K.: Cloning of internal Ca pump from rabbit stomach smooth muscle. Nucleic Acids res. **18:** 4026,1990.
36. Mearow, K.M., Thilander, B.G., Khan, I., Garfield, R.E., Grover, A.K.: In situ hybridization and immunocytochemical localization of SERCA2 encoded Ca^{2+} pump in rabbit heart and stomach. Mol. & Cell. Biochm. **121:** 155,1993.
37. Zderic, S.A., Wein, A.J., Rohrnmann, D., Gong, C., Nigro, D., Haugaard, N., Levin, R.M.: Mechanisms of bladder smooth muscle hypertrophy and decompensation: lessons from normal development and the response to outlet obstruction. World J. Urol. 1998, in press.
38. Zderic, S.A., Rohrmann, D., Gong, C., Snyder, H.M., Duckett, J.W., Wein, A.J., Levin, R.M.: The decompensated detrusor II: evidence for loss of sarcoplasmic reticulum following bladder outlet obstruction in the rabbit. J. Urol. **156:** 587, 1996.
39. Sullivan, M.P., Yalla, S.V.: Detrusor contractility and compliance characteristics in adult male patients with obstructive and non obstructive voiding patterns. J. Urol. **155:** 1995, 1996.
40. Djavan, B., Madersbacher, S., Klingler, C., Marberger, M.: Urodynamic assessment of patients with acute urinary retention: is treatment failure after prostactectomy predictable? J. Urol. **158:** 1829,1997.
41. O'Connor, L.T., Vaughan, E.D., Felsen, D.: In vivo cystometric evaluation of progressive bladder outlet obstruction in rats. J. Urol. **158:** 631,1997.
42. Arai, M., Matsui, H., Periasamy, M.: Sarcoplasmic reticulum gene expression in cardiac hypertrophy and heart failure. Circ. Res. **74:** 555, 1995.
43. Komuro, I., Kurabayashi, M., Shibazaki, Y., Takaku, F., Yazaki, Y.: Molecular cloning and characterization of a Ca Mg dependent ATP ase from rat cardiac sarcoplasmic reticulum. Regulation of its expression by pressure overload and developmental stage. J. Clin. Invest. **83:** 1102,1989.
44. Feldman, A.M., Weinberg, E.O., Ray, P.E., Lorell, B.H.: Selective changes in cardiac gene expression during compensated hypertrophy and the transition to cardiac decompensation in rats with chronic aortic banding. Circ. Res. **73:** 184, 1993.
45. Briggs, F.N., Lee, K.F., Feher, J.F., Wechsler, A.S., Ohlendieck, K., Campbell, K.: Ca-ATPase isozyme expression is altered by chronic stimulation of skeletal muscles. FEBS Letters **259:** 269,1990.
46. Schulte, L.M., Navarro, J., Kandarian, S.C.: Regulation of sarcoplasmic reticulum gene expression by hind limb unweighting. Am. J. Phys. **264:** C1308, 1993.
47. Grover, A.K., Samson, S.E., Fomin, V.P., Werstiuk, E.S.: Effects of peroxide and superoxide on coronary artery: ANGII response and sarcoplasmic reticulum Ca2+ pump. Am. J. Physiol. **269:** C546,1995.
48. Yoshida, K., Inui, M., Harada,K., Saido, T.C., Sorimachi, Y., Ishihara, T., Kawashama, S., Sobue,K.: Reperfusion of rat heart after brief ischemia induces proteolysis of calspectin by calpain. Circ. Res. **77:** 603,1995.
49. Levin, R.M., Goldman, M., Wein, A.J.: Effect of isoproterenol and EGTA on the volume pressure relationship of the in vitro whole bladder preparation. Neurourology & Urodynamics **3:** 133,1984.
50. Coplen, D., Macarak, E.J., Levin, R.M.: Developmental changes in fetal bovine whole bladder physiology. J. Urol. **151:** 1391,1994.

13

CYCLOOXYGENASE-2

A Key Regulator of Bladder Prostaglandin Formation

John M. Park,[1] Jürgen B. Schnermann,[2] and Josephine P. Briggs[3]

[1]Department of Urology
Children's Hospital and Harvard Medical School
Boston, Massachusetts 02115
[2]Department of Physiology
University of Michigan Medical School
Ann Arbor, Michigan
[3]NIDDK
National Institutes of Health
Bethesda, Maryland

1. INTRODUCTION

Since Felix Hoffman first synthesized acetylsalicylic acid (aspirin) in 1897, prostaglandin synthesis inhibitors have been widely used as effective anti-inflammatory, analgesic and anti-pyretic agents. Much of our present understanding regarding prostaglandins (PGs) in fact derives from observing the effects of aspirin and non-steroidal anti-inflammatory drugs (NSAIDs) which inhibit PG synthesis by interfering with one of the rate-limiting enzymes, cyclooxygenase. In addition to regulating immune responses, pain, and fever, PGs are now implicated as playing an important role as autocrine/paracrine modulators of cellular responses such as mitogenesis and apoptosis.

Much of the early investigative efforts has focused on studying the role of PGs in reproductive, respiratory, vascular and gastrointestinal systems. A surge of interest about the significance of PGs in the lower urinary tract function can be traced back to early 1970s, after the report of observation that bladder distention released a PG-like substance into circulation.[1] Subsequent studies, both clinical and experimental, established that bladder was an independent site of PG synthesis.[2] There is evidence to suggest that PGs might be involved in the regulation of bladder tone and spontaneous contractile activity.[3] These findings led to a hypothesis that locally generated PGs may serve as an endogenous regulator of bladder's storage and micturition function.[4] It was also postulated that urinary PGs

Advances in Bladder Research, edited by Baskin and Hayward.
Kluwer Academic / Plenum Publishers, New York, 1999.

might play a role in urinary epithelial cytoprotection, analogous to their established roles in gastric mucosa.[2]

Recently, there has been a wave of new investigation exploring the regulatory mechanism of cyclooxygenase (also known as prostaglandin endoperoxide synthase). Until several years ago, only one isoform of cyclooxygenase had been known. It is now well established that cyclooxygenase (COX) exists in two differentially regulated isoforms.[5] This discovery has stimulated an interest in studying the regulatory mechanisms of PG synthesis via COX isoforms in diverse biological situations such as inflammation, development, and carcinogenesis. In this brief review, we summarize the current information regarding bladder PGs and explore the potential impact of recent development regarding COX isoforms on understanding the role of bladder PG regulation.

2. THE BLADDER AS A SITE OF LOCAL PROSTAGLANDIN SYNTHESIS

The first evidence that bladder could synthesize its own PGs was provided by Gilmore and Vane in 1971. They reported that when the bladder was distended with fluid in anesthetized dogs, the level of circulating PGE_2 increased. Since then, other experimental observations followed, establishing that bladder increases its local PG production in response to various stimuli including distention, pelvic nerve stimulation, mechanical mucosal trauma , increased urine osmolarity, cholinergic receptor stimulation, inflammatory cytokinesand bacterial endotoxin.[6–8] The type of PGs produced by the bladder appears to vary with species, but PGE_2 and PGI_2 appear to be the predominant ones in most mammalian bladders.[2] There is a quantitative difference in PG production between different regions of the bladder. In rabbit, for example, the body of the bladder appears to synthesize a greater amount of PGE_2 than the base.[9] There is evidence that both smooth muscle and uroepithelial cells are capable of synthesizing PGs.[8,10]

3. THE FUNCTION OF BLADDER PROSTAGLANDINS

Early experiments documented that addition of exogenous PGs produced a slow, tonic contraction of the bladder muscle strips in vitro.[3] Pre-treatment of muscle strips with PG synthesis inhibitors (such as indomethacin) markedly diminished the tone and contractility, whereas addition of PGs to the bladder wall strips that had been previously treated with indomethacin quickly restored the tone and contractility.[3] This effect of PG synthesis inhibitors could be reproduced by selective PG-receptor antagonists.[11] Topical application of PGE_2 onto the dome of partially filled rat bladders in vivo produced strong bladder contractions.[12] In a small number of trials involving human subjects, intravesical instillation of PGE_2 or $PGF_{2\alpha}$ appeared to lower the volume threshold for the appearance of uninhibited bladder contractions, thereby reducing the overall capacity.[3,13] Slow-filling cystometrogram in urethane-anesthetized rats that had been pre-treated with indomethacin and ketoprofen demonstrated an increase in bladder capacity and compliance.[14] Based on these findings, it was postulated that locally produced bladder PGs may be an endogenous regulator of smooth muscle tone and contractility.

In addition, there is compelling evidence to suggest that PGs may directly modulate bladder's storage and micturition function by affecting afferent signals to the central nervous system.[4] Urinary PGs appear to activate capsaicin-sensitive C-fibers in vivo.[4,12] At the somatic level, there is ample evidence that PGs can directly sensitize and stimulate capsaicin-sensitive afferent nerves.[15] PGs' ability to trigger bladder contractions in vivo is abolished when the animals are pre-treated with a systemic dose of capsaicin, a neurotoxic agent that selectively destroys C-fibers.[16] In human studies, the pre-requisite requirement for the previously mentioned in vivo PG effects on the bladder seems to be the presence of intact neural pathways to the central nervous system, further suggesting that PGs are involved in the activation of neural target cells.[13]

PGs' role in gastric mucosal cytoprotection has been well established.[17] There are several clinical and experimental observations that urinary PGs might also play a similar cytoprotective role for bladder uroepithelium.[2] Intravesical instillation of PGE_2 and $PGF_{2\alpha}$ have been shown to provide beneficial effects in cystitis of various etiology including chemical agent (e.g., cyclophosphamide), infection (e.g., schistosomiasis), and radiation.[2,18,19] The mechanism of mucosal cytoprotection by PGs is not fully understood, but cAMP mediated increase in glycosaminoglycan production may be involved.[20]

4. THE BIOLOGY OF CYCLOOXYGENASE ISOFORMS

In PG biosynthesis, the committed, rate-limiting enzymatic step occurs at cyclooxygenase (COX), which catalyzes the conversion of arachidonic acid into the unstable intermediate precursor, prostaglandin endoperoxide (PGH_2). The final PGs produced depend on the type of synthase present in different cell types (Figure 1). Before 1989, only one isoform of cycloxygenase had been known, the isozyme now called COX-1 (for cyclooxygenase-1) or PGHS-1 (for prostaglandin endoperoxide H synthase-1). Discovery has been made of another COX isoform,[21] which is now referred as COX-2 or PGHS-2. Subsequent studies have demonstrated that although COX-2 is similar to COX-1 in catalytic site structure and pharmacokinetics, it differs substantially from COX-1 with respect to its pattern of expression, regulatory mechanism and biological function.[5] COX-1 is often called the 'constitutive' isoform. Its expression is seen in most tissues at basal states and has been associated with regulation of various "house-keeping" functions such as platelet aggregation and gastric mucosal cytoprotection. Its expression level does not appear to change significantly except during development. COX-2, in contrast, has been termed the 'inducible' isoform. It is normally undetectable in most mammalian tissues (except for brain, spleen, testes and renal macula densa) but is rapidly induced by serum, growth factors, cytokines, tumor promoters, and bacterial endotoxin. COX-2 expression can be inhibited by anti-inflammatory steroids such as dexamethasone. It has been associated with regulation of inflammation, fever, pain, tissue proliferation, and carcinogenesis. COX-1 and COX-2 are encoded by separate genes located on different chromosomes.[22,23] Although their catalytic efficiency in terms of cyclooxygenase activity is similar, enough differences exist in their enzymatic regulatory sites, allowing differential sensitivity to various NSAIDs. All current NSAIDs in general clinical use inhibit both COX-1 and COX-2. Several COX-2 specific inhibitors have been recently developed and remain the focus of active investigation.[24] Experimental studies suggest that COX-2 specific inhibitors may result in improved anti-inflammatory effects and reduced side effects such as gastric ulcers and bleeding, thought to be mediated by COX-1.[25]

5. COX ISOFORM REGULATION IN BLADDER

Given the significance of biological roles bladder PGs might play and the new information regarding differential regulation of PG synthesis by COX-1 and COX-2, we inves-

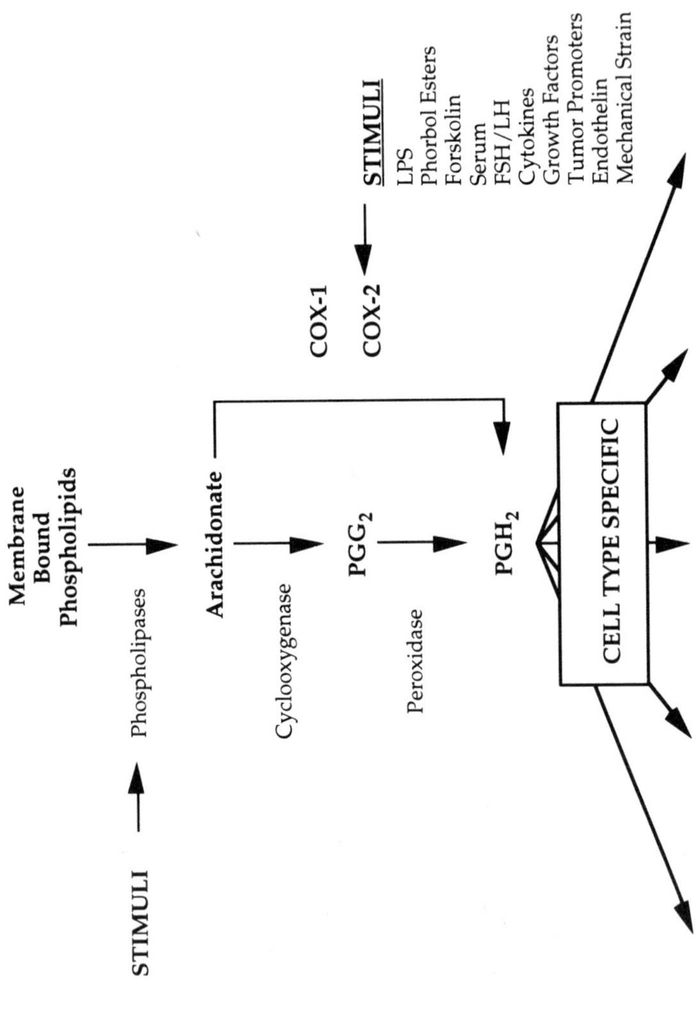

Figure 1. Arachidonate metabolism via cyclooxygenase (COX) pathway. LPS, lipopolysaccharide; FSH, follicle stimulating hormone; LH, luteinizing hormone.

tigated the regulatory pattern of COX isoforms in bladder during fetal development and after acute outlet obstruction.[26]

Bladder progenitor tissues were obtained by microscopic dissection from developing murine embryos between gestational days 11.5 and newborn. Using semi-quantitative reverse transcription–polymerase chain reaction (RT-PCR) assay with normalization to β-actin expression, COX-1 and COX-2 levels were assessed. COX-1 was uniformly undetectable in bladder during fetal development until the time of birth. COX-2, in contrast, was highly expressed during the early gestation (11.5 days), and its expression level declined progressively at subsequent gestational time points, reaching the low basal level seen in adult bladders by birth (Figure 2). This pattern of COX-2 expression suggests an important developmental role. The gestational time point at which the COX-2 level was maximal corresponds to the beginning of bladder formation, wherein cloaca becomes divided into urogenital sinus anteriorly and rectum posteriorly by urorectal septum. At this stage, although the urogenital sinus cavity (the future bladder) is morphologically distinguishable, the epithelium and the surrounding mesenchyme have not yet differentiated into transitional epithelium and smooth muscle cells. Mice lacking COX-2 have been produced by homologous recombination.[27,28] These mice were not reported to demonstrate obvious bladder abnormality, suggesting COX-2 is not essential for bladder development. These observations do not completely exclude a functional role of COX-2 in bladder formation, however, mainly due to biological redundancy with alternate biochemical factors. Interestingly, one of the consistent anomalies observed in these mice lacking COX-2 was the increased incidence of post-natal renal failure, the cause of which is unknown.

We also studied the pattern of COX isoform expression in murine bladders after complete bladder outlet obstruction. Under ketamine and xylozine anesthesia, bladders of adult female CD-1 mice were completely obstructed by ligating their urethra, and bladder distention was allowed to occur with animals' own urine for variable duration up to 24 hours. Bladder COX-1 level was low at basal states and remained unchanged after obstruction. COX-2, in contrast, was rapidly induced in response to outlet obstruction. (Figures 3 and 4) COX-2 induction was confirmed at both mRNA and protein levels. By

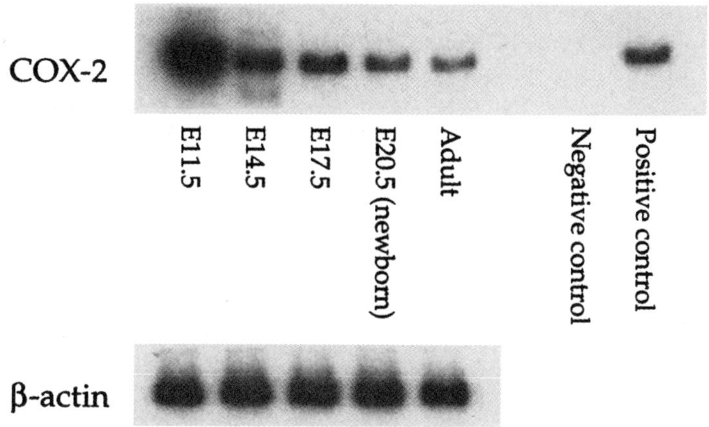

Figure 2. COX-2 expression in bladder during fetal development. Bladder progenitor tissues were obtained by micro-dissection from murine embryos at various gestational time points indicated and were assessed for COX-2 mRNA levels by semi-quantitative RT-PCR analysis. Corresponding β-actin levels are shown. cDNA prepared from murine kidneys served as positive control, and RT(-) reaction was used as negative control.

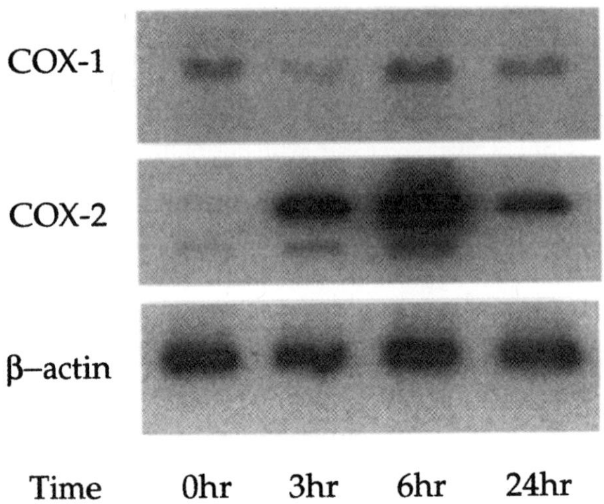

Figure 3. COX-1 and COX-2 isoform expression after varying duration of bladder outlet obstruction (RT-PCR) with corresponding β-actin levels.

semi-quantitative RT-PCR with normalization to β-actin, COX-2 mRNA levels increased rapidly, as early as after 3 hours of obstruction, reached maximal levels after 6 hours, and decreased progressively thereafter. This temporal pattern of rapid gene activation and subsequent down-regulation is consistent with that of immediate-early gene family. These genes are activated rapidly (usually within hours) and transiently in response to stimuli without requiring other protein synthesis, and are thought to play an important role in the regulation of other gene transcription.[29]

In order to localize the site of COX-2 activation, semi-quantitative RT-PCR assays were performed in microscopically dissected uroepithelial and stromal (predominantly smooth muscle cells) layers (Figure 5). We found that COX-2 activation occurred specifically in the stromal layer after outlet obstruction and suggested that the most likely source was the smooth muscle cells. This finding is consistent with the pattern of COX-2 expression seen in other organs composed of smooth muscle cells. COX-2 was found to be increased in rat aortic smooth muscle cells after balloon distention in vivo and serum stimulation in vitro.[30] When exposed to cyclical stretch/relaxation in the absence of serum or other hormonal factors, COX-2 expression increased in renal mesangial cells, also thought to be derived from smooth muscle cell lineage.[31]

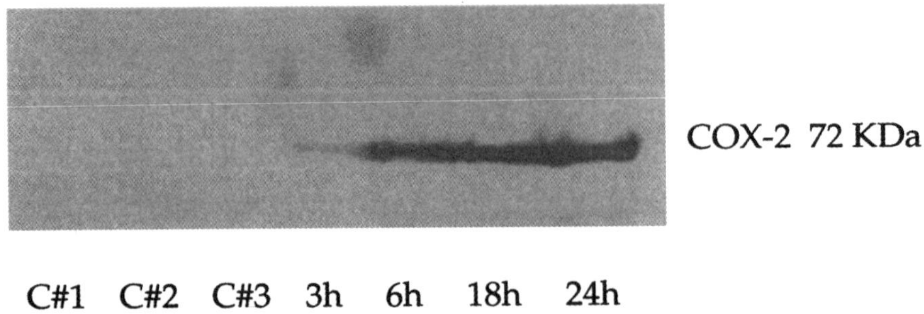

Figure 4. COX-2 immunoblot assay after varying duration of bladder outlet obstruction. C#1–3 represent three independent samples of non-obstructed control bladder tissues.

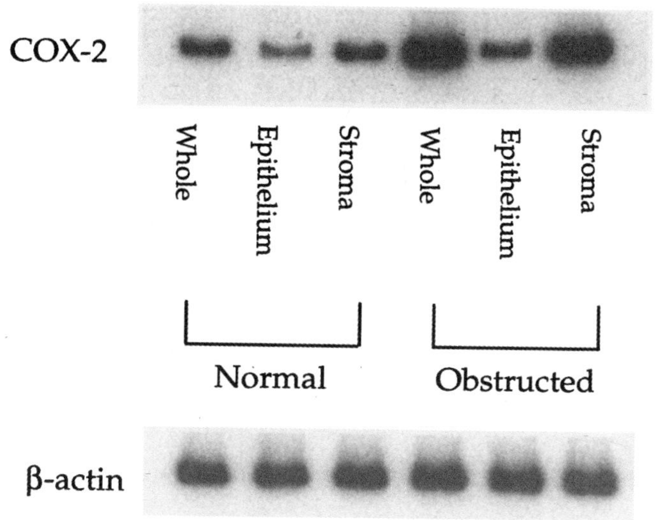

Figure 5. Localization of COX-2 activation after 6-hour bladder outlet obstruction. COX-2 mRNA levels were assessed in micro-dissected epithelial and stromal (predominantly muscle) layers by semi-quantitative RT-PCR.

We proposed a hypothesis that local mechanical stretch was the primary stimulus for COX-2 induction in bladder outlet obstruction. To test this, we used an in vivo model of regional bladder wall distention.[32] After performing a complete urethral obstruction, the lower half of the bladder was encased in a polyethylene tube (4 mm width, 8 mm length) to limit the distention and subsequent wall stretch to the upper half. (Figure 6, top) When regional COX-2 levels were assessed by RT-PCR, COX-2 activation was confined to the distended upper half of the bladder (Figure 6, bottom). This model separates the effect of local wall stretch from that of intravesical pressure elevation. Since upper and lower halves shared the common lumen within the same bladder, the effect of intraluminal pressure should be the same for both, while the degree of local stretch varied. Based on this finding, we postulated that in bladder outlet obstruction, the primary stimulus for changes in gene expression was increase in local cellular stretch rather than luminal pressure elevation. This hypothesis is consistent with the bladder's unique ability to store a wide range of fluid volumes at low pressures. Although elevated storage pressure is commonly associated with chronic obstruction, this is a late event and a result of end-stage decompensation. In terms of early changes in gene expression, increase in local mechanical stretch may be the initial stimulus. Establishing mechanical stretch as the primary stimulus is important in that it provides a framework to further explore the specific mechanism of COX-2 gene activation in bladder smooth muscle cells.

6. BLADDER COX-2: A NEW THERAPEUTIC TARGET?

It was hypothesized early on that various syndromes of voiding dysfunction may be treated by modulating local PG levels. Several studies suggested beneficial effects of intravesical instillation of PGs in patients with urinary retention of various etiology.[33] A small number of clinical trials also investigated the potential therapeutic efficacy of COX inhibitors such as indomethacin in patients with detrusor instability and painful bladder syndromes.[13] Although these studies did not provide conclusive evidence for a true improvement due to the drug tested, the majority of patients reported a significant relief of their symptoms. It became evident, however, that these agents were ill suited for long-term use, primarily due to high incidence of gastric side effects.

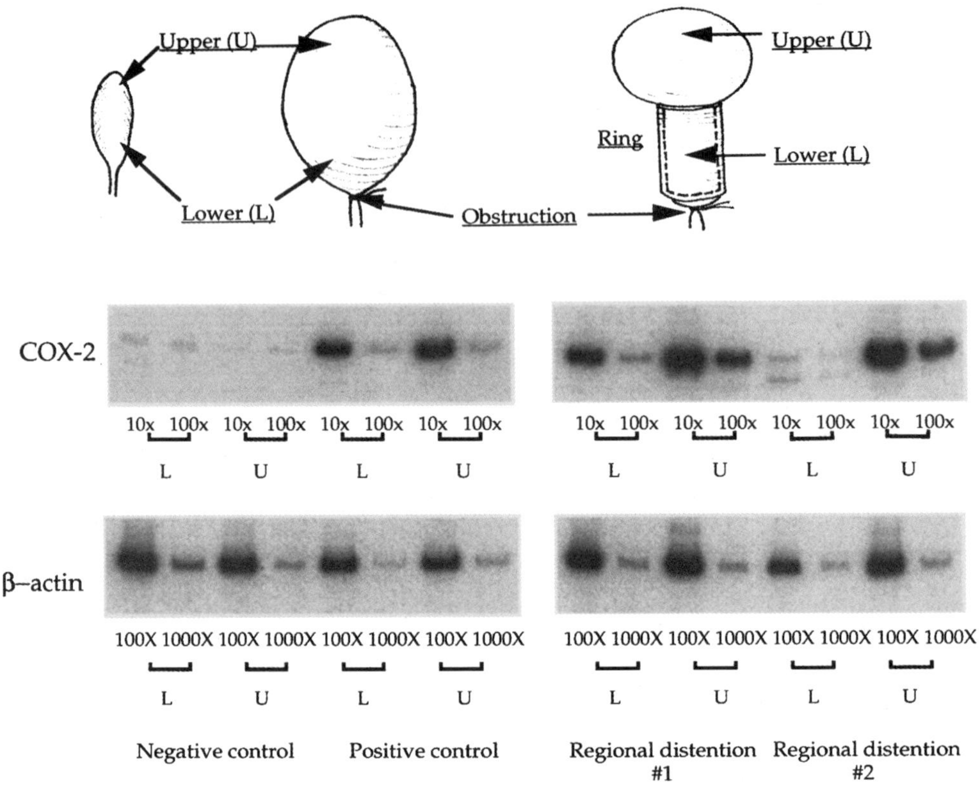

Figure 6. (Top) Regional bladder distention model. (Bottom) Regional COX-2 mRNA levels were assessed between distended upper half and non-distended lower half within the same bladder by semi-quantitative RT-PCR. To account for any endogenous regional variation, similar analysis was performed in non-obstructed (negative control) and globally distended (positive control) bladders. RT-PCR was performed at two log-scale dilutions to demonstrate appropriate quantitative comparisons between samples.

The fact that PG biosynthesis occurs by two distinctly regulated isoenzymes provides a new opportunity to investigate the utility of COX inhibitors in the treatment of bladder dysfunction. Most of the current NSAIDs are non-specific COX inhibitors; that is, they block both COX-1 and COX-2.[24] As mentioned above, COX-1 inhibition leads to decreased PG production in gastric mucosa, and is thought to result in a higher incidence of peptic ulcer disease. Several COX-2 specific inhibitors have now been developed, and at least in experimental settings, these agents demonstrate a potent anti-inflammatory property without gastric side effects.[34] It is conceivable that COX-2 specific inhibitors may in fact be a "better-and-improved aspirin" in terms of therapeutic efficacy versus side effects ratio. Based on our recent experimental findings regarding COX-2 regulation in bladder, we believe that COX-2 specific inhibitors may also be useful in the treatment of bladder dysfunction. In conditions of outlet obstruction such as benign prostatic hyperplasia and posterior urethral valves, increased COX-2 expression may be responsible for elevated local PG levels, which in turn may activate detrusor instability by affecting smooth muscle tone and capsaicin-sensitive C-fiber mediated afferent signaling to the central nervous system. If this hypothesis is correct, COX-2 inhibitors might be able to reduce detrusor instability with minimal side effects, making them suitable for long-term use.

Much remains to be learned about the biological significance of COX isoforms. Most cells express COX-1 at basal states, yet they have the capacity to activate COX-2 if given a proper stimulus. What might be the reason for the presence of both COX-1 and -2 isoforms? Accumulating evidence suggests that COX isoforms regulate at least partially independent PG biosynthetic systems.[5,35] COX-1 proteins are localized on the luminal aspect of endoplasmic reticulum (ER) and are thought to regulate the production of extracellular PGs that function as local autocrine/paracrine hormones. A subpopulation of COX-2 proteins are also localized within ER lumen and may serve to augment the function of COX-1 by increasing secretory PG production. We believe that this mechanism is probably responsible for elevated urinary PG levels after bladder outlet obstruction. Another subpopulation of COX-2 proteins are localized on the luminal surface of the nuclear envelope, and their PG products may possess nuclear activity, regulating processes such as cell differentiation and replication.[36] COX-2 expression is induced by mitogens and tumor promoters, and has been shown to be activated in the setting tissue growth, repair and carcinogenesis. Constitutive over-expression of COX-2 has been well documented in colonic adenocarcinoma,[37,38] and it appears that elevated COX-2 expression is closely associated with these cancer cells having an increased resistance to apoptosis[39] as well as a greater maligant potential.[40] COX-2 products may also participate in the regulation of fibroblast proliferation and tissue fibrosis.[41] It remains to be seen whether COX-2 activation might play a role in modulation of local tissue remodeling and cell proliferation after bladder outlet obstruction.

REFERENCES

1. Gilmore, N. J., and J. R. Vane.: Hormones released into the circulation when the urinary bladder of the anesthetized dog is distended. Clin. Sci., **41:** 69, 1971.
2. Mikhailidis, D. P., J. Y. Jeremy, and P. Dandona.: Urinary bladder prostanoids - their synthesis, function and possible role in the pathogenesis and treatment of disease. J. Urol., **137:** 577. 1987.
3. Bultitude, M. I., N. H. Hills, and K. E. L. Shuttleworth.: Clinical and experimental studies on the action of prostaglandins and their synthesis inhibitors on detrusor muscle in vivo and in vitro. Br. J. Urol., **48:** 631, 1976.
4. Maggi, C. A.: Prostanoids as local modulators of reflex micturition. Pharmacol. Res., **25:** 13, 1992.
5. Smith, W. L., R. M. Garavito, and D. L. DeWitt.: Prostaglandin endoperoxide H synthases (cyclooxygenases) -1 and -2. J. Biol. Chem., **271:** 33157, 1996.
6. Giuliani, S., P. Santicioli, M. Tramontana, P. Geppetti, and C. A. Maggi.: Peptide N-formyl-methionyl-leucyl-phenylalanine (FMLP) activates capsaicin-sensitive afferents in guinea pig atria and urinary bladder. Br. J. Pharmacol., **102:** 730, 1991.
7. Khalaf, I. M., J. G. Lehoux, L. A. Elshawarby, and M. M. Elhilali.: Release of prostaglandins into the pelvic venous blood of dogs in response to vesical distention and pelvic nerve stimulation. Invest. Urol., **17:** 244, 1979.
8. Downie, J. W., and M. Karmazyn.: Mechanical trauma to bladder epithelium liberates prostanoids which modulate neurotransmission in rabbit detrusor muscle. J Pharmacol Exp. Ther., **230:** 445, 1984.
9. Leslie, C. A., A. J. Pavlakis, J. S. Wheeler, M. B. Siroky, and R. J. Krane.: Release of arachidonate cascade products by the rat urinary bladder: neurophysiological significance? J. Urol., **132:** 376, 1984.
10. Brown, W. W., Z. T. V., and B. B. Davis.: Prostaglandin E2 production by rabbit urinary bladder. Am. J. Physiol., **239:** F452, 1980.
11. Maggi, C. A., S. Giuliani, R. Patacchini, B. Conte, M. Furio, P. Santicioli, P. Meli, L. Gragnani, and A. Meli.: The effect of SC-19220, a prostaglandin antagonist, on the micturition reflex in rats. Eur. J. Pharmacol., **152:** 273, 1988.
12. Maggi, C. A., S. Giuliani, B. Conte, M. Furio, P. Santicioli, P. Meli, L. Gragnani, and A. Meli.: Prostanoids modulate reflex micturition by acting through capsaicin-sensitive afferents. Eur. J. Pharmacol., **145:** 105, 1988.

13. Cardozo, L. D., and S. L. Stanton.: Comparison between bromocriptine and indomethacin in the treatment of detrusor instability. J. Urol., **123:** 399, 1980.
14. Maggi, C. A., P. Santicioli, M. Furio, and A. Meli.: The effect of cyclooxygenase inhibitors on the low filling rate cystometrogram in urethane anesthetized rats. J. Urol., **134 :** 800, 1985.
15. Gold, M. S., D. B. Reichling, M. J. Shuster, and J. D. Levine.: Hyperalgesic agents increase a tetrodotoxin-resistant Na+ current in nociceptors. Proc. Natl. Acad. Sci. USA., **93:** 1108, 1996.
16. Maggi, C. A., F. Santicioli, and A. Meli.: Evidence for the involvement of a capsaicin-sensitive innervation in the afferent branch of micturition reflex in rats. Acta Physiologica Hungarica., **69:** 425, 1987.
17. Wallace, J. L., and D. N. Granger.: The cellular and molecular basis of gastric mucosal defense. F.A.S.E.B. J., **10:** 731, 1996.
18. Mohiuddin, J., H. G. Prentice, S. Schey, H. Blacklock, and P. Dandona.: Successful treatment of cyclophosphamide-induced hemorrhagic cystitis with intravesical infusion of prostaglandin E2. Ann. Int. Med., **101:** 142, 1984.
19. El-Gendi, M. A., S. H. Nassar, M. D. Toppozada, and F. Abdel-Raheem.: Pharmacotherapeutics of prostaglandin E2 and 15(S)-15-methyl prostaglandin F2a in chronic schistosomal bladder ulcer: a clinicoendoscopic study. Prostaglandins, **24:** 97, 1982.
20. Green, R. M., V. I. MacAndrew, and M. R. Lloyd.: Stimulation of glycosaminoglycan synthesis by cyclic AMP. Biochem. Biophys. Res. Commun., **107:** 232, 1982.
21. Simmons, D. L., D. B. Levy, Y. Yannoni, and R. L. Erikson.: Identification of a phorbol ester repressible v-src-inducible gene. Proc. Natl. Acad. Sci. USA. ,**86:** 1178, 1989.
22. Yokoyama, C., and T. Tanabe.: Cloning of human gene encoding prostaglandin endoperoxide synthase and primary structure of the enzyme. Biochem. Biophys. Res. Commun., **165:** 888, 1989.
23. Kosaka, T., A. Miyata, H. Ihara, H. S., T. Sugimoto, O. Takeda, E. Takahashi, and T. Tanabe.: Characterization of the human gene (PTGS2) encoding prostaglandin endoperoxide synthase 2. Eur. J. Biochem., **221:** 889, 1994.
24. Smith, W. L.: Pharmacology of prostaglandin endoperoxide synthase isoenzyme 1 and 2. Ann. N.Y. Acad. Sci., **714:** 136, 1994.
25. Seibert, K., J. L. Masferrer, P. Needleman, and D. Salvemini.: Pharmacological manipulation of cyclo-oxygenase-2 in the inflamed hydronephrotic kidney. Br. J. Pharmacol., **117:** 1016, 1996.
26. Park, J. M., T. Yang, L. J. Arend, S. A. M., J. B. Schnermann, and J. B. Briggs.: Cyclooxygenase 2 is expressed in bladder during fetal development and stimulated by outlet obstruction. Am. J. Physiol., **273:** F538, 1997.
27. Morham, S. G., R. Langenbach, C. D. Loftin, H. F. Tiano, N. Vouloumanos, J. C. Jennette, J. F. Mahler, K. D. Kluckman, A. Ledford, C. A. Lee, and O. Simithies.: Prostaglandin synthase 2 gene disruption causes severe renal pathology in the mouse. Cell, **83:** 473, 1995.
28. Dinchuk, J. E., B. D. Car, R. J. Focht, J. J. Johnston, B. D. Jaffe, M. B. Covington, N. R. Contel, V. M. Eng, R. J. Collins, P. M. Czerniak, S. A. Gorry, and J. M. Trzaskos.: Renal abnormalities and an altered inflammatory response in mice lacking cyclooxygenase II. Nature, **378:** 406, 1995.
29. Herschman, H. R.: Primary response genes induced by growth factors and tumor promoters. Annu. Rev. Biochem., **60:** 281, 1991.
30. Pritchard, K. A., M. K. O'Bannon, J. M. Miano, N. Vlasic, U. G. Bhatia, D. A. Young, and M. B. Stemerman.: Induction of cyclooxygenase-2 in rat vascular smooth muscle cells in vitro and in vivo. J. Biol. Chem., **269:** 8504, 1994.
31. Akai, Y., T. Homma, K. D. Burns, T. Yasuda, K. F. Badr, and R. C. Harris.: Mechanical stretch/relaxation of cultured rat mesangial cells induces protooncogenes and cyclooxygenase. Am. J. Physiol., **267:** C482, 1994.
32. Yamamoto, M., S. C. Harm, W. A. Grasser, and M. A. Thiede.: Parathyroid hormone related protein in the rat urinary bladder: a smooth muscle relaxant produced locally in response to mechanical stretch. Proc. Natl. Acad. Sci. USA.,**89:** 5326, 1992.
33. Pavlakis, A. J., M. B. Siroky, C. A. Leslie, and R. J. Krane.: Prostaglandins in lower urinary tract. Neurourol. Urodyn., **2:** 105, 1983.
34. Seibert, K., J. Masferrer, Y. Zhang, K. Leahy, S. Hauser, J. Gierse, C. Koboldt, G. Anderson, M. Bremer, S. Gregory, and et al.: Expression and selective inhibition of constitutive and inducible forms of cyclooxygenase. Adv. Prostagl. Thromb. Leukotr. Res., **23:** 125, 1995.
35. Williams, C. S., and R. N. DuBois.: Prostaglandin endoperoxide synthase: why two isoforms? Am. J. Physiol., **270:** G393, 1996.
36. Morita, I., M. Schindler, M. K. Regier, J. C. Otto, T. Hori, D. L. DeWitt, and W. L. Smith.: Different intracellular locations for prostaglandin endoperoxide H synthase-1 and -2. J. Biol. Chem., **270:** 10902, 1995.

37. Kargman, S., G. O'Neill, P. Vickers, J. Evans, J. Mancini, and S. Jothy.: Expression of prostaglandin G/H synthase-1 and -2 protein in human colon cancer. Cancer Res., **55:** 2556, 1995.
38. Eberhart, C. B., R. J. Coffey, A. Radhika, F. M. Giardiello, S. Ferrenbach, and R. N. DuBois.: Up-regulation of cyclooxygenase-2 gene expression in human colorectal adenomas and adenocarcinomas. Gastroenterology, **107:** 1183, 1994.
39. Tsujii, M., and R. N. DuBois.: Alteration in cellular adhesion and apoptosis in epithelial cells overexpressing prostaglandin endoperoxide synthase 2. Cell, **83:** 493, 1995.
40. Tsujii, M., S. Kawano, and R. N. DuBois.: Cyclooxygenase-2 expression in human colon cancer cells increases metastatic potential. Proc. Natl. Acad. Sci. USA., **94:** 3336, 1997.
41. Wilborn, J., L. J. Crofford, M. D. Burdick, S. L. Kunkel, R. M. Strieter, and M. Peters-Golden.: Cultured lung fibroblasts isolated from patients with idiopathic pulmonary fibrosis have diminished capacity to synthesize prostaglandin E2 and to express cyclooxygenase-2. J. Clin. Invest., **95:** 1861, 1995.

14

ROLE OF ANGIOTENSIN II IN BLADDER SMOOTH MUSCLE GROWTH AND FUNCTION

Earl Y. Cheng, Robert S. Decker, and Chung Lee

Division of Urology, Children's Memorial Hospital
Departments of Urology and Medicine
Northwestern University Medical School
Chicago, Illinois 60611

1. INTRODUCTION

Outlet obstruction of the developing urinary bladder may result in pathologic changes in the bladder wall that can adversely affect lower urinary tract dynamics and worsen compliance. When obstruction is not recognized and persists, upper tract deterioration and renal function compromise may ensue. This process occurs in several pediatric urologic conditions such as posterior urethral valves or urethral atresia. Bladder neck obstruction may also be neurogenic in origin as is the case in many children with myelomeningocele (spina bifida). The exact cellular mechanisms and pathways that regulate obstructive changes in the developing bladder have yet to be elucidated. Viewing this process simplistically, physiologic impairment of urinary flow likely causes increased bladder wall stress from retained urine and over distension of bladder muscle. This increased mechanical stress then causes a cascade of cellular events which results in secondary changes in bladder wall architecture. The most prominent histopathologic findings in the obstructed bladder are muscular hypertrophy/hyperplasia and increased extracellular matrix protein deposition.[1–3] Resultant changes in muscle and collagen content, alterations in cellular architecture, and changes in bladder compliance have adverse effects on the bladder's ability to function as a low pressure reservoir for urinary storage. Also, bladder wall compliance changes may in turn produce further problems with increased wall stress that can then lead to a viscous cycle of events which can result in a small non functional bladder in need of surgical reconstruction. The specific factors which transmit the signal of increased wall stress to induce biologic remodeling of the obstructed bladder are poorly understood.[4] Even less is known about the cellular and physiologic determinants of reversibility of these obstructive changes. It is clear that if we are to improve upon our current clinical management of patients with a neurogenic bladder that has resulted from obstruc-

Advances in Bladder Research, edited by Baskin and Hayward.
Kluwer Academic / Plenum Publishers, New York, 1999.

tion, we need to have a better understanding of the cellular factors which regulate bladder smooth muscle growth and collagen production under both normal and pathologic conditions.

Angiotensin II is an octapeptide which has been well described with regard to its systemic actions in regulating vascular tone, blood volume, and sodium homeostasis via the systemic renin angiotensin system. Recent work in the cardiovascular system has demonstrated that angiotensin II is also a locally produced cellular factor in the wall of the heart that is active in the modulation of smooth muscle growth and collagen production following pathologic insult.[5] Additionally, angiotensin II has been found to cause the up regulation of several secondary growth factors (e.g. transforming growth factor beta, platelet derived growth factor, basic fibroblast growth factor, etc.) which are well described with regard to their autocrine and paracrine effects on cell proliferation and function.[6–11] Production of angiotensin II which acts in a "growth factor like" fashion is hypothesized to come from a local renin–angiotensin system in the wall of the heart. Similar local renin–angiotensin systems have also been identified in the brain and kidney.[12] Interestingly, the heart responds to increased load in a similar fashion as does the bladder. As is the case in the bladder, muscle hypertrophy and excess production of fibrillar collagens are histologic hallmarks of the obstructed myocardium. Numerous in vivo and in vitro studies have now provided an abundance of evidence that indicates angiotensin II is a key regulator of these obstructive changes on a cellular level. In cultured cardiac cells, angiotensin II administration induces a hyperplastic and/or hypertrophic response,[13] as well as the excess production of collagen.[14] In addition, pharmacologic inhibition of angiotensin II production lessens obstructive changes in the heart and has reparative effects on already damaged myocardium.[15,16] Given the similarities that exist between the heart and the bladder in their response to increased load, it is tempting to speculate that angiotensin II may have a similar regulatory role in muscle growth and collagen production in both organs.

Current knowledge regarding the specific physiologic effects of angiotensin II on bladder function is lacking. Available information thus far has demonstrated that angiotensin II has direct effects on bladder muscle tone and contraction. Several studies have shown that angiotensin II causes contraction of human, canine, and rabbit bladder muscle in-vitro[17–20] which is mediated by the type 1- subtype (AT-1) receptor.[21] The significance of these angiotensin II induced contractions under normal conditions is not known although the contractile response is altered in the human neurogenic bladder.[22] If angiotensin II is indeed important in bladder function, it is important to know where it comes from and what regulates its production. Angiotensin II is certainly available to most organs in the body via its presence in the blood. Levels of angiotensin II in the blood are regulated by the systemic renin angiotensin system, which is in turn dependent on several factors including blood pressure and vascular tone. However, as mentioned earlier, angiotensin II may also be derived from local renin angiotensin systems which act on a cellular level and have their own separate and independent regulation. Preliminary evidence now supports that certain elements of a local renin angiotensin system exist in the bladder. Sullivan et al.,[23] utilizing bladder tissue from fetal sheep, have shown both immunohistochemical and functional evidence of locally produced angiotensin II in the bladder wall. Further investigation is certainly necessary to unequivocally document the existence of a local renin angiotensin system in the bladder. Nevertheless, the functional significance of locally produced angiotensin II in the bladder requires further study.

In our laboratory, we are presently pursuing the hypothesis that angiotensin II is a trophic factor in the wall of the bladder that is in part responsible for the regulation of smooth muscle growth and collagen production under both normal and pathologic states. To date, using animal models of bladder neck obstruction, we have looked at the in vivo effects of pharmacologic agents which inhibit angiotensin II production and/or block angiotensin II receptors on the obstructive changes in the bladder. Additionally, we have also utilized in vitro cell culture techniques to investigate the effects of angiotensin II and other related growth factors on primary cultured bladder stromal cells.

2. IN VIVO ANIMAL STUDIES

The effects of captopril (angiotensin converting enzyme inhibitor) on the obstructed neonatal rabbit bladder were determined.[24] Angiotensin II production is dependent on its conversion from angiotensin I by angiotensin converting enzyme. Thus, the amount of bioavailable angiotensin II and its subsequent cellular effects are decreased by captopril. Partial bladder neck obstruction (BNO) was induced in a group of 2 day old rabbits (n= 8) by placing a loose 2-0 silk ligature around the vesicourethral junction. A second group of rabbits subjected to the identical partial BNO procedure (n=8) was concomitantly administered captopril (1 mg/kg/day) via a subcutaneously placed osmotic pump. 12 days post-obstruction, bladders from the above animals along with paired controls (n=8) were harvested for routine histology and assayed for total protein, DNA, and collagen content. Examination of bladder histology revealed that rabbits with partial BNO which did not receive captopril had obstructive changes in the bladder which are similar to those described in the adult rabbit; namely a 2–3 fold increase in wet weight of the bladder, muscular hypertrophy in the muscularis propria, prominent smooth muscle cell proliferation in the serosa (serosal hyperplasia), and increased transmural deposition of collagen.[25,26] In contrast, obstructed rabbits which were administered captopril had reduction in all of these parameters. Most notably there was a marked reduction of the serosal hyperplasia and the amount of interfascicular and intrafascicular collagen (see Figure 1). These histologic findings were also consistent with biochemical studies which showed a significant reduction in the amount of total DNA and decrease in the total collagen in captopril treated animals. These observations suggest that angiotensin II is intimately involved in the regulation of smooth muscle proliferation and collagen production in the obstructed neonatal rabbit bladder. Despite this, it is important to note that captopril did not completely abolish the obstructive changes in the rabbit bladder since histologic and biochemical evidence of muscle hypertrophy was still present in the captopril treated animals. Many possible reasons exist to explain this observation. First and foremost, it is likely that multiple cellular factors are involved in the regulation of bladder smooth muscle cell growth and collagen deposition. Angiotensin II may only be one of these many regulatory factors. Also, it has been shown that angiotensin II conversion from angiotensin I in the human bladder may occur by an enzyme other than angiotensin converting enzyme, which is likely a serine protease.[27] Given this, captopril would not be expected to completely eliminate the local formation of angiotensin II and therefore not entirely obliterate its effects on smooth muscle. Additionally, the effects of captopril in the rabbit may be species specific since serosal hyperplasia is a unique finding in the obstructed rabbit bladder that has not been described in other animals or the human. This certainly raises concerns with regard to the utility of the obstructed rabbit bladder model as an applicable animal model to the human condition. Lastly, it is possible that captopril may have antiproliferative and antifibrotic effects which are separate from its effect on angiotensin II production. Some investigators have suggested that the antifibrotic activity of captopril may represent a non-specific sulfahydryl

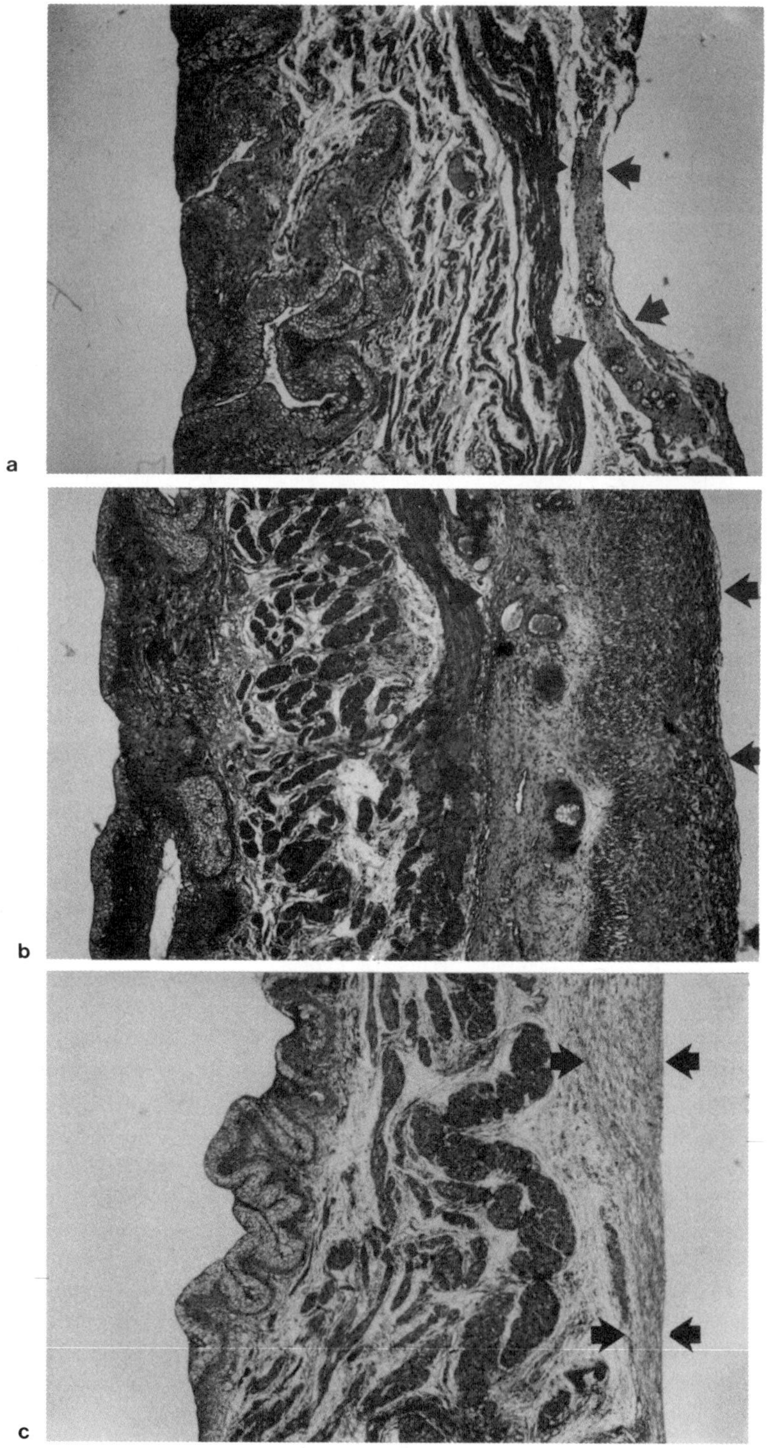

Figure 1. Trichrome stained bladder sections from a) controls, b) partial bladder neck obstruction, and c) obstruction with captopril. Obstruction results in significant muscle hypertrophy, serosal hyperplasia (serosa marked by arrows) and transmural collagen deposition. Captopril administration results in diminution in all of these changes, especially in the amount of serosal hyperplasia and collagen deposition. Of note, a moderate degree of muscular hypertrophy in the muscularis propria still persists in the captopril treated animals. Magnification 40 X. (Reprinted with permission from Cheng, E.Y. et al.,[24] Elsevier Science Inc.).

effect related to its chemical structure, as opposed to its effect on angiotensin converting enzyme activity.[28] If so, this may help explain our observations that captopril has a greater inhibitory effect on collagen deposition and smooth muscle hyperplasia, than on the development of muscle hypertrophy.

To address many of these concerns, we decided to repeat a similar study in a model of partial BNO in the rat.[29] In comparison to the rabbit, the rat bladder is more similar to the human bladder with respect to architecture (normal and obstructed) and the changes in bladder wall dynamics that occur in response to obstruction. For example, hydronephrosis occurs rarely in the acutely or chronically obstructed rabbit bladder whereas upper tract dilation not uncommonly develops with prolonged BNO in both rats and humans. In this study partial BNO was once again induced for 2 weeks. Separate groups of obstructed rats were given either captopril (50 mg/kg/day) or losartan (angiotensin subtype AT-1 receptor antagonist, 30 mg/kg/day) via the drinking water. A higher dose of captopril was used in the rats to see if a greater inhibitory effect on the obstructive bladder changes could be achieved than those observed in the rabbit model. Also, captopril is much better tolerated in rats than in rabbits. Additionally, losartan was utilized to address the concerns that the inhibitory effects of captopril may be due to its unique chemical structure and not necessarily due to its suppression of angiotensin II production. The results of this study were somewhat surprising in that neither captopril nor losartan significantly ameliorated the histologic or biochemical features of partial BNO in the rat. Prominent muscle hypertrophy and collagen deposition occurred in all groups of animals, including those treated with captopril and losartan. Although not statistically significant on biochemical evaluation, captopril appeared to have a modest effect on the amount of excess transmural collagen present in the bladder. The lack of an inhibitory effect of losartan on the obstructed rat bladder is also supported by a recent similar study by Persson et al.[30] Why should there be such disparity between the results of these studies in the rat and those in the rabbit? As alluded to earlier, it may simply be that these are species specific observations. Since muscle hypertrophy was not substantially inhibited in either the rabbit or rat, it is also possible that angiotensin II is more active in promoting cellular proliferation and collagen production, as opposed to cellular hypertrophy. Another intriguing possible explanation revolves around the theory that the cellular effects of angiotensin II are mediated though contraction and/or cell stretch mechanisms. Angiotensin II is known to induce a relatively weak bladder contraction in the rat in comparison to other species, including humans.[30] If the cellular effects of angiotensin II are indeed linked to its influence on bladder tone, then the lessened effects of both captopril and losartan on the obstructive rat bladder may be in part related to angiotensin II=s relatively weak effects on rat bladder contractile activity. Nevertheless, further elucidation of the exact mechanisms in which angiotensin II exerts its different effects in the rat and rabbit bladder is needed to clarify and explain these preliminary observations.

3. IN VITRO CELL CULTURE STUDIES

The in vitro effects of angiotensin II on bladder stromal cell mitogenesis were investigated utilizing primary stromal cell cultures developed from 3 day old rabbit bladders. Since angiotensin II is known to upregulate secondary growth factors with well defined mitogenic properties such as basic fibroblast growth factor (βFGF), the effects of βFGF were also examined.[31] Stromal cells were plated at a density of 30,000 cells/ml in standard 12 well tissue culture plates and allowed to grow for 24 hrs. Subconfluent cells (80–90%

confluent) were growth arrested in serum deficient (0.25% NCS) or serum free media for 24 hrs. Cells were then stimulated with angiotensin II (10^{-7}M) or βFGF (10ng/ml) for an additional 48 hrs. Cell counts and ^{3}[H] thymidine incorporation were measured. Angiotensin II and βFGF both individually stimulated neonatal bladder stromal cell proliferation and ^{3}[H] thymidine incorporation under serum deficient conditions. On average, angiotensin II demonstrated a 26% increase in cell number ($p<0.001$) and a 35% increase in ^{3}[H] thymidine incorporation ($p<0.001$) when compared to paired control values. βFGF was an even more potent mitogen with a 47% increase in cell number ($p<0.001$) and 180% increase in ^{3}[H] thymidine incorporation ($p<0.001$) as compared to controls. Contrastly, angiotensin II and βFGF both individually failed to have significant stimulatory effects under serum free conditions. This suggests that there is a synergistic factor contained in serum that is necessary for both angiotensin II and βFGF to exert their stimulatory effects.

These in vitro results utilizing stromal cells from the neonatal rabbit, in conjunction with the observations from the in vivo captopril study in the obstructed rabbit bladder, strongly suggest that angiotensin II is involved in the stimulation of cellular proliferation in the obstructed bladder. Once again, the exact mechanisms in which angiotensin II stimulates bladder stromal cell division requires further investigation.

4. FUTURE DIRECTION

As is many times the case with any research, the results of our initial studies raise more questions than they answer. One of the primary questions that needs to be addressed is whether the observations in animals supporting angiotensin II's involvement in regulating smooth muscle growth and collagen production in the developing bladder have relevance to the human condition. This is especially important given the apparent disparate effects of angiotensin II on the bladders of different animal species. Current efforts in our laboratory now involve the use of primary stromal and epithelial cell cultures from the pediatric human bladder. Utilization of human cells will hopefully eliminate some of the concerns regarding differences between animals and humans, although one must always be careful when extrapolating in-vitro cell culture observations to the intact in-vivo state. Nonetheless, we have thus far identified the immunohistochemical presence of the AT-1 receptor in cultured human bladder stromal cells and preliminary results suggest that angiotensin II is involved in the regulation of collagen production in the human bladder. Utilizing northern blot analysis, human bladder stromal cells were found to increase both collagen types I and III gene expression in response to the individual in vitro administration of angiotensin II (10^{-7}M) and transforming growth factor beta 1 (TGFβ ng/ml). Further work is now being conducted to investigate whether these actions of angiotensin II on collagen gene production are mediated through the secondary up regulation of TGFβ Along these same lines, we are also investigating whether angiotensin II has a mitogenic effect on human stromal cells as it does in the rabbit and whether this effect is via the up regulation of secondary growth factors such as βFGF. Additionally, we are intrigued by the notion that the effects of angiotensin II on cell proliferation and collagen production may be linked to cell stretch. Studies utilizing vascular smooth muscle cells have demonstrated that angiotensin II induced mitogenesis was enhanced over 3 fold by the synergistic application of mechanical strain.[32] Similar experiments are now being performed to see if synergistic effects of mechanical strain and angiotensin II are present in human bladder stromal cells. Lastly, all of our studies have thus far focused on the effects of angiotensin II in the stromal compartment of the bladder wall. The possible important contribution of bladder epithelium must not be overlooked. The significance of stromal epithelial interactions in the development and function of numerous organs has recently come to be realized and better understood. With respect to the bladder, it appears that smooth muscle differentiation and proliferation is dependent on the presence of epithelial cells or factors secreted by them.[33] It is quite conceivable that angiotensin II may be produced by not only stromal cells in the bladder but also by epithelial cells. This possibility is now the focus of ongoing studies utilizing primary cell cultures of human urothelium. As this work progresses, it is our hope that further knowledge will be gained as to the precise actions that angiotensin II has in the human bladder in regulating smooth muscle growth and collagen production under both normal and pathological conditions.

5. SUMMARY

Preliminary studies by our group and others indicate that angiotensin II may have an important role in the cellular regulation of smooth muscle growth and collagen production in the bladder. The exact mechanisms in which angiotensin II elicits its cellular effects are not known. Given the available information thus far, we hypothesize the following (see Figure 2): 1) Outlet obstruction of the bladder causes increased cell stretch/strain which in turn induces the local production of angiotensin II. Angiotensin II may also influence cell stretch/strain via its direct effects on bladder tone. 2) Angiotensin II then acts as a trophic factor in the bladder wall to cause smooth muscle cell hypertrophy/hyperplasia and increased collagen production via an autocrine and/or paracrine pathway. 3) The cellular effect(s) of angiotensin II may be mediated by secondary growth factors such as bFGF and TGFb Much more extensive research is certainly needed to reveal whether some part, or all of this hypothesis is correct. If angiotensin II is indeed active in regulating muscle and collagen changes in the pathologic bladder, then the clinical implications are extremely exciting since numerous pharmacologic agents are now available which can either inhibit angiotensin II production and/or block receptor mediated events. These agents may prove to be extremely useful in the clinical management of the neurogenic bladder in which obstructive changes may be prevented and potentially reversed. Despite this, caution must be exercised with regard to the potential use of any medications which alter the systemic renin- angiotensin system in the pediatric population since some research has suggested that an intact system may be necessary for the normal development of some organs, including the kidney.[34]

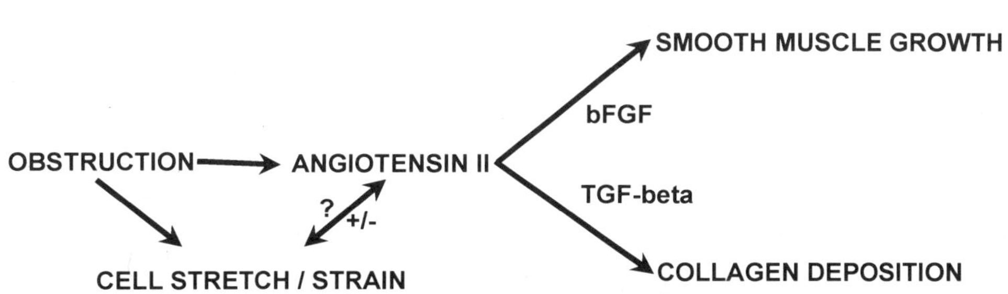

Figure 2. Hypothesis of the role of angiotensin II in the obstructed bladder. Outlet obstruction of the bladder results in the local production of angiotensin II which in turn induces smooth muscle growth and collagen production. These cellular effects of angiotensin II may be direct or via secondary growth factors such as βFGF or TGF-beta (see text).

ACKNOWLEDGMENT

Material in this chapter has been adapted in part from previous articles published on this topic by Earl Y. Cheng. See references # 4, 24, 29, and 31.

REFERENCES

1. Kim K.M., Kogan B.A., Massad C.A., and Huang Y.C.: Collagen and elastin in the obstructed fetal bladder. J. Urol., **146:** 528, 1991.
2. Peters C.A., Vasavada S., Dator D., Carr M., Shapiro E., Lepor H., McConnell J., Retik A.B., and Mandell J.: The effect of obstruction on the developing bladder. J. Urol., **148:** 491, 1992.
3. Kato K., Monson F.C., Longhurst P.A., Wein A.J., Haugaard N., and Levin R.M.: The functional effects of long-term outlet obstruction on the rabbit urinary bladder. J. Urol., **142:** 600, 1990.
4. Cheng, E.Y., Baskin L.S., DiSandro M., and Zderic S.A.: Bladder development and function: cellular factors. Dialogues in Pediatric Urology, **20:** 1, 1997.
5. Dzau V.J.: Local contractile and growth modulators in the myocardium. Clin. Cardiol., **16 :** 5, 1993.
6. Gibbons G.H., Pratt R.E., and Dzau V.J.: Vascular smooth muscle hypertrophy vs. hyperplasia: autocrine transforming growth factor-beta 1 expression determines growth response to angiotensin II. J. Clin. Invest., **90:** 456, 1992.
7. Naftilin A.J., Pratt R.E., and Dzau V.J.: Induction of platelet-derived growth factor A-chain and c-myc gene expressions by angiotensin II in cultured rat vascular smooth muscle cells. J. Clin. Invest., **86:** 1690, 1990.
8. Chua C.C., Diglio C.A., Siu B.B., and Chua B.H.: Angiotensin II induces TGF-beta 1 production in rat endothelial cells. Biochimica et Biophysica Acta., **1223:** 141, 1994.
9. Fischer T.A., Ungureanu-Longrois D., Singh K., de Zengotita J., DeUgarte D., Alali A., Gadbut A.P., Lee M.A., Balligand J.L., Kifor I., Smith T.W., and Kelly R.A.: Regulation of βFGF expression and ANG II secretion in cardiac myocytes and microvascular endothelial cells. Am. J. Physiol., **272:** H958, 1997.
10. Koibuchi Y., Lee W.S., Gibbons G.H., and Pratt R.E.: Role of transforming growth factor -beta 1 in the cellular growth response to angiotensin II. Hypertension, **21:** 1046, 1993.
11. Itoh H., Mukoyama M., Pratt R.E., Gibbons G.H., and Dzau V.J.: Multiple autocrine growth factors modulate vascular smooth muscle response to angiotensin II. J. Clin. Invest., **91:** 2268, 1993.
12. Dzau V.J. and Pratt R.E.: Cardiac, vascular and intrarenal renin angiotensin system in normal physiology and disease. In: The Renin–Angiotensin System. Edited by J.I.S. Robertson and M.G. Nicholls. London, England: Gower Medical Publishing: chapter 42.1–42.11, 1993.
13. Sadoshima J. and Izumo S.: Molecular characterization of angiotensin II induced hypertrophy of cardiac myocytes and hyperplasia of cardiac fibroblasts. Critical role of the AT-1 receptor subtype. Circ. Res., **73:** 413, 1993.
14. Brilla C.G., Reams G.P., Maisch B., and Weber K.T.: Renin-angiotensin system and myocardial fibrosis in hypertension: regulation of the myocardial collagen matrix. Eur. Heart. J., 14 suppl J:57, 1993.
15. Rossi M. and Peres L.C.: Effect of captopril on the prevention and regression of myocardial cell hypertrophy and interstitial fibrosis in pressure overload cardiac hypertrophy. Am. Heart J., **124:** 700, 1992.
16. Weber K.T., Brilla, C.G., and Janicki, J.S.: Myocardial fibrosis: functional significance and regulatory factors. Cardiovasc. Res., **27:** 341, 1993.
17. Andersson K.E., Hedlund H., and Stahl M.: Contractions induced by angiotensin I, angiotensin II, and bradykinin in isolated muscle from the human detrusor. Acta Physiologica Scandinavica, **145:** 253, 1992.
18. Erspamer V., Ronzoni G., and Falconieri-Erspamer G.: Effects of active peptides on the isolated muscle of the human urinary bladder. Invest. Urol., **18:** 302, 1981.
19. Steidle C.P., Cohen M.L., and Neubauer B.L.: Bradykinin-induced contractions of canine prostate and bladder: effect of angiotensin-converting enzyme inhibition. J. Urol., **144:** 390, 1990.
20. Anderson, G.F., Barraco, R.A., Normile, H.J., and Rosen, T.D.: Evidence for angiotensin II receptors in the urinary bladder of the rabbit. Can. J. Physiol. Pharmacol., **62:** 390, 1984.
21. Tanabe N., Ueno A., and Tsujimoto G.: Angiotensin II receptors in the rat urinary bladder smooth muscle: type 1 subtype receptors mediate contractile responses. J. Urol., 150:1056, 1993.
22. Saito, M., Kondo A., Kato T., and Miyake K.: Response of the human bladder to angiotensins: a comparison between neurogenic and control bladders. J. Urol., **149:** 408, 1993.
23. Sullivan M.P., Peters C.A., Williams G.H., Yalla S.B., and Kifor I.: Immunohistochemical and functional evidence of angiotensin II in the fetal sheep bladder. J. Urol., **157:** (4) (suppl.: Abst. #680)

24. Cheng E.Y., Lee C., Decker R., Sensibar J., Lang S., Kaplan W.E., Maizels M., and Firlit C.F.: Captopril (an inhibitor of angiotensin converting enzyme) inhibits obstructive changes in the neonatal rabbit bladder. Urol., **50:** 465, 1997.
25. Malkowicz S.B., Wein A., Elbadawi A., Van Arsdalen K., Ruggieri M., and Levin R.: Acute biochemical and functional alterations in the partially obstructed rabbit urinary bladder. J. Urol., **136:** 1324, 1986.
26. Monson F.C., Goldschmidt M.H., Zderic S.A., Ruggieri M.R., Levin R.M., and Wein A.J.: Use of a previously undescribed elastic lamina of the serosa to characterize connective tissue hypertrophy of the rabbit bladder wall following partial outlet obstruction. Neurourology and Urodynamics, **7:** 385, 1988.
27. Lindberg B.F., Nillson L.G., Hedlund H., Stahl M., and Andersson K.E.: Angiotensin I is converted to angiotensin II by a serine protease in human detrusor smooth muscle. Am. J. Physiol., **266:** R1861, 1994.
28. Nguyen L., Ward W., Ts'ao C., and Molteni A.: Captopril inhibits proliferation of human lung fibroblasts in culture: a potential antifibrotic mechanism. Soc. Exp. Bio. and Med., **205:** 80, 1994.
29. Palmer L.S., Lee C., Decker R.S., Lang S., Kaplan W.E., Firlit C.F., and Cheng E.Y.: The effect of angiotensin converting enzyme inhibition and angiotensin II receptor antagonism on obstructed rat bladder. J. Urol., **158:** 1100, 1997.
30. Persson K., Pandita K.R., Waldeck K., and Andersson K.E.: Angiotensin II and bladder obstruction in the rat: influence on hypertrophic growth and contractility. Amer. J. Physiol., **271:** A1186, 1996.
31. Cheng E.Y., Grammatopoulos T., Lee C., Decker R., Sensibar J., Kaplan W.E., Maizels M., and Firlit C.F.: Angiotensin II and bFGF induce neonatal bladder stromal cell mitogenesis. J. Urol., **156:** 593, 1996.
32. Krishnankutty S., Wilson E., Chatterjee K., and Ives, H.E.: Mechanical strain and collagen potentiate mitogenic activity of angiotensin II in rat vascular smooth muscle cells. J. Clin. Invest., **92:** 3003, 1993.
33. Baskin L.S., Hayward S.W., Young P., and Cunha G.R.: role of mesenchymal-epithelial interactions in normal bladder development. J. Urol., **156:** 1820, 1996.
34. Chung K.H., Gomez R.A., and Chevalier R.L.: Regulation of renal growth factors and clusteri.n by AT1 receptors during neonatal ureteral obstruction. Am. J. Phys., **286:** F1117, 1995.

15

NEW CONCEPTS ON THE NORMAL AND ABNORMAL DEVELOPING BLADDER

Ellen Shapiro

New York University School of Medicine
Department of Urology
540 First Avenue, Suite 10 U
New York, New York 10016

1. INTRODUCTION

In the mid-1980s, a large number of children with myelomeningocele (MMC) were undergoing bladder augmentation for neurogenic bladder dysfunction leading to renal deterioration and incontinence. At that time, a small portion of the bladder was excised and submitted to pathology. We noted that pathology reports indicated "acute and chronic inflammation" but there was no mention of any other bladder wall abnormalities. This description did not explain the radiographic findings of trabeculation nor did it address the appearance of the bladder at cystoscopy or at the time of augmentation. On reviewing these bladders histologically, we noted that there was a significant increase in connective tissue and marked disruption of bladder muscle bundle formation by dense infiltrating collagen or extracellular matrix (ECM) protein.[1] Some of the muscle cells appeared to be encased within collagen. These observations provided an explanation as to why these bladders develop significant problems with compliance and why they do not contract normally . We then embarked on investigations to determine when these bladder wall changes occurred since it was assumed at that time that the bladders were normal at birth and then deteriorated due to recurrent infection or hypercontractility and/or detrusor sphincter dyssynergia.

Elucidating the timing of the morphological changes in the MMC bladders requires the acquisition of human fetal specimens. Quantitative morphometric studies using color image analysis demonstrated that bladders of human fetuses with MMC are abnormal as early as 16–21 weeks of gestation.[1] We also observed a paucity of smooth muscle (SM) in the bladder wall and a threefold increase in connective tissue or extracellular matrix (ECM) protein disrupting muscle bundles and surrounding individual muscle cells in fetuses with MMC when compared to control fetal bladders. In the control bladders from

Advances in Bladder Research, edited by Baskin and Hayward.
Kluwer Academic / Plenum Publishers, New York, 1999.

age-matched from patients who died of nonurological causes and underwent autopsy, or who served as organ donors, well organized muscle bundles were apparent with only small amounts of interlacing connective tissue. Gross histologic analysis of the bladder specimens of the normal stillborn fetuses revealed a network of organized muscle bundles that were imbedded in loose connective tissue of primitive mesenchyme. These findings suggested that structural changes occurred in the developing fetus with MMC, and the changes were not a result of the recurrent urinary tract infections or chronic detrusor sphincter dyssynergia.

Based upon these observations, we hypothesized that an intact innervation may be one of the prerequisites for normal morphogenesis of bladder SM. Since human fetal MMC spinal cords and peripheral nerves may be morphologically intact, other factors may contribute to the maldevelopment of the fetal MMC bladder.[2] We will review our most recent research investigations on new aspects of normal bladder development. These findings have underscored the abnormalities observed in MMC bladder development which will be discussed at length at the end of this chapter.

2. METHOD AND RESULTS

We have serially cross-sectioned and sagittally sectioned (3 micron thick) the entire lower urinary tract of one 13-week and two 16.5 week male fetuses. Every 20th section was stained with Masson's trichrome. Immunohistochemical (IHC) staining was performed with antibodies specific for SM and skeletal muscle (SKM) actin. We confirmed that the posterior longitudinal muscle of the detrusor extends caudally to the base of the prostate as the deep trigone and the anterior detrusor extends to the mid-prostate. As the ureter enters the detrusor, a fine layer of muscle fibers that appears to be detrusor in origin (Waldeyer's sheath), joins the inner and outer aspect of the longitudinal detrusor musculature. There is a definite space between the sheath and the ureteral wall muscle (Waldeyer's space). In cross-section, the ureters, approach the hiatus and are bridged by a thick muscle layer which becomes attenuated at each of the ends, which then becomes associated with the ureteral sheath. This layer most likely represents the interureteric ridge. The submucosal region distal to interureteric ridge has a paucity of SM. The definitive muscle of the superficial trigone (ST) is best visualized on sagittal sections stained with SM actin and suggest that the ST originates from ureteral wall muscle and extends distally forming a thick SM unit which becomes the definitive bladder neck. This SM unit extends caudally to the ejaculatory duct entrance into the urethra and contributes to the SM encircling the ejaculatory duct and utricle. Our anatomical observations in the fetus are consistent with the neurophysiology of the lower urinary tract. The ureter, ST/bladder neck region/stroma and smooth muscle of the ejaculatory duct and stroma of the prostate appear to be of mesonephric duct origin. Their SM activity is mediated by adrenergic (sympathetic) nerves, whereas the bladder SM is mediated by muscarinic cholinergic (parasympathetic) nerves. The reciprocal innervation of the smooth muscle containing tissues is explained by their unique embryological origins.

We have recently examined the early development and innervation of the human fetal bladder and external urinary sphincter (EUS) musculature. The bladder and external urinary sphincter of six fetuses, ages 9.5–25 weeks, were serially cross-sectioned. IHC staining was performed using specific antibodies to SM and SKM actin. IHC staining for S-100 proteins for localizing Schwann cells and for neurofilament protein for the detection of axons was performed.[3] The 9.5 week bladder and EUS showed staining for SM actin. The bladder also stained for SKM actin at this time, while no SKM was detected in the EUS until 13.5 weeks. By 16 weeks, the bladder no longer expressed SKM actin while the EUS demonstrated increased SKM actin with persistence of SM actin until 20 weeks. Double immuno-labeling confirmed co-expression of SM and SKM during early development. Only SM was found in the bladder at 16 weeks, whereas the composition of the EUS at 20 weeks was primarily SKM. These studies show that the bladder and EUS in the male fetus transiently express a mixed phenotype during early development. These observations suggest that developmentally programmed transdifferentiation, which is a relatively rare phenomena, occurs in the development of the lower urinary tract. This phenomenon is characterized by cells of one differentiated type and function changing to another differentiated cell type.[4] The co-expression of SM and SKM markers in these structures precede the conversion of SM in the bladder and SKM in the external urinary sphincter.

IHC staining for S-100 protein and neurofilament protein showed no staining of the bladder or EUS at 9.5 weeks. Neural staining developed in a cephalad to caudad direction. Neural staining was present in the bladder as early as 9.5 weeks, while no innervation was detectable in the EUS until 16 weeks.(Figure 1 A and B) In addition to earlier differentiation of the bladder, the bladder also develops innervation several weeks prior to the EUS. Delayed or altered muscular differentiation and innervation of the bladder and/or EUS may affect function adversely early in development.[5] If the innervation of the bladder and external urinary sphincter do not function in a coordinated function, then the external urinary sphincter may cause a functional obstruction while the bladder is capable of contracting or subsequently developing hypercontractility. This presumed discoordination in function of the bladder and EUS may explain the gross bilateral reflux observed in some

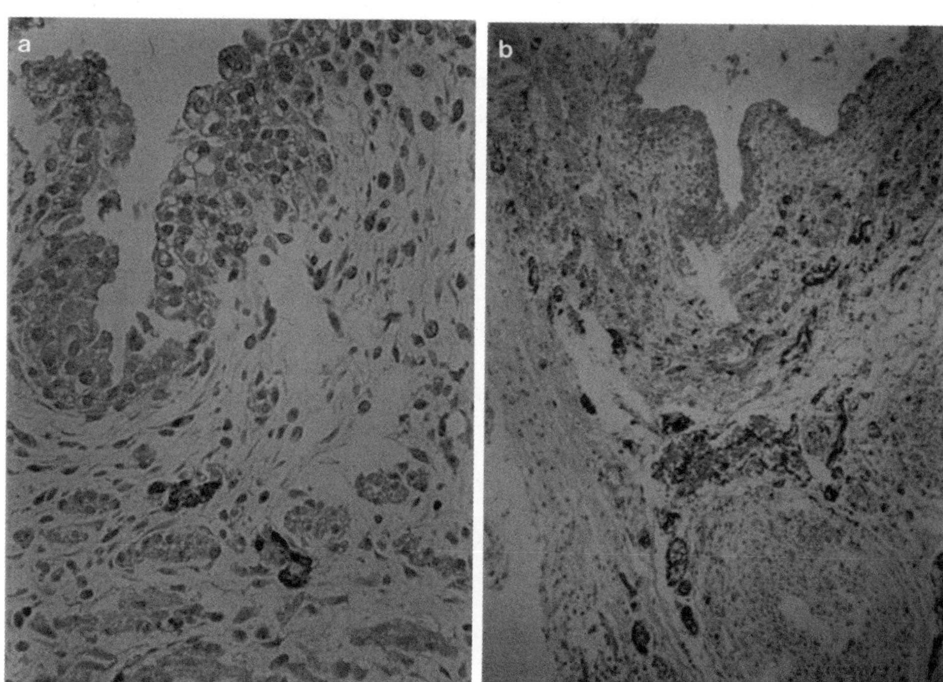

Figure 1. S-100 protein staining of the control fetal bladder at: (A) 9.5 weeks (40 x); (B) 11 weeks (20x).

male neonates. Similar studies in females are needed to further delineate the sex differences in the development of gross bilateral reflux.

We have also studied uroplakins (UPs) in the developing human fetal bladder.[6] UPs are specialized membrane proteins constituting the asymmetric unit membrane plaque of the urothelium and represents specific molecular markers of urothelial differentiation.[7] We examined UP expression in the urothelium of both the prostate and bladder in the developing fetus. We have shown that UPs are present as early as 9.5 weeks in bladder and prostatic urethral urothelium. UPs were not expressed in the distal ureter until 11 weeks (Figure 2 A and B). Early expression of UPs in the prostatic urethral urothelium is of interest since the prostatic urothelium is also capable of expressing androgen receptor (AR) ((9.5 weeks) and 5 alpha reductase (11 weeks). As the prostatic ducts elongate and extend through the stroma of the prostate to the periphery of the gland, UPs and AR continued to be expressed. UP expression was diminished or absent distal to the region of ductal budding where acini form and become cannulated. AR continued to be expressed in these newly formed glands. This, again, may represent a form of programmed transdifferentia-

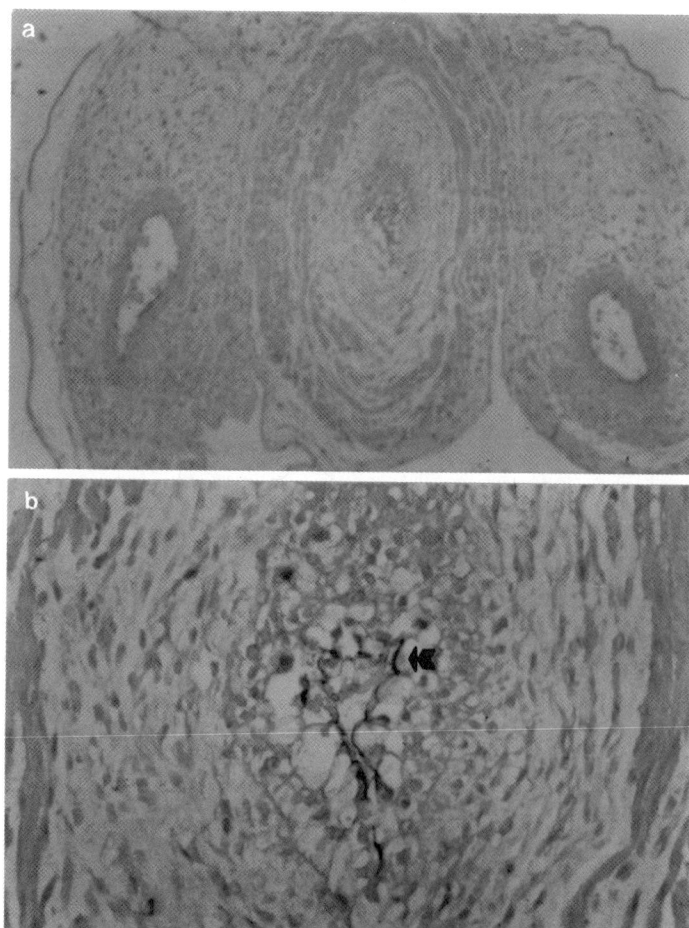

Figure 2. Uroplakin expression (arrow head) in the bladder urothelium at: (A) 9.5 weeks (10 x); (B) 9.5 weeks (40 x).

tion since the prostatic urethral urothelium has the capability of changing from one differentiated cell type (urothelium) to another differentiated cell type (glandular epithelium) during the process of human prostatic development.

In order to provide insights into epithelial differentiation, we have also studied keratin (K) expression during human fetal lower urinary tract development.[8] Six human fetal prostates and bladders (9.5–23 weeks) were serially cross-sectioned. Control BPH and bladder tissues were obtained from 3 patients undergoing open prostatectomy. Sections from the bladder and apex, mid-gland, and base of the prostate were stained with antibodies to KA4 (K14,15, 16, and 19) and 10.11 (K8 and 18) which stain simple epithelium (luminal cells) and antibodies to 34ßE12 (K5) which stain stratified epithelium (basal cells).

Staining of prostatic urothelial (PU) and bladder urothelial (BU) luminal cells was seen at 9.5 weeks while prostatic glandular epithelial (PGE) luminal cell staining was present at 11.5 weeks which corresponds to the appearance of glandular budding. With increasing fetal age, the PGE staining intensity was greater centrally than in the noncannulated, peripheral glands. Staining of PU and BU basal cells was not present until 11.5 and 16 weeks, respectively, while PGE basal cell staining was present as early as 13 weeks. BPH tissue showed staining for basal and luminal cells within the glands while control bladder stained for luminal cells with only weak expression of K5 in the basal cells.

These studies show that PGE cytokeratin expression is similar in the fetus and BPH. K5 stains basal cells of the fetal PU and BU was found to be a "fetal urothelial marker" since it is present in only trace amounts in normal adult urothelium. Urothelial cells grown in culture synthesize these fetal keratins and undergo a "reverse differentiation", expressing a more primitive phenotype in cell culture."Reverse differentiation" may play a role in the development of bladder cancer, since poorly differentiated squamous cell and transitional cell carcinomas also express the fetal urothelial marker, keratin 5.[8]

With this new knowledge of normal bladder development, we examined 7 male fetuses with MMC at 20 weeks gestational age.[9] Meuli et al reported morphological studies on these human MMC spinal cords and showed that some neuroplates were morphologically normal and capable of differentiation despite the absence of normal neurulation, suggesting that the primary malformation may mesodermal and not neural. Neurological dysfunction was postulated to result from a traumatic phenomena with destruction of the neural tissue during gestation, labor and delivery, since the cord is not protected by the normal meningeal, vertebral and cutaneous coverings.

Recent experimental evidence in a fetal sheep model suggests that exposure of the normal lumbar spinal cord to the intrauterine environment in mid-gestation results in progressive secondary destruction of the neural tissue leading to paraplegia at birth.[10] Furthermore, in utero surgical repair preserves neurological function since all survivors were able to stand and walk. Although the bladders in this model were not examined, the animals were reported to have normal excretory function.[11–13] Other in utero surgical repair studies in a pig and a rabbit model have also shown similar neurological results, but in those studies, the bladders of operated and control animals showed no histological differences.[14–17] These studies in animal models provide the rationale for in utero intervention in human MMC cases. Clinical observations suggest the feasibility of in utero intervention. Normal lower extremity movements have been observed on ultrasound at 17 weeks in fetuses with MMC.[18] Since open prenatal intervention is currently being performed as early as the 18th week of gestation, in utero surgical repair with coverage of the exposed spinal cord may prevent further destruction of neural tissue and salvage neurological function in humans with MMC.[19]

In our most recent investigations, we examine the lower GU tract in human fetal MMC specimens previously examined and reported by Meuli et al providing the unique opportunity to correlate the findings in the lower GU tract with the severity of the spinal cord lesions.[20] Acquisition of these specimens allowed us to investigate the hypothesis that the mesodermal and/or neural defects leading to MMC results in a global smooth muscle maldevelopment of the lower GU tract and that secondary injury to the spinal cord from amniotic fluid and subsequent birth trauma may lead to further anatomical and physiological dysfunction due to autonomic nerve injury.[21] The bladders were examined immunohistochemically from male MMC and control fetuses in order to determine whether nerve density, SM and SKM are altered in the MMC GU tract. Fetal surgery with coverage of the spinal cord in selected cases may prevent further injury to the somatic nervous system during the second half of gestation. Our study characterizes the extent of the early injury to the lower GU structures and the potential for prenatal intervention to impact favorably on autonomic function of the bladder.

IHC staining was performed using S-100 protein and neurofilament protein, the peripheral innervation of the bladder was found to be markedly decreased in MMC. Trichrome and SM actin staining also demonstrated that SM in control specimens was well differentiated when compared to the MMC specimens. A paucity of SM was observed in MMC bladder and bladder neck with an excess of collagen in an inter- and intrafascicular distribution. Double immunofluorescent staining revealed persistent co-expression of SM and SKM actin by myocytes of MMC detrusor while controlled bladders at 20 weeks show only SM as early as 16 weeks. The SKM component of structures in MMC fetuses, including the external sphincter, was morphologically similar to controls. These studies show that a global defect in the development of SM exists in MMC lower genitourinary tract. Peripheral nerve density is diminished, suggesting that an intact nervous system is important in the development of normal SM.[9]

ACKNOWLEDGMENT

A portion of this work was funded by NIH R01 DK 47503-02.

REFERENCES

1. Shapiro, E., Becich, M. J., Perlman, E. and Lepor, H.: Bladder wall abnormalities in myelodysplastic bladders: a computer assisted morphometric analysis. J. Urol., **145:** 1024, 1991.
2. Meuli, M., Meuli-Simmen, C., Hutchins, G. M., Seller, M. J., Harrison, M. R. and Adzick, N. S.: The spinal cord lesion in human fetuses with myelomeningocele: implications for fetal surgery. J. Ped. Surg., **32:** 448, 1997.
3. Nakajima, T., Watanabe, S., Sato, Y., Kameya, T. and Shimosato, Y.: An immunoperoxidase study of S-100 protein distribution in normal and neoplastic tissues. Am. J. Surg. Pathol., **6:** 715, 1982.
4. Patapoutian, A., Wold, B. J., and Wagner, R. A.: Evidence for developmentally programmed transdifferentiation in mouse esophagus. Science, **270:** 1818, 1995.
5. Shapiro, E.: Programmed Transdifferentiation: Insights into the development of lower genitourinary and rectal wall musculature. The American Urological Association, New Orleans, Louisiana, 1997.
6. Shapiro, E. and Wu, X-R.: Uroplakin expression in the lower genitourinary tract of the human male fetus. Societe Internationale D'Urologie, Montreal, Canada, 1997.
7. Moll, R., Wu, X-R, Lin, J-H, and Sun, T-T.: Uroplakins, specific membrane proteins of urothelial umbrella cells, as histological markers of metastatic transitional cell carcinomas. Am. J. Pathol. **147:** 1383, 1995.
8. Shapiro, E., Gitlin, J. S., Sun, T-T., and Wu, X-R.: Expression of cytokeratins in human fetal prostate and bladder. Abstract. AUA Annual Meeting, 1998.

9. Shapiro, E., Seller, M. J., Steiner, M. S., Raghow, S., Lepor, H., Kalousek, D. K., Hutchins, G. M., Perlman, E. J., and Meuli, M.: Altered smooth muscle development and growth factor expression in the lower genitourinary and gastrointestinal tract of the human male fetus with myelomeningocele. Submitted J. Urol. 1998.
10. Meuli, M., Meuli-Simmen, C., Yingling, C. D., Hutchins, G. M., Hoffman, K. M., Harrison, M. R. and Adzick, N. S.: Creation of myelomeningocele in utero: a model of functional damage from spinal cord exposure in fetal sheep. J. Ped. Surg., **30:** 1028, 1995.
11. Meuli, M., Meuli-Simmen, C., Hutchins, G. M., Yingling, C. D., Hoffman, K. M., Harrison, M. R. and Adzick, N. S.: In utero surgery rescues neurological function at birth in sheep with spina bifida. Nat. Med., **1:** 342, 1995.
12. Meuli, M., Meuli-Simmen, C., Yingling C. D., Hutchins, G. M., Timmel, G. B., Harrison, M. R. and Adzick, N. S.: In utero repair of experimental myelomeningocele saves neurological function at birth. J. Ped. Surg., **31:** 397, 1996.
13. Meuli-Simmen, C., Meuli, M., Hutchins, G. M., Harrison, M. R., Buncke, H. J., Sullivan, K. M. and Adzick, N. S.: Fetal reconstructive surgery: experimental use of the latissimus dorsi flap to correct myelomeningocele in utero. Plast. Reconstr. Surg., **96:** 1007, 1995.
14. Heffez, D. S., Aryanpur, J., Hutchins, G. M. and Freeman, J. M.: The paralysis associated with myelomeningocele: clinical and experimental data implicating a preventable spinal cord injury. Neurosurg., **26:** 987, 1990.
15. Heffez, D. S., Aryanpur, J., Rotellini, N. A., Hutchins, G. M. and Freeman, J. M.: Intrauterine repair of experimental surgically created dysraphism. Neurosurg., **32:** 1005, 1993.
16. Housley, H. T., Graf, J. L., Lipshultz, G. S., Calvano, C. J., Harison, M. R. and Jennings, R. W.: Creation of myelomeningocele in the fetal rabbit. The American Urological Association, New Orleans, Louisiana, 1997.
17. Personnel communication-L. S. Baskin.
18. Korenromp, M. J., Gool, J., Bruinese, H. and Kriek, R.: Early leg movements in Myelomeningocele. Lancet, **8485:** 917, 1986.
19. Hodgen, G. D.: Antenatal diagnosis and treatment of fetal skeletal malformations: with emphasis on in utero surgery for neural tube defects and limb bud regeneration. JAMA, **246:** 1079, 1981.
20. Meuli, M., Meuli-Simmen, C., Hutchins, G. M., Seller, M. J., Harrison, M. R. and Adzick, N. S.: The spinal cord lesion in human fetuses with myelomeningocele: implications for fetal surgery. J. Ped. Surg., **32:** 448, 1997
21. Somasundaram K. and Prathap, K.: The effect of exclusion of amniotic fluid on intra-uterinehealing of skin wounds in rabbit foetuses. J. Pathol., **107:** 127, 1972.

16

BIOCHEMICAL AND PHYSIOLOGICAL CHARACTERIZATION OF THE URINARY BLADDER IN EHLERS-DANLOS SYNDROME

Catherine M. Deveaud,[1] William A. Kennedy II,[2] Stephen A. Zderic,[2] and Pamela S. Howard[1,2]

[1]Department of Anatomy and Histology
School of Dental Medicine
University of Pennsylvania
[2]Division of Urology
Children's Hospital of Philadelphia
Philadelphia, Pennsylvania 19104

1. INTRODUCTION

Ehlers-Danlos syndrome (EDS) is a heterogenous group of disorders characterized by inherited abnormalities of connective tissue.[1,2] The main clinical manifestations of this disease are skin fragility, skin hyperextensibility and joint hypermobility. More than 10 subtypes of EDS have been defined based on clinical, genetic and biochemical criteria. Recent molecular studies of EDS have identified genetic defects responsible for several subtypes of this disease; however, for others, the molecular defect remains unknown.

Clinical pathologies associated with EDS may include the presence of diverticuli in different organs. Diverticuli of the gastro-intestinal tract,[3,4] as well as the genito-urinary tract[5–11] have been reported. Non-EDS diverticuli of the urinary bladder have frequently been associated with urinary tract infections, dysuria, hematuria, and vesicoureteral reflux. As a result, diverticulectomy has often been advocated as the treatment of choice. The durability of this treatment plan is often questioned, since the EDS associated bladder diverticuli form a distinct group characterized by postoperative recurrence.[12] Levard et al.[8] have speculated that the recurrence of these diverticuli supports the theory that there is an intrinsic anomaly of the vesical wall in certain subtypes of EDS.

In this study, we report the results of a detailed investigation of the native bladder wall and bladder diverticular wall of a child who has been diagnosed clinically with EDS. Native bladder wall and congenital diverticular wall of a child who did not present with EDS were used for comparison.

Advances in Bladder Research, edited by Baskin and Hayward.
Kluwer Academic / Plenum Publishers, New York, 1999.

2. MATERIALS AND METHODS

2.1. Patients

A twelve year old girl with a prior clinical diagnosis of Ehlers Danlos syndrome presented to the Urology Clinic with a history of new onset urinary incontinence and urinary frequency. She initially presented to our institution at birth with skeletal abnormalities, bilateral hip and knee dislocations, joint hypermobility and lax eyelids. Investigation at that time led to the diagnosis of EDS (type VII). Neither parent reported symptoms of EDS.

More recently the patient reported a voiding history characterized by urinary frequency of one hour intervals, and day and nighttime urinary incontinence. Her voiding was characterized by straining and the need to apply manual compression to the suprapubic region to empty her bladder. She reported occasional suprapubic pain. Urine analysis was normal and urine culture was with no growth of microorganisms. Urological imaging consisted of a renal ultrasonograph with two normal kidneys, and a voiding cystourethrogram (VCUG) with a greater than 700ml capacity, trabeculation, and a large left-sided diverticulum which did not empty with voiding. Urodynamic studies demonstrated a bladder capacity of 1150ml, a residual of 775ml, and a voiding pattern consistent with abdominal valsalva maneuvers.

Treatment options of clean intermittent catheterization and diverticulectomy were discussed with the patient and family, with diverticulectomy chosen. Postoperatively the patient has done well, voiding to completion with resolution of urinary frequency and incontinence. Postoperative urodynamic studies revealed a bladder capacity of 350ml, a residual of 75ml, and a voiding pattern characterized by a coordinated and sustained detrusor contraction.

A four and a half year old boy with no prior medical history, presented to the Urology Clinic with a febrile urinary tract infection and a left pyelonephritis. The patient's urological history was only significant for decreased force of his urinary stream. His evaluation included a renal ultrasonograph which was normal, and a VCUG which showed left vesicoureteral reflux and a left paraureteral diverticulum. The VCUG also demonstrated a dysfunctional voiding pattern with the diverticulum enlarging with voiding and a large post void residual urine. His treatment plan consisted of a bladder diverticulectomy and left ureteral reimplantation. Postoperatively the patient has done well with an improved urinary stream and no further febrile urinary tract infections.

Tissue from both the diverticulum and native bladder wall (non-diverticulum) were removed from both patients at the time of diverticulectomy. Pieces of tissue were then further divided for histologic, immunofluorescent, biochemical, molecular and physiologic analyses, as described below.

2.2. Histology and Immunofluorescence Histochemistry

All tissues were embedded in freezing compound (OTC) and 5 micrometer thick frozen sections were cut on a Slee cryostat (London, U.K.) maintained at -28°C. All specimens were stained utilizing Masson's trichrome technique[13] for routine histologic examination. For immunohistochemistry, sections were incubated with collagen type-specific antibodies, elastin and fibrillin-1 antibodies as previously published.[18] Slides were examined with a Zeiss microscope equipped with epifluorescence optics and also, with a confocal laser scanning immunofluorescent microscope.

The human type III collagen monoclonal antibody was prepared and characterized in our laboratory[14] and used undiluted. The type IV and type I collagen monoclonal antibodies were the gift of Drs. J. Scheinman[15] and J. Werkmeister,[16] respectively, and used at 1:200 dilutions. A monospecific antibody against a synthetic peptide sequence of human fibrillin-1 and a polyclonal antibody against elastin were the gifts of Dr. J. Rosenbloom. For controls, tissue sections were incubated : 1) with a primary antibody and no secondary antibody, 2) with a primary antibody which had been absorbed with the immunogen, and 3) PBS, followed by a secondary antibody. In all instances, control staining was either minimal or not detectable.

2.3. Collagen RNA Analysis

Small pieces of native and diverticular wall from the EDS bladder were frozen in liquid nitrogen and pulverized. Total RNA was extracted by the method of Chomczynski and Sacchi,[17] quantified by absorbance at 260nm, and electrophoresed in an agarose gel containing ethidium bromide to determine the integrity of the 28S and 18S ribosomal RNA. Dilutions of RNA (0.2, 0.15, 0.1 and 0.05μg) were denatured and applied to nylon membranes (Hybond, Amersham), using a Slot-blot microfiltration apparatus (Hoefer PR 648 Filtration Manifold). The immobilized RNA was UV irradiated for 2 minutes. (Due to the limited amount of native bladder tissue available for study from the control patient, no RNA was extracted from this tissue). Hybridization with ^{32}P-radiolabeled human cDNA inserts for the $\alpha 1$ chains of type I (Hf677) and type III (Hf934) collagens were carried out as previously described[18] and filters were washed under stringent conditions to eliminate the possibility of cross-hybridization. The filters were exposed to X-ray film (X-Omat AR, Kodak) at -70^{0}C and quantified by densitometric scanning (Molecular Dynamics, ImageQuant V3.1.1) and linear regression analysis. No statistical analysis was carried out due to the limited number of patients.

2.4. Quantitative Analysis for Total Collagen Concentration

Quantitative amino acid analysis was performed on pulverized, lyophilized crude bladder tissue from both native and diverticular walls of the EDS and control patient as described previously.[18,20] Bladder tissue amino acids were identified and quantified by comparison to a standard consisting of a commercial amino acid mixture (Sigma) supplemented with hydroxyproline, extra proline and hydroxylysine. Total collagen concentration was determined by multiplying the total hydroxyproline content by 7.46.[19]

2.5. Indirect Enzyme-Linked Immunosorbent Assay (ELISA)

Cyanogen bromide (CNBr) peptides were prepared from the pulverized, dried crude bladder samples as previously described.[18,20] Type III collagen was quantified by an indirect ELISA using a human type III collagen polyclonal antibody (DMI, Kennebunk, ME). This antibody was determined to react only against type III collagen peptides, and not type I collagen peptides, by mixing human collagen types I and III CNBr-peptides in different ratios and assaying by indirect ELISA (data not shown). The concentration of type III collagen in the CNBr-solubilized bladder tissue samples was analyzed in triplicate and determined directly from the standard inhibition curve of human type III collagen CNBr peptides. Collagen ratio analysis was determined as previously described.[20] The quantity of type I collagen was derived by subtraction of the type III concentration from the total collagen concentration, based upon the fact that types I and III collagen comprise 97% of the collagen types in the bladder.[18]

A similar ELISA analysis was carried out utilizing a monospecific antibody against a synthetic peptide (287) sequence of human fibrillin-1. Elastin concentration was determined from the cyanogen bromide insoluble tissue residue, since elastin contains no methionine residues and is not digested by cyanogen bromide.

2.6. Smooth Muscle Physiology

Muscle strips of 1×12mm were cut from fresh operative native bladder wall and diverticular wall specimens. 4–0 silk sutures were attached at both ends giving a total muscle length of 10 mm between the silk ties. All strips were suspended in organ baths perfused with Tyrode's physiological solution and oxygenated with 95% oxygen mixed with 5% carbon dioxide at 37°C. One silk tie fixed the strip at the base of the bath while the second tie was attached to an isometric force transducer with input to a Grass polygraph. All strips were equilibrated for 1 hour at slack length. Active compliance (muscle tone intact) was determined as follows: the strips were stretched in 5mm increments, followed by a quick 2.5mm release to a new equilibrium length. After 10 minutes, field stimulation was applied at 80V and 32Hz. All strips were stretched progressively to the length for optimal force generation. At this point, the Tyrode solution in the baths was exchanged for calcium-free Tyrode's containing 1 mM EGTA. Field stimulation was then repeated at 2-minute intervals until no further active force could be generated. The time at which 50% of the starting force was lost was recorded, the buffer was then changed back to a normal Tyrode solution, and the strips were allowed a 30-minute period to recover the initial starting tension in response to field stimulation. All data were collected from the Grass polygraph recording strip and expressed as a percentage of the initial starting tension with only normal Tyrode's solution in the baths. Passive compliance (muscle tone ablated) was determined in the same fashion following overnight incubation in calcium-free Tyrode's buffer in the presence of 5mM EGTA and 10mM sodium azide.[23]

3. RESULTS

3.1. Histology and Immunohistochemistry

The architecture of the EDS and control bladder diverticuli is shown in Figure 1. The control bladder diverticular wall (A) was characterized by a urothelial layer a few cells thick (U), lamina propria connective tissue of variable density (lp), and small, compact muscle bundles surrounded by dense connective tissue (m). In contrast, the EDS diverticular wall (B) had a thickened urothelium (U), less dense connective tissue in the lamina propria (lp) and a muscle layer comprised of rudimentary smooth muscle cells and loose connective tissue (m). When stained with an antibody against type III collagen, there was strong reactivity within rudimentary muscles bundles of EDS diverticulum (Figure 1D), whereas reactivity was restricted to the perimysial connective tissue of the control diverticulum (Figure 1C). Type I collagen staining was present in the perimysial regions with no endomysial reactivity [Data not shown]. Type IV collagen was localized to the basement membranes surrounding individual muscle cells in both control (E) and EDS (F) diverticuli. The EDS diverticular bladder wall was further characterized by a lack of detrusor smooth muscle cells organized into bundles or fasicles (Figure 2B) as compared to control diverticulum (Figure 2A). Both diverticular bladder walls contained numerous blood vessels which were extensively dilated (Figures 2 A, B).

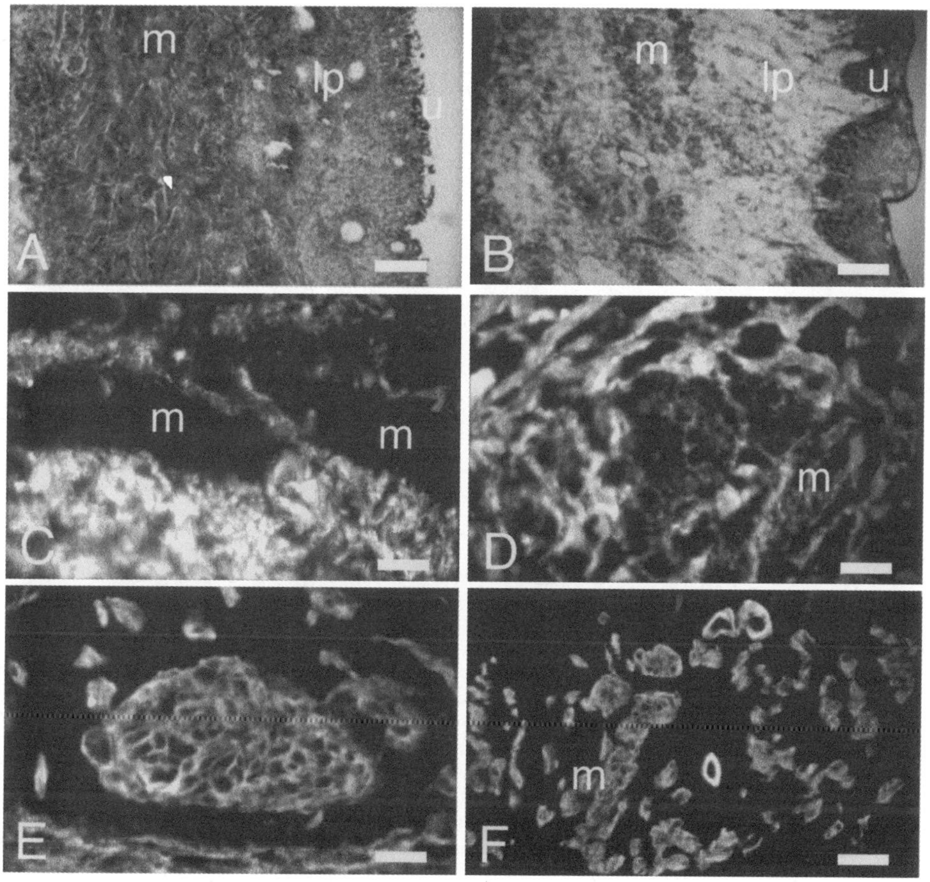

Figure 1. Histological and immunocytochemical characterization of diverticular walls from congenital control (A,C,E) and EDS patient (B,D,F). Masson's trichrome staining of diverticula walls. (A) Control and (B) EDS. (u: urothelium; lp: lamina propria; m: detrusor muscle). Note dense staining of connective tissue in lp of control and increased muscle bundles in detrusor region (m). Magnification (A,B) = 54 X, bar = 200 um. Immunofluorescent staining with an antibody against collagen III (C,D); immunofluorescent staining with antibody against collagen IV delineating individual muscle cells and blood vessels. (E,F). Note the infiltration of type III collagen within the muscle bundles of the EDS diverticulum and the absence of type III collagen in control muscle bundles (m). Magnification (C,D,E,F) = 218X, bar = 45 μm.

In order to determine whether extracellular matrix abnormalities were found only in the diverticular walls, we also examined native bladder tissue (non-diverticulum) from both EDS and control bladders. While the architecture of muscles bundles from the native detrusor region of both the control and EDS bladders were relatively similar, Masson's trichrome stained sections showed a mild degree of connective tissue infiltration within the smooth muscle bundles of native EDS bladder compared with no infiltration in native control bladder (data not shown). This infiltration in the native EDS bladder was also confirmed to be specific for type III collagen (Figure 3D). Conversly, staining for type III collagen in the native control detrusor (Figure 3C) localized only to the perimysial connective tissue around the muscle bundles. Sections stained with an antibody against type IV collagen, and visualized by scanning confocal microscopy, delineated in more detail an investment of basement membrane surrounding individual muscle cells within the

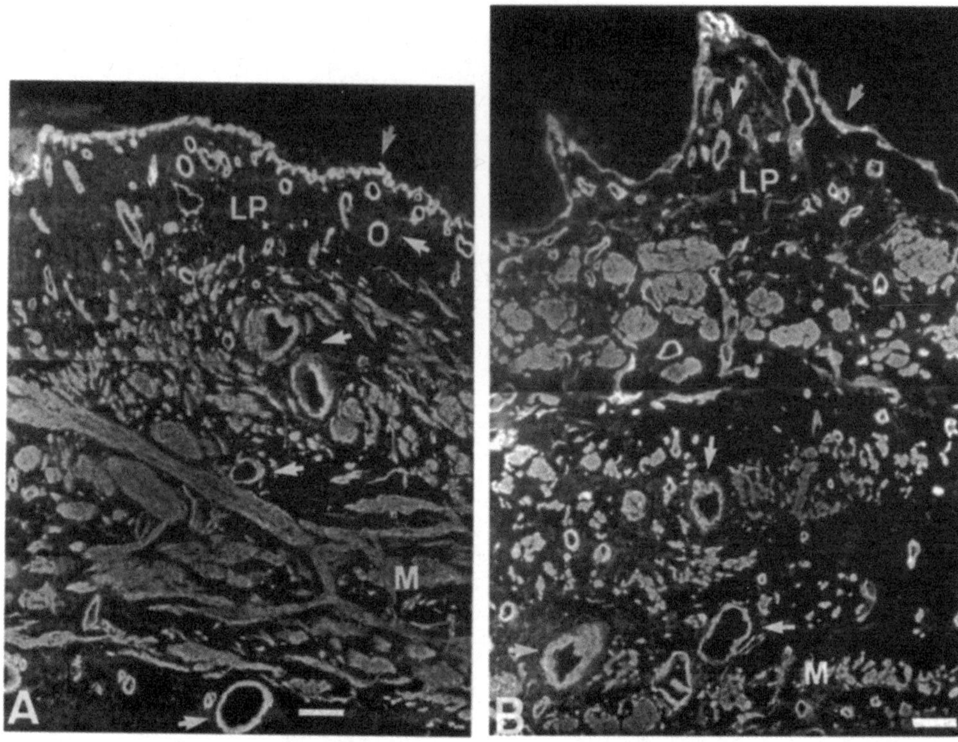

Figure 2. Comparison of vasculature and muscle cell arrangement of diverticula from control (A) and EDS (B) bladders delineated by staining with an antibody against type IV collagen. Note the dilated blood vessels in both diverticular walls (arrowheads). Note the lack of organized muscle bundles in the outer detrusor layer of the EDS diverticulum (B) compared to the diverticulum of the congenital control. Magnification = 75X, bar = 100μm.

bundles (Figure 3A,B). This investment was smooth and uniform in the native control bladder (Figure 3A), but appeared thickened and irregular around the native EDS muscle cells (Figure 3B).

Since the diverticuli were highly expansive, we analyzed the tissue sections for the localization of elastic fiber components, elastin and the elastin associated microfibrillar protein, fibrillin-1. Elastin appeared as fibers in the matrix and associated with smooth muscle cells in both native and diverticular walls of both patients (Figure 4). The native bladder segments of both the control (A) and EDS (B) showed the presence of wavy elastin fibers associated with smooth muscle cells of the detrusor layer. Although not quantitative, there appeared to be more elastin present in the native EDS bladder sections compared to controls. Comparison of the diverticular walls with elastin antibody staining revealed that the elastin fibers appeared associated with and connecting between muscle fascicles (Figure 4C) in the control diverticular wall. In the EDS diverticular wall, elastin fibers appeared thicker and no direct inter-association of muscle fibers could be delineated (Figure 4D).

In a similar fashion, fibrillin-1 was associated with detrusor smooth muscle cells of both the control diverticulum and EDS diverticulum (data not shown). Fibrillin-1 localized heavily to the urothelial cells of both the control and EDS diverticuli. However, little staining of the control diverticular lamina propria was detected, in contrast to staining of fine fibers throughout the lamina propria of the EDS diverticulum.

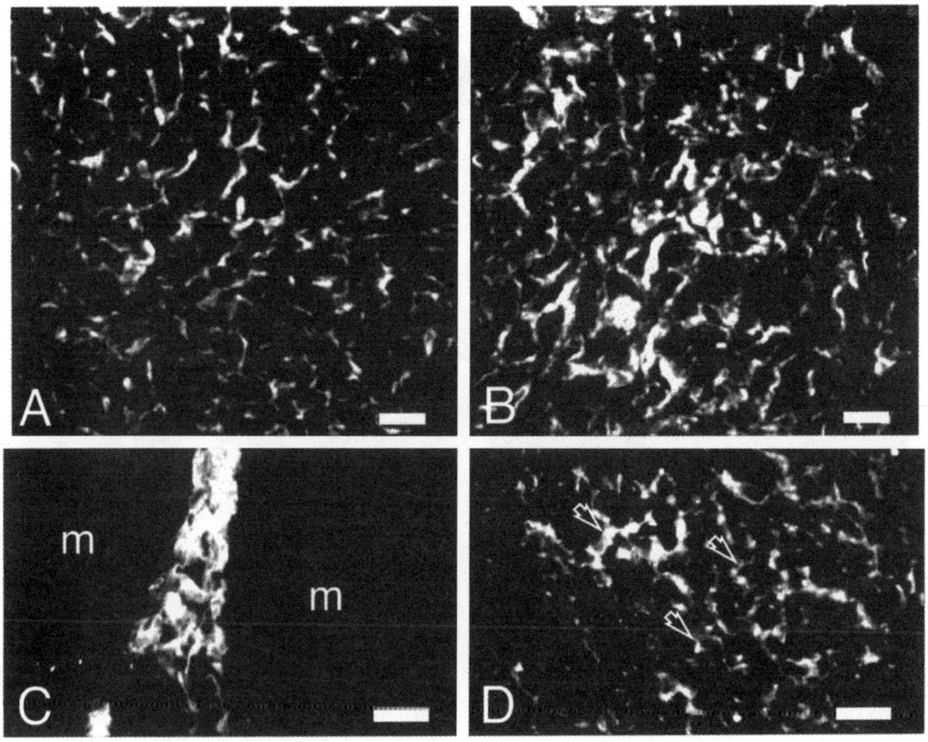

Figure 3. Comparison between native control detrusor (A, C) and native EDS detrusor (B, D). Confocal micrograph of native control bladder (A) and native EDS bladder (B) detrusor muscle stained with antibody against type IV collagen (0.7 μm thick). Note the thickened and irregular pattern of collagen IV in EDS native bladder (B) compared with control native bladder (A). Magnification = 740 X, bar = 100 μm. Immunofluorescent staining with antibody against collagen III (C, D). Note the presence of infiltration of collagen within the muscle bundles of EDS native bladder (B and D, arrows) and minimal, if any collagen type III in control muscle m (A and C). Magnification = 218 X, bar = 45 μm.

3.2. Steady-State RNA Analysis

Linear regression analysis of steady-state mRNA levels for the $\alpha 1$ chains of both type I and type III collagen revealed that the EDS diverticulum was characterized by a modest, coordinate increase of 41% and 32% in α-1(I) and α-1(III) steady-state mRNA levels, respectively, compared to the native EDS bladder segment. RNA analysis was not carried out on the native bladder control sample due to the paucity of available tissue.

3.3. Matrix Protein Analysis

Quantitation of hydroxyproline, an imino acid found almost exclusively in collagen, was carried out to determine the total collagen concentration in both the diverticuli and native bladder wall samples from both patients. The analysis showed that there was an increase in total collagen content in both diverticuli, between 52–64%, compared to their respective native bladder wall segments (Table 1). When quantified by indirect ELISA, a similar 50% increase in Type I collagen was also found in both diverticuli compared to their native segments. When type III collagen was analyzed, there was a similar 50% in-

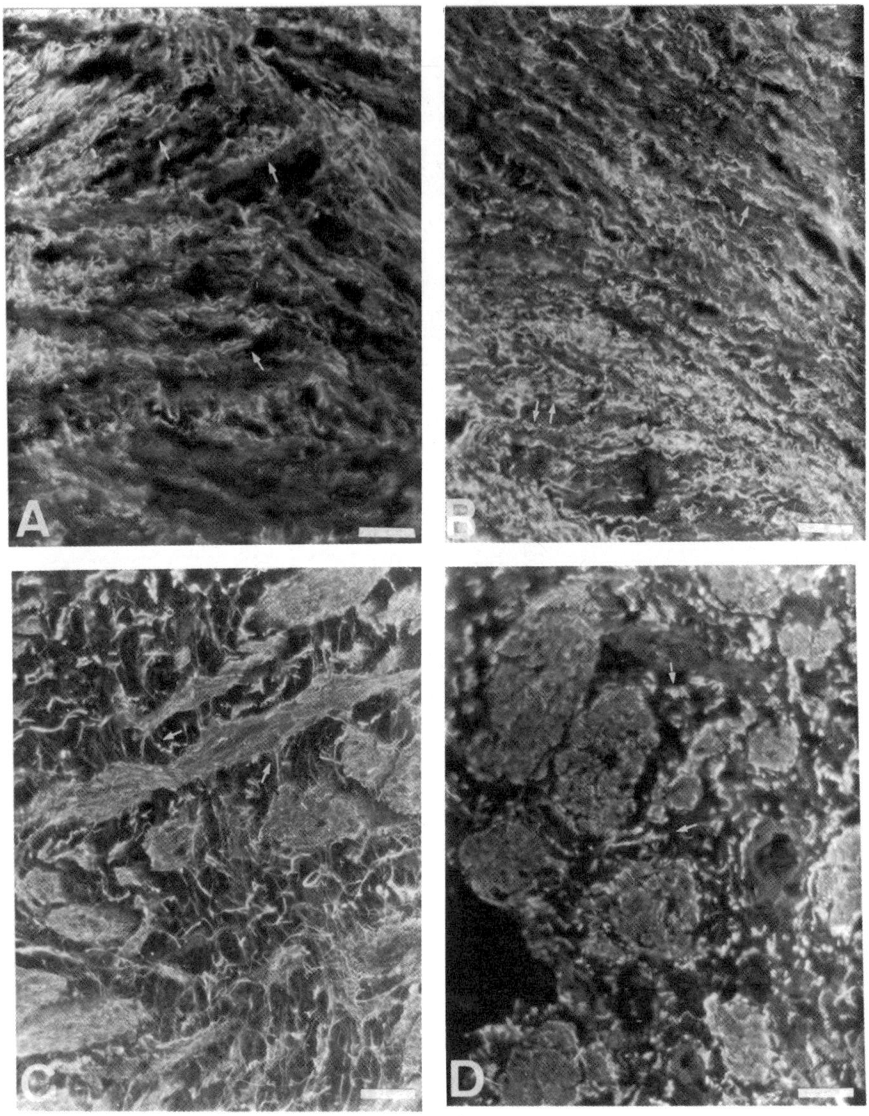

Figure 4. Immunofluorescent staining with an antibody against elastin of detrusor muscle from native bladder wall segments (A=control, B=EDS) and diverticular wall segments (C=control, D=EDS). Note the increased staining of fine wavy fibers in (B). In the diverticular walls, not that the elastin fibers appear "connected to the muscle bundles in the control (C), whereas the elastin fibers in the EDS diverticular wall appear to be thicker, but not connected to other structures. Magnification = 187.5 X, bar = 50 μm.

crease in the control diverticulum compared to its native segment. However, the type III collagen values increased by 115% in the EDS diverticulum compared to its native segment, resulting in an increase in the type III : type I collagen protein ratio.

The EDS diverticulum contained 157.4 mg/g dry weight of insoluble elastin, compared to only 44.6 mg/g dry weight for the native EDS bladder segment. The elastin residue fraction of the control bladder segments was similar to the native EDS segment: control native bladder – 47.6 mg/g dry weight, control diverticulum – 50.6 mg/g dry

Table 1. Quantification of matrix protein (mg/gram dry weight)

	Total collagen	Collagen I	Collagen III	Elastin	Fibrillin-1
Native control	384.2	311.2	73	47.6	ND
Tic control	582.2	473.9	108.3	50.6	4.1
% change Tic control	52%	52%	49%	6%	
Native EDS	211.3	174.3	37	44.7	1.5
Tic EDS	346.4	266.7	79.7	157.4	5.0
% change Tic EDS	64%	53%	115%	252%	233%

ND: not determined.

weight. The EDS diverticulum was therefore characterized by an increase in elastin of 252%.

Analysis of bladder wall tissue by indirect ELISA with a monospecific antibody produced against a synthetic peptide sequence of fibrillin-1 revealed that the EDS diverticulum was characterized by an increase in soluble fibrillin-1 concentration, 5.0 mg/g dry weight compared to the native EDS bladder, 1.5 mg/g dry weight (Table 1), an increase of 233%. When the ratio of elastin to fibrillin concentration was determined for each tissue segment, there was nochange in these values (Table 2). Due to paucity of tissue sample, no fibrillin-1 data was generated for the native control bladder wall.

Comparison of the ratios of matrix proteins (Table 2) showed that there was an increase in the ratios of elastin and fibrillin-1 to both collagens I and III in the EDS diverticulum compared to its native segment, however, the fibrillin-1 to elastin ratio remained constant.

3.4. Physiological Data

In order to investigate the compliance or elasticity of these tissues, stress-strain analyses of muscle strips were performed with each tissue sample (diverticulum and native bladder walls of both control and EDS). Comparison of diverticular wall segment from the control patient with its respective native bladder wall tissue showed that the diverticulum was more compliant (Figure 5). A similar comparison of segments from the EDS patient showed that the compliance properties were equivalent to one another. However, both EDS bladder segments were more compliant than their respective control bladder samples. The same experiments were then performed in presence of sodium azide (NaN_3), to eliminate the smooth muscle cell contribution to bladder wall compliance (Figure 6). Results of these analyses showed that, in the absence of muscle tone, both control and EDS native bladder wall segments became more compliant than their respective diverticular segments. While the native EDS bladder and its diverticulum were still more com-

Table 2. Ratio of matrix proteins

	Control		EDS	
	Native	Diverticulum	Native	Diverticulum
Fibrillin-I:elastin	ND	0.08	0.03	0.03
Elastin:collagen I	0.15	0.11	0.26	0.59
Elastin:collagen III	0.65	0.47	1.21	1.97
Fibrillin-I:collagen I	ND	0.01	0.01	0.02
Fibrillin-I:collagen III	ND	0.04	0.04	0.06

ND: not determined.

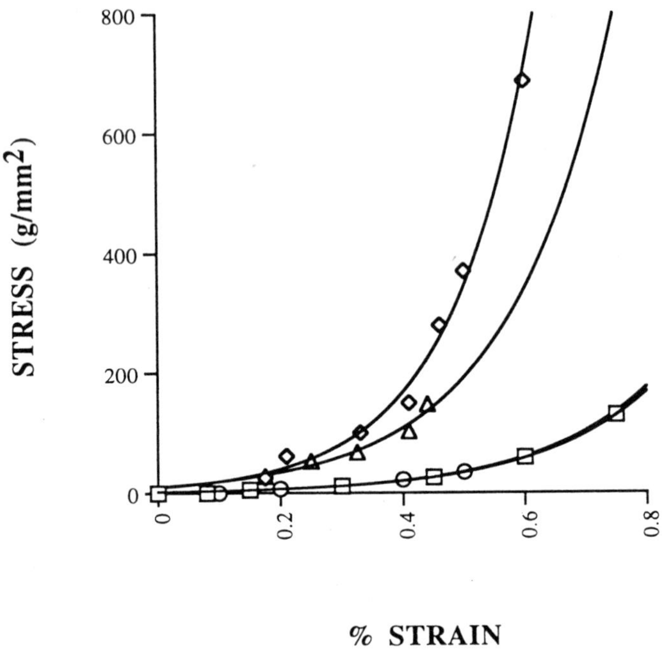

Figure 5. Stress-strain relationships of muscle strips in absence of sodium azide. Diamond: Native control bladder; triangle: Diverticula Control; square: Native EDS bladder; circle: Diverticula EDS.

pliant respectively than the control bladder and its diverticulum, there was little change in the compliance values of both diverticuli in the presence or absence of sodium azide. These data indicate that there is little contribution of smooth muscle tone to the compliance properties of the bladder diverticular walls.

4. DISCUSSION

The presence of bladder diverticulum associated with Ehlers-Danlos syndrome has been previously reported.[5–11] EDS is a congenital connective tissue disorder characterized by abnormalities of collagen structure and synthesis. Although the association between the occurrence of bladder diverticuli and EDS has been characterized through case reports in the literature, a molecular and biochemical basis for the diverticular pathology and its ability to recur has not been established. Several investigators have hypothesized as to the pathophysiology of this phenomenon. Zalis and Roberts[9] and Breivik et al.[10] suggested that the increased extensibility of the connective tissue was the underlying factor in the genesis of diverticuli. Levard et al.[8] and Rabin et al.[11] noted, after histologic examination of their patient's tissue, that the diverticular walls had a reduced amount of smooth muscle. Levard et al.[8] also observed that the muscle bundles were disrupted by connective tissue sclerosis. Rabin et al.[11] established that collagen fibers in their patient's bladder wall had features previously described for patients with type IX EDS.

The initial diagnosis of the EDS patient in this study was that of EDS type VII. EDS VII is characterized by a heterozygosity of autosomal dominant mutations in the type I

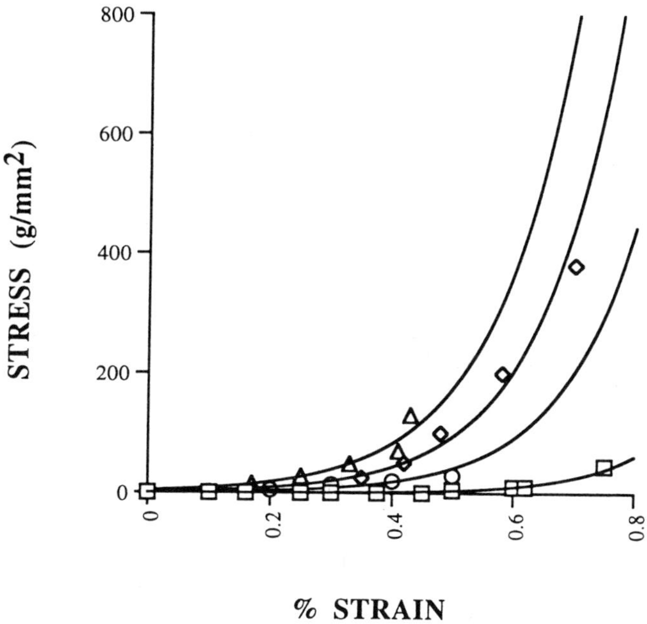

Figure 6. Stress-strain relationships of muscle strips in presence of sodium azide. Triangle: Diverticula Control bladder; diamond: Native Control bladder; circle: Diverticula EDS; square: Native EDS.

collagen genes (EDS VII A and B). A more rare form, EDS VII C, is characterized by a deficiency of procollagen N-proteinase, which fails to cleave the amino-propeptide of type I collagen. However, bladder diverticuli have also been observed in patients with EDS IX, a rare X-linked recessive form, which is characterized by abnormal copper utilization, which results in a decrease in lysyl oxidase activity required for collagen and elastin cross-linking. Further genetic analysis will be required to characterize the specific defect in this EDS patient.

The findings of our study represent the first comprehensive report of the connective tissue composition and physiological characterization of the native bladder and diverticular wall associated with Ehlers-Danlos syndrome. Similar to the studies referenced above, we also observed differences in the amount, size and arrangement of the smooth muscle component in the diverticular walls. The EDS bladder was further characterized by an infiltration of type III collagen among rudimentary muscle bundles and Type III collagen was quantitatively increased in the EDS diverticulum compared to its native bladder segment. Type III collagen is normally found in tissues and organs which are extensible, as in normal bladder. It has also been shown to be a component of heterotypic collagen fibrils, i.e. collagen fibrils composed of more than one genetic collagen type. However, an alteration in the concentration of type III collagen, a change in the ratio of type III:I present in heterotypic fibrils, and an alteration in the arrangement of the fibers with one another, with other matrix components and with the cellular components could be responsible for changes in the functional capabilities of the tissue or organ.

We have previously shown that type III collagen was increased in both anatomically and functionally obstructed human bladders, the majority of which were characterized by urodynamic studies as having low- or non-compliant (i.e., stiffer) bladders.[18] In the present study the bladder diverticuli were also characterized by an increase in type III collagen; however, by stress-strain analysis, they were more compliant (i.e., less stiff). Is there a common mechanism which could be responsible for these similarities in type III collagen accumulation and differences in compliance? We have previously speculated that a mechanical stimulus could be the "signalling mechanism" for increased type III collagen synthesis: a "stretch" stimulus in the case of the diverticuli, and a "modified stretch"—rapid uninhibited contractions coupled with increased smooth muscle tension in the case of the "non-compliant" bladders. One major difference between these groups was the detrusor smooth muscle layer. In the diverticuli, there were fewer muscle cells, and a lack of organized muscle bundles or fascicles. Conversely, in the non-compliant bladders, the muscle was characterized by hypertrophy/hyperplasia, trabeculation and patchy, type III collagen infiltration of some, but not all detrusor smooth muscle bundles. We have previously postulated that the detrusor layer acts to "gird" or limit the distensibility or overexpansion of the lamina propria connective tissue matrix (the capacitance layer).[21,22] This mechanism would explain the thickened bladder wall and non-compliant nature of the obstructed bladders. In the diverticuli, lack of a normal detrusor layer would allow for expansion or "uncoiling" of the collagen fibers with increased filling and capacitance. "Stretching" of the existing muscle cells may increase the synthesis of connective tissue proteins. The increase in steady state collagen mRNA levels and also accumulated collagenous proteins in the EDS diverticulum compared to the native bladder wall supports this hypothesis. Further, we have previously shown that in vitro stretching of bladder smooth muscle cells results in an increase in the synthesis of types I and III collagen.[24]

Type IV collagen normally forms a cocoon-like investment surrounding individual smooth muscle cells. Immunostaining with an antibody against type IV collagen showed a thickened and irregular pattern of basement membrane investment around smooth muscle cells of both native and diverticular walls of only the EDS patient. It is tempting to speculate that the additional basement membrane between muscle cells along with infiltrated type III collagen could interfere with normal signalling between smooth muscle cells, and result in decreased contribution of muscle tone.

The histological and immunohistological data demonstrated that the diverticulum as well as normal part of the bladder of the EDS patient displayed large abnormalities of the wall architecture. These latter correlate with the changes in compliance observed for EDS bladder and diverticular walls which had increased elasticity compared respectively to control bladder and its diverticulum. Assessment of the role of elastin and elastin-associated microfibrillar protein, fibrillin-1, showed that, quantitatively, the ratio of fibrillin-1 to elastin did not change in the EDS diverticulum compared to the EDS native segment. However, in the diverticulum, there was an approximately 250% increase in both matrix proteins, which could be directly related to the increased expansion and compliance of the diverticulum.

Although bladder tissue hyperelasticity in EDS diverticuli has been suspected,[8] this report documents for the first time direct evidence of hyperelasticity of EDS bladder wall associated with a disorganization of bladder wall architecture and changes in extracellular matrix composition. This suggests that the composition and/or arrangement of both the structural and cellular elements of the EDS bladder are different than that of the control bladder. The ability of the EDS native bladder and diverticulum to remain more compliant when muscle tone was ablated suggests that these compliance changes may be attributed to the matrix components of these tissues. The reduced amount of smooth muscle in the diverticulum in association with increased concentrations of type III collagen, elastin and fibrillin-1 may account for the hyperextension of the EDS diverticular wall.

Further observations that both diverticuli were 1) more compliant than their respective native portions and, 2) showed no change in the stress/strain relationship in the presence or absence of sodium azide suggest that a common element between both diverticuli could be the decreased muscle content and organization in the diverticuli and/or a lack of the appropriate linker connections between the muscle cells and the matrix. Since the compliance curves of both sodium azide treated native bladder segments "shifted to the right" compared to untreated native segments, smooth muscle tone (active compliance) appears to play a role in the overall compliance properties of the native bladder walls. Active tension did not appear to contribute to the compliance properties of either diverticuli. Therefore the differences in the stress-strain relationships between the congenital control diverticulum and EDS diverticulum, the latter more compliant, can only be attributed to the passive complinace elements—the extracellular matrix proteins.

ACKNOWLEDGMENTS

This paper was supported by grants DK48215 and R29-DK44144 from the National Institutes of Health. The authors wish to acknowledge the assistance of Dr. Peter Bannerman from the Neurologic Department of Children's Hospital of Philadelphia in obtaining confocal pictures of anti-collagen IV immunostained frozen sections, Drs. Dorothea Rohrmann and Gregory Dean for performing muscle strips experiments, Dr. Umberto Kucich for ELISA assays, and Dr. William Abrams for amino acid analyses which were performed in the Biopolymer Analysis Laboratory, University of Pennsylvania Dental School.

REFERENCES

1. Yeowell H.N., and Pinnell S.R.: The Ehlers-Danlos syndromes. Seminars in Dermatology, **12:** 229–240, 1993.
2. Byers P.H. Disorders of collagen biosynthesis and structure, Chap. 115, in Scriver, CL:The Metabolic Basis of Inherited Diseases, New York, McGraw-Hill, 1989.
3. Toyohara T., Kanako T., Araki H., Takahashi K., and Nakamura T.: Giant epiphrenic diverticulum in a boy with Ehlers-Danlos syndrome. Pediatric Radiology, **19**: 437, 1989.
4. Junquera F.F., Santos V.J., Molero R.J., and Vilaseca M.J.: Ehlers-Danlos syndrome with giant duodenal diverticulum as a cause of cholangitis. Gastroenterologia Hepatologia, **18:** 84, 1995.
5. Cuckoro P.M., Blackhall R.J., and Mouriquand P.D.,: Huge bladder diverticula associated with Ehlers-Danlos syndrome. Journal of the Royal Society of Medicine, **87:** 290, 1994.
6. Bade J.J., Ypma A.F., Van Elk P., and Mensink H.J.: A pelvic mass: bladder diverticulum with haemorrage in Ehlers-Danlos syndrome patient. Scandinavian Journal of Urology and Nephrology, **28:** 319, 1994.
7. Stage K.H., and Tank E.S.: Primary congenital bladder diverticula in boys. Urology, **40:** 536, 1992.
8. Levard G., Aigrain Y., Ferkadji L., Elghoneimi A., Pichon J., and Boureau M.: Urinary bladder diverticula and the Ehlers-Danlos syndrome in children. Journal of Pediatric Surgery, **24:** 1184, 1989.
9. Zalis F.G., and Roberts D.C.: Ehlers-Danlos syndrome with a hypoplastic kidney, bladder diverticulum and diaphragmatic hernia. Arch. Dermatol., **96:** 540, 1967.
10. Breivik N., Refsum S. Jr., Oppedal B.R., and Vesterhus P.: Ehlers-Danlos syndrome and diverticula of the bladder. Z. Kinderchir., **40:** 243, 1985.
11. Rabin J.M., Hirschfield L., and Badlani G.H.: Type IX Ehlers-Danlos syndrome: Bladder diverticula with transitional cell carcinoma. Urology, **38:** 563, 1991.
12. McKusick V.A.: Connective tissue diseases in man, in: Catalogues of AutosomalDominant, Autosomal recessive, and X.Linked Phenotypes,. Baltimore, JohnsHopkins Univ Press, 1983, pp 1004–1026.
13. Humason G.L.: Animal tissue techniques. San Francisco, W.H. Freeman Co., pp 121, 1962.
14. Macarak E.J., Howard P.S., and Lally E.T.: Production and characterization of a monoclonal antibody to human type III collagen. J. Histochem. Cytochem., **34:** 1003, 1986.
15. Scheinman J., and Tsai C.: Monoclonal antibody to type IV collagen with selective basement membrane localization. Lab. Invest., **50:** 101, 1984.

16. Werkmeister J., Ramshaw, and Ellender G.: Characterization of a monoclonal antibody against native human type I collagen. Eur. J. Biochem., **187:** 439, 1990.
17. Chomczynski P., and Sacchi N.: Single-step method of RNA isolation by acid guanidinium thiocyanate-phenol-chloroform extraction. Anal Biochem., **162:** 156, 1987.
18. Deveaud C.M., Macarak E.J., Kucich U., Ewalt D.H., Abrams W.R., and Howard P.S.: Molecular analysis of collagens in bladder fibrosis. J. Urol., In Press, 1998.
19. Neuman R and Logan M: The determination of collagen and elastin in tissues. J. Biol. Chem., **186:** 549, 1950.
20. Baskin L.S., Constantinescu S.C., Howard P.S., McAninich J.W., Ewalt D.H., Duckett J.W., Snyder H.M., and Macarak E.J.. Biochemical characterization and quantitation of the collagenous components of urethral stricture tissue. J. Urol., **150:** 642, 1993.
21. Ewalt D.H., Howard P.S., Blyth B., Snyder H.M., Duckett J.W., Levin R.M. and Macarak E.J. Is lamina propria matrix responsible for normal bladder compliance? J. Urol., **148:** 544, 1992.
22. Coplen D.E., Macarak E.J. and Levin R.M.. Developmental changes in normal fetal bovine whole bladder physiology. J. Urol., **151:** 1391, 1994.
23. Dean G.E., Cargill R.S., Macarak E.J., Snyder H.S., Duckett J.W. and Levin R.M. Active and passive compliance of the fetal bovine bladder. J. Urol., **158:** 1094, 1997.
24. Baskin L.S., Howard P.S. and Macarak E.J.. Effect of physical forces on bladder smooth muscle and urothelium. J. Urol., **150:** 601, 1993.

17

THE ROLE OF COLLAGEN IN BLADDER FILLING

Edward J. Macarak and Pamela S. Howard

University of Pennsylvania and The Division of Pediatric Urology
Children's Hospital of Philadelphia
Philadelphia, Pennsylvania 19104

1. INTRODUCTION

Under normal circumstances, the bladder fills with little or no increase in intravesical pressure. When the volume capacity is reached, afferent signals, likely from stretch receptors associated with nerve endings, trigger the micturition reflex. The bladder then systematically empties and a new filling cycle begins. During each cycle, there must be a close integration of tension generating (muscle cells or "active" elements) with tension bearing and elastic elements (collagen and elastin), that are passive structural components.

Despite the intense study of the active components of the bladder wall, relatively little is known with respect to the linkages that exist between active and passive elements. In addition, there is little understanding of the structural architecture of the bladder wall and how it accommodates during bladder filling. While it can be demonstrated that the passive elements are required for normal function, there is little appreciation regarding the actual mechanism by which structural macromolecules contribute to overall function. Information currently available suggests that compositional changes in the visco-elastic or passive elements of the bladder occur in response to a variety of pathologic conditions. In most instances, these changes result from some form of obstructive uropathy. If there is increased resistance at the bladder neck, the bladder must correspondingly increase the pressure within the bladder to open the bladder sphincter and expel the urine. If the bladder is forced to operate under these conditions for extended periods of time, compensatory changes occur with respect to physiologic function and matrix composition.[1,2,3,4,5,6 and references therein]

Outflow obstruction can occur as a result of a physical obstruction at the bladder neck or proximal urethra. In some instances, these obstructions can be of an acute and congenital nature, e.g., posterior urethral valves. Obstruction can also occur chronically. Most males over the age of 40 experience symptoms related to mild obstruction as a result of benign prostatic hyperplasia. Other forms of outlet obstruction can best be described as

Advances in Bladder Research, edited by Baskin and Hayward.
Kluwer Academic / Plenum Publishers, New York, 1999.

functional obstructions, e.g., detrusor - sphincter dyssynergia. In this instance, the obstruction results not from a physical obstruction but rather from a functional inability to coordinate the actions of the sphincter and detrusor muscle. In effect, it is a form of obstruction similar to anatomic ones because in all cases, the bladder is forced to work against an increased outlet resistance. Thus, while the etiology of these conditions may be different, the end results are quite similar in that bladder function is adversely affected.

A major clinical symptom associated with outlet obstruction of any type is some form of bladder instability.[7] Bladder instability can manifest itself in the form of involuntary contractions that alters the normal cycling regimen of the bladder. With increasing instability, the bladder wall is subjected to a different regimen of cycling, i.e., it may be forced to spend greater portions of the filling cycle at higher than normal pressures. This integral of time and pressure is an important measure of normal cycling. In addition to sustaining higher intravesical pressures, the wall may be experiencing involuntary contractions subjecting the constituent cells to a variety of abnormal mechanical signals. Whatever the underlying cause, the cells within the bladder wall begin receiving abnormal "cues" to which they respond. Stimulation of cells can occur by autocrine or paracrine secretion of cytokines, neurotransmitters, hormones and/or other "effector" molecules.[8,9] In addition, increased or prolonged mechanical "stretch" of the bladder wall can alter channels within the cell membranes of nerve or other bladder wall cells.[10] Ultimately these signals, regardless of their origin or associated pathology, alter the phenotype of bladder wall cells. It is likely that the altered phenotypic expression of bladder wall cells leads to some of the physical and compositional changes that occur in response to outlet obstruction.

Initially these changes are largely reversible. If the obstruction is relieved before the bladder undergoes "decompensation", both structural and functional changes almost completely disappear.[11] Thus there appears to be a "window" during which intervention is beneficial to the long term outcome of the patient. Animal studies from a number of laboratories have demonstrated that the initial response of the bladder wall to obstruction can best be described as a compensatory response. That is, the bladder muscle responds to the increased work load caused by the obstruction by increasing its muscle mass. Biologically, the detrusor is functionally compromised and must increase its contractile force to overcome the increased resistance at the bladder neck. In response to this functional challenge, the detrusor compensates by increasing the overall amount of contractile protein. This can be observed biochemically as the level of actin and myosin increases as well as other associated contractile proteins.

In addition to changes in contractile proteins, alteration in the connective tissue composition of the bladder wall also occurs. While these changes are substantial, they are not as great as those in contractile proteins. Measurements of collagen usually show a significant absolute increase in collagen; however, relative to the more substantial increase in overall mass resulting from detrusor hypertrophy, such changes are usually not significant.

2. RESULTS

Changes in collagen manifest themselves in two ways: 1) a change in the distribution and 2) a change in the concentration. To evaluate these changes, our strategy has been to employ immunohistochemical, biochemical and molecular biological approaches. Alteration in collagen architecture is best studied using immunohistochemistry with collagen type-specific antibodies to evaluate the distribution of collagens in normal and pathologic bladder tissue.

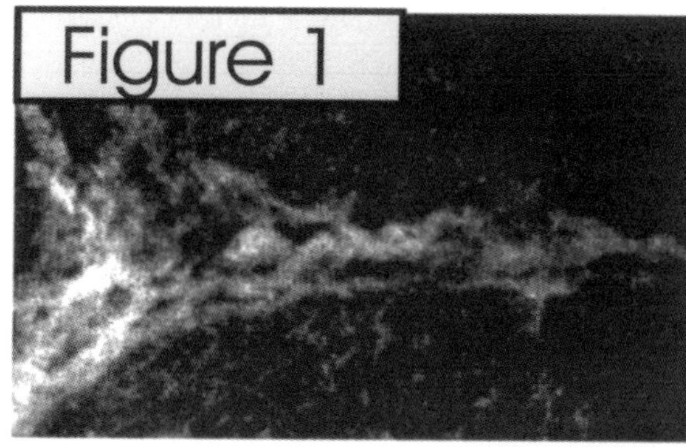

Figure 1. Type III collagen localization in a detrusor muscle bundle from a bladder with normal compliance. Localization is largely confined to the interfascicular regions of the detrusor.

Previous studies from our laboratory as well as those from other laboratories have clearly demonstrated that changes in collagen composition accompany various types of outlet obstruction. Ewalt et al. showed that neurogenic bladders have an altered distribution of type III collagen.[12] In Figure 1, a section from a normal bladder was stained with an antibody to type III collagen. The section shows several muscle bundles with localization of the type III collagen confined largely to the perimysial region surrounding the detrusor muscle bundles (MB). Little or no type III collagen is found within muscle bundles. In contrast, type III collagen in a subset of neurogenic bladders is found not only in the perimyseal location but also within muscle bundles (endomysial location, Figure 2). These changes appear to be specific for type III collagen since type I collagen shows no similar change in distribution (data not shown). Thus sections through noncompliant bladders characteristically show altered distribution of type III collagen within some but not all muscle bundles.

Thus immunohistochemistry can demonstrate changes in a particular collagen type and location in selected patient tissue samples. While such information is of obvious value, it is difficult to characterize these changes without some means to objectively evaluate them. Therefore, it is necessary to biochemically evaluate these same changes by applying techniques that lend themselves to their quantification. To evaluate and quantify changes in collagen, tissue samples are solubilized using cyanogen bromide. This reagent cleaves proteins containing methionine making the resulting peptides soluble. When bladder tissue is treated with this reagent, approximately 95% of the tissue weight is solubilized. The small residue that remains undissolved is largely elastin that does not contain the amino acid methionine and is therefore resistant to cyanogen bromide digestion. This approach was used to study a group of patients with non-compliant bladders.[13]

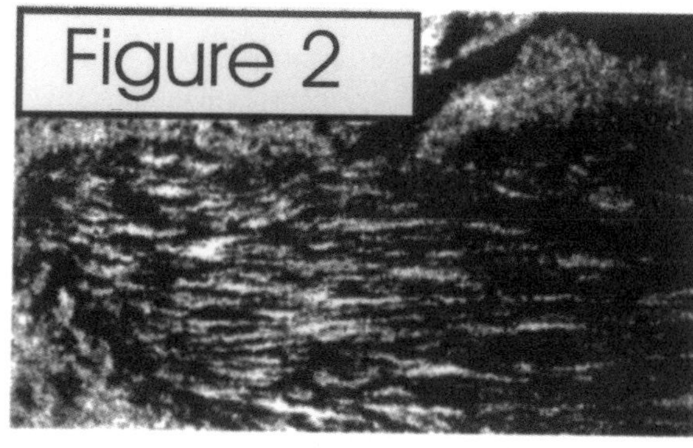

Figure 2. Type III collagen localization ina detrusor muscle bundle form a noncompliant bladder. Note the increased pericellular localization of type III collagen within the muscle bundles (intrafascicular region).

Table 1. Collagen quantification

	Collagen (mg/gm dry wt.)	% Type I	% Type III	Type I (mg/gm dry wt.)	Type III (mg/gm dry wt.)
Normal	263 ± 60	76.2 ± 3	23.8 ± 3	200.7 ± 49	62.6 ± 15
Non-compliant	289 ± 52	67.2 ± 11	32.8 ± 11	193.6 ± 53	93.1 ± 36
Change	9.8%	12%	37.8%	−3.67	48.7%

Tissue samples from pediatric patients with either functional (largely neurogenic bladders) or anatomic (largely posterior urethral valves), were biochemically analyzed to evaluate changes in matrix composition. Results show an overall increase in hydroxyproline content in dysfunctional bladders; although, not surprisingly, the change is not significant when normalized to overall bladder weight. However, analysis of the absolute increase in collagen, that is the increase in collagen mass relative to that found in controls, is significant. Regardless of whether either neurogenic and nonneurogenic bladders are analyzed separately or combined as a group, type III collagen increases by approximately 48% (Table 1).[13] Thus the biochemical and immunohistochemical results are in agreement and demonstrate that both functional and anatomic obstruction can lead to an alteration in not only the location but also the quantity of type III collagen.

Since the major fibrillar collagens found in the bladder wall are types I and III collagen, a consequence of these changes is an alteration in the type III:type:I collagen ratio (Table 2). In normal bladders, the collagen composition is approximately 25% type III collagen and 75% Type I collagen (Type III:I ratio = 1:3) In bladders with progressive fibrotic changes, this ratio shifts to approximately 33% type III and 66% Type I collagen (Type III:I ratio = 1:2). While little is currently known about the functional consequences of such changes, there are several possibilities that likely relate to the mechanical properties of the tissue.

These data as well as those from other studies support the idea that abnormal cycling secondary to obstruction results in a progressive fibrosis of the bladder wall. This fibrosis is characterized by infiltration of some detrusor muscle bundles with increased quantities of type III collagen. Changes at the protein level are generally preceded by alterations in gene activity of the respective genes that encode for these proteins. These changes likely have potential functional significance although how tension is actually borne by collagen fibers in the bladder wall is not well-understood. For example, if alteration in type III collagen occurs and is associated with changes in the functional compliance properties of the bladder wall, what role do these fibers play in the architecture of the bladder wall and in tension transfer

To begin to answer this question, we have analyzed three sets of fetal bovine bladders at gestational ages of 2nd trimester, 3rd trimester and term or late gestation. Gestational ages were determined by measurement of crown-rump lengths. Fetal bladders from all gestational ages were filled to either (a) 0%, (b) 25%, (c) 50%, (d) 75% or (d) 100% of their respective capacities. Total capacity was defined as the point in the cystometrogram

Table 2. Type III: type I ratio

Non-compliant	Control
32.8%:67.2%	23.8%:76.2%

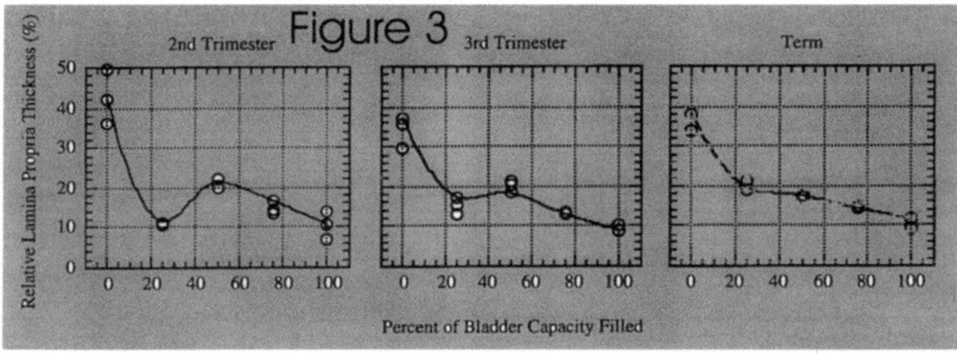

Figure 3. Relative lamina propria thickness at different gestational ages. Bladders were filled to different levels of their respective total capacities with OTC freezing compound and frozen. Sections were taken and stained with hematoylin and eosin and used for measurments of relative lamina propria thickness. Using a calibrated eyepiece on a dissection microscope, the relative thicknesses of the lamina propria and detrusor were measured. The thickness of the lamina proprai was divided by the total thickness of the bladder wall and multiplied by 100 to derive the relative percentage thickness of the lamina propria. Measurements were made on three fetal bladders at each point in the filling cycle.

at which a sharp rise in intravesical pressure occurred. After filling, bladders were allowed to equilibrate for approximately 15 minutes before being snap frozen and sectioned through the bladder dome. Sections were stained with antibodies to type III collagen and the position and tortuosity of type III collagen fibers noted.

In addition, some sections were stained with conventional histologic stains and the thicknesses of the lamina propria and detrusor measured. For these measurements, sections were viewed using a dissecting microscope fitted with a calibrated eyepiece. The thickness of the lamina propria was divided by the total thickness of the bladder wall and multiplied by 100 to derive the relative percentage of the total thickness contributed by the lamina propria.

The results of these measurements show that the relative thickness of the lamina propria to the overall bladder wall thickness generally decreased as the volume within the bladder increased at all developmental stages studies (Figure 3). In the three gestational ages studied, the lamina propria comprised approximately 40% of the total bladder thickness. In addition, as filling increased from 50 to 100% of capacity, bladders from all gestational ages demonstrated a progressive and linear decrease in the relative contribution of the lamina propria to the overall wall thickness. Thus, during this portion of the filling cycle, the lamina propria is thinning more rapidly than is the detrusor layer.

Distinct differences, however, were observed in the filling of 2nd and 3rd trimester bladders. These bladders exhibited three distinct phases as determined by changes in the relative contributions of the lamina propria and detrusor to overall thickness. In 2nd and 3rd trimester bladders, the first phase coincided with filling from 0–25% during which time there was a significant decrease in the relative thickness of the lamina propria. A second distinct phase occurred as bladders were filled from 25–50% of capacity. During this portion of the filling cycle, there was a relative increase in the thickness of the lamina propria, or more accurately, the detrusor was thinning as a faster rate than was the lamina propria. The final phase occurred between 50 and 100% of capacity. During this phase, the lamina propria once again thinned at a faster rate than did the detrusor. Thus for 2nd and 3rd trimester bladders, the lamina propria thins at a faster rate than does the detrusor during all portions of the filling cycle except when the bladder is filled from 25 to 50% of capacity.

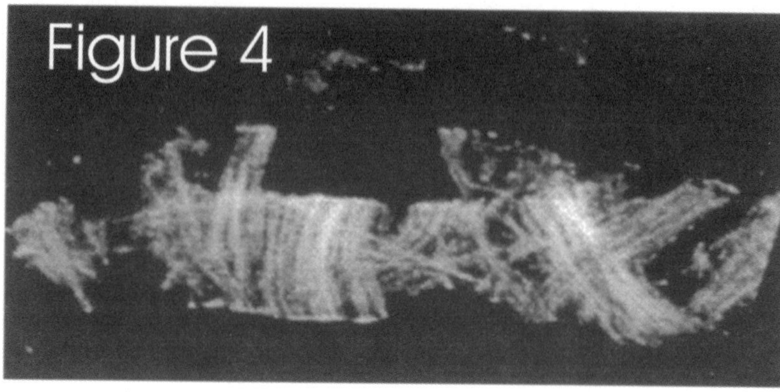

Figure 4. Confocal microscopy of the type III collagen fibers in the detrusor layer showinh extension of the fibers. To the left, the fiber is tightly coiled while to the right the fibers are extended or more taut. (bar = 10mM).

These data are in contrast to those from the term bladders that showed a consistent decrease in the relative lamina propria thickness during all phases of the filling cycle (Figure 3). These data suggest that during the 2nd and 3rd trimesters of development, tension is sustained largely by the lamina propria except when the bladder is filled to 25–50% of capacity. During this portion of the filling cycle, tension is borne by the detrusor layer since it is thinning at a faster rate then is the lamina propria.

To correlate these changes with the actual morphology of collagen fibers, sections stained with antibody to type III collagen were examined immunohistochemically. While the fibers containing type III collagen were clearly visualized, it was difficult to evaluate their internal conformation (data not shown). To more clearly resolve these structures, additional observations were performed using a laser scanning confocal microscope.[14]

Using the confocal microscope, the internal conformation of the fibers can clearly be visualized.[13] As shown in Figure 4, fibers containing type III collagen exhibit a characteristic "coiled configuration" that can be complex. When tissue from bladders at different levels of filling were examined, it was observed that the tortuosity of the coils was varied depending upon the state of filling. For example, in empty bladders, coils of type III collagen were compact and showed maximum tortuosity (Figures 5 A and B). In partially filled bladders, type III collagen fibers were less tortuous and had an expanded coil-structure. As shown in Figures 5C and 5D, bladders filled to 25% of capacity exhibit expanded coils that were thinner than the coils observed in the empty bladder. The drawings to the right of the photographs in Figure 5 represent our perception of how the coil structure expands during filling to become less tortuous.

Finally, when type III fibers in full bladders were examined, the collagen coils showed the least tortuosity and had the most expanded coil structure (Figure 5e). Note that the diameter of the coil is considerable less than that seen for fibers in Figure 5. Thus, the period of the collagen coiling (the distance from the "peak" of one coil to the "peak" of the next coil) is smallest in the undistended bladder (greatest tortuosity) and is the largest in the bladders filled to capacity (least tortuosity).

3. DISCUSSION

Observations from our laboratory over the past seven years suggest that various forms of outlet obstruction result in a form of fibrosis within the detrusor muscle layer.

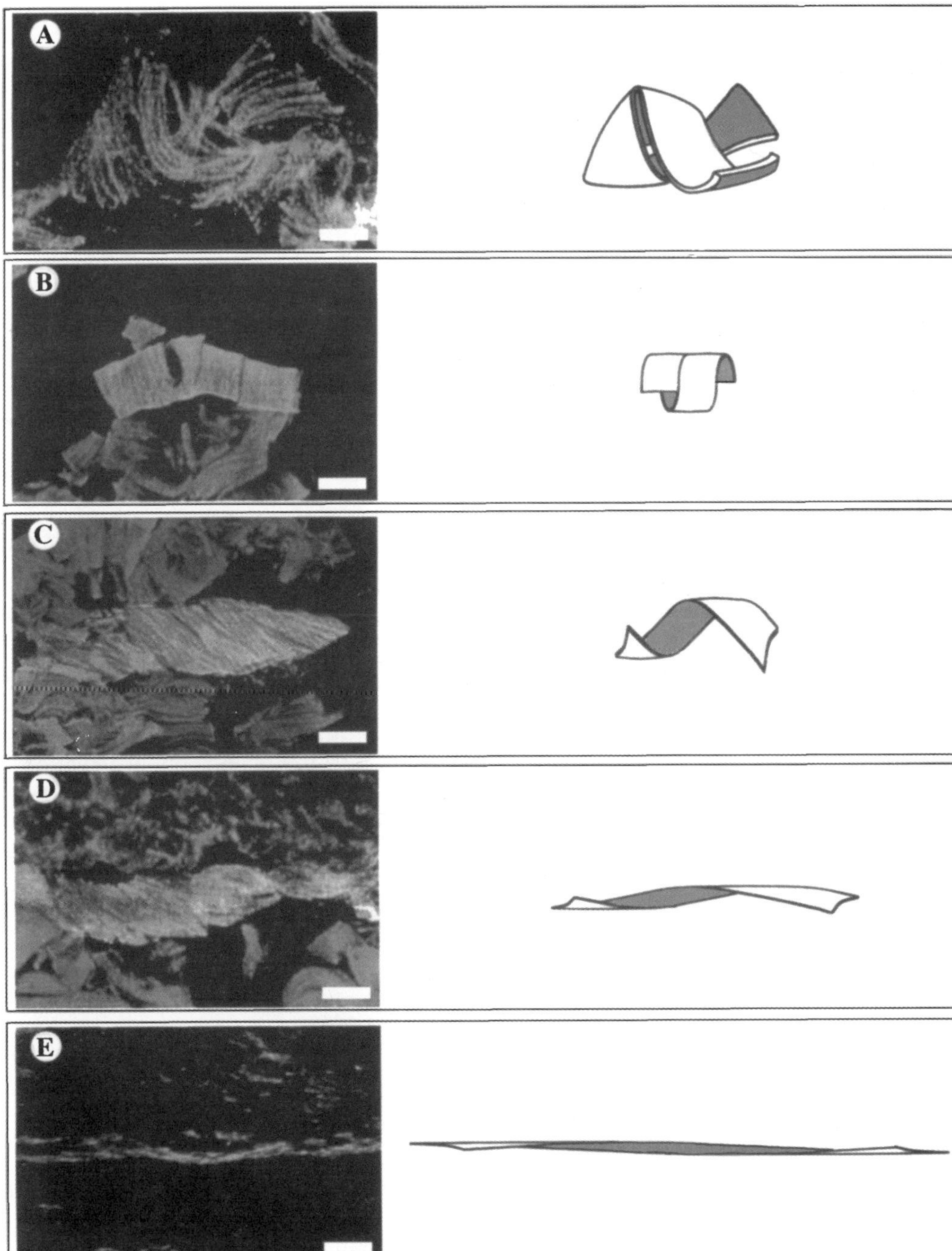

Figure 5. Confocal microscopy of type III collagen fibers in the detrusor muscle of 3rd trimester fetal bovine bladders. To the right of the photpgraphs are 3-D schematic diagrams of the collagen fibers (bar = 10mM). A. Section showing the random orientation of the relaxed type III collagen fiber bundle in the 0% distended fetal bovine bladder. The fiber bundle possesses a "wavy" characteristic and coiling is evident. B. Section showing the tightly coiled fiber bundle in the 25% distended fetal bovine bladder. C. Section showing the early stages of type III collagen fiber expansion or extention in the 25% distended fetal bovine bladder. D. Section showing the later stages of type III collagen fiber extention in the 25% distended fetal bovine bladder. E. Section showing the linearization of type III collagen fibers in the 100% distended fetal bovine bladder.

The fibrosis is the result of an increase in type III collagen that accumulates within and possibly around some muscle bundles. This increased and altered expression of type III collagen appears to functionally disrupt the ability of the bladder to cycle as a low compliance structure although the exact mechanism by which increased type III may alter compliance is unknown. One possibility is that increased synthesis of type III collagen by bladder wall cells may alter the mechanical properties of the fibers that transfer tension within the wall. It has previously been demonstrated that collagen fiber diameter can be altered by changing the concentration of type III and type I collagen.[15] Thus, if one assumes that mechanical strength or some other physical property is related to fibril diameter, it is obvious, based upon our biochemical data, that such changes do occur as a result of obstruction. The actual mechanism responsible for these changes is currently unknown; however, it is possible that similar mechanisms operate during development.[16]

This, however, raises the question of how volume accommodation actually occurs in the bladder. If type III collagen is a tension bearing element within the bladder wall, then, as the bladder fills, the collagen fibers should change their shape to accommodate the increased tension that develops. If the tendon is used as an analogy, it can be appreciated that, as the tendon is "loaded", its length stretches to a new length that is approximately 4–6% longer. The use of the word stretch is actually incorrect here since the collagen fibers do not actually stretch but more correctly "straighten out" as they begin to carry tension. In the unloaded state, tendon fibers have a very slightly coiled conformation that, upon loading, is lost. The increased length thus results from a loss of the slight waviness of tendon collagen fibers rather than their stretching.

In contrast to the tendon, the bladder wall undergoes a much greater expansion than 4–6%. How then does the bladder wall and the collagen fibers that comprise its structural architecture actually sustain such large swings in distension? Based upon our observations with the confocal microscope, we suggest that bladder wall tension increases during bladder filling. In fact, this tension may be a necessary factor in the normal development of the bladder. This tension is dissipated within the structure of the bladder wall by expansion of the coil structure of type III collagen. Fibers in an empty and undistended bladder are highly tortuous with each individual coil directly adjacent to another. In contrast, a full bladder has collagen fibers whose coil structure is extended or expanded and is much less tortuous that in the empty bladder. Such coils potentially could accommodate large tensional changes with little or no danger of its fibers being extended to such an extent that they "plastically or irreversibly" deform and fracture. The role of type I collagen must also be considered.

Although we have focused on the role of Type III collagen in this study, Type I collagen is the major structural collagen found in the bladder. Recent evidence suggests that the presence of Type III is critical for Type I fibrillogenesis in the cardiovascular systems and other organs as well.[17] Thus there is a need for more information about the interaction between these two important structural proteins in the bladder.

3. SUMMARY

While this model is speculative, it is attractive in that it can account for the physiologic properties of the bladder. It also implies that connections must exist between the tension generating elements, i.e., the smooth muscle cells, and the other components of the bladder. In bladders that become noncompliant, it is likely that there is some interference with the ability of the collagen fibers to elastically and reversibly alter their tortuosity. This, predictably, would reduce total bladder capacity. Further studies will be required to establish the relationship between compliance changes and the passive mechanical elements of the bladder wall that comprise its structural protein matrix.

ACKNOWLEDGMENTS

It is a pleasure to acknowledge the contributions of the individuals who made substantial contributions to this work. Especially noteworthy were the efforts of Steven Chang who was primarily responsible for carrying out these experiments on bladder filling including the confocal microscopy. In addition to Steven, we are grateful also to J.S. Chung, Catherine Deveaud, David Ewalt, Harry P. Koo, and M.K. Yeung.

REFERENCES

1. Mallory, B., Steers, W. and de Groat, W.C.: Electrophysiological study of micturition refelxes in the rat. Amer. J. Physiol., **257:** R410, 1989.
2. Cass, S.C. and Hinman, F.: Constant urethral flow in female dog. II Effect of consrtiction of vesical neck and external meatus. J. Urol., **99:** 447, 1976.
3. Uvelius, B., Peterson, L. and Mattiasson, A.: Smooth muscle hypertrophy and hyperplasia in the rat detrusor after short-term infravesical outflow obstruction. J.Urol., **131:** 173, 1984.
4. Levin, R. M., High, H. and Wein, A.J.: The effect of short-term obstruction on urinary bladder function in the rabbit. J. Urol., **132:** 789, 1984.
5. Kim, K.M., Kogan, B.A., Massad, C.A. and Hyang, Y-C.: Collagen and elastin in the obstructed fetal bladder. J. Urol., **146:** 528, 1991.
6. Nielsen, K.K., Andersen, C.B., Petersen, L.K., Oxlund, H. and Nordling, J.: Morphological, stereological and biochemical analysis of the mini-pig urinary bladder after chronic outflow obstruction and after recovery from obstruction. Neurourology and Urodynamics., **14:** 269, 1995.
7. Elbadawi, A.: Pathology and pathophysiology of detrusor in incontinence. Urologic Clinics of North American., **22:** 499, 1995.
8. Levin, R.L.: Developmental aspects of urinary bladder physiology and pharmacology. In: Dialogues in Pediatric Urology, Ed Richard Ehrlich, William J. Miller Assoc.,Inc, Pearl River, New York, pp2, 1996.
9. Sachs, F.: Mechanical transduction by ion channels: how forces reach the channel. Society of General Physiologists Series., **52:** 209, 1997.
10. Chen, M-W., Levin, R.M. and Buttyan, R.: Peptide growth factors in normal and hypertrophied bladder. World J. Urol., **13:** 344, 1995.
11. Gabella, G. and Uvelius, B.: Reversal of muscle hypertrophy in the rat urinary bladder after removal of urethral obsrtuction. Cell Tissue Res., **277:** 333, 1994.
12. Ewalt, D.H., Howard, P.S., Blyth, B., Snyder, HM 3d., Duckett, J.W., Levin R.M. and Macarak E.J.: Is lamina propria matrix reponsible for normal bladder compliance? J. Urol., **148:** 544, 1992.
13. Deveaud, C.M., Macarak, E.J., Kucich, U., Ewalt, D.H., Abrams, W.R. and Howard, P.S.: Molecular analysis of collagens in bladder fibrosis. J. Urol. In press, 1998.
14. Chang, S.L., Howard, P.S. and Macarak, E.J..: Role of type III collagen in bladder filling. Neurology and urodynamics. in press, 1998.
15. Keene, D.R., Sakai, L. Y., Bachinger, H.P. and Burgeson, R.E.: Type II collagen can be present on banded collagen fibrils regardless of fibril diameteer. J. Of cell Biol., **105:** 2393, 1987.
16. Close, C.E., Carr, M. C., Burns, M. W.and Mitchell, M.E.: Lower urinary tract changes after early valve ablation in neonates and infants: is early diversion warranted? J. Urol., **157:** 984, 1997.
17. Liu, X., Wu, H., Byrne, M., Krane, S. and Jaenisch, R.: Type III collagen is crucial for collagen I fibrillogenesis and for normal cardiovascular development. Proc. Nat. Acad. Sci. USA., **94:** 1852, 1997.

DISCUSSION

Muscle and Extracellular Matrix in the Bladder

Q & A with Steve Zderic, M.D.

Bob Levin, Albany, New York.

Q: One interesting phenomenon you didn't discuss, which I think is very important is compliance and its regulation by the Sarcoplasmic reticulum (SR). In the work that we've done together in the fetal bladder that has very low SR, compliance is regulated a lot by the extracellular calcium concentration, and with the development of the SR, then the intercellular calcium is regulated to a much finer degree, and actually you get a much more compliant bladder in that case. So when we get the dysregulation of the SR during obstruction and other things, that can also directly relate to the changes in compliance so that alterations in compliance may have an active process, in that the higher the base level of calcium, the lower compliance. So I think it may be a very important clinical correlate to the demonstrations that you've shown, both for the neonatal development and for the response to obstruction.

A: I knew I was going to be sorry that I threw that slide out, but, as I said, I wanted to limit it to the active. I think all of us would agree that the cytosolic calcium contributes to the resistance to deformation, and we're beginning to look at that.

Barry Kogan, Albany, New York.

Q: The question always comes up, are the changes that we're seeing reversible, number one, and are the changes that you're seeing beneficial or harmful in the long-term?

A: Again, now that we're able to work at a PCR level, we should be able to go on a micro level. For example, when you go and reverse the obstruction, biopsy the bladder, and instead of just simply performing the histochemistry that we did, actually be able to do a PCR analysis at the time of reversal, and the subsequently after a period of recovery to see which of these changes may well predict which bladders will come back and which ones won't. But we're very interested in the reversal model.

Mike Watson, Research Triangle Park, South Carolina.

Q: It is interesting that you bring up the cardiovascular aspects of this, since for the last 17 years my work has focused on the cardiovascular sarcoplasmic reticulum. In the decompensated bladder smooth muscle cells, what is the increase in basal calcium concentration?

A: In answer to your question, I'm hoping to acquire a tracer imaging system that would let us begin to answer that - what are the basal calcium concentrations in the group?

Q: Right, because obviously, if there's a proteolysis taking place, there's a lot of calcium-activated proteases that are at least on the drawing board for congestive heart failure. So, I would assume a similar sort of situation in smooth muscle also.

A: Bob Levin has started to accumulate some evidence that degradation is a very important mediator of some of these bladder changes, but the truth is, there are probably a number of pathways that get activated, including the lysosomal enzymes I was studying back when I was a medical student. It may well be that those flood gates also open up, dumping out the contents of lysosomal enzymes to create fairly widespread proteolysis and damage to the cells' crucial structures.

Q: Is there any information about the FK blinding proteins associated with sarcoplasmic reticulum in smooth muscle.

A: I've seen the references to FK-506 looking at association with SR. I just, I don't know what relationship they'd have in the bladder.

Q & A with John McConnell, M.D.

Bill Steers, Charlottesville, Virginia.

Q: John, that was a wonderful and very thorough review of the hypotrophy literature. You mentioned the real eloquent new studies on Karl Eric Anderson on the calcinurin, suggesting the signaling of the nucleus with calcium being involved, and Steve Zderic was talking about CERCA and intracellular calcium stores. Several years ago, in our laboratory we looked at the ligated rat model and gave verapamil. We were completely able to abolish hypertrophy with verapamil. Do you think that was due to changes with loading of intracellular stores? Steve may want to comment on why, if you block potentially voltage gated channels, and the same has been shown in the cardiac literature, I believe as well. With cardiac hypertrophy you can almost completely block hypertrophy in that scenario as well.

A: I don't think that's a paradox. To me, that fits with the hypothesis that actually, I think we have to give Bob Levin credit for, for many years ago suggesting that intracellular calcium abnormalities, in fact, may be the initiating signal here. The real question, especially when you're using verapamil, is what you're doing to block the ability of these cells to generate more active tension or more load, if you will, and that's in fact what's preventing the hypertrophy. Or whether it's actually the reduction in the calcium. If I was still actively working in this area, I wouldn't operate on any more rabbits, because I think what we need to be doing is developing these models in mice, so we can take care of modern knock-out type models, that's the only way we're gonna answer these questions. You can obstruct mice.

Barry Kogan Albany, New York.

Q: John, you mentioned a question about whether the animal models were too abrupt. Do you have any human data, or do you have any hypotheses about what differences you might see in preparations that were slower in developing?

A: The problem in the human model, at least as an adult urologist, is that the tissues that we see, and we of course learned this from Gosling years ago, is that when you biopsy these bladders or analyze cystectomy specimens from patients who have had their bladders removed for cancer, and you look at the noncancerous part of the bladder, you know you're looking at extracellular matrix disease. You cannot find smooth muscle hypertrophy in most of these older male bladders. So, I don't think we're gonna learn too much, frankly, from looking at that. It might be more revealing to do longitudinal followup in some of the pediatric cases where you expect to have significant changes over time, and that might be a population that might yield more information.

Q: Also John, you commented on gender differences, particularly with aging. Do you have any data on gender differences in animal populations.

A: I don't personally know. I think that there has been some work with obstruction of the bladder in the female rat. I am not aware of any significant differences there, but you know, if you just look at the histologic level, many of the changes in the bladder wall that we've blamed, as a prostatologist on the prostate you see in women with aging, and so I would think it's the effects of denervation, effects of all of aging just on the function of smooth muscle cells. Some of that may be vascular insufficiency, and I think that's another important area of investigation.

Peter Juenemann, Germany.

Q: You gave some very critical points about the obstruction part and the obstruction model in all the animal models that are presented and published in the literature. When I come for a clinical point of view, I am always wondering how the bladder is able to recover, not only in a young patient, even in an adult patient, let's say 40, 50 years of age. For example, a quadriplegic patient who receives or undergoes dorsal rhizotomy, and the normal bladder function, at least the bladder architecture reoccurs even though he has this problem for many, many years, let's say 10 to 15 years, so I'm wondering whether one can compare a mechanical obstruction, which all the animal models are based on, with a functional obstruction, which we probably see, for example, in those patients with a spinal cord lesion or, or in pediatric patients with the Hinman syndrome where you also see this hypertrophy. Do you have any data or any idea whether there is a difference between a functional obstruction versus a mechanical obstruction?

A: No, although I think, and I don't know if Sam's gonna talk about this, but I think the Penn Group, over the years, has compared different models, different ways of obstructing the rabbit. Sort of slow degrees of obstruction, and you tend to get more of a ramp up of the increase in muscle mass, I think there's less ECM. Is that not correct Bob? At first, yes (Bob). So, you know, I can't answer your question. But the point you're making is important. I think that neurologic sort of disorders of the bladder is really a totally different issue, and I think we need to look at animal models of denervation to specifically answer those questions. Now, the one thing I'll quibble with you about though, is when you say the bladder architecture is returning to normal, I think what you mean is that these bladders look pretty good on x-ray.

Q: Even functionally, if you put a ventral root stimulator on those neurofibers, let's say on S3, S4 nerve fibers, the Brinkley stimulator, for example, and you stimulate them, and they respond quite properly.

A: But what we don't really know though, is what those bladders are really like at the cellular levels. So, I think unless you biopsy those patients, you don't really know what's going on.

Tom Lue, UCSF.

Q: You discussing about the female rats, and we actually did some studies recently, just the normal aging female rat. We actually looking at all the electromicroscopic features of urethra different ages after pregnancy, after delivery, and all that stuff. But anyway, just to summarize, you are asking whether the bladder in older rats, whether they have hypertrophy. And the answer is yes. They are much bigger. They become sort of like a square shape instead of a nice polygonal shape, so they don't have obstruction, so there's something else going on in the older rats that's created this kind of different picture.

A: Did you give those rats Viagra?

Q & A with Ellen Shapiro, M.D.

David Thomas, Leeds, U.K.

Q: One gets the impression that in tissue culture, the smooth muscle cells revert to a more fetal form. Do smooth muscle cells in tissue culture co-express to any extent?

A: I don't know. But I do know that Lynn Wilson, also in our department, has developed cultures of prostatic smooth muscle, so it may be important to look at this phenomeon in this culture system.

Barry Kogan, Albany, New York.

Q: Can you tell us some of the functional implications of your findings?

A: A pediatric surgeon, had accumulated these specimens. He was very interested in knowing whether fetal intervention would be possible and its impact on lower exrtreemity function. Prior to18 weeks, they often would note fetal leg movement on sonography in patients that were known to have spina bifida. It was thought that if the cord was covered, this may prevent any further damage to the cord.

Our work has shown that probably the cat's out of the bag by 18 weeks, and that the bladders are already abnormal. Covering the cord, which is possible by fetal intervention between 18 and 20 weeks, may avert any further damage, which I think would have long-term implications for all of our patients. Anything that can be done to save function would be useful. So, we were hoping to be able to correlate the amount of damage seen in some of the spinal cord specimens with the findings in the bladder, but it seemed like most of the bladders demonstrated a significant degree of abnormality, and really, none were normal per se.

Simon Chacko, Pennsylvania.

Q: It has been shown that dystrophin is important for skeletal muscle myogenesis, and I wonder whether, have you looked at the expression of dystrophin at stages where you see striated muscle and smooth muscle, and later when you don't see the striated muscle, to see maybe the expression of dystrophin isn't there? If you have dystrophin, a number of cells will become striated muscle cell, including smooth muscle, or even fibroblasts.

A: We've not looked at that, but the comment has been made by those who've examined the limb development in these fetuses, as well as observing the pelvic floor. In many of them the striated muscle appears normal. Whether it expresses all of its normal phenotypic components, I don't know, but it would certainly be an important question to answer.

Barry Kogan, Albany, New York.

Q: Ellen, you emphasize you look primarily at male fetuses, is that because of Herb's interest in prostate, or is there some other reason?

A: No. We do have female fetuses, and I just had this discussion with Pam Howard earlier. We're really interested in looking at overall smooth muscle development. We know it's abnormal in the prostate. In the Manx cat model in which some of the fetuses have a neurogenic bladder, some have also been observed to have a neurogenic uterus. So, not many people have looked at the uterine development. So, we're not only interested at looking at normal and abnormal uterine development if it indeed occurs abnormally in the spina bifida female fetuses, which we do have four fetuses to examine, but also examine the development of the urethral musculature since so little is known about female continence.

Steve Zderic, Philadelphia.

Q: The question I have, is if you look at the newborn patient with spina bifida, at their bladders, what kind of myosins are they expressing? Are those continued expressions of striated myosin continuing, or do you just not have those?

A: Currently, I don't have any specimens that are newborn. All of these specimens are 20-week, but because they were abortus'.

Q: Newborn vesicostomies?

A: I haven't done a vesicostomy in a spina bifida for a long time.

The other implication of that, is that if that really is persisting, it leads you to wonder what the quality of regeneration will be into the grafts that are being proposed.

Q & A with Edward Macarak, Ph.D.

Tom Lue, UCSF.

Q: I'm just curious about, have you studied the elastic fibers?

A: We've studied some elastic fibers, but the data on the elastic fibers is not as clear cut as it is with the collagen. It's much more difficult to biochemically quantitate elastin than collagen, and so for that reason, we really have not carried those studies as far along as perhaps we might.

Q: Have you looked at the structure under electron microscopy? We have been doing that, and actually, the structure of even just the collagen fibers and the ground matrix, actually look quite different in a normal state versus a diseased state. They seem to pack together much more tightly.

A: No, we've not done any electron microscopy on this cohort of patient samples.

Saso Jezermik, Denmark.

Q: I would like to ask you, how did you measure the thickness of the collagen layer? Did you measure at several points and take the average around the bladder or at one point?

A: It was measured at multiple points in the section. We used actually the thickest point of the bladder between all of the samples to give us the detrusor thickness, and then we subtracted that from the total thickness to give us the difference of the mucosal thickness, and then it was averaged by making several measurements across the thickness of the bladder.

Barry Kogan, Albany, New York.

Q: I'm still not sure why we had such disparate findings regarding Type I and Type III. But I guess my real question is, in disease states, I'm sure you've done your confocal microscopy where Type III is greatly increased and Type I is somewhat increased, what do these collagen fibers look like and how are they different, particularly if you compare the noncompliance with the compliance?

A: Sadly, we've really not done any of these studies on any of the patient samples, but I can say that in some studies that'll be presented at the AUA by Joel Hutcheson, in obstructed sheep, we find a similar pattern in terms of the expression of Type III collagen. I think that, in terms of the difference, surely, it's now known that if you look at animals that have a genetic defect in the production of Type III collagen, that tissues that contain fibers that have inadequate amounts of Type III, the collagen fibers are actually very fat; however, I think that the one thing that may have been a problem in the way you actually did your quantitation, is that some of these fibers that you indicated were only Type I, probably had considerable amounts of Type III with them.

Q: I'd be interested, do you have studies of the configuration of the mixed actual collagen fibers of Type III and Type I mixed together?

A: No, we don't. In fact, that's a good question because some of these fibers that we've stained with the antibody to Type III collagen also most likely contain Type I collagen. The thing that we don't understand is that there's not a really good way to determine the mechanics of a pure tissue containing only Type III collagen. You can do that for Type I in a tendon, but there's no comparable structure that contains essentially, exclusively Type III collagen that we could actually mechanically test and evaluate and determine an elastic constant for.

Q: Just as a final comment, it might be really fascinating in the noncompliant bladders, one might guess that somehow this collagen doesn't nicely stretch like a telephone cord as you showed. It might be really fascinating to look at those in some of the bad bladders.

A: I think there's lots of possibilities, but certainly, one is that for some reason you get abnormal tethering of muscle bundles, or some way that the actual normal pattern of interaction between these muscle bundles is altered by the over expression of Type III.

Q & A with John Park, M.D.

Bob Levin, Albany, New York.

Comment: The work that we've done really supports the idea that stretch is a major factor in the response to outlet obstruction. As we compared over distention in vivo with the response to outlet obstruction, and in general, they both do the same thing. You get virtually the same type of genetic response to dilation as you do to obstruction. You get about the same type of stimulation of tritium uptake in DNA synthesis after over distention as you do with obstruction. The difference of course is the time course, the response to over-distention occurs very quickly and recovers very quickly; whereas, the response to obstruction takes longer time to develop and does not recover to the same extent or the same time course. But the idea that distention is the initiating factor for a lot of the gene activations in the sequences is a very reasonable theory, and a model of looking at stretch-induced changes is a very nice model.

Earl Chang, Chicago.

Q: How do your in vitro observations of angiotensin II and HB-EGF in rat smooth muscle cells relate to the in vivo finding, on the muscular changes following obstruction in the rat model of bladder neck obstruction? Have you looked at the mitogenic affects of angiotensin II, as well as HB-EGF in your culture system?

A: In your first question, I assume that you're referring to the study by Dr. Karl Erik Andersson's group, where they created partial bladder obstruction in rats and treated these animals with losartan. There was no appreciable difference in rats treated with losartan. I think the lesson to be remembered when thinking about the inhibition of a molecule and its functions having no obvious in vivo effect is that of gene knock out mice. Some of the molecules that are considered to be critically important often do not demonstrate obvious morphological changes after they have been knocked out . That doesn't necessarily mean that these molecules are not important. The opposite may be true, in that they are so important there exist alternate mechanisms for rescue. Having said that, I think equating the importance of angiotensin II as the only mediator of cell response after stretch is becoming outdated. When cardiovascular researchers initially came out with findings implicating angiotensin II as a critical mediator of cell response after stretch, it drew a lot of interest because it supported the idea of a local tissue renin-angiotensin system; however, subsequent works have now demonstrated that angiotensin II by itself may not bethe sole mediator of cell response after stretch. There are other factors that are at play here.

Sudip Bandyopadhyay, Cleveland Clinic Foundation.

Q: What transcription factor drives this HB-EGF?

A: We're working on that right now. I showed the results of 1.7kb promoter reporter trasfection experiment today. Since then, we have narrowed the cis-acting region down to a couple hundred base pairs. We're in the process of specifically nailing down the stretch responsive element, which we think will be of a great importance not just to urological research, but to cardiovascular research which has a lot of interest in the stretch-regulated gene expression. We're not quite there yet.

Q: Is there any known binding element, signal transducer and activator of transcription?

A: The 1.7kb region that we tested contains several consensus transcription factor binding sites, among which are a couple of AP-1 binding sites. I believe that some of the stretch-responsive genes have been implicated to involve AP-1 sites. We're in the process of narrowing it down.

Avraham Raz, Detroit, Michigan.

Q: Something we missed is the specificity. Is this stretching effect specific to this particular muscle cell? Is this specific to this particular gene or is it a general phenomenon?

A: Dr. Jeremy Tuttle, who was earlier in the audience, along with several other investigators are really the driving force behind the research interest in this mechanical stretch phenomenon. It has been stated that all cells respond to mechanical stretch although the response itself is cell type specific. We too were worried about the specificity of this response. You take the cells out of an organ, and then you subject them to an arbitraily created stretch. We wondered whether we were watching some sort of a nonspecific response that occurs in all cells. We have now identified several genes that do change with the stretch in the bladder smooth muscle cells while not in other cell types. For example, HBEGF, which is known to be expressed in renal tubular cells, does not change when renal cells are subjected to mechanical stretch. So I think as far as we can tell, it's a specific response by the bladder smooth muscle cells.

Q & A with Sam Chacko, DVM, Ph.D.

John McConnell, Dallas.

Q: All of these isoforms, as you mentioned, really are derived by alternate splicing. Has your group done any work looking at whether or not the hypertrophy or the obstruction may induce changes in the actual alternate splicing technique, and this really has nothing to do with specific genes, but is a generalized defect in the cell with alternate splicing.

A: That's a very important question. I think that one of the fascinating things about this model is that whatever hypertrophy causes turns on the splicing of one type of isoform as opposed to the other type of isoform in a reversible manner. One of the problems with the smooth muscle myosin is that we don't know anything about the transcriptioal regulation of smooth muscle myosin. This cannot be achieved without a good cell culture system for phenotypically differentiated smooth muscle cells, and that's why we do need a good in vitro system. We can only study that in an in vitro system there's a lot that we need to know, and we don't have a perfect in vitro system to study it. There are a lot of factors, you can talk about; chronic stretching of the bladder wall and overdistensiion probably releases something that turn on the gene. Another possibility is that it could be some sort of hypoxic changes that turn on the expression of new myosin isoforms, or it could be neurogenic, anything that you're talking about may have an effect. For example, what is the effect of nerves? In skeletal muscle, changing the innervation changes a slow muscle to a fast muscle we've shown that take the nerve and put it in the fast muscle, and the muscle becomes fast. So there may be changes in response to the nerve. So there's a lot of factors. So there is some evidence that the nuerogenic factors can alter the expression of myosin isoforms in muscles, in general. We have a lot that we need to know. And I think the problem is that we don't know enough about the transcription or regulation of myosin; therefore, we need good biological study using, unfortunately, we do need cell culture system to study this. Without that, it's difficult to study. So your question is extremely important, and I think with time I'm sure We'll know more about it.

Barry Kogan, Albany, New York.

Q: Sam, have you looked at different parts of the bladder, in particular, the trigone and bladder neck? I wonder, we think that continence in some circumstances may be maintained by tonic smooth muscle contractions. So I wonder if you see different types of myosin isoforms in the bladder neck versus the fundus of the bladder.

A: Yes. We have looked at the bladder body and dome, and also the urethra and the neck region. As we come towards the neck and the urethra, the smooth muscle has more of the myosin that is found in the tonic type of smooth muscle with the tonic type of insert. Again, I must say that we have begun to look at the human, and we see similar changes and also in the mouse. This is not just something unique to rabbit, and this is a general phenomenon.

Section III

NERVES AND PHARMACOLOGY IN THE BLADDER

OVERVIEW OF NERVES AND PHARMACOLOGY IN THE BLADDER

Robert M. Levin

Division of Basic and Pharmaceutical Sciences
Albany College of Pharmacy
Albany, New York 12208

The first presentation in this section was by Dr. Karl Eric Andersson on 'The Pathways for relaxation of the detrusor smooth muscle'. One common problem that can be related to a number of specific pathologies is spontaneous bladder contractions that occurs during bladder filling. This can be a symptom of outlet obstruction secondary to BPH, neurogenic bladder dysfunction, or it can occur in association with aging. Also associated with aging, and with a variety of dysfunctions, is a general decrease in bladder compliance. Dr. Andersson discussed several of the cellular mechanisms that can result in a relaxation of the detrusor muscle including activation of the cyclic AMP system by both stimulation of adenylate cyclase, or inhibition of phosphodiesterase (PDE), the enzyme that catalyzes the degradation of cyclic AMP. There is considerable interest in the identification of the specific isoforms of PDE present in the bladder detrusor, urethra, and cavernosal tissue with the goal of developing tissue-selective PDE inhibitors. Dr. Andersson discussed in detail the relationship between nitric-oxide stimulated relaxation and cyclic GMP.

As discussed by Dr. Andersson, ATP, in addition to providing biological energy to a variety of cellular processes, has been shown to have specific neurotransmitter functions in the lower urinary tract. ATP is capable of stimulating a contraction via the activation of P2x receptors and relaxation by stimulating P2y receptors. The last major section of Dr. Andersson's presentation related to the pharmacology of ion channels, as targets for the control of bladder tension. Agents that block calcium channels, or open potassium channels result in the relaxation of bladder smooth muscle, and the reduction of spontaneous contractile activity. Dr. Andersson ended his presentation with information showing that there are unknown factors that may be released from the urothelium that can modulate bladder smooth muscle tone, and that these factors may play an important role in the mechanisms by which sensory information may be transmitted to the smooth muscle elements of the bladder.

The second presentation was by Dr. William C. de Groat. Dr de Groat discussed 'Maturation of bladder reflex pathways during postnatal development'. As discussed by

Advances in Bladder Research, edited by Baskin and Hayward.
Kluwer Academic / Plenum Publishers, New York, 1999.

Dr. de Groat, the mechanisms involved in storage and periodic elimination of urine undergo marked changes during prenatal development. In the young fetus prior to maturation of the nervous system, urine is eliminated from the bladder by nonneural mechanisms; however at later stages of development, voiding is regulated by primitive reflex pathways organized in the spinal cord. As the central nervous system matures during the postnatal period, reflex voiding is brought under voluntary control which originates in the higher centers of the brain. Thus the function of the lower urinary tract are regulated by a hierarchical control system consisting of primitive components which are intrinsic to the bladder and urethra, spinal cord, as well as the more complex components in the cerebral cortex. In adults, injuries or diseases of the nervous system can lead to the re-emergence of primitive reflexes that were prominent early in development, but suppressed during postnatal maturation. Therefore, developmental studies in neurourology are likely to provide insights into the mechanisms underlying neurogenic disorders of bladder functions in adults as well as an understanding of problems of the developing nervous system encountered in pediatric urology. Dr. de Groat's presentation reviewed recent studies by himself and others in which tract tracing techniques and electrophysiological methods were used in experimental animal models to examine the synaptic changes underlying developmental plasticity in the parasympathetic reflex pathways to the urinary bladder. For example, in many species (e.g., rats and cats) voiding in neonates is dependent on an exteroceptive somato-bladder reflex mechanism triggered when the mother licks the perineal region of the young animal. This exteroceptive reflex which is organized in the sacral spinal cord disappears in older animals but reappears in adult animals after spinal cord injury. In summary neuroanatomical and electrophysiological techniques have provided new insights into the organization of spinal cord circuitry and neurotransmitter mechanisms involved in primitive voiding reflexes in neonatal animals. It is clear that developmental and spinal cord injury induced plasticity in parasympathetic reflex pathways is due in part to alterations in glutamatergic transmission between INTs and PGNs. Thus synaptic remodeling in the sacral parasympathetic nucleus is likely to be an important factor in the postnatal maturation of voiding reflexes.

The third presentation was by Dr. Robert Levin on 'Etiological factors in the progression of bladder dysfunction secondary to BPH'. Bladder dysfunction secondary to BPH is a major affliction associated with human aging. It is clear that bladder dysfunction secondary to BPH is a slow progressive disease. In many cases, medical treatment is not sought until the dysfunction is relatively severe. In man, it is difficult to investigate the cellular mechanisms by which progressive bladder dysfunction occurs. However, many of the functional changes associated with human bladder pathology can be induced in experimental animal model systems and this has been demonstrated prominently in a rabbit model of bladder outlet obstruction. Dr. Levin presented evidence from his work and that of others that virtually all bladder cellular elements are involved in the response to partial outlet obstruction. The urothelium undergoes an early, rapid and substantial period of hyperplasia; the connective tissue component undergoes rapid changes in localization and content; the smooth muscle elements undergo both hyperplasia and hypertrophy; and fibroblasts differentiate into functional smooth muscle cells. In general, partial outlet obstruction induces a rapid and marked remodeling of the bladder wall which includes compartment-specific cellular hypertrophy (growth and enlargement of individual cells), hyperplasia (cell division), and re-organization (structural relationship between connective tissue and smooth muscle elements). Dr. Levin has identified in the rabbit model of obstruction, three major cellular alterations which are induced by partial outlet obstruction and directly result in the reduction in the ability of the hypertrophied bladder to generate pressure and empty: 1) Progressive denervation; 2) Selective dysfunction of the sarcoplasmic reticulum to store and release calcium; and 3) Selective mitochondrial dysfunction. Dr. Levin presented the following hypothesis concerning the etiology of the above three cellular dysfunctions is as follows: 1) Partial outlet obstruction results in bladder hypertrophy. 2) Bladder hypertrophy induces localized and cyclic hypoxia / ischemia. 3) Hypoxia / ischemia induces a net release of calcium from the sarcoplasmic reticulum.

4) High intracellular calcium activates calcium-activated hydrolytic enzymes including calpain (calcium-activated protease) and phospholipase A2 (calcium-activated lipase); and results in the generation of free radicals and the subsequent activation of lipid peroxidases. Nerve membranes, the sarcoplasmic reticulum, and mitochondrial membranes are selective targets for the activated hydrolytic enzymes (includingcalpain, phospholipase A2 and free-radical stimulated lipid peroxidation). Thus, the above sequence of events can explain the specific functional, contractile, and biochemical effects of partial outlet obstruction.

This theory is based on the activation of specific hydrolytic enzymes that result in cellular and subcellular membrane dysfunction. Rather than publish a review article based on his presentation, which have been published previously, he has chosen to publish an original article on the 'Subcellular distribution of free fatty acids, phospholipids, and endogenous lipase activity of rabbit urinary bladder smooth muscle and mucosa'. It is Dr Levin's view that a detailed study of the changes in the lipid and phospholipid content of specific cellular and subcellular membranes that accompany specific experimental pathologies such as ischemia, diabetes, ageing, and partial outlet obstruction can give important novel information on the cellular mechanisms of the functional and biochemical dysfunctions that are induced by each pathological (or physiological) process. The current article provides background information on the distribution of free fatty acids and phospholipids in normal rabbit bladder smooth muscle and mucosa.

Dr. Nishizawa presented an excellent talk on 'Effects of Glutamate Receptor Antagonists on Lower Urinary Tract Function in Conscious Unanesthetized Rats'. Glutamic acid which functions as an excitatory neurotransmitter at various sites in the central nervous system acts on at least two subtypes of ionotropic receptors: NMDA and AMPA/kainate (AMPA) receptors. NMDA and AMPA receptors have been implicated in the control of micturition in the rat. The precise role of glutamate receptor on the spinal control of micturition remains unknown. Dr. Nishizawa's presentation described the results of his experimental studies concerning the effects of the intrathecal administered glutamate receptor antagonists on the bladder and urethral activities during isovolumic bladder contraction in conscious normal and chronic spinal rats.

Dr. Nishizawa explained that in normal rats, isovolumic bladder contractions were developed concomitant with spasmodic urethral EMG activities at constant intervals before drug administration. MK-801 (non-competitive NMDA receptor antagonist) produced a decrease in bladder contraction pressure and urethral EMG activity with dose dependent manner. LY 293558 (competitive AMPA receptor antagonist) produced a decrease in bladder contraction pressure and urethral EMG activity with dose dependent manner.

Detrusor-sphincter dyssynergia was developed in chronic spinal rats. MK-801 and LY 293558 reduced bladder pressure during contraction and markedly decreased urethral contractions concomitant with bladder contraction. From these studies, Dr. Nishizawa concluded that in both normal rats and chronic spinal rats, two subtypes of glutamate receptors-NMDA and AMPA receptors- were involved in the control of bladder and urethral activities at the level of the spinal cord.

The final presentation was by Dr. William Steers on 'Observations from the spontaneously hypertensive rat: Insight into NGF regulation and nonadrenergic hyper-innervation in the lower urinary tract'. Dr. Steers explained that there is an association between hypertension and the symptoms from BPH. Approximately 1/3 of men being treated for symptomatic BPH are hypertensive. Thus, the prevalence for hypertension in men with symptomatic BPH is statistically higher than the incidence in age-matched control men with no symptoms. Because of the close relationship between hypertension and sympathetic nerve activity, there may be a relationship between sympathetic overactivity, symptomatic BPH, and hypertension. In order to study this possibility, Dr. Steers and co-workers have evaluated the voiding behavior of the spontaneously hypertensive rat (SHR), and correlated the voiding behavior with a number of factors including the density of sympathetic innervation, the synthesis and release of nerve growth factor (NGF), and the morphology of the spinal micturition centers.

Awake cystometry of the rats demonstrated that the SHR rats had an increased voiding frequency, smaller micturition volume, and increased incidence of unstable bladder contractions during filling. Dr. Steers presented detailed evidence that there was a markedly increased sympathetic innervation in the bladder of the SHR rats, and this increased sympathetic innervation correlated with increased expression and synthesis of NGF.

Data presented by Dr. Steers provide excellent evidence that there is a causal relationship between the increased sympathetic innervation induced by the increased expression and synthesis of NGF, the increased voiding frequency and hyperactive bladders of the SHR rats, and the spontaneous hypertension. However, the mechanisms by which these occur is the subject of further investigation.

18

PATHWAYS FOR RELAXATION OF DETRUSOR SMOOTH MUSCLE

Karl-Erik Andersson

Department of Clinical Pharmacology
Lund University Hospital
Lund, Sweden
and The Department of Urology
University of Virginia
Charlottesville, Virginia 22902-0422

1. INTRODUCTION

To fullfill its functions, the bladder has to contract during voiding and relax during the filling phase to accommodate increasing volumes of urine. During filling of the normal bladder, there is no significant rise in intravesical pressure until the bladder is near its functional capacity.[1] The mechanisms of bladder contraction have been well established,[2,3] but those responsible for bladder relaxation during urine storage remain unknown. Both myogenic and neural factors have been discussed.[4] Even though some investigators have suggested that viscoelastic properties of the smooth muscle itself may facilitate bladder compliance, and that bladder adaptation is independent of nervous influences,[5,6] at least in animals, a neural factor cannot be excluded.[4] During filling, the detrusor maintains an active, myogenic tone. Modulation of this tone by the smooth muscle cell itself, or by extrinsic influences, may be one of the decisive factors determining the ability of the bladder to store urine.

Increasing bladder capacity and eliminating urge, without interfering with the ability to completely empty the bladder, is one of the goals of pharmacologic treatment of bladder overactivity. It has been proposed[7] that there is an increased smooth muscle (re)activity, resulting in a decreased bladder compliance and an increased afferent response to distention in many cases of bladder overactivity. If this is correct, it should be possible to treat bladder overactivity by reducing the myogenic activity of the detrusor. Theoretically, this may be done without interference with the contractile strength of the muscle during nervous activation.

Myogenic activity can be influenced by both endogenous and exogenous factors causing stimulation or inhibition. Such factors may act by blocking excitatory stimuli, or

Advances in Bladder Research, edited by Baskin and Hayward.
Kluwer Academic / Plenum Publishers, New York, 1999.

by activating relaxation-mediating pathways in the detrusor. In the present review, some of the pathways for "active" relaxation of detrusor smooth muscle are briefly discussed with the aim of identifying targets for future drugs that can increase bladder capacity without interfering with the emptying of the bladder.

2. THE CYCLIC AMP PATHWAY

Lower urinary tract muscles can be relaxed by drugs increasing the intracellular concentrations of cyclic AMP (cAMP) or cyclic GMP (cGMP). Multiple types of adenyl cyclases exist, which catalyze the formation of cAMP.[8] In general, increases in cAMP seem to have a main role in bladder relaxation, whereas cGMP is important for urethral relaxation.[9–11] A decrease of cAMP concentration may increase bladder tone. Thus, stimulation of M_2 muscarinic receptors, which are negatively coupled to adenylyl cyclase, was found to contribute to muscarinic receptor-mediated rat bladder contraction both in vitro and in vivo.[12] Several agents, acting through different receptors linked to adenylyl cyclase, or directly stimulating the enzyme, have been shown to relax the bladder and simultaneously to increase the intracellular concentrations of cAMP.

2.1. β-Adrenoceptor Agonists

Detrusor smooth muscle seems to lack a significant inhibitory innervation. In the human bladder, fluorescence histochemical studies localizing catecholamines have either failed to demonstrate noradrenaline-containing nerves, or an extremely sparse innervation has been found.[13] However, release of noradrenaline can be demonstrated on electrical stimulation of nerves,[2] and noradrenaline relaxes the detrusor.[14,15] This effect can be due to both pre- and postjunctional actions. Stimulation of α_2-adrenoceptors on cholinergic neurons may lead to a decreased release of acetylcholine, and stimulation of postjunctional β-adrenoceptors to direct relaxation of the detrusor muscle.[2] Since β-adrenoceptors dominate over α-adrenoceptors postjunctionally in the bladder,[16] noradrenaline should relax the detrusor through stimulation of β-adrenoceptors.

The existence of subtypes of β-adrenoceptors (β_1 and β_2) is well established. Isoprenaline, which non-subtype selectively stimulates β-adrenoceptors has a pronounced inhibitory effect on the human bladder, an effect considered to be mediated by stimulation of adenylyl cyclase and an increase in cyclic AMP levels.[2] Functional studies revealed that the β-adrenoceptors of the human bladder were "atypical", since they could be antagonized by propranolol or bupranolol, effectively blocking both β_1- and β_2-adrenoceptors, but not by drugs selectively blocking β_1- (practolol or metoprolol) or β_2-(butoxamine) adrenoceptors.[17–19] Dobutamine or procaterol, selectively stimulating β_1- and β_2 adrenoceptors, respectively, had little relaxant effect.[19] On the other hand, β_2-adrenoceptor agonists such as terbutaline or salbutamol can inhibit bladder activity in humans.[20] Responses mediated via "atypical" β-adrenoceptors have recently been shown to be mediated by a β_3-adrenoceptor, which may explain some previous findings with drugs acting selectively on β_1- and β_2 adrenoceptors.[21,22]

The human detrusor is able to express β_1-, β_2-, as well as β_3-adrenoceptors.[19,23] Igawa et al.[19] showed that CGP-12177A, a β_3-adrenoceptor partial agonist with β_1/β_2-adrenoceptor blocking properties, effectively relaxed isolated detrusor preparations. These findings suggest that stimulation of both β_2- and β_3-adrenoceptors can relax the bladder. Whether or not the β_3-adrenoceptor is a good target for drugs meant for treatment of the overactive detrusor remains to be established. There seems to be a decrease with age both in density of β-adrenoceptors,[24] and in forskolin-activated adenylyl cyclase activity,[25] and it has been suggested that different β-adrenoceptor subtypes may be differently affected by age.[24] Possible age-dependent changes of the β_3-adrenoceptor of the bladder should be explored.

2.2. Vasoactive Intestinal Polypeptide (VIP)

The lower urinary tract of humans as well as animals is supplied with nerves containing VIP.[13] The majority of these nerves also contain immunoreactivity to Nitric Oxide Synthase (NOS)[26–28] and the vesicular acetylcholine transporter (VAChT), which is a specific marker for cholinergic neurons.[29] VIP receptors (types 1 and 2), linked via a stimulatory G-proteins to adenylyl cyclase, mediate the actions of the peptide.[30] The stimulatory effect on adenylyl cyclase leads to an increase in cAMP, which in turn activates cAMP-dependent protein kinase. However, VIP may increase not only cAMP but also cGMP concentrations in various smooth muscles.[31]

There is little evidence that VIP is utilized by postganglionic efferent neurons as a direct relaxant transmitter in the bladder. The action of VIP may vary from species to species. For example, in isolated guinea-pig bladder, the peptide produces contraction,[32–33] wheras in rabbit bladder, it causes relaxation and inhibition of spontaneous action potential discharge.[34,35] In the human bladder, the peptide has little effect on contractions induced by muscarinic receptor stimulation or by electrical stimulation of nerves.[36] However, VIP can effectively inhibit spontaneous contractile activity in isolated detrusor muscle from several animal species and from humans.[2] If this myogenic activity is of importance for afferent nerve activity, direct inhibitory effects of VIP on bladder smooth muscle could influence bladder capacity. Since VIP most probably is contained within cholinergic nerves, and since there is normally no ongoing sacral parasympathetic outflow during the filling phase,[37] the importance of VIP as a direct detrusor relaxant seems doubtful. However, several investigators have reported that VIP-levels are markedly reduced in bladders from patients suffering from idiopathic detrusor instability,[38,39] or detrusor hyperreflexia.[40] These findings were interpreted to suggest that lack of VIP contributes to some forms of detrusor overactivity.

The physiological and pathophysiological role of VIP in the bladder and its therapeutic potential remains to be established. VIP given i.v. to patients in a dose causing increases in heart rate, had no effect on cystometric parameters.[41] Thus, available information does not suggest that VIP receptors should be a promising target for drugs producing bladder relaxation.

2.3. Forskolin

Forskolin, a naturally occurring diterpene and plant alkaloid, is believed to stimulate directly the catalytic subunit of the adenylyl cyclase enzyme and thus increase the intracellular concentration of cAMP. However, the drug may also have effects that are not mediated via cAMP.[42] Forskolin caused a concentration-dependent relaxation of rabbit[43] and guinea-pig[44] detrusor muscle strips, and was shown also to effectively relax both porcine and human detrusor muscle.[45–48] Forskolin and isoproterenol were equipotent in relaxing rabbit detrusor strips, and produced the same amount of relaxation. However, forskolin caused a much higher increase in cAMP levels than isoproterenol, suggesting compartmentalization of cAMP in the detrusor.[49] Forskolin has been used mostly as a tool for stimulation of adenylyl cyclase in various tissues. Recently, however, the drug was given to men with erectile dysfunction with reported positive effects.[50]

Whether or not forskolin or any of its analogues[51,52] may have a clinically useful effect in the treatment of bladder disorders remains to be demonstrated.

2.4. Cyclic AMP Phosphodiesterase Inhibitors

Cyclic nucleotide phosphodiesterases (PDEs) catalyze the degradation of cAMP and cGMP, and the cellular levels and effects of these nucleotides are determined in part by the rate of degradation. Presently, seven different families of PDE are known which differ in their specificity for cAMP and cGMP, cofactor requirements and kinetic properties.[53,54] Inhibitors of PDE can influence the intracellular cyclic nucleotide metabolism and thus contraction and relaxation of detrusor smooth muscle. In both porcine[47,48] and human[45,46] detrusor, several PDE isoenzyme families have been demonstrated. Longhurst et al.[44] found that inhibition of cyclic AMP specific PDE4 isozymes by rolipram significantly reduced the contractile response of guinea-pig bladder strips to field stimulation, but that inhibition of cyclic GMP-specific PDE5 isozymes by zaprinast had no effect. Kinetic characteristics together with functional in vitro studies suggest that PDE1 may be of importance in the intracellular regulation of human detrusor tone. The PDE1 inhibitor vinpocetine was more effective than other selective PDE inhibitors in relaxing carbachol-contracted detrusor strips.[46]

Truss et al.[55] reported preliminary, promising experiences with vinpocetine in patients with urge incontinence and low compliance bladder. Whether this means that vinpocetine or any other selective inhibitor of PDE1 will be useful in the treatment of bladder overactivity remains to be established in controlled clinical trials.

3. THE CYCLIC GMP PATHWAY

Many endogenous agents and as well as drugs have cGMP as their common mediator in eliciting different physiological responses. There are multiple types of soluble and particulate guanylyl cyclases which catalyze cGMP synthesis.[56] Soluble guanylyl cyclase is the target of NO, which binds to its heme moiety and activates the enzyme with resulting increase in cGMP. cGMP, in turn, regulates protein phosphorylation, ion channel conductivity and phosphodiesterase activity. Wheeler et al.[57] investigated guanylyl cyclase activities in urinary tract smooth muscle, and found the rank order of the soluble form to be: urethra=ureter>bladder dome. Atrial natriuretic peptide stimulated particulate guanylyl cyclase activity was at least three times higher in the ureter than in the bladder or urethra, illustrating regional differences that may be physiologically important.

3.1. Nitric Oxide

L-arginine-derived NO has been shown to be a major inhibitory and relaxation-mediating transmitter of nonadrenergic, noncholinergic (NANC) nerves in several parts of the body, including lower urinary tract smooth muscles.[58] NOS has been shown to be localized to nerve fibres in the detrusor, but is most abundant in the nerves of the trigone and urethra. NOS seems to be contained within cholinergic nerves together with acetylcholine, VIP, and neuropeptide Y (NPY), and when these nerves are activated and NO released, the possibility of co-release and concerted action of transmitters should be considered.[59]

Considering its function in the stomach, where NO has been suggested to act as a mediator of adaptive relaxation to accommodate food and fluid,[60] it could well be conceived that NO has a similar role in the bladder. However, the occurrence and importance of relaxant responses in the detrusor, attributable to NO, seem to be controversial. James et al.[61] reported relaxation of human detrusor in response to electrical field stimulation and suggested that this response was partly mediated via the production of NO. These results were confirmed by Williams et al.,[62] who also found that the relaxant response was unchanged in hyper-reflexic bladders. However, the relaxations did not occur in all strips, were sometimes inconsistent, and were reduced, but not abolished by Nω-nitro L-arginine methyl ester (L-NAME). Since the responses were tetrodotoxin-resistant, it is unlikely that the NO producing the effect was derived from nerves. It has not been possible to convincingly produce nerve-mediated relaxations in human, pig, or rat detrusor[58], and the source of NO involved in relaxant responses to nerve-stimulation remains to be established.

If NO, released from nerves or some other source in the detrusor, were one of the factors keeping the bladder relaxed during filling, the detrusor smooth muscle cells should be the likely target. However, the detrusor smooth muscle has a low sensitivity to NO and agents acting via the cGMP system; nevertheless, relaxant effects can be obtained with NO-donors in some species.[58] However, in the guinea pig bladder, neither sodium nitroprusside, nor atrial natriuretic factor, stimulating soluble and particulate guanylyl cyclase, respectively, produced any relaxation.[57] In the human bladder, sodium nitroprusside was even found to produce contraction,[63] and it was suggested that the NO effects were mediated by the interstitial cells, staining for cGMP, which have been demonstrated in human and guinea-pig detrusor.[27] These findings make it unlikely that NO has a role as a direct relaxant neurotransmitter in the detrusor, but does not exclude that NO may have important functions as a modulator of both afferent nerve function, and of the effects of other transmitters.

3.2. ATP

There is presently good evidence suggesting that ATP is a NANC excitatory transmitter in animal as well as human detrusor, and that it produces contraction by stimulating P_{2x} receptors. However, the contractile potency of ATP in the bladder is low.[2] Boland et al.[64] suggested that in the mouse bladder, this might be due to ATP stimulating not only contraction-mediating P_{2x} receptors, but also relaxation-mediating P_{2y} receptors. Bolego et al.[65] arrived at the same conclusion in the rat bladder, and showed that the ATP-induced relaxation was abolished by the G-protein blocking agent, guanosine 5'-O-(3-thiophosphate). Boland et al.[64] hypothetized that the P_{2y} receptors counteracted contraction or induced smooth muscle inhibition during bladder filling, thereby explaining in part the high compliance of the detrusor smooth muscle.

The physiological importance of P_{2y} receptor-mediated relaxation is unclear. However, the effect points to the possibility of obtaining bladder relaxation by selective stimulation of the P_{2y} receptor.

4. ION CHANNELS

4.1. Potassium Channels

In detrusor muscle strips, K+channel blocking drugs increase spontaneous mechanical activity, and drugs opening these channels reduce it.[2,66] There is much evidence suggesting that several types of K+channels may be involved both in determining the membrane potential and in action potential depolarization. In the guinea-pig detrusor, at least three types of K^+ channel have been demonstrated: ATP-sensitive (K_{ATP}) channels; small conductance and calcium dependent (SK_{Ca})channels; and large conductance and calcium-dependent (BK_{Ca}) channels. ATP-sensitive K^+-channels have been found also in human detrusor muscle, together with calcium-activated K^+-channels (BK_{Ca}), which seem to be involved in the control of basal tension and membrane potential.[67] Openers of ATP-sensitive K+ channels have been shown to be an effective way of decreasing bladder activation.[2,66] For example, cromakalim reduced the spike frequency in isolated guinea-pig bladder, abolished spikes, and produced a concentration- dependent hyperpolarization of the cell membrane.[68,69] In the guinea-pig detrusor, Grant and Zuzack[70] found marked stimulatory effects of charybdotoxin and iberiatoxin, which were used as probes for investigation of the large conductance, Ca^{2+} activated K^+ channels, and concluded that these channels may be involved in the control of basal tension and possibly the membrane potential of detrusor cells.

Studies on isolated human detrusor muscle, and on bladder tissue from several animal species have shown that K^+ channel openers effectively reduce or abolish spontaneous myogenic contractions. They also depress contractions induced by electrical stimulation, carbachol, and low, but not high external K^+ concentrations.[2] The K^+-channel openers were particularly effective in hypertrophic rat bladder muscle in vitro[71] and effectively suppressed bladder hyperactivity in rats with bladder outflow obstruction without affecting bladder emptying.[72]

Openers of ATP-sensitive K^+-channels, such as cromakalim and pinacidil, were found to be up to 200 times more potent as inhibitors of vascular than of detrusor smooth muscle,[73,74] and in the clinical trials performed with these drugs, no bladder effects have been found at doses reducing blood pressure.[75,76] However, new drugs with K_{ATP}-channel opening properties have been described, which may be useful for treatment of bladder overactivity.[77–79] Zeneca ZD6169 and its analogs were shown to activate K_{ATP}-channels in guinea pig detrusor strips as well as in human bladder cells[80–81], and given orally, ZD6169 was claimed to have an in vivo selectivity, significantly reducing micturition frequency in rats at doses producing no cardiovascular effects.[77] Selectivity for bladder tissue was, however, not obvious in vitro.[74]

Pinacidil given intravesically significantly reduced bladder hyperreflexia induced by penile ligation in anesthetized rabbits.[82] Yu et al.,[83] giving ZD6169 intravesically, found the drug to increase the intercontraction interval in normal, but not in capsaicin-treated rats. Intravesical ZD6169 also suppressed C-fos gene expression in the spinal cord induced by acetic acid instilled into the bladder, and it was concluded that intravesical ZD6169 can influence bladder capacity by suppressing capsaicin-sensitive C-fiber afferents in the bladder.[83] ZD6169, given i.a. or orally, increased threshold pressure, but had otherwise little effect on volume-induced micturition. However, the drug markedly reduced or prevented PGE_2-induced bladder activity when given i.a.; it was also effective when given orally.[84]

K^+ channel opening is an attractive way of treating bladder overactivity, since it would theoretically make it possible to eliminate undesired bladder contractions without affecting normal micturition. However, the relaxant effects of levocromakalim and YM934, a novel benzoxazin opener of ATP-sensitive K+ channels, were similar in normal and hyper-reflexic bladders, suggesting that there were no appreciable changes in K_{ATP} channel function in the unstable bladder.[85] The site of action of K+ channel openers that may be useful for bladder relaxation, be it smooth muscle cells, nerves, or both, remains to be established.

4.2. Calcium Channels

Action potentials in detrusor muscle from for example guinea pigs are blocked by L-type Ca^{2+} channel blockers.[86] There is evidence from patch-clamp studies that the upstroke of the action potential also in the human bladder may involve dihydropyridine-sensitive Ca^{2+} channels.[87,88] Detrusor muscle cells possess stretch-activated channels, which are non-selective cation channels with some permeability for calcium ions.[89] Spontaneous intracellular Ca^{2+} transients have been demonstrated in isolated human bladder smooth muscle cells,[90] and may be the origin of myogenic contractile activity. Such activity is, however, dependent on extracellular Ca^{2+}, since Ca^{2+} antagonists can abolish spontaneous bladder contractions.

Activation of detrusor muscle, both through cholinergic, and NANC pathways, require influx of extracellular Ca^{2+} through L-type calcium channels, as well as mobilization of intracellular Ca^{2+}.[2] If there is an increased Ca^{2+} influx in bladder overactivity, one would expect changes in, e.g., the binding or functional properties of the voltage-dependent Ca^{2+} channels. In bladders from children with myelodysplasia, in whom detrusor hyperreflexia is a common phenomenon, this was not found.[91]

Considering the importance of Ca^{2+} influx for contractile activation, blockade of L-type Ca^{2+} channels seems to be an attractive way of inhibiting spontaneous bladder contraction. However, irrespective of the mode of activation, nifedipine inhibited detrusor contraction in isolated human detrusor preparations,[92] and nifedipine also abolished the NANC mediated contractile component in rabbit[93] and rat detrusor,[94,95] and the atropine-resistant contractile component in hypertrophic human detrusor muscle.[96] Therefore, Ca^{2+} antagonists could be expected to depress bladder contractility ncessary for normal bladder emptying. On the other hand, if the myogenic contractions are exclusively dependent on Ca^{2+} influx through L-type channels, and the micturition contraction is mainly dependent on inositol trisphosphate (IP_3)-initiated Ca^{2+} release, it may be possible to reduce bladder overactivity without interfering with bladder emptying.

5. PARATHYROID HORMONE-RELATED PROTEIN (PTHRP)

Distension of the bladder wall can cause the synthesis of agents that may increase bladder compliance. One such factor is parathyroid hormone-related protein (PTHrP).[97,98] Yamamoto et al.[97] demonstrated that PTHrP mRNA levels changed in response to stretch of the rat bladder wall. PTHrP mRNA increased substantially with distension of the bladder, and in vitro PTHrP-1(1–34)-NH_2 relaxed carbachol-induced contractions in strips from bladders kept empty in vivo. Persson et al.[98] showed an increase in PTHrP secretion from rat bladder base and urethra in response to stretch, and suggested that PTHrP may facilitate bladder accommodation during filling by a direct effect on the smooth muscle, and that the protein may have a paracrine effect on vessels regulating blood flow during bladder distension or modulate the threshold for afferent nerve firing. The mechanism for PTHrP-induced bladder relaxation does not seem to have been clarified, nor whether or not PTHrP has a role in the normal micturition cycle, increasing compliance during bladder filling.

6. UNKNOWN FACTORS

It has been speculated that factors released form the urothelium should be able to modify the bladder response to for example stress,[99] similar to what is found in the endothelium and vascular smooth muscle.[100] The bladder can produce a relaxant factor when subjected to mechanical stress.[101] Fovaeus et al.,[101] using a co-axial bioassay system with endothelium-free, noradrenaline-contracted, rat aortic preparations mounted within urothelium-intact urinary bladder, showed that carbachol caused a concentration-dependent relaxation of the vessel preparation. The relaxant effect of carbachol was lost if the urinary bladder segment was removed. However, the relaxation was affected neither by removal of the urothelium, nor by bladder segment inversion. It was resistant to inhibition of the L-arginine/nitric oxide and cyclo-oxygenase pathways, and unaffected by propranolol. The effect was interpreted as being caused by an unidentified relaxant factor released from the bladder tissue.[101] The nature of the relaxant factor and the source of release are the subject of further investigations.

ACKNOWLEDGMENT

This work was supported by the Swedish Medical Research Council, grant no. 06837.

REFERENCES

1. Coolsaet, B.: Bladder compliance and detrusor activity during the collection phase. Neurourol. Urodyn., **4:** 263, 1985.
2. Andersson, K.-E.: Pharmacology of lower urinary tract smooth muscles and penile erectile tissues. Pharmacol. Rev., **45:** 253, 1993.
3. Turner, W.H. and Brading, A.F.: Smooth muscle of the bladder in the normal and the diseased state: pathophysiology, diagnosis, treatment. Pharmacol. Ther., **75:** 77, 1997.
4. Vaughan, C.W. and Satchell, P.M.: Urine storage mechanisms. Progr. Neurobiol., **46:** 215, 1995.
5. Tang, P.C. and Ruch, T.C.: Non-neurogenic basis of bladder tonus. Am. J. Physiol., **181:** 249, 1969.
6. Klevmark, B.: Motility of the urinary bladder in cats during filling at physiological rates. II. Effects of extrinsic bladder denervation on intramural tension and intravesical pressure patterns. Acta Physiol. Scand., **101:** 176, 1977.
7. Brading, A.F. and Turner, W.H.: The unstable bladder: towards a common mechanism. Br. J. Urol., **73:** 3, 1994.
8. Sunahara, R.K., Dessauer, C.W. and Gilman, A.G.: Complexity and diversity of mammalian adenylyl cyclases. Annu. Rev. Pharmacol. Toxicol., **36:** 461, 1996.
9. Morita, T., Tsujii, T., and Dokita, S.: Regional difference in functional roles of cAMP and cGMP in lower urinary tract smooth muscle contractility. Urol. Int., **49:** 191, 1992.
10. Dokita, S., Smith, D., Nishimoto, T., Wheeler, M.A. and Weiss, R.M.: Involvement of nitric oxide and cyclic AMP in rabbit urethral relaxation. Eur. J. Pharmacol., **269:** 269, 1994.
11. Persson, K. and Andersson, K.-E.: NANC-mediated relaxation of the rabbit lower urinary tract smooth muscle and associated levels of cGMP and cAMP. Eur. J. Pharmacol., **268:** 159, 1994.
12. Hegde, S.S., Choppin, A., Bonhaus ,D., Briaud, S., Loeb, M., Moy, T.M., Loury, D. and Eglen, R.M.: Functional role of M_2 and M_3 muscarinic receptors in the urinary bladder of rats in vitro and in vivo. Br. J. Pharmacol., **120:** 1409, 1997.
13. Lincoln, J. and Burnstock, G.: Autonomic innervation of the urinary bladder and urethra. In: The Autonomic Nervous System. Vol. 6, Chapter 8, Nervous Control of the Urogenital System, ed by CA Maggi. Harwood Academic Publishers, London, UK, 1993; pp. 33–68.
14. Perlberg, S. And Caine, M.: Adrenergic response of bladder muscle in prostatic obstruction Urology, **20:** 524, 1982.
15. Åmark, P., Nergårdh, A. and Kinn, A.C.: The effect of noradrenaline on the contractile response of the urinary bladder. Scand. J. Urol. Nephrol., **20:** 203, 1986.
16. Levin, R.M. and Wein, A.J.: Quantitative analysis of alpha and beta adrenergic receptor densities in the lower urinary tract of the dog and the rabbit. Invest. Urol., **17:** 75, 1979.
17. Nergårdh, A., Boréus, L.O. and Naglo, A.S.: Characterization of the adrenergic beta-receptors in the urinary bladder of man and cat. Acta Pharmacol. Toxicol., **40:** 14, 1977.
18. Larsen, J.J.: α- and β-Adrenoceptors in the detrusor muscle and bladder base of the pig and β-adrenoceptors in the detrusor of man. Br. J. Urol., **65:** 215, 1979.

19. Igawa, Y., Yamazaki, Y., Takeda, H., Hayakawa, K., Akahane, M., Ajisawa, Y., Yoneyama, T. and Nishizawa, O.: The role of β3-adrenoceptors in normal and neurogenic detrusors. Neurourol. Urodyn., **16:** 363, 1997.
20. Andersson, K.-E.: Current concepts in the treatment of disorders of micturition. Drugs, **35:** 477, 1988.
21. Lipworth, B.J.: Clinical pharmacology of β_3-adrenoceptors. Br. J. Clin. Pharmacol., **43:** 291, 1996.
22. Strosberg, D. and Pietri-Rouxel. F.: Function and regulation of the β_3-adrenoceptor. Trends Pharmacol. Sci., **17:** 273, 1997.
23. Takeda, M., Mizusawa, T., Obara, K., Koizumi, T., Tsutsui, T., Hatano, A., Kanai, T. and Takahashi, K.: Adrenergic β_1-, β_2-, and β_3-receptor subtypes in the detrusor of human urinary bladder. - Evaluation by mRNA expression and isometric contraction. Neurourol. Urodyn., **16:** 365, 1997.
24. Nishimoto, T., Latifpour, J., Wheeler, M.A., Yoshida, M., Weiss, R.M.: Age-dependent alterations in b-adrenergic responsiveness of rat detrusor smooth muscle. J. Urol., **153:** 1701, 1995.
25. Wheeler, M.A., Pontari, M., Nishimoto, T., Weiss, R.M.: Changes in lipid composition and isoproterenol- and ethanol-stimulated adenylate cyclase activity in aging Fischer rat bladders. J. Pharmacol. Exp. Ther., **254:** 277, 1990.
26. Persson, K., Alm, P., Johansson, K., Larsson, B. and Andersson, K.-E.: Co-existence of nitrergic, peptidergic and acetylcholine esterase-positive nerves in the pig lower urinary tract. J. Auton. Nerv. Syst., **52:** 225, 1995.
27. Smet, P.J., Jonavicius, J., Marshall, V.R. And De Vente, J.: Distribution of nitric oxide synthase-immunoreactive nerves and identification of the cellular targets of nitric oxide in guinea-pig and human urinary bladder by cGMP immunohistochemistry. Neuroscience, **71:** 337, 1996.
28. Smet, P.J., Edyvane, K.A., Jonavicius, J. and Marshall, V.R.: Neuropeptides and transmitter-synthesizing enzymes in intrinsic neurons of the human urinary bladder. J. Neurocytol., **25:** 112, 1996.
29. Arvidsson, U., Reidl, M., Elde, R. and Meister, B.: Vesicular acetylcholine transporter (VAChT) protein: A novel and unique marker for cholinergic neurons in the central and peripheral nervous systems. J. Comp. Neurol., **378:** 454, 1997.
30. Fahrenkrug, J.: Transmitter role of vasoactive intestinal peptide. Pharmacol. Toxicol., **72:** 354, 1993.
31. Chakter, S. and Rattan, S.: Involvement of cAMP and cGMP in relaxation of internal anal sphincter by neural stimulation, VIP, and NO. Am. J. Physiol., **264:** G702, 1993.
32. Johns, A.: The effect of VIP on the urinary bladder and taenia coli of the guinea-pig. Can. J. Physiol. Pharmacol., **56:** 106, 1979.
33. Mackenzie, I. and Burnstock, G.: Neuropeptide action on the guinea-pig bladder; a comparison with the effects of field stimulation and ATP. Eur. J. Pharmacol., **105:** 85, 1984.
34. Callahan, S.M. and Creed, K.E.: Non-cholinergic neurotransmission and the effects of peptides on the urinary bladder of guinea-pigs and rabbits. J. Physiol., **374:** 103, 1986.
35. Levin, R.M. and Wein, A.J.: Effect of vasoactive intestinal peptide on the contractility of the rabbit urinary bladder. Urol. Res., **9:** 217, 1989.
36. Sjögren, C., Andersson, K.-E. and Mattiasson, A.: Effects of vasoactive intestinal polypeptide on isolated urethral and urinary bladder smooth muscle from rabbit and man. J. Urol., **133:** 136, 1985.
37. de Groat, W.C., Booth, A.M. and Yoshimura, N.: Neurophysiology of micturition and its modification in animal models of human disease. In: The Autonomic Nervous System. Vol. 6, Chapter 8, Nervous Control of the Urogenital System, ed by CA Maggi. Harwood Academic Publishers, London, UK, 1993; pp. 227–289.
38. Gu, J., Restorick, J.M., Blank, M.A., Huang, W.M., Polak, J.M., Bloom, S.R. and Mundy, A.R.: Vasoactive intestinal polypeptide in the normal and unstable bladder. Br. J. Urol., **55:** 645, 1983.
39. Chapple, C.R., Milner, P., Moss, H.E. and Burnstock, G.: Loss of sensory neuropeptides in the obstructed human bladder. Br. J. Urol., **70:** 373, 1992.
40. Kinder, R.B., Restorick, J.M. and Mundy, A.R.: Vasoactive intestinal polypeptide in the hyperreflexic neuropathic bladder. Br. J. Urol., **57:** 289, 1985.
41. Klarskov, P., Holm-Bentzen, M., N¨rgaard, T., Ottesen, B., Walter, S. and Hald, T.: Vasoactive intestinal polypeptide concentration in human bladder neck smooth muscle and its influence on urodynamic parameters. Br. J. Urol., **60:** 113, 1987.
42. Laurenza, A., McHugh-Sutkowski, E.M. and Seamon, K.B.: Forskolin: a specific stimulator of adenylyl cyclase or a diterpene with multiple sites of action. Trends Pharmacol. Sci., **10:** 442, 1989.
43. Morita, T., Wheeler, M.A. and Weiss, R.M.: Relaxant effect of forskolin in rabbit detrusor muscle: role of cyclic AMP. J. Urol., **135:** 1293, 1986.
44. Longhurst, P.A., Briscoe, J.A.K., Rosenberg, D.J. and Leggett, R.E.: The role of cyclic nucleotides in guinea-pig bladder contractility. Br. J. Pharmacol., **121:** 1665, 1997.

45. Truss, M., Ückert, S., Stief, C.G., Forssmann, W.G. and Jonas, U.: Cyclic nucleotide phosphodiesterae (PDE) isoenzymes in the human detrusor smooth muscle. II. Effect of various PDE inhibitors on smooth muscle tone and cyclic nucleotide levels in vitro. Urol. Res., **24:** 129, 1996a.
46. Truss, M., Ückert, S., Stief, C.G., Kuczyk, M. and Jonas, U.: Cyclic nucleotide phosphodiesterae (PDE) isoenzymes in the human detrusor smooth muscle. I. Identification and characterization. Urol. Res., **24:** 123, 1996b.
47. Truss, M., Ückert, S., Stief, C.G., Kuczyk, M., Schulz-Knappe, P., Forssmann, W.-G. and Jonas, U.: Effects of various phosphodiesterase-inhibitors, forskolin, and sodium nitroprusside on porcine detrusor muscle tonic responses to muscarinic stimulation and cyclic nucleotide levels in vitro. Neurourol. Urodyn., 1996c; **15:** 59, 1996.
48. Truss, M., Ückert, S., Stief, C.G., Kuczyk, M., Schulz-Knappe, P., Forssmann, W.-G. and Jonas, U.: Porcine detrusor cyclic nucleotide phosphodiesterase isoenzyme : characterization and functional effects of various phosphodiesterase inhibitors in vitro. Urology, **45:** 893, 1995.
49. Morita, T., Ando, M., Kihara, K., Oshima, H., Kondo, S. and Tashima, Y.: Dissociation of magnitude of relaxation from cyclic AMP levels in rabbit urinary bladder. J. Smooth Muscle Res., **28:** 121, 1992.
50. Mulhall, J.P., Daller, M., Traish, A.M., Gupta, S., Park, K., Salimpour, P., Payton, T.R., Krane, R.J. and Goldstein, I.: Intracavernosal forskolin: role in management of vasculogenic impotence resistant to standard 3-agent pharmacotherapy. J. Urol., **158:** 1752, 1997.
51. Laurenza, A., Robbins, J.D. and Seamon, K.B.:. Interaction of aminoalkylcarbamates of forskolin with adenylyl cyclase: synthesis of an iodinated derivative of forskolin with high affinity for adenylyl cyclase. Mol. Pharmacol., **41:** 360, 1992.
52. McHugh Sutkowski, E., Tang, W.-J., Broome, C.W., Robbins, J.D. and Seamon, K.B.: Regulation of forskolin interactions with type I, II, V, and VI adenylyl cyclases by G_{sa}. Biochemistry, **33:** 12852, 1994.
53. Beavo, J.A.: Cyclic nucleotide phosphodiesterases: functional implications of mutiple isoforms. Physiol. Rev., **75:** 725, 1995.
54. Polson, J.B. and Strada, S.J. :Cyclic nucleotide phosphodiesterases and vascular smooth muscle. Annu. Rev. Pharmacol. Toxicol., **36:** 403, 1996.
55. Truss, M., Stief, C.G., Ückert, S., Becker, A.J., Forssmann, W.-G. and Jonas, U.: Initial clinical experience with the selective phosphodiesterase (PDE)-I isoenzyme inhibitor vinpocitine in the treatment of urge-incontinence and low compliance bladder. J. Urol., **157:** 187, 1997.
56. Schmidt, H.H.H.W., Lohmann, S.M. and Walter, U.: The nitric oxide and cGMP signal transduction system: regulation and mechanism of action. Biochim. Biophys. Acta, **1178:** 153, 1993.
57. Wheeler, M.A., Pontari, M., Dokita, S., Nishimoto, T., Ho Cho, Y., Whan Hong, K., Weiss, R.M.: Age-dependent changes in particulate and soluble guanylyl cyclase activities in urinary tract smooth muscle. Mol. Cell. Biochem., **169:** 115, 1997.
58. Andersson, K.-E. and Persson, K.: Nitric oxide synthase and the lower urinary tract: possible implications for physiology and pathophysiology. Scand. J. Urol. Nephrol., **29(Suppl 175):** 43, 1995.
59. Lundberg, J.M.: Pharmacology of cotransmission in the autonomic nervous system: integrative aspects on amines, neuropeptides, adenosine triphosphate, amino acids and nitric oxide. Pharmacol. Rev., **48:** 113, 1996.
60. Desai, K.M., Sessa, W.C., and Vane, J.R.: Involvement of nitric oxide in the reflex relaxation of the stomach to accommodate food or fluid. Nature, **351:** 477, 1991.
61. James, M.J., Birmingham, A.T. and Hill, S.J.: Partial mediation by nitric oxide of the relaxation of human isolated detrusor strips in response to electrical field stimulation. Br. J. Clin. Pharmacol., **35:** 366, 1993.
62. Williams, P., Datta, S., German, K., Stephenson, T.P. and Bedwani, J.R.: Comparison of relaxation responses of detrusor strips from neuropathic and control patients. Br. J. Urol., **76:** 726, 1995.
63. Moon, A., Pickard, R.S., Gillespie, J.I. and Neal, D.E.: Contractile responss to sodium nitroprusside and L-arginine in isolated human detrusor. J. Urol., **157:** 258, 1997 (abstract 1012).
64. Boland, B., Himpens, B., Paques, C., Casteels, R. and Gillis, J.M.: ATP induced-relaxation in the mouse bladder smooth muscle. Br. J. Pharmacol., **108:** 749, 1993.
65. Bolego, C., Pinna, C., Abbracchio, M.P., Cattabeni, F. and Puglisi, L.: The biphasic response of rat vesical smooth muscle to ATP. Br. J. Pharmacol., **114:** 1557, 1995.
66. Andersson, K.-E.: Clinical pharmacology of potassium channel openers. Pharmacol. Toxicol., **70:** 244, 1992.
67. Trivedi, S., Potter-Lee, L., Li, J.H., Yasay, G.D., Russell, K., Ohnmach, C.J., Empfield, J.R., Trainor, D.A. and Kau, S.T.: Calcium dependent K-channels in guinea pig and human urinary bladder. Biochem. Biophys. Res. Commun., **213:** 404, 1995.
68. Fujii, K., Foster, C.D., Brading, A.F. and Parekh, A.B.: Potassium channel blockers and the effects of cromakalim on the smooth nuscle of the guinea-pig bladder. Br. J. Pharmacol., **99:** 779, 1990.

69. Foster, C.D., Speakman, M.J., Fujii, K. and Brading, A.F.: The effects of cromakalim on the detrusor of human and and pig urinary bladder. Br. J. Urol., **63:** 284, 1989.
70. Grant, T.L. and Zuzack, J.S.: Effects of K^+ channel blockers and cromakalim (BRL 34915) on the mechanical activity of guinea pig detrusor smooth muscle. J. Pharmacol. Exp. Ther., **259:** 1158, 1991.
71. Malmgren, A., Andersson, K.-E., Andersson, P.-O., Fovaeus. M. And Sjögren, C.: Effects of cromakalim (BRL 34915) and pinacidil on normal and hypertrophied rat detrusor in vitro. J. Urol., **143:** 828, 1990.
72. Malmgren, A., Andersson, K.-E., Sjögren, C. and Andersson, P.-O.: Effects of pinacidil and cromakalim (BRL 34915) on bladder function in rats with detrusor instability. J. Urol., **142:** 1134, 1989.
73. Edwards, G., Henshaw, M., Miller, M. and Weston, A.H.: Comparison of the effects of several potassium-channel openers on rat bladder and rat portal vein in vitro. Br. J. Pharmacol., **102:** 679, 1991.
74. Barras, M., Van der Graf, P.H., Christophe, P. and Itzchak, A.: Relaxant efficacy of potassium channel openers in rabbit isolated bladder and mesenteric artery. Eur. J. Urol., **30 (Suppl 2):** 240, 1996 (abstract 894).
75. Hedlund, H., Mattiasson, A. and Andersson, K.-E.: Effects of pinacidil on detrusor instability in men with outlet obstruction. J. Urol., **146:** 1345, 1991.
76. Komersova, K., Rogerson, J.W., Conway, E.L., Lim, T.C., Brown, D.J., Krum, H., Jackman, G.P., Murdoch, R. and Louis, W.J. :The effect of levcromakalim (BRL 38227) on bladder function in patients with high spinal cord lesions. Br. J. Pharmacol., **39:** 207, 1995.
77. Howe, B.B., Halterman, T.J., Yochim, C.L., Do, M.L., Pettinger, S.J., Stow, R.B., Ohnmacht, C.J., Russel, K., Empfield, J.R., Trainor, D.A., Brown, F.J. and Kau, S.T.: ZENECA ZD6169: A novel KATP channel opener with in vivo selectivity for urinary bladder. J. Pharmacol. Exp. Ther., **274:** 884, 1995.
78. Masuda, N., Uchida, W., Shirai, Y., Shibasaki, K., Goto, K. and Takenaka, T.: Effect of the potassium channel opener YM934 on the contractile response to electrical field stimulation in pig detrusor smooth muscle. J. Urol., **154:** 1914, 1995.
79. Hashitani, H., Suzuki, H. and Kumazawa, J.: Effects of Y26763, a novel K-channel opener, on electrical responses of smooth muscle in the guinea-pig bladder. J. Urol., **155:** 1454, 1996.
80. Li, J.H., Yasay, G.D., Zografos, P., Kau, S.T., Ohnmacht, C.J., Russell, K., Empfield, J.R., Brown, F.J., Trainor, D.A., Bonev, A.D., Heppner, T.J. and Nelson, M.T.: ZENECA ZD6169 and its analogs from a novel series of anilide tertiary carbinols: in vitro K_{ATP} channel opening activity in bladder detrusor. Pharmacology, **52:** 33, 1995.
81. Trivedi, S., Stetz, S.L., Potter-Lee, L., McConville, M., Li, J.H., Empfield, J.R., Ohnmacht , C.J., Russell, K., Brown, F.J., Trainor, D.A., Kau, S.T.: K-channel opening activity of ZD6169 and its analogs: Effect on ^{86}RB efflux and ^{3}H-P1075 binding in bladder smooth muscle. Pharmacology, **50:** 388, 1995.
82. Levin, R.M., Hayes, L., Zhao, Y. and Wein, A.J.: Effect of pinacidil on spontaneous and evoked contractile activity. Pharmacology, **45:** 1, 1992.
83. Yu, Y.B., Fraser, M.O. and de Groat, W.C.: Effect of intravesical administration of ZD6169 on the micturition reflex and on C-fos expression in the spinal cord induced by noxious bladder stimulation in the rat. Society for Neuroscience, **22:** 93, 1996 (abstract 45.8).
84. Pandita, R.K., Persson, K. and Andersson, K.-E.: Effects of the K^+ channel opener, ZD6169, on volume and PGE_2-stimulated bladder activity in conscious rats. J. Urol., **158:** 2300, 1997.
85. Martin, S.W., Radley, S.C., Chess-Williams, R., Korstanje, C. and Chapple, C.R.: Relaxant effects of potassium-channel openers on normal and hyperreflexic detrusor muscle. Br. J. Urol., **80:** 405, 1997.
86. Mostwin, J.L.: The action potential of guinea pig bladder smooth muscle. J. Urol., **135:** 1299, 1986.
87. Montgomery, B.S. and Fry, C.H.; The action potential and net membrane currents in isolated human detrusor smooth muscle cells. J. Urol., **147:** 176, 1992.
88. Gallegos, C.R. and Fry, C.H.: Alterations to the electrophysiology of isolated human detrusor smooth muscle cells in bladder disease. J. Urol., **151:** 754, 1994.
89. Wellner, M.C. and Isenberg, G.: Stretch-activated nonselective cation channels in urinary bladder myocytes: importance for pacemaker potentials and myogenic response. Experientia, **66:** 93, 1993.
90. Chambers, P., Neal, D.E. and Gillespie, J.I.: Ca^{2+} signalling in cultured smooth muscle cells from human bladder. Exp. Physiol., **81:** 553, 1996.
91. Shapiro, E., Tang, R., Rosenthal, E. and Lepor, H.: The binding and functional properties of voltage dependent calcium channel receptors in pediatric normal and myelodysplastic bladders. J. Urol., **146:** 520, 1991.
92. Forman, A., Andersson, K.-E., Henriksson, L., Rud, T. and Ulmsten, U.: Effects of nifedipine on the smooth muscle of the human urinary tract in vitro and in vivo. Acta Pharmacol. Toxicol., **43:** 111, 1978.
93. Andersson, K.-E., Fovaeus, M., Morgan, E. and Mclorie, G.: Comparative effects of five different calcium channel blockers on the atropine resistant contraction in electrically stimulated rabbit urinary bladder. Neurourol. Urodyn., **5:** 579, 1986.

94. Bo, X. and Burnstock, G.: The effects of BAY K 8644 and nifedipine on the responses of rat urinary bladder to electrical field stimulation, β-γ–methylene ATP and acetylcholine. Br. J. Pharmacol., **101:** 494, 1990.
95. Zar, M.A., Iravani, M.M. and Luheshi, G.N.: Effect of nifedipine on the contractile responses of the isolated rat bladder. J. Urol., **143:** 835, 1990.
96. Sjögren, C., Andersson, K.-E., Husted, S., Mattiasson, A. and M¯ller-Madsen, B.: Atropine resistance of the transmurally stimulated isolated human bladder. J. Urol., **128:** 1368, 1982.
97. Yamamoto, M., Harm, S.C., Grasser, W.A. and Thiede, M.A.: Parathyroid hormone-related protein in the rat urinary bladder: A smooth muscle relaxant produced locally in response to mechanical stretch. Proc. Natl. Acad. Sci, USA, **89:** 5326, 1992.
98. Persson, K., Bruns, M.E.H., Bruns, D.E., Tuttle, J.B. and Steers, W.D.: Stretch increases secretion of parathyroid hormone related protein by cultured bladder smooth muscle cells. Neurourol. Urodyn., **13:** 406, 1994.
99. Levin, R.M., Wein, A.J., Krasnopolsky, L., Atta, M.A. and Ghoniem, G.M.: Effect of mucosal removal on the response of the feline bladder to pharmacological stimulation. J. Urol., **153:** 1291, 1995.
100. Mombouli, J.-V. and Vanhoutte, P.M.: Endothelium-derived hyperpolarizing factor(s): updating the unknown. Trends Pharmacol. Sci., **18:** 252, 1997.
101. Fovaeus, M., Fujiwara, M., Högestätt, E.D., Persson, K. and Andersson, K.-E.: A non-nitrergic smooth muscle relaxant factor released from the contracting rat urinary bladder. Acta Physiol. Scand., **162:** 115, 1998.

19

MATURATION OF BLADDER REFLEX PATHWAYS DURING POSTNATAL DEVELOPMENT

William C. de Groat and Isao Araki

Department of Pharmacology, Medical School
University of Pittsburgh
Pittsburgh, Pennsylvania 15261

1. INTRODUCTION

The mechanisms involved in storage and periodic elimination of urine undergo marked changes during prenatal and postnatal development.[1–3] In the young fetus prior to maturation of the nervous system, urine is presumably eliminated from the bladder by nonneural mechanisms; however at later stages of development voiding is regulated by primitive reflex pathways organized in the spinal cord. As the central nervous system matures during the postnatal period, reflex voiding is eventually brought under voluntary control which originates in the higher centers of the brain. Thus the functions of the lower urinary tract are regulated by a hierarchical control system consisting of very primitive components which are intrinsic to the target organs and spinal cord as well as more complex components involving neural circuitry in the cerebral cortex.

In adults, injuries or diseases of the nervous system can lead to the reemergence of primitive functions that were prominent early in development but then suppressed during neural maturation.[1–4] Therefore, developmental studies in neurourology are likely to provide insights into the mechanisms underlying neurogenic disorders of bladder function in adults as well as an understanding of problems of the developing nervous system encountered in pediatric urology.

This paper will focus on alterations in spinal reflex mechanisms that underlie the maturation of voiding function during the early postnatal period in experimental animals. In many species (e.g., rats and cats) voiding in neonates is dependent on an exteroceptive somato-bladder reflex mechanism triggered when the mother licks the genital or perineal region of the young animal.[1–3,5–10] The exteroceptive reflex is organized in the sacral spinal cord and has an afferent limb in the pudendal nerve and an efferent limb in the pelvic nerve. Similar reflexes have been identified in human infants.[11,12]

Advances in Bladder Research, edited by Baskin and Hayward.
Kluwer Academic / Plenum Publishers, New York, 1999.

In newborn kittens and rats the exteroceptive perineal-to-bladder reflex is essential for survival because isolation of the newborn from its mother leads to urinary retention.[5,7,8,13,14] This indicates that an adult form of reflex voiding which is induced by bladder distension is not functional in neonatal animals. The latter type of voiding, which is mediated by a spinobulbospinal bladder-to-bladder reflex pathway and by a micturition center in the brain stem,[1,3,15,16] only emerges several weeks after birth.[1–3,7] During this same period of time the perineal-to-bladder reflex becomes progressively weaker and eventually disappears. Thus postnatal maturation of voiding function is associated with a prominent reorganization of synaptic connections in bladder reflex pathways leading to downregulation of primitive spinal mechanisms and upregulation of mature supraspinal mechanisms. It seems likely that this developmental switching mechanism is dependent upon competition between brain and spinal pathways because spinal cord injury in adult animals and humans which interrupts brain-spinal cord connections causes the reemergence of the neonatal perineal-to-bladder reflex.[1,3,14,17] This paper will review recent studies in which tract tracing techniques and electrophysiological methods were used to examine the synaptic changes underlying developmental neuroplasticity in the parasympathetic pathways to the urinary bladder.

2. ANATOMICAL STUDIES

The reflex circuitry controlling micturition consists of four basic components: spinal efferent neurons, spinal interneurons, primary afferent neurons and neurons in the brain that modulate spinal reflex pathways (Figure 1). New research methodologies including transneuronal virus tracing[18–24] and measurements of gene expression[25,26] have provided insights into the morphology and interconnections of these components. Neurotropic viruses such as pseudorabies virus (PRV) have been particularly useful because after they are injected into a target organ (urinary bladder, urethra or urethral sphincter)[20,22–24] they move intra-axonally from the periphery to the central nervous system where they replicate and then pass retrogradely across synapses to infect second and third-order neurons in neural pathways (Figure 2).[18,19] Because PRV can be transported across many synapses, it could sequentially infect all of the neurons that connect directly or indirectly to the target organ.

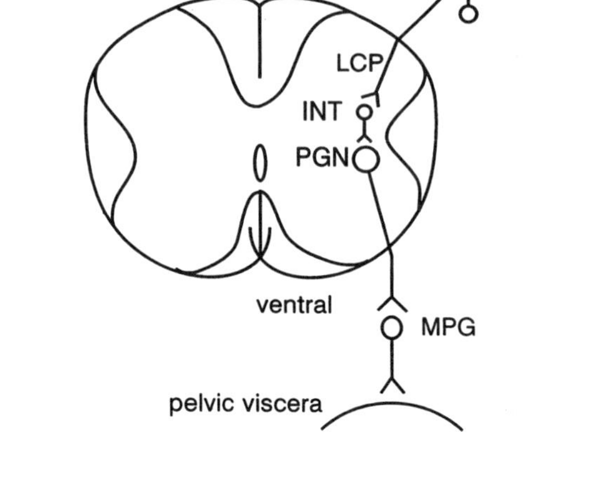

Figure 1. Schematic diagram for the neural circuitry underlying spinal segmental parasympathetic reflexes in the lumbosacral spinal cord. PGN, Parasympathetic preganglionic neurons; INT, dorsal interneurons; LCP, lateral collateral pathway of primary afferents; MPG, major pelvic ganglion.

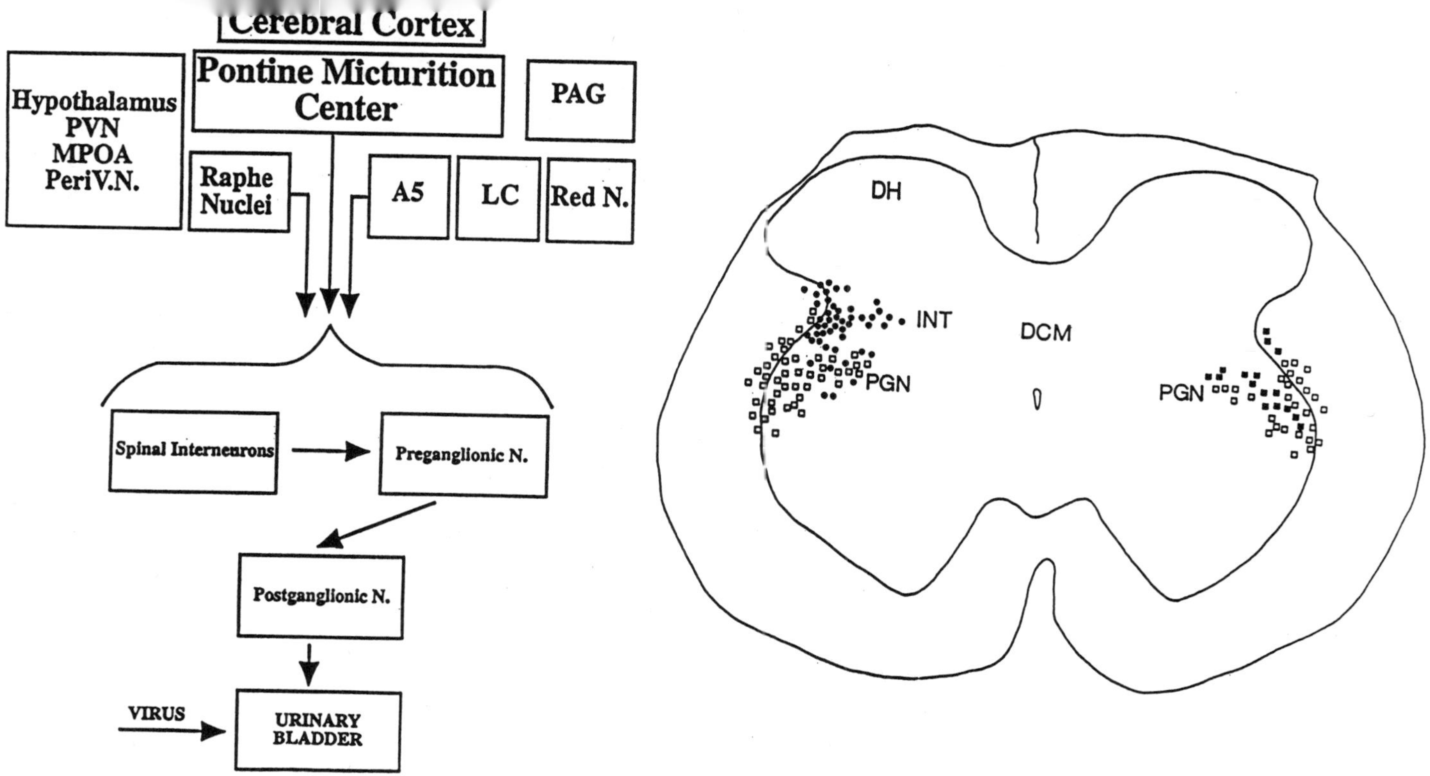

Figure 2. Transneuronal virus tracing of the central pathways controlling the urinary bladder of the rat. Injection of pseudorabies virus into the wall of the urinary bladder leads to retrograde transport of virus and sequential infection of postganglionic neurons, preganglionic neurons and then various central neural circuits synaptically linked to the preganglionic neurons. At long survival times virus can be detected with immunocytochemical techniques in neurons at specific sites throughout the spinal cord and brain extending to the pontine micturition center in the pons (ie., Barrington's nucleus or the laterodorsal tegmental nucleus) and to the cerebral cortex. Other sites in the brain labeled by virus are: (1) the paraventricular nucleus (PVN), medial preoptic area (MPOA) and periventricular nucleus (Peri V.N.) of the hypothalamus, (2) periaqueductal gray (PAG), (3) locus coeruleus (LC) and subcoeruleus, (4) red nucleus (Red N), (5) medullary raphe nucleus and (6) the noradrenergic cell group designated A5. L_6 spinal cord section showing on the left side the distribution of virus labeled parasympathetic preganglionic neurons (✸) and interneurons (●) in the region of the parasympathetic nucleus as well as interneurons in the dorsal commissure and the superficial laminae of the dorsal horn 72 hrs after injection of the virus into the bladder. On the left side the entire population of preganglionic neurons (PGN) was labeled by axonal tracing with fluorogold injected into the pelvic ganglia. The right side shows the distribution of bladder PGN (■) among the entire population of PGN (✸). Composite diagram of neurons in 12 spinal sections (42µm).

2.1. Pathways in the Spinal Cord

The spinal cord grey matter is divided into three general regions: (1) the dorsal horn which contains interneurons that process sensory input, (2) the ventral horn which contains motoneurons and (3) the intermediate region located between the dorsal and ventral horns which contains interneurons and preganglionic neurons. These regions are further subdivided into layers or laminae that are numbered starting with the superficial layer of the dorsal horn (lamina I) and extending to the ventral horn (lamina IX) and the commissure (lamina X) connecting the two sides of the spinal cord (Figure 3D).

Anatomical studies in neonatal[21,24] and adult rats[20] have revealed a similar general organization of the spinal micturition circuitry. Parasympathetic preganglionic neurons (PGN) are located in lateral lamina VII in the sacral parsympathetic nucleus (SPN) (Figure 2). Bladder afferent axons project through lateral lamina I of the dorsal horn to the SPN and also through the intermediate region to the dorsal commissure (lamina X) (Figure 3A).[27] Activation of bladder afferent pathways stimulates immediate early gene expression (c-fos) in neurons in the SPN, lamina I of the dorsal horn and the dorsal commissure indicating direct or indirect connections from afferents to neurons in these regions (Figure 3B).[25,26] In these same areas PRV tracing revealed interneurons that are involved in the control of bladder,[20,21,24] urethra[23] and urethral sphincter function[22] (Figure 3C).

2.2. Pathways in the Brain

Although voiding in neonatal rats does not depend on neural mechanisms in the brain, a large number of neurons at various sites in the brain were labeled by PRV injected into the urinary bladder of 2 and 10 day old rat pups.[24] Indeed, the distribution of PRV-infected neurons was somewhat broader than in adult rats. At the earliest survival times (72 hr) the most prominent labelling was in the pontine micturition center (PMC, Barrington's nucleus). Other neuronal populations that were labelled at slightly longer times (78–84 hr) included the nucleus raphe magnus, A5 and A7 regions, parapyramidal reticular formation, the periaqueductal gray, locus coeruleus, the lateral hypothalamus, medial preoptic area and the frontal cortex (Figure 2). Many of these sites are labeled by injection of tracer substances into the PMC[28] and correspond to sites where electrical stimulation facilitates or inhibits bladder activity in adult animals.[29] Thus even in neonatal animals some supraspinal mechanisms may already be functioning, possibly in an inhibitory manner, to suppress the spinobulbospinal micturition reflex pathway allowing micturition to be controlled by primitive spinal reflex mechanisms (see below).

3. IN VIVO PHYSIOLOGICAL STUDIES

Measurements of bladder activity as well as voiding function in anesthetized or unanesthetized neonatal rats and kittens have revealed strong reflex activation of the parasympathetic outflow to the bladder in response to tactile stimulation of the perineal region.[3,7–10,13,14] The reflex activity is mediated by pathways in the spinal cord because it is unaffected by transection of the cord rostral to the lumbosacral segments. In rats the voiding response to perineal stimulation is depressed by either NMDA or nonNMDA glutamatergic receptor antagonists[30,31] indicating that glutamic acid, the principal excitatory transmitter in the central nervous system, plays an essential role in the spinal mechanisms involved in the perineal-to-bladder reflex.

Although the spinobulbospinal micturition reflex is nonfunctional in young kittens and rat pups, the pathway seems to be "wired-up" and can be activated under certain conditions.

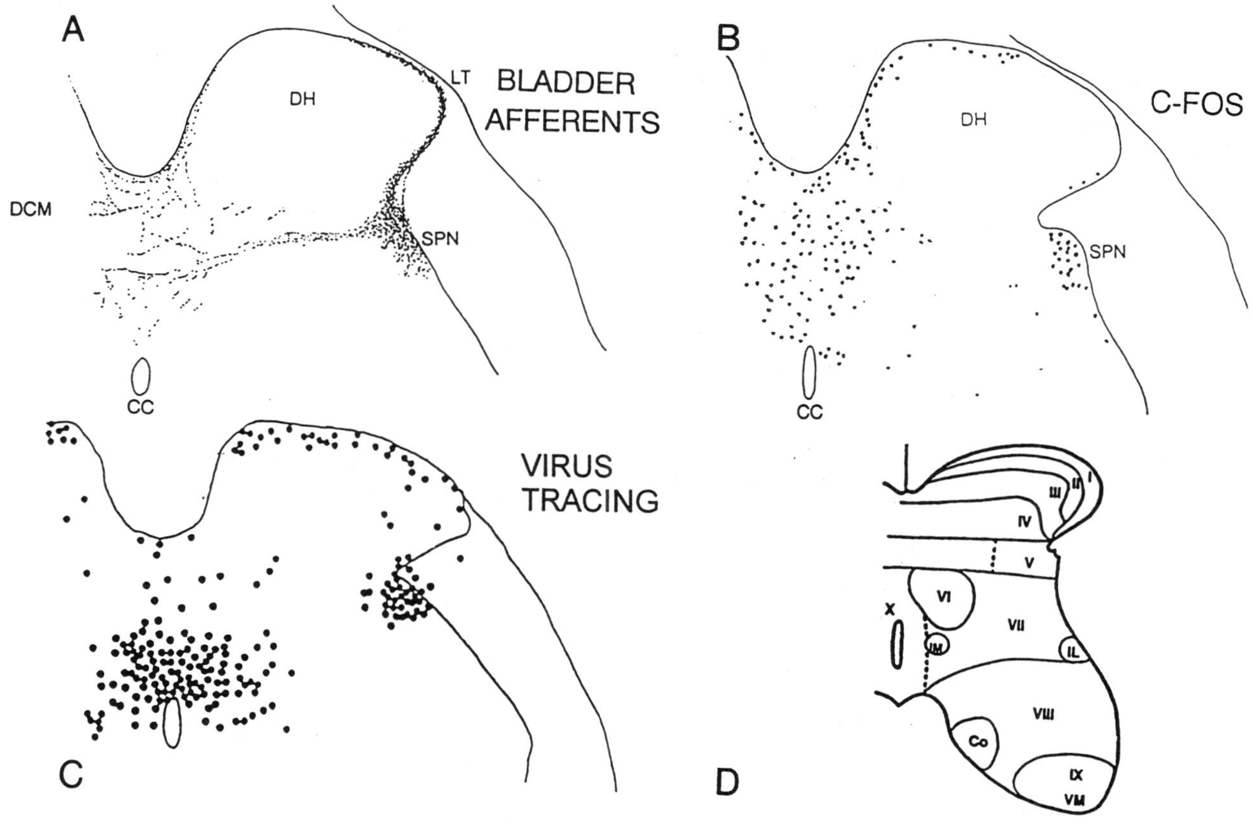

Figure 3. Comparison of the distribution of bladder afferent projections to the L_6 spinal cord of the rat (A) with the distribution of c-fos positive cells in the L_6 spinal segment following chemical irritation of the lower urinary tract of the rat (B) and the distribution of interneurons in the L_6 spinal cord labeled by transneuronal transport of pseudorabies virus injected into the urinary bladder (C). Afferents labelled by WGA-HRP injected into the urinary bladder. C-fos immunoreactivity is present in the nuclei of cells. DH, dorsal horn; SPN, sacral parasympathetic nucleus; CC central canal. Calibration represents 500 μm. Modified from de Groat et al.: Neurophysiology of micturition and its modification in animal models of human disease, In Maggi CA, ed: The Autonomic Nervous System, vol. 3, Nervous Control of the Urogenital System, London, 1993, Harwood Academic Publishers, 227–290.D, drawing shows the laminar organization of the cat spinal cord.

For example in 50% of 2 day old rats and 100% of 6–9 day old rats, removal of the forebrain (decerebration) unmasked reflex bladder activity in response to bladder distension.[13] The activity was abolished by transection of the spinal cord rostral to the lumbar level, indicating that it was mediated by a supraspinal pathway. In kittens similar reflexes could be evoked after the administration of certain anesthetics such as chloralose or ketamine.[3,7] Under general anesthesia the immature spinobulbospinal reflex in kittens exhibits a very long central delay (225–325 msec) in comparison to the delay in adult cats (60–75 msec).[1,3] This long delay is presumably due to the slow conduction velocity of immature central axons in the newborn animal. The latency of the reflex progressively shortens during the first eight postnatal weeks reaching the adult latency by 2–3 months of age. These findings are consistent with the results of PRV tracing and indicate that supraspinal bladder-to-bladder reflexes are organized relatively soon after birth but are tonically suppressed by inhibitory mechanisms in the brain. Suppression of these inhibitory mechanism by decerebration or anesthetics in young animals or by the natural process of neural maturation in older animals leads to the emergence of the latent reflex pathway and the appearance of mature voiding mechanisms.

4. IN VITRO PHYSIOLOGICAL STUDIES

4.1. In Vitro Brain Stem-Spinal Cord-Bladder Preparation

Perineal-to-bladder reflexes have also been studied in vitro using preparations consisting of: (1) the spinal cord or the brain stem-spinal cord, (2) the bladder and perineal region, (3) the relevant peripheral neural pathways.[30,32] Reflexly active preparations have been obtained from rat pups ranging from 4 to 10 days of age. As noted in vivo, tactile stimulation of the perineal region in these preparations induces neurally evoked contractions of the bladder that are blocked by the administration of ganglionic blocking agents or by nonNMDA glutamatergic receptor antagonists as noted in vivo.[31] Transection of the spinal cord above the lumbar segments did not block but commonly enhanced the reflex indicating that input from the brain stem tonically inhibits the spinal reflex pathway. Excitatory projections from the brain to the parasympathetic outflow to the bladder were also demonstrated using electrical stimulation. In these experiments bladder contractions were elicited by microstimulation of a narrow fiber tract running rostrocaudally through the medulla.[32] This finding is consistent with in vivo and anatomical experiments showing that the descending limb of the spinobulbospinal micturition reflex is functional in neonatal animals.

4.2. In Vitro Spinal Cord Slice Preparation

The mechanisms underlying the developmental suppression of the perineal-to-bladder reflex and its re-emergence after spinal cord injury have been studied using patch clamp recordings in identified parasympathetic preganglionic neurons (PGN) in thin slice spinal cord preparations obtained from 6 to 28 day old neonatal rats (Figure 4F).[33–35] PGN visually identified by the presence of a fluorescent tracer transported from the peripheral nervous system were studied with whole cell recording techniques. Interneurons located within 100 μm of the PGN were stimulated with extracellular micropipettes to elicit postsynaptic currents in the PGN (Figure 4F).

Two types of interneurons were identified: (1) interneurons located dorsal as well as medial to the PGN that elicited excitatory postsynaptic currents (EPSCs)and (2) interneurons located medial to the PGN that elicited inhibitory postsynaptic currents (IPSCs)(Figure 4A,F). Inhibitory interneurons were relatively uncommon representing approximately 15% of the neurons studied.[34] Dorsal interneurons were smaller in size than

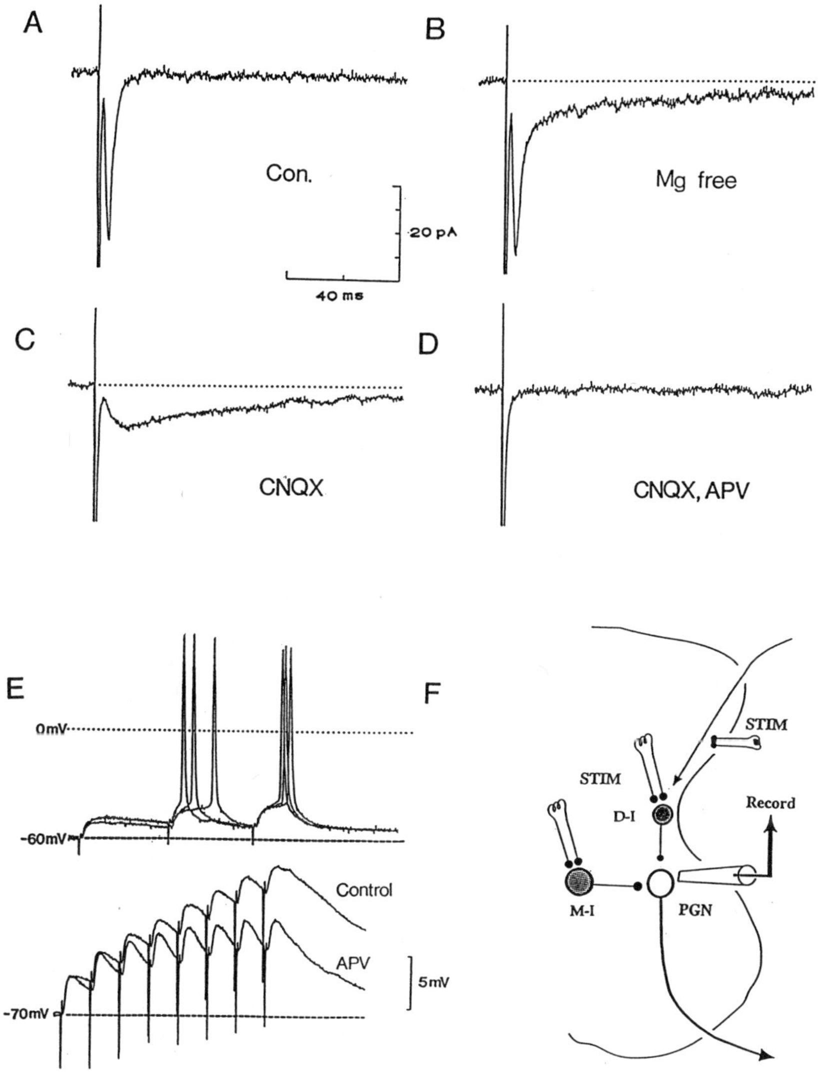

Figure 4. A–D, N-methyl-D-aspartate (NMDA) and non-NMDA receptor-mediated components of evoked EPSCs elicited in a preganglionic neuron (PGN) by electrical stimulation of a dorsal interneuron (INT). Averaged responses of 20–30 EPSCs recorded in the presence of 1 μM strychnine and 10 μM bicuculline to block glycinergic and $GABA_A$ inhibitory mechanisms. Recording were made in physiological saline (1 mM Mg) (A) or Mg-free external solution (B). C and D: EPSCs in Mg-free external solution were recorded in the presence of 5 μM 6-cyano-7-nitro-quinoxaline-2–3-dione (CNQX), an AMPA receptor antagonist (C) or 5 μM CNQX and 50 μM 2-amino-5-phosphonovalerate (APV), an NMDA receptor antagonist (D). Note Mg-free solution unmasks a slow EPSC (B) and CNQX blocks the fast EPSC (C) and APV blocks the slow EPSC (D). Holding potential -60 mV. (E). Repetitive stimulation of an interneuron (INT) (3 pulses at 50-ms intervals) induced action potentials in a preganglionic neuron. Note that the potentials were induced by the 2nd and 3rd excitatory postsynaptic potentials (EPSPs) but not by the 1st EPSP indicating temporal facilitation. Dashed horizontal lines: membrane potentials. Averaged records of 15–20 synaptic responses (bottom) evoked in another PGN by high frequency stimulation (40 Hz, 8 pulses) of an interneuron before (Control) and during application of 50 μM APV. Note APV blocked a part of the temporal facilitation of the EPSPs. Action potentials were suppressed by application of hyperpolarizing current. (F). Diagram showing stimulation and recording methods for the patch slice preparation. Stimulation of dorsal INT (D-I) always elicited excitatory inputs to PGN whereas stimulation of medial INT (M-I) could elicit excitatory or inhibitory inputs. Dorsal IN were excited by electrical stimulation on the lateral edge of the dorsal horn, an area containing primary afferent axons.

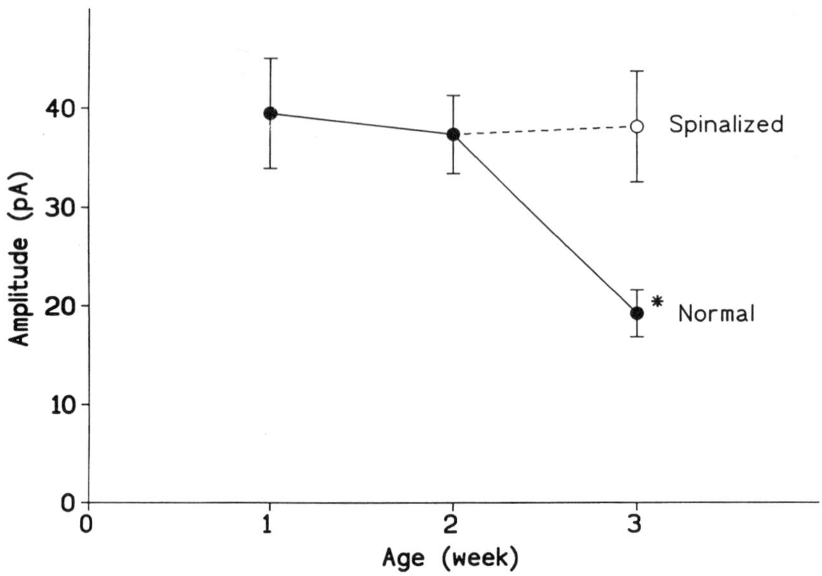

Figure 5. Developmental change in the efficacy of excitatory transmission at interneuronal- PGN synapses and its modification by chronic spinal transection. Mean peak amplitudes of unitary EPSCs recorded from PGNs in intact (●) or spinalized (○) rats are plotted against ages. Mean peak amplitudes were measured from averaged responses of >30 consecutive EPSCs (including failures). The holding potential was -60 mV. Each point represents the mean ± SEM (vertical bar) from 14 or 15 cells. The asterisk represents significant difference from the values observed in other preparations. ($p < 0.006$).

PGN (10–15 μm versus 15–25 μm) and received excitatory synaptic inputs from axonal pathways on the lateral edge of the dorsal horn, an area known to contain primary afferent projections to the parasympathetic nucleus. The EPSCs in PGN evoked by dorsal interneurons consisted of fast and slow components that were blocked by respectively by nonNMDA (CNQX) and NMDA (APV) glutamatergic receptor antagonists (Figure 4A-D). Repetitive stimulation of dorsal interneurons facilitates the amplitude of EPSCs and elicits action potentials in PGN (Figure 4E). Because dorsal interneurons are exclusively excitatory to PGN and receive excitatory inputs from presumptive afferent axons it is reasonable to speculate that these interneurons are part of a disynaptic parasympathetic reflex pathway (Figure 4F) that might underlie the perineal-to-bladder reflex. Thus the effect of postnatal maturation on the efficiency of dorsal interneuron-to-PGN synaptic transmission was evaluated[35] to determine if changes in transmission could account for the downregulation of the reflex pathway during postnatal development.

In preparations removed from 1 and 2 week old rats the dorsal interneuron-evoked EPSCs were large and of constant amplitude (mean 40 pA) (Figure 5). However, in preparations from 3 week old rats, an age when the spinobulbospinal micturition has emerged, the EPSCs were reduced in amplitude by 50% (20 pA) (Figure 5). Quantal analysis of unitary EPSCs indicated that this reduction in amplitude was attributable to a decrease in the presynaptic release of glutamic acid and was not due to a change in the density or sensitivity of postsynaptic glutamatergic receptors.[35] Transection of the spinal cord between 1 and 2 weeks of age prevented the reduction in synaptic transmission that occurred at 3 weeks of age (Figure 5).

As illustrated in Figure 6, these observations are most reasonably explained by anatomical changes in the synapses in the sacral parasympathetic nucleus. It seems likely that

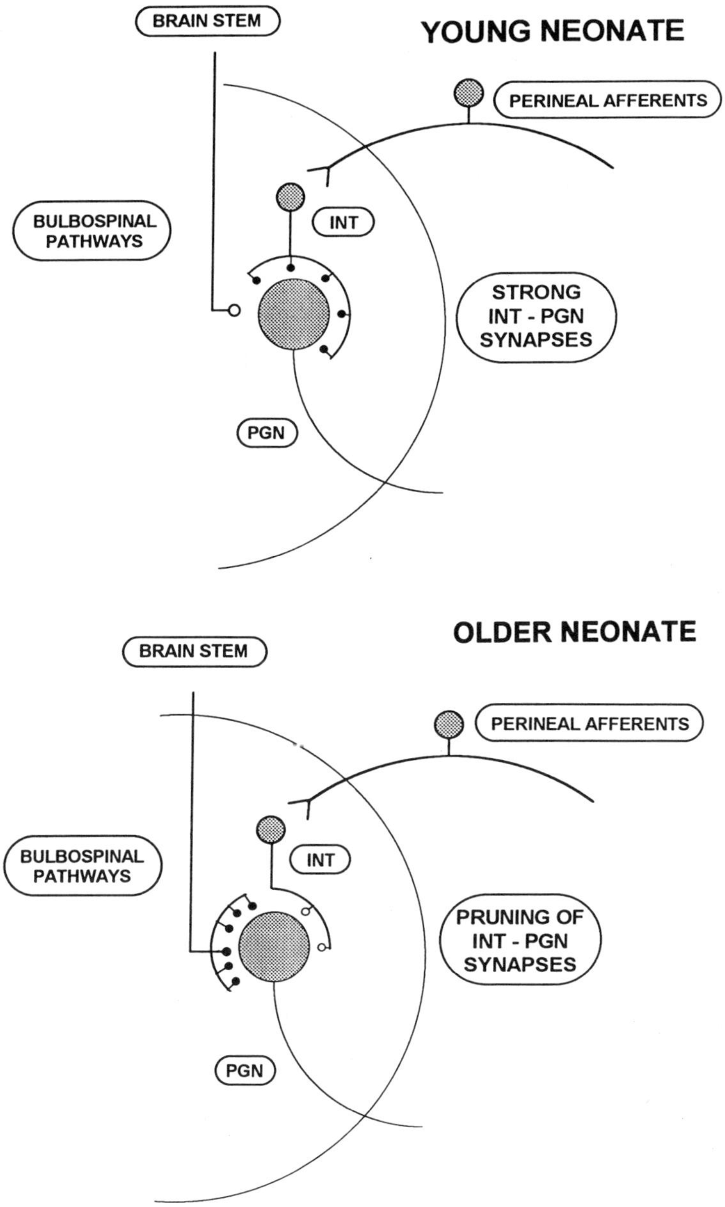

Figure 6. Diagram showing possible anatomical changes at synapses on preganglionic neurons (PGN) in the sacral parasympathetic nucleus(SPN) of the neonatal rat that might underlie the alteration of bladder reflexes during postnatal development. In young neonates (birth to 2 weeks of age) interneurons (INT) located immediately dorsal to the SPN make strong synaptic connections with the PGN. These INTs receive synaptic inputs from primary afferent neurons some of which are likely to innervate the perigenital region. It is hypothesized that the perigenital-to-bladder reflex that mediates reflex voiding in the neonatal rat is mediated by this disynaptic reflex pathway. Inputs to the SPN from bulbospinal axons arising in the pontine micturition center (PMC) make weak synaptic connections with the PGN in young neonates. However, in older neonates (3–4 weeks of age) bulbospinal inputs to the PGN expand at the time when the spinobulbospinal micturition reflex emerges. At the same time segmental interneuronal synapses on the PGN are eliminated (synaptic pruning) contributing to the downregulation of the perigenital-to-bladder reflex. Quantal analysis in the spinal slice preparation of unitary EPSCs evoked in PGN by stimulation of single interneurons revealed that there is a 50% decrease in the number of quanta of the excitatory transmitter, glutamic acid, released at INT-PGN synapses during the third postnatal week. This change was prevented by spinal cord transection during the second postnatal week suggesting that it is due to the influence of descending pathways from the brain on the INT-PGN synapses.

maturation of the bulbospinal pathways to the PGN during the second to third postnatal week, at the same time when the mature bladder-to-bladder spinobulbospinal reflex is emerging, downregulates the interneuronal-PGN excitatory synapses, possibly by synapse elimination in response to competition for synaptic space between boutons of interneurons and descending axons. Transection of the spinal cord which causes degeneration of the descending axons blocks this synaptic reorganization and maintains the primitive neonatal reflex pathway. A similar degeneration of descending axons in adult animals could stimulate axonal sprouting in interneuronal pathways and cause the re-emergence of the neonatal perineal-to-bladder reflex in chronic spinal animals.

5. SUMMARY

Neuroanatomical and electrophysiological techniques have provided new insights into the organization of the spinal cord circuitry and the neurotransmitter mechanisms involved in primitive voiding reflexes in neonatal animals. In addition, studies of unitary synaptic transmission in spinal cord slice preparations indicate that developmental and spinal cord injury induced plasticity in sacral parasympathetic reflex pathways is due in part to alterations in glutamatergic excitatory transmission between interneurons and parasympathetic preganglionic neurons. It is proposed that these synaptic changes are due to competition between segmental and supraspinal inputs. Thus synaptic remodeling in the sacral parasympathetic nucleus is likely to be an important factor in the postnatal maturation of voiding reflexes.

REFERENCES

1. de Groat, W.C., Booth, A.M., Yoshimura, N.: Neurophysiology of micturition and it modification in animal models of human disease, in: Maggi, C.A.(Ed.), The Autonomic Nervous System. In: Nervous Control of the Urogenital System, vol. **3**, Harwood Academic, London, 227, 1993.
2. de Groat, W.C., Araki, I., Vizzard, M.A., Yoshiyama, M. Yoshimura, N., Sugaya, K., Tai, C. Roppolo, J.R.: Developmental and injury induced plasticity in the micturition reflexpathway, Behavioral Brain Research, 127, 1999.
3. de Groat, W.C., Nadelhaft, I., Milne, R.J., Booth, A.M., Morgan, C., Thor, K.: Organization of the sacral parasympathetic reflex pathways to the urinary bladder and large intestine, J. Auton.Nerv. Syst. **3:** 135, 1981.
4. de Groat, W.C.: Mechanisms underlying the recovery of lower urinary tract function following spinal cord injury, Paraplegia, **33:** 493 1995.
5. Capek, K., Jelinek, J.: The development of the control of water metabolism. I. The excretion of urine in young rats, Physiol. Bohemoslov., **5:** 91, 1956.
6. de Groat, W.C., Douglas, J.W., Glass, J., Simonds, W., Weimer, B., and Werner, P.: Changes in somatovesical reflexes during postnatal development in the kitten, Brain Res., **94:** 150, 1975.
7. Thor, K.B., Blais, D.P., de Groat, W.C.: Behavioral analysis of the postnatal development of micturition in kittens, Dev. Brain Res., **46:** 137, 1989.
8. Thor, K.B., Kawatani, M. and de Groat, W.C.: Plasticity in the reflex pathways to the lower urinary tract of the cat during postnatal development and following spinal cord injury. Development and Plasticity of Mammalian Spinal Cord. Fidia Research Series, Vol. **3**: 65, 1986.
9. Maggi, C.A., Santicioli, P., and Melli, A.: Postnatal development of micturition reflex in rats. Am. J. Physiol., **250:** R926, 1986.
10. Thor, K.B., Hisamitsu, T. and de Groat, W.C.: Unmasking of a neonatal somatovesical reflex in adult cats by the serotonin autoreceptor agonist 5-methoxy-N, N-dimethyltryptamine. Devel. Brain Res., **54:** 35, 1990.

11. Beach, F.A.: Ontogeny of 'coitus-related' reflexes in the female guinea pig. Proc. Natl. Acad. Sci. U.S.A., **56:** 526, 1966.
12. Boehm, J.J. and Haynes, J.L.: Bacteriology of 'midstream catch' urines. Am. J. Dis. Child., **111:** 366, 1966.
13. Kruse, M.N., de Groat, W.C.: Micturition reflexes in decerebrate and spinalized neonatal rats, Am. J. Physiol., **258:** R1508, 1990.
14. Kruse, M.N., de Groat, W.C.: Spinal pathways mediate co-ordinated bladder/urethral sphincter activity during reflex micturition in normal and spinal cord injured neonatal rats, Neurosci. Lett. **152:** 141, 1993.
15. de Groat, W.C.: Nervous control of the urinary bladder of the cat, Brain Res., **87:** 201, 1975.
16. Mallory, B., Steers, W.D., de Groat, W.C.: Electrophysiological study of micturition reflexes in rats, Am. J. Physiol., **257:** R410, 1989.
17. Cheng, C., Ma, C., de Groat, W.C.: Effect of capsaicin on micturition and associated reflexes in chronic spinal rats, Brain Res., **678:** 40, 1995.
18. Card. J.P., Rinaman, L., Schwaber, J.S., Miselis, R.R., Whealy, M.E., Robbins, A.K., Enquist, L.W.: Neurotropic properties of pseudorabies virus: uptake and transneuronal passage in the rat central nervous system, J. Neurosci., **10:** 1974, 1990.
19. Loewy A.D.: Pseudorabies virus: A transneuronal tracer for neuroanatomical studies. In: Viral Vectors, Kaplitt, M.G. Loewy, A.D. (Eds) Academic Press: San Diego. pp. 349–366, 1995.
20. Nadelhaft, I., Vera, P.L.: Central nervous system neurons infected by pseudorabies virus injected into the rat urinary bladder following unilateral transection of the pelvic nerve, J. Comp. Neurol., **359:** 443, 1995.
21. de Groat, W.C., Roppolo, J.R., Yoshimura, N., Sugaya, K.: Neural control of the urinary bladder and colon, in : Taché, Y., Wingate, D., Burks, T., (Eds.), Proceedings of the Second International Symposium on Brain-Gut Interactions, CRC Press, Boca Raton, Fl, 167, 1993.
22. Nadelhaft, I., Vera, P.L.: Neurons in the rat brain and spinal cord labeled after pseudorabies virus injected into the external urethral sphincter, J. Comp. Neurol., **375:** 502, 1996.
23. Vizzard, M.A., Erickson, V.L., Card, J.P.,Roppolo, J.R., and de Groat, W.C.: Transneuronal labeling of neurons in the adult rat brainstem and spinal cord after injection of pseudorabies virus into the urethra, J. Comp. Neurol., **355:** 629, 1995.
24. Sugaya, K., Roppolo, J.R., Yoshimura, N., Card, J.P., de Groat, W.C.: The central neural pathways involved in micturition in the neonatal rat as revealed by the injection of pseudorabies virus into the bladder, Neurosci. Lett. **223:** 197, 1997.
25. Birder, L.A., de Groat, W.C.: Increased *c-fos* expression in spinal neurons after chemical irritation of the lower urinary tract of the rat, J. Neurosci., **12:** 4878, 1992.
26. Birder, L.A., de Groat, W.C.: Introduction of *c-fos* gene expression in spinal neurons in the rat by nociceptive and non-nociceptive stimulation of the lower urinary tract, Am. J. Physiol., **265:** R643, 1993.
27. Steers, W.D., Ciambotti J., Etzel, B., Erdman, S., de Groat, W.C.: Alterations in afferent pathways from the urinary bladder of the rat in response to partial urethral obstruction, J. Comp. Neurol., **310:** 401, 1991.
28. Sugaya, K., Mori, S., Tsuchida, S.: Input and output neuronal structures of the pontine micturition center. I. Mainly input neuronal structure , Jpn. J. Urol., **79:** 1210, 1988.
29. Torrens, M.J., Morrison, J.F.B.: The Physiology of the Lower Urinary Tract, Springer, Berlin, 1987.
30. Sugaya, K., de Groat, W.C.: Effects of MK-801 and CNQX, glutamate receptor antagonists on bladder activity in neonatal rats, Brain Res., **640:** 1, 1994.
31. Yoshiyama, M., Erickson, K.A., Erdman, S.L. and de Groat, W.C: AMPA and NMDA glutamatergic mechanisms involved in the voiding reflex in the neonatal rat. Neuroscience (in press).
32. Sugaya, K., de Groat, W.C.: Micturition reflexes in the in vitro neonatal rat brainstem-spinal cord-urinary bladder preparation, Am. J. Physiol. , **266:** R658, 1994.
33. Araki, I.: Inhibitory postsynaptic currents and the effects of GABA on visually identified sacral parasympathetic preganglionic neurons in neonatal rats, J. Neurophysiol., **72:** 2903, 1994.
34. Araki, I. and de Groat, W.C.: Unitary excitatory synaptic currents in preganglionic neurons mediated by two distinct groups of interneurons in neonatal rat sacral parasympathetic nucleus, J. Neurophysiol., **76:** 215, 1996.
35. Araki, I. and de Groat, W.C: Developmental synaptic depression underlying reorganization of visceral reflex pathways in the spinal cord, J. of Neurosci.,**17:** 8402, 1997.

SUBCELLULAR DISTRIBUTION OF FREE FATTY ACIDS, PHOSPHOLIPIDS, AND ENDOGENOUS LIPASE ACTIVITY OF RABBIT URINARY BLADDER SMOOTH MUSCLE AND MUCOSA

Laura J. O'Connor, Tamar Nicholas, and Robert M. Levin

Division of Basic and Pharmaceutical Sciences
Albany College of Pharmacy
106 New Scotland Ave.
Albany, New York, 12208, and
Division of Urology, Albany Medical College, Albany, New York
Stratton VA Medical Center, Albany, New York

1. ABSTRACT

1.1. Objectives

The urinary bladder wall can be separated into two major compartments: the urothelium (mucosa) and the detrusor smooth muscle. Specific dysfunctions of both layers have been linked to ischemia, which may induce significant cellular and subcellular membrane damage via the activation of selective calcium dependent and independent hydrolytic enzymes. Preliminary to investigating changes in cell membrane composition induced by ischemia, we measured the free fatty acid (FFA) and phospholipid (PL) content of normal rabbit bladder muscle and mucosal cellular and subcellular membranes, and characterized the endogenous lipase activity.

1.2. Methods

Rabbit bladders were excised and the muscle and mucosal layers separated; each layer was homogenized, then fractionated by differential centrifugation. Endogenous lipase activity of the homogenates, and FFA and PL concentrations of the homogenates and subcellular fractions were measured.

Advances in Bladder Research, edited by Baskin and Hayward.
Kluwer Academic / Plenum Publishers, New York, 1999.

1.3. Results

(1) The basal FFA concentration of the mucosal homogenates was 5 times that of the muscle homogenates. **(2)** The basal PL concentrations of the two tissues were similar. **(3)** Subcellular studies: FFA concentration was greatest in the mitochondrial fraction of both compartments. In the mucosa, PL concentration was significantly greater in the mitochondria and microsomes than in the other fractions; in the smooth muscle, the PL concentration was highest in the mitochondria. **(4)** The maximal endogenous lipase activity was 10 times higher in the mucosal homogenates than in the muscle homogenates.

1.4. Conclusions

These results are consistent with those of previous studies which indicate that the mucosa is metabolically more active than the resting smooth muscle, which may cause the mucosa to be significantly more sensitive than the muscle to hypoxic/ischemic damage.

2. INTRODUCTION

The urinary bladder can be divided into two components, the mucosal epithelium (urothelium) and the underlying detrusor smooth muscle. The urothelium provides a non-adherent surface that discourages bacterial adhesion to the bladder wall[1,2] and acts as a permeability barrier to the penetration of solutes[3,4], thus protecting the smooth muscle elements from damage by urine consituents and changes in pH[3–6]. The symptoms of several urological disorders (interstitial cystitis, recurrent infections, incontinence) have been ascribed to defects in mucosal function, and may be related directly to destruction of the antiadherent properties of the urothelium and its function as a permeability barrier[5–7].

Micturition is mediated through the coordinated contraction of bladder smooth muscle elements[8–10]. Bladder dysfunction secondary to benign prostatic hyperplasia (BPH) has been ascribed to progressive changes in innervation and smooth muscle contraction[11,12]. In animal models of BPH (partial outlet obstruction), the associated bladder contractile dysfunction is mediated by three processes: **(1)** progressive denervation, **(2)** mitochondrial damage and the associated decreased ability to generate ATP, and **(3)** progressive reduction in the ability of the sarcoplasmic reticulum to store and release calcium[13,14].

Studies indicate that disorders of mucosal function (*i.e.*, interstitial cystitis and recurrent infection) and smooth muscle function (*i.e.*, bladder contractile failure secondary to BPH and hyperreflexia) are related to cyclical ischemia and reperfusion injury[15–17]. Ischemia and reperfusion may activate degradative enzymes that cause cellular damage through hydrolysis of neuronal, cellular, and subcellular membranes. Very little is known about the effect of ischemia / reperfusion on urinary bladder cell membrane structure and function. Preliminary to investigating changes in cell membrane composition induced by ischemia, we measured the free fatty acid (FFA) and phospholipid (PL) content of normal rabbit bladder muscle and mucosal cellular and subcellular membranes, and characterized the endogenous lipase activity.

3. METHODS

3.1. Tissue Preparation

Frozen tissue was weighed, thawed on ice in 100 mM glycylglycine buffer, pH 8.0, then homogenized at a concentration of 100 mg/ml with a Polytron homogenizer for 10 sec. The homogenate was filtered through gauze, then centrifuged at 2000 x g for 10 min. to pellet plasma membranes/nuclei. The supernatant was centrifuged at 13,300 x g for 15 min. to isolate mitochondria. Microsomes were pelleted by centrifugation at 105,000 x g for 60 min.; the supernatant (soluble fraction) was saved. Each pellet was then resuspended in the original buffer. Aliquots of each fraction were assayed for FFA and PL.

3.2 Basal FFA and PL Assays

These enzymatic colorimetric methods were used to quantitate endogenous FFA and PL. Separate muscle and mucosa fractions in glass test tubes were heated at 90°C for 5 min. to inactivate endogenous lipase in the samples. Then, 50 ul aliquots were incubated at 37°C with NEFA C or Phospholipids B color reagent kits (Wako Chemicals USA, Inc., Richmond, VA), following recommended manufacturer instructions. The optical density of all samples was read at 550 nm for FFA and 505 nm for PL. Results were expressed as nmols FFA (or PL) / mg protein.

3.3. FFA Generation Assay (Endogenous Lipase Activity)

Homogenates from muscle and mucosa were kept on ice and diluted 1:6 in buffer (0.01 M glycylglycine, 3 mM $CaCl_2$, 1 % Tween-20, 6.67 % methanol, pH 8.0). The same buffer with the addition of 0.67% lecithin as a substrate was used for initial enzyme characterization. Samples were incubated at 37°C for alloted time increments. The reaction was stopped by incubating tubes at 90°C for 5 min. The reaction tubes were centrifuged at 1000 x g for 5 min. to remove protein precipitates and FFA concentration was determined spectrophotometrically using a NEFA C test kit as described above.

4. RESULTS

The basal concentrations of free fatty acids (FFA) and phospholipids (PL) of the muscle and mucosal homogenates are presented in figure 1. The FFA concentration of the mucosal homogenates was 5 times that of the muscle homogenates; the PL concentrations of the two tissues were similar. The basal concentrations of muscle and mucosa PL were approximately 100-fold and 1000-fold lower than their respective FFA concentrations.

The subcellular distribution of FFA is shown in figure 2. The FFA concentrations of the mucosal nuclear and mitochondrial pellets were significantly greater than those of the respective muscle pellets; the FFA concentrations of both tissues' microsomal pellet and soluble fraction were similar. The subcellular distribution of PL is shown in figure 3. The PL concentrations of the mucosal mitochondrial and microsomal pellets were significantly greater than those of the respective muscle pellets; the PL concentrations of both tissues' nuclear pellet and soluble fraction were similar.

The endogenous lipase activities of the mucosa and the muscle are displayed in figures 4–6. Lipase activities were linear with time and with tissue concentration (figures 4 and 5). Maximal lipase activity of the mucosal homogenate was 10 times higher than that of the muscle homogenate (figure 6). Enzyme pH sensitivity is displayed in figure 7. The enzymes of both tissues exhibited maximal activity between pH 7 and 9.

Additional studies demonstrated that the lipase activities of both tissues were not affected by either pre-incubation with 1 mM EGTA, or incubation in the presence of various concentrations of calcium up to 10 mM. In addition, the enzyme activities were not affected by the addition of exogenous substrates, including lecithin (up to 1 mM).

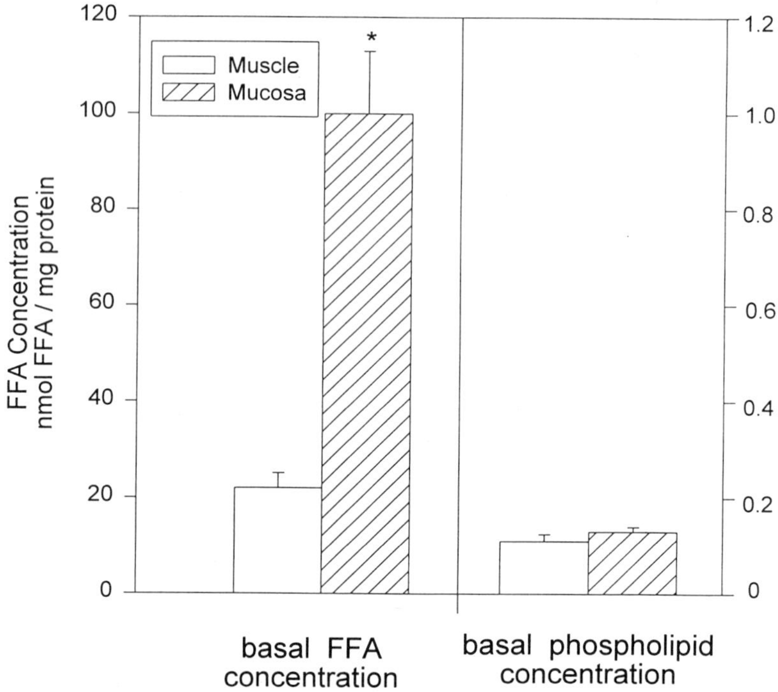

Figure 1. FFA and phospholipid concentrations of rabbit bladder muscle and mucosa. Each bar is the mean ± SEM of 6–8 individual preparations. * = significantly different from the muscle ($p < 0.05$).

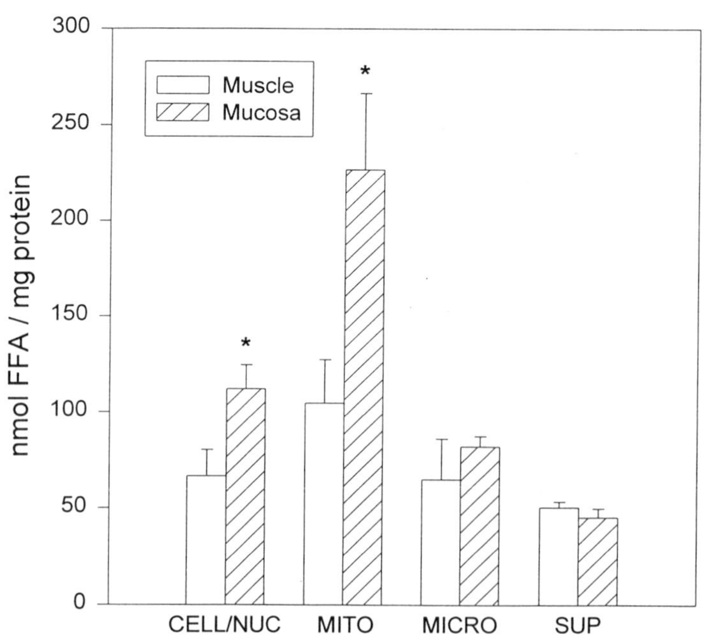

Figure 2. Subcellular distribution of FFA in rabbit bladder muscle and mucosa. Each bar is the mean ± SEM of 6–8 individual preparations. Cell/Nuc is the low speed particulate fraction (plasma membrane / nuclei); Mito = mitochondria; Micro = microsomes; Sup = soluble fraction. * = significantly different from the muscle ($p < 0.05$).

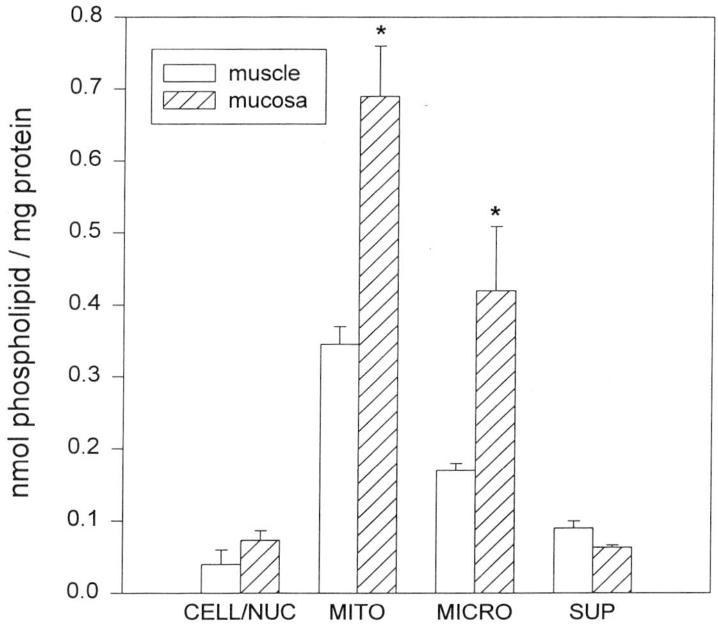

Figure 3. Subcellular distribution of phospholipids in rabbit bladder muscle and mucosa. Each bar is the mean ± SEM of 6–8 individual preparations.Cell/Nuc is the low speed particulate fraction (plasma membrane / nuclei); Mito = mitochondria; Micro = microsomes; Sup = soluble fraction. * = significantly different from the muscle ($p < 0.05$).

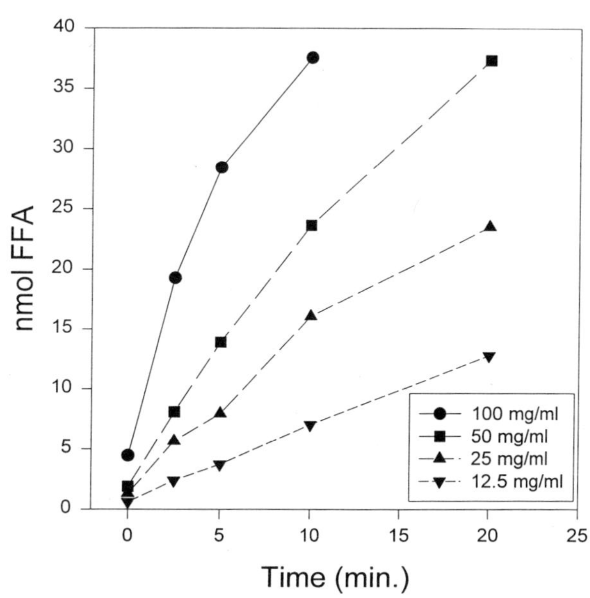

Figure 4. Calcium independent FFA formation (endogenase lipase activity) in rabbit bladder mucosa: Each point is the mean ± SEM of 6–8 individual preparations. * = significantly different from the muscle ($p < 0.05$).

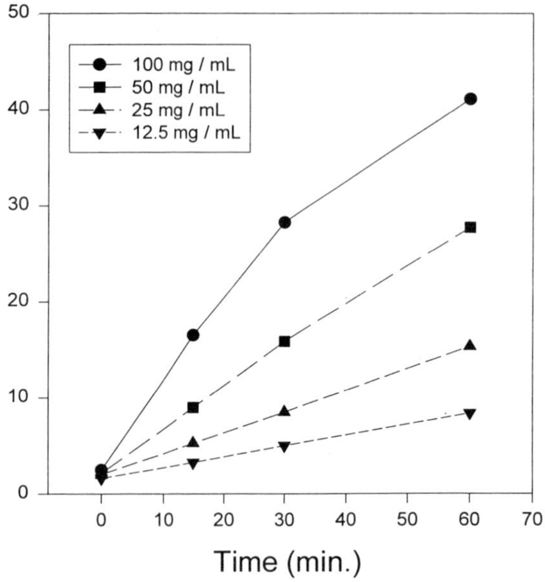

Figure 5. Calcium independent FFA formation (endogenase lipase activity) in rabbit bladder muscle: Each point is the mean ± SEM of 6–8 individual preparations. * = significantly different from the muscle ($p < 0.05$).

5. DISCUSSION

The urothelium serves two major functions[1–7]: It protects the bladder from infection via a non-adherent glycosaminoglycan surface[1,2] and from penetration by solutes in the urine[3–6]. Thus, both anti-adherence and the permeability barrier are dependent upon urothelial integrity. Disruption of these functions have been implicated in a variety of disease states including recurrent infection, interstitial cystitis, non-bacterial cystitis, and incontinence[3,6]. Hypoxia and ischemia have been as etiological factors for these diseases, which are believed to be mediated by direct damage to the anti-adherence and permeability barrier functions of the urothelium[3–7,15].

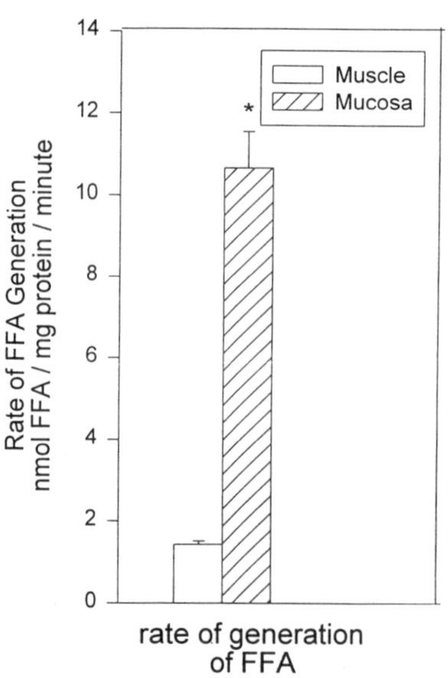

Figure 6. Comparative endogenous lipase activity of bladder muscle and mucosa: Each bar is the mean ± SEM of 6–8 individual preparations. * = significantly different from the muscle ($p < 0.05$).

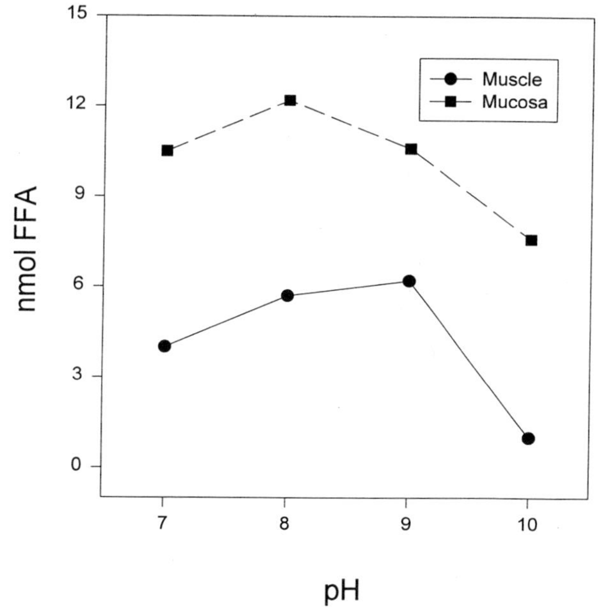

Figure 7. Calcium independent FFA formation (endogenase lipase activity) in rabbit bladder muscle and mucosa as a function of pH: Each bar is the mean of 4 individual preparations.

Experimentally, both hypoxia (in vitro), and ischemia (in vivo) result in a marked increase in bacterial adherence to the surface of the urothelium and a significant increase in permeability through the urothelium into the smooth muscle compartment[3,4,18]. In addition, hypoxia (in vitro) has been shown to significantly reduce urethral resistance, which would be an etiological factor for incontinence[15]. It also causes a loss of tonic contraction of bladder smooth muscle and dissociation of intracellular calcium and tonic tension[15,16].

Several investigators have shown, in animals, that there is a brief phasic reduction in blood flow and tissue oxygen tension (ischemia) during micturition[17–20]. In the presence of experimental partial outlet obstruction, there is a significant increase in the severity and duration of the ischemia induced by micturition, and a significant delay in the recovery of tissue oxygenation during bladder filling[17–21]. Thus, obstruction significantly and substantially increases the severity and duration of the cyclical ischemia, which is followed by a period of reperfusion[17,20].

In several uropathologies that affect men (obstructive uropathy), women (interstitial cystitis) and both genders (decreased voiding efficiency with age), a primary etiological factor may be ischemia / reperfusion and the associated activation of calcium-dependent and independent hydrolytic enzymes. Our working hypothesis is that there is a progressive loss of compliance (increased wall stiffness) with age, and when the bladder contracts during micturition, there is a decrease in local blood flow, *i.e.*, phasic ischemia. During filling, blood flow returns to normal and reperfusion occurs. The cyclical ischemia mediates a phasic release of intracellular calcium which activates a number of degradative calcium-dependent enzymes including calpain and phospholipase A_2; free radicals generated during reperfusion stimulates lipid peroxidation. The result is the increased hydrolysis of cellular and subcellular membranes and alterations in the structure and properties of these membranes. This process is exacerbated in the presence of prostatic enlargement and obstructive uropathy. In men with BPH induced bladder dysfunction, the result of the increased bladder wall thickness (hypertrophy) is decreased compliance and increased wall tension during micturition.

It is clear from previous studies that the mucosa has a higher metabolic rate than resting smooth muscle[21,22]. The mucosa has a significantly greater rate of glycolysis than detrusor smooth muscle, a higher rate of oxidative metabolism, and higher mitochondrial enzyme activities[21]. Although the ATP concentrations of muscle and mucosa are similar, the creatine phosphate levels are significantly greater in the muscle than in the mucosa. Finally, the mucosa is significantly more sensitive to hypoxia than is the muscle[21,22]; hypoxia results in a signficantly greater rate of high energy phosphate (ATP and CP) degradation in the mucosa than in detrusor.

The reported study compared FFA and PL content, and endogenous lipase activity of rabbit urinary bladder mucosa and smooth muscle. Higher lipase activity of the mucosa is consistent with the significantly greater FFA content. The greater lipolytic activity would also be consistent with the greater sensitivity to hypoxia-induced cellular damage observed in the mucosa.

Future studies will correlate the effect of ischemia and partial outlet obstruction on lipid metabolism and FFA and PL concentrations of urinary bladder muscle and mucosa cellular and subcellular membranes.

ACKNOWLEDGMENTS

Supported in part by grants from the VAMC and NIH grant DK 26508

REFERENCES

1. Parsons,CL; Greenspan,C; Moore,SW; Mulholland,SG (1977): Role of surface mucin in primary antibacterial defense of bladder. Urology IX, 48–52.
2. Ruggieri, M., Hanno, P. and Levin, R. M.: Effects of heparin on the adherence of five species of urinary tract pathogens to urinary bladder mucosa. Urological Research, 12: 199-203, 1984.
3. Levin, R. M., Whitmore, K, Monson, F. C., Witkowski, B. A. and Ruggieri, M.: Trypan blue as an indicator of urothelial integrity, Neurourol Urodyn., 9: 269-280, 1990.
4. Monson, F. C., Wein, A. J., McKenna, B. A., Whitmore, K. and Levin, R. M.: Indigocarmine as a quantitative indicator of urothelial integrity. J. Urol., 145: 842-845, 1991.
5. Parsons,CL; Boychuk,D; Jones,S; Hurst,R; Callahan,H (1990): Bladder surface glycosaminoglycans: an epithelial permeability barrier. J. Urol. 143, 139–142.
6. Lilly,JD; Parsons,CL (1990): Bladder surface glycosaminoglycans is a human epithelial permeability barrier. Surg. Gynecol. Obstet. 171, 493–496.
7. Parsons,CL; Lilly,JD; Stein,P (1991): Epithelial dysfunction in nonbacterial cystitis (interstitial cystitis). J. Urol. 145, 732–735.
8. Steers, W.D. Physiology of the urinary bladder. In Cambells Urology, (eds. Walsh, P.C., Retik, A.B., Stamey, T.A., and Vaughan, E.D. Jr.) Saunders, Phila., 1992, pp142–176.
9. Wein AJ, Levin RM, Barrett DM: Voiding function: Relevant anatomy, physiology, and pharmacology. In Gillenwater JY, Grayhack JT, Howards SS, Duckett JD (eds): Adult and Pediatric Urology, 2nd ed. Chicago, Year Book Medical Publishers, pp. 933–999, 1991.
10. Anderson, K.E. Pharmacology of lower urinary tract smooth muscles and penile erectile tissue. Pharmacol. Rev. 45: 253–307, 1993.
11. Grayhack, J.T. and Kozlowski, J.M. Benign Prostatic Hyperplasia. In Adult and Pediatric Urology (Gillenwater, J.Y., Grayhack, J.T., Howards, S.S., and Duckett, J.W. eds.) Year Book Med. Pub. Inc. (Chicago), pp 1062-1126, 1987.
12. Sterling, A.M., Ritter, R.C., and Zinner, N.R. The physical basis of obstructive uropathy, in Benign Prostatic Hypertrophy, Springer-Verlag, N.Y. 1983 (pp 433-442)
13. Levin, R. M., Longhurst, P. A., Monson, F. C., Haugaard, N., and Wein, A. J.: Experimental studies on bladder outlet obstruction. In: Prostate Diseases. Edited by H. Lepor MD, R.K. Lawson MD. Philadelphia: W.B. Saunders Co., pp. 119–130, 1993.

14. Levin, R. M., Longhurst, P. A., Monson, F. C., Kato, K., and Wein, A. J.: Effect of bladder outlet obstruction on the morphology, physiology, and pharmacology of the bladder. Prostate, Supp. **3:** 9–26, 1990.
15. Levin, R. M., Levin, S. S. and Wein, A. J.: Etiology of Incontinence: A review and hypothesis. Scandanavian J. Urol., Nephrol. 30, Supp. 179: 15–25, 1996.
16. Levin, R.M., Levin, S.S., Zhao, Y., and Buttyan, R. Cellular and molecular aspects of bladder hypertrophy. Eur. Urol. 32: (supp) 15–21, 1997.
17. Greenland,JE; Hvistendahl,JJ; Andersen,H; Jorgensen,TM; Constantinou,CE; McMurray,G; Brading,AF; Frokiær,J (1997): Detrusor and kidney blood flow is reduced in response to early bladder outlet obstruction in pigs. J. Urol. 157(4, Suppl), 172.
18. Ruggieri, M. R., Hanno, P. M., Samadzdeh, S., Johnson, E. W. and Levin, R. M.: Heparin inhibition of increased bacterial adherence following overdistention, ischemia, and partial outlet obstruction of the rabbit urinary bladder. J. Urology, 136: 132–135, 1986.
19. Lin,AT; Chen,MT; Yang,CH; Chang,LS (1995): Blood flow of the urinary bladder: effects of outlet obstruction and correlation with bioenergetic metabolism. Neurourol. Urodynam. 14, 285–292.
20. Brading, A.F. Alterations in the physiological properties of urinary bladder smooth muscle caused by bladder emtpying against an obstruction. Scand. J. Urol. Nephrol. 184: 51–58, 1997.
21. Hypolite, J. A., Longhurst, P. A., Gong, C., Briscoe, J., Wein, A. J. and Levin, R. M.: Metabolic studies on rabbit bladder smooth muscle and mucosal epithelium. Mol. Cell. Biochem., 125: 35–42, 1993
22. Levin, R. M., Hypolite, J. A., Haugaard, N. and Wein, A. J.: Comparative response of rabbit bladder smooth muscle and mucosa to anoxia. Neurorurol. Urodyn., 15: 79–84, 1996.

21

EFFECTS OF GLUTAMATE RECEPTOR ANTAGONISTS ON LOWER URINARY TRACT FUNCTION IN CONSCIOUS UNANESTHETIZED RATS

Oamu Nishizawa, Yasuhiko Igawa, Tomoya Satoh, Seiji Yamashiro, and Kimio Sugaya

Department of Urology
Shinshu University School of Medicine
3-1-1 Asahi
Matsumoto City 390, Japan

ABSTRACT

Objectives. To study the effects of the intrathecal administered glutamate receptor antagonists on the bladder and urethral activities during isovolumetric bladder contraction in conscious normal and chronic spinal rats.

Methods. Twenty-eight female Wistar rats with and without previous spinal cord transection were used. Before and after intrathecal administration of glutamate receptor antagonist, urodymamic parameters under isovolmetric condition of the bladder were analyzed.

Results. In normal rats, MK-801(noncompetitive N-methyl-D-aspartate [NMDA] receptor antagonist) and LY 293558 (competitive AMPA receptor antagonist) produced a decrease in bladder contraction pressure and urethral activity with dose dependent manner. In chronic spinal rats, detrusor-sphincter dyssynergia (DSD) was developed before drug administration. MK-801 and LY 293558 partially inhibited bladder contraction pressure, and markedly depressed urethral contraction concomitant with bladder contraction. LY 293558 produced urethral relaxation concomitant with bladder contraction.

Conclusions. In both normal rats and chronic spinal rats, two subtypes of glutamate receptors (NMDA and AMPA receptors) in the spinal cord were involoved in the control of bladder and urethral activities. The AMPA receptor in the spinal cord seems to take an important role in the development of DSD.

Advances in Bladder Research, edited by Baskin and Hayward.
Kluwer Academic / Plenum Publishers, New York, 1999.

1. INTRODUCTION

Glutamic acid which functions as an excitatory neurotransmitter at various sites in the central nervous system acts on at least two subtypes of ionotropic receptors -NMDA and AMPA/kainate (AMPA) receptors. NMDA and AMPA receptors have been implicated in the control of micturition in the rat.[1,2] The precise role of the glutamate receptor in the spinal control of micturtition is unknown. The present paper describes the results of our experimental studies on the effects of the intrathecal administered glutamate receptor antagonists on the bladder and urethral activities during isovolumetric bladder contraction in conscious normal and chronic spinal rats.

2. MATERIALS AND METHODS

2.1. Animals

Twenty-eight female Wistar rats (8–10 weeks after birth, weighing 120–170 g) with and without previous spinal cord transection were used in this study.

2.2. Procedures

2.2.1. Spinal Cord Transection. Rats were anesthetized with inhalation anesthesia of sevoflurane. The laminectomy was done at the 7–8th thoracic vertebraes. The dura was opened and the spinal cord was transected at the middle of 7–8th thoracic spinal cord under direct vision. After spinal cord transection the bladder was emptied by crede manuver twice a day. After 4 weeks spinal rats were used for experiment.

2.2.2. Intrathecal Catheter Implantation. A polyethlene catheter (PE-10) was inserted into the subarachinoid space through the hole made in the dura at the level of 5th lumbar vertebrae in normal rats and at the level of 7–8th thoracic vertebrae in spinal rats. The tip of catheter was located at the level of 6th lumbar spinal cord segments for intrathecal administration of drugs.

2.3. Urodynamic Examination

Figure 1 shows the experimental setup. The bladder and urethral activities were monitored by measuring bladder pressure, urethral pressure and external urethral sphincter electromyogram (EMG) under isovolumetric condition of the bladder. Under inhalation anesthesia of sevoflurane, animals were fixed at prone position, abdomen was opened through a midline laparotomy. Both ureters were cut and distal ends were ligated to drain the urine and thus allow bladder volume maintenance by external infusion only. A polyethylene catheter (PE-60) was inserted through the dome into the bladder for recording bladder pressure and for filling the bladder with warm saline. The bladder neck was ligated to produce isovolmetric condition of the bladder. A 3 Fr. microtip catheter transducer was inserted into the urethra through the external meatus to measure urethral pressure. Two wire electrodes were inserted into the urethral external sphincter to measure urethral EMG. All experiments were performed at fixed prone position in the conscious state after the recovery from anesthesia. The bladder contraction pressure, urethral pressure and EMG activity were analyzed before and after drug administration. The paired t-test was used to compair

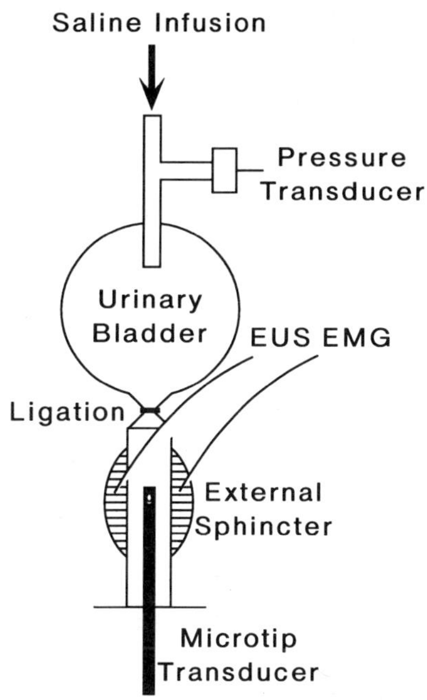

Figure 1. Experimental setup.

the urodynamic parameters before and after drug administration, and $p<0.05$ expressed significant difference.

2.4. Administration of Drugs

MK-801 (noncompetitive NMDA receptor antagonist) and LY293558 (competitive AMPA receptor antagonist) were administered by microsyringe via an intrathecal catheter at the 6th lumbar spinal cord segment. Both drugs were dissolved with physiological saline, and a volume of 1–10μl were used.

3. RESULTS

3.1. Normal Rats

In normal rats isovolumetric bladder contractions were developed concomitant with spasmodic urethral EMG activities at constant interval before drug administration. MK-801(n=7) was administered at 6 doses of 0.1, 0.3, 1, 3, 10, 30μg in 7 rats and LY 293558 was administered at 5 doses of 0.01, 0.03, 0.1, 0.3, 1.0 μg in 7 rats. Figure 2A shows a representative recording before and after MK-801. MK-801 produced a decrease in bladder contraction pressure and urethral EMG activity with dose dependent manner(Figure 3 and 4). Figure 2B shows a representative recording before and after LY 293558. LY 293558 produced a decrease in bladder contraction pressure and urethral EMG activity with dose dependent manner (Figure 3 and 4).

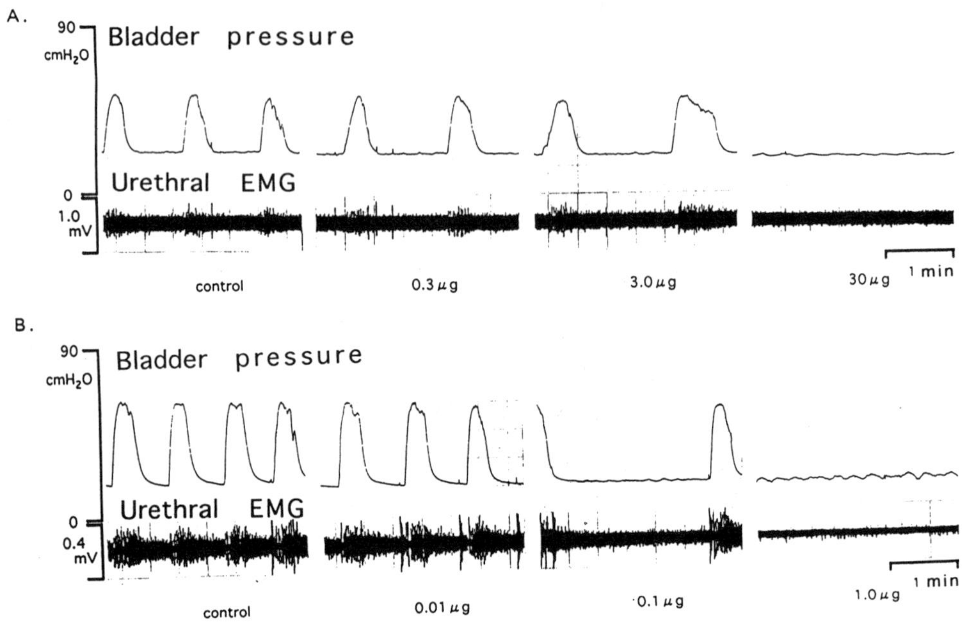

Figure 2. Representative recording of bladder pressure and urethral EMG. A: Before and after MK-801, B: Before and after LY 293558.

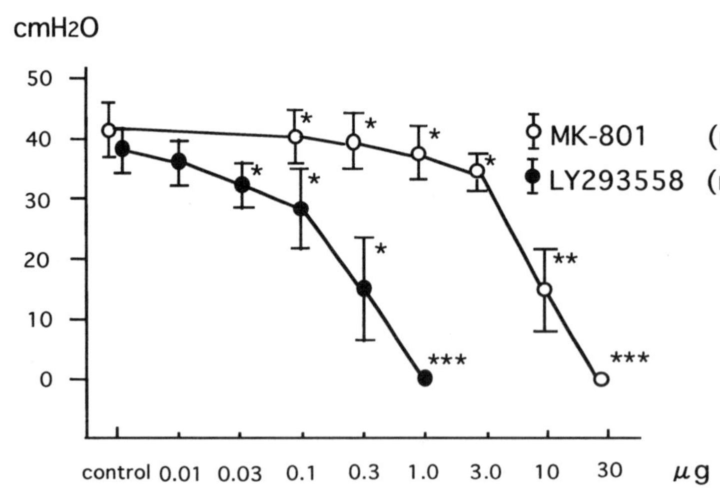

Figure 3. Bladder contraction pressure before and after drug administration in the normal rats (mean ± S.E., *P <0.05; **P <0.01; ***P <0.001).

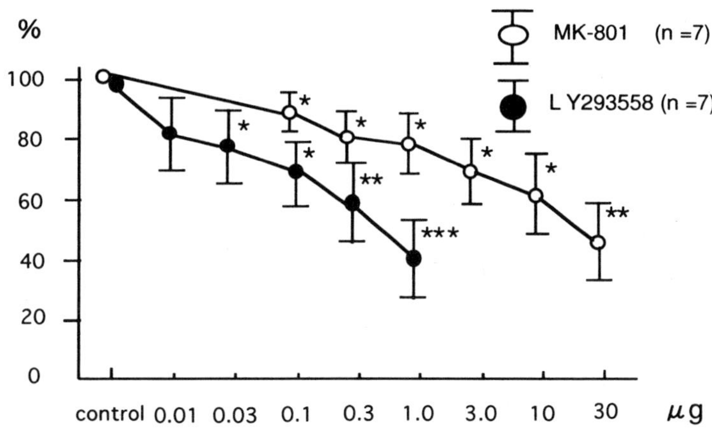

Figure 4. Urethral EMG activity before and after drug administration in the normal rats (mean ± S.E., *P <0.05; **P <0.01; ***P <0.001).

3.2. Chronic Spinal Rats

In all chronic spinal rats, DSD (Detrusor-sphincter dyssynergia) was developed and the amplitude of urethral contraction comcommitant with bladder contraction was 6.8± cmH20 (mean ± S.E.) before drug administration.

MK-801 and LY 293558 were administered separately at 6 doses of 0.3, 1, 3, 10, 30, 100 μg in 7 rats. MK-801 and LY 293558 partially inhibited bladder contraction pressure (Figure 5). MK-801 and LY 293558 markedly decreased urethral contraction concomitant with bladder contraction (Figure 6). 100 μg of Mk-801 abolished urethral contraction concomitant with bladder contraction. The prominant finding was that 3 μg of LY293558

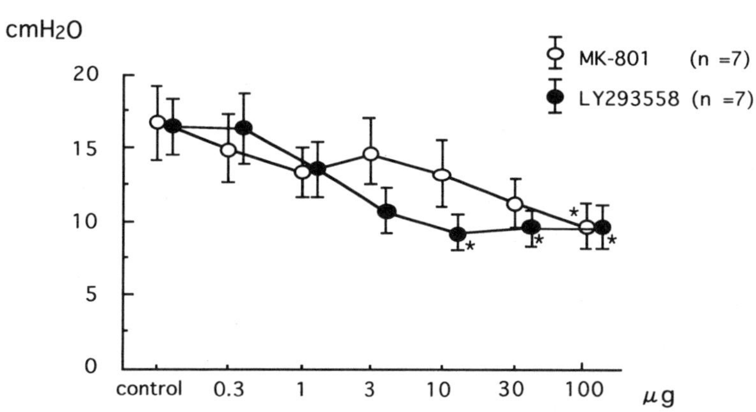

Figure 5. Bladder contraction pressure before and after drug administration in the chronic spinal rats (mean ± S.E., *p <0.05).

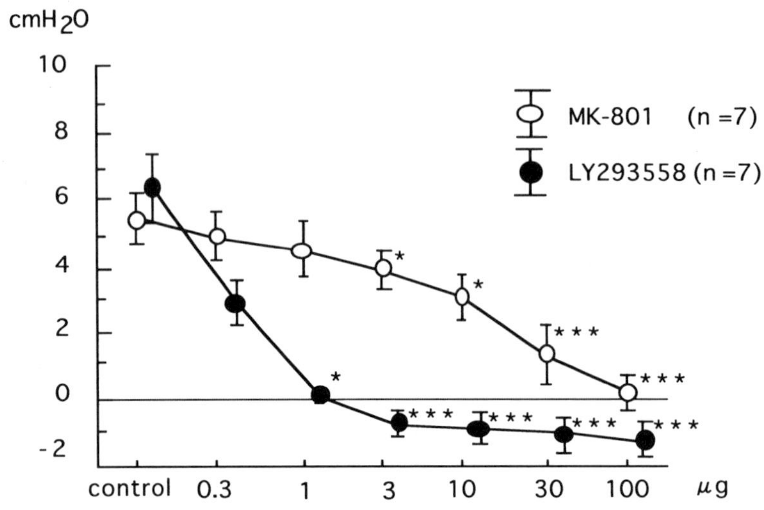

Figure 6. Urethral pressure before and after drug administration in the chronic spinal rats (mean ± S.E., *p <0.05; ***p <0.001).

abolished urethral contraction and 10 μg of LY293558 produced uretral relaxation concomitant with bladder contraction. The amplitude of urethral relaxation was -1.2± 0.6cmH20.

4. CONCLUSIONS

In both normal rats and chronic spinal rats, two subtypes of glutamate receptors, NMDA and AMPA receptors, were involoved in the control of bladder and urethral activities at the level of the spinal cord. The intrathecal administration of glutamate receptor antagonist inhibited bladder and urethral activities of normal rats and chronic spinal rats. The intrathecal administration of glutamate receptor antagonist seem to be useful drug therapy for treating overactive bladder.

Between normal rats and chronic spinal rats, there is difference in the effective dose of MK-801 and LY293558 needed to inhibit the bladder and urethral activities. In normal rats 30μg of MK-801 and 1μg of LY293558 abolished bladder contraction pressure, and in chronic spinal rats 100μg of MK-801 or LY293558 did not abolish bladder contraction pressure. The lower urinary tract function of chronic spinal rats is only modulated with glutamate receptor antagonist stronger than that of normal rats. It appears that the degree of modulation of glutamate receptor on the bladder and urethral activities is influenced by spinal cord transection.

LY 293558 was more effective than MK-801 in inhibiting the urethral activities of chronic spinal rats. 3μg of LY 293558 abolished urethral contraction concomitant with bladder contraction and 10μg of LY 293558 produced urethral relaxation concomitant with bladder contraction which is detrusor–sphincter synnergia (DSS) and is developed in the nerve intact normal rats. 10μg of LY293558 restored the pathological condition -DSD- developed before drug administration to normal condition -DSS-. The AMPA receptor in the spinal cord seems to take an important role in the development of DSD. It is possible that AMPA receptor antagonist may become an useful drug for treatment of DSD.

REFERENCES

1. Yoshiyama, M., Roppolo, J. R, de Groat, W.C. :Alteration by urethane of glutamatergic control of micturition. Eur. J. Pharmacol., **264:** 417, 1994.
2. Yoshiyama, M., Roppolo, J. R, de Groat, W. C. : Effects of CYKI 52466 and CNQX, AMPA/kainate receptor antagonists, on the micturition reflex in the rat. Brain Res., **691:** 185, 1995.

22

OBSERVATIONS FROM THE SPONTANEOUSLY HYPERTENSIVE RAT

Insight into NGF Regulation and Noradrenergic Hyper-Innervation in the Lower Urinary Tract

William D. Steers, David B. Clemow, Katarina Persson, Todd Sherer, Karl-Erik Andersson, and J. B. Tuttl

Departments of Urology and Neuroscience
University of Virginia School of Medicine
Charlottesville, Virginia 22902-0422
and Department of Clinical Pharmacology
Lund University

1. INTRODUCTION

An association between hypertension (HTN) and symptoms from benign prostatic hyperplasia (BPH) has been reported.[1–4] Nearly a third of men undergoing treatment of prostatism are hypertensive. This increased prevalence of HTN exceeds that for age-matched cohorts. Such an association between these two seemingly independent disorders is intriguing because it raises the possibility of a common underlying mechanism- namely over-activity of the sympathetic nervous system and/or aging. In line with this reasoning, young anxious males can exhibit episodic rises in systemic blood pressure as well as the voiding symptoms urgency, frequency and slow urinary stream similar to men with prostatism. These anxiety-related symptoms have also been attributed to sympathetic nerves and adrenergic mechanisms.

Several hypotheses can be proposed linking HTN and the symptoms of BPH. The most obvious explanation is that there exists a global increase in sympathetic output leading to heightened vascular tone as well as increased urethral resistance that contributes to a functional obstruction in BPH. Alternatively, an age-related up-regulation of α_1-adrenoceptor function may exist in both conditions producing enhanced contractile responses to norepinephrine in vascular and prostatic smooth muscles.

An association between HTN and BPH may shed light on the pathogenesis of hyperactive voiding behavior and changes in autonomic function in both conditions. If altered

Advances in Bladder Research, edited by Baskin and Hayward.
Kluwer Academic / Plenum Publishers, New York, 1999.

neural function underlies voiding behavior in prostatism, this finding could explain the surprising lack of an association between the severity of voiding symptoms and degree of urethral obstruction, prostatic size or efficacy of treatments that don't relieve obstruction.[5,6] Moreover, the relatively mediocre results of uro-selective α_{1A}- adrenoceptor antagonists could be predicted based on neural plasticity in afferent (sensory) or spinal mechanisms in patients with HTN or BPH.[7]

To obtain insight into an association between hypertension and voiding behavior we examined the Spontaneously Hypertensive Rat (SHR) and related cross-bred strains. Aside from changes in innervation exhibited in the resistance vessels, enhanced cardiovascular and renal reflexes develop in the SHR. It is postulated that the observed increased innervation and enhanced ganglionic and reflex activity in the SHR are the result of overproduction or altered metabolism of Nerve Growth Factor (NGF).

Do SHR demonstrate a defect in NGF production in non-vascular smooth muscles such as the bladder? If so, do SHR bladders exhibit increased noradrenergic innervation and altered adrenergic pharmacology? Although SHR demonstrate increased levels of catecholamine-containing synaptosomes in the bladder, it is unknown whether this finding is due merely to changes in vascular innervation.[8] Could changes in NGF and noradrenergic innervation or function influence voiding behavior and urodynamics? Does altered bladder function correlate with systemic blood pressure in cross bred strains from SHR? SHR exhibit both hypertension and motor hyperactivity.[9] These two phenotypes can be selected for by cross breeding WKY and SHR creating hypertensive (WKHT) and hyperactive (WKHA) strains. Thus it should be possible to elucidate whether voiding behavior is correlated with blood pressure. Herein we present evidence for an altered genetic regulation of NGF in the vasculature and urinary bladder of the SHR that leads to alterations in the autonomic innervation of these tissues with the resultant physiological changes of hypertension and hyperactive voiding.

2. BLOOD PRESSURE AND URINARY FREQUENCY: INSIGHTS FROM HIGH AND LOW BLOOD PRESSURE F2 RATS

A clinical association between hypertension and voiding dysfunction exists. To investigate the possible genetic linkage between hypertension and hyperactive voiding, SHRs have been crossbred with WKYs and the offspring randomly interbred to produce a gene-segregating F2 population. Traits from the two strains used to develop a F2 population will randomly mix throughout the population allowing for genetic correlates to be made.[10] In a F2 population study, a positive correlation exists when two distinct traits are linked to each other. We found a positive correlation (linear regression: r=0.75) between voiding frequency and blood pressure in a WKY:SHR-derived F2 population.[11] F2s with high blood pressure (High BP F2s) void at a significantly higher frequency than F2s with low blood pressure (Low BP F2s). High BP F2s void lower volumes per void then Low BP F2s with no difference in total voided volume between the two groups. This finding could have significant implications in the understanding of bladder overactivity in patients with HTN and BPH.

3. VOIDING BEHAVIOUR

Kawabe suggested that SHR exhibit an altered 24 hr urinary.[12] Indeed, SHR void at least 3 times more frequently and with smaller quantities of urine than WKY rats during the same light cycle as measured using a metabolic chamber. Since SHR weigh less than WKY rats, bladders from SHR also weigh less. Increased voiding frequency even exceeded young (4 wk) WKY rats of comparable size to adult SHR. Yet the ratio of bladder to body weights in SHR and WKY rats were identical, No changes in urine osmolality were found. Nor was there an increase in total urine output for SHR. These latter two findings suggest a difference in voiding frequency was not merely due to increased urine output.

4. CYSTOMETRY

Awake cystometry confirmed that SHR void at a smaller volume than WKY rats implying a lowered afferent threshold, since bladder compliance appeared similar in both strains. Likewise, bladder capacities were lower in SHR. However, micturition pressures from SHR are significantly greater than WKY rats. This increased voiding pressure, in the face of increased α-adrenoceptor response in vitro, may result from increased outlet resistance. Detrusor instability, manifested as non-voiding bladder contractions in between micturitions, were far more frequent in SHR than WKY rats.

5. BLADDER INNERVATION IN SHR

The urinary bladder receives efferent input from predominately noradrenergic sympathetic and cholinergic parasympathetic nerves originating from the L1/L2 and L6/S1 spinal levels in the rat.[13] Neuropeptides and other substances such as adenosine triphosphate (ATP) and nitric oxide (NO) are probably co-released with norepinephrine and acetylcholine, and influence neurotransmission.[14] Afferents relaying information from the bladder to the spinal cord contain a variety of neuropeptides as well as NO.[15]

Examination of noradrenergic innervation in the SHR bladder using glyoxylic acid revealed profuse catecholamine-fluorescent varicosities throughout the bladder far exceeding that observed in WKY controls.[16] Although most fibers coursed below the level of the urethral orifices, dense fluorescent varicosities were found as rostral as the bladder dome which did not receive significant noradrenergic input in the WKY rat. Catecholamine fluorescent fibers were detected throughout the bladder smooth muscle bundles and not limited to the vasculature. Corroborating this finding, high performance liquid chromatography (HPLC) measurements revealed significantly higher norepinephrine content in bladders from SHR than WKY rats. Using retrograde axonal tracing methods, neuronal cross sectional area profiles for bladder afferents in the L6/S1 dorsal root ganglia (DRG) were significantly larger in SHR than WKY animals.[17] Together these findings support the notion of increased noradrenergic and altered afferent innervation in the SHR. Further studies are required to quantify total nerve content and specifically afferent innervation in SHR bladders. Electrophysiological methods are still needed to examine neurotransmission in the pelvic ganglia and spinal/supraspinal reflexes in the SHR which would help pinpoint any alterations in neural function.

6. IN VITRO DETRUSOR SMOOTH MUSCLE BATH STUDIES

Electrical field stimulation (EFS) of isolated bladder strips from SHR and WKY rats evoked frequency-dependent, tetrodotoxin-sensitive contractions. The contractile response in SHR was significantly lower than in WKY rats at frequencies ≥16Hz. The decreased response persisted after treatment with the muscarinic antagonist scopolamine. The β-adrenoceptor antagonist, propranolol, and α_2-adrenoceptor antagonist, rauwolscine, failed to affect EFS evoked contractions. In WKY, norepinephrine produced a contraction-dependent inhibition of EFS contractions by as much as 83%. This inhibition was not observed in SHR. This suggests a relative decrease in prejunctional α_2 adrenoceptor function in the cholinergic nerves of SHR. Facilitatory α_1-adrenoceptors on cholinergic nerves have been demonstrated in the rat bladder.[18] In line with this, phenylephrine, selective for α_1-adrenoceptors caused concentration-dependent enhancement of electrically-induced bladder contractions; the maximal increase was more prominent in SHR than in WKY rats.

Contractile responses of bladder smooth muscle to the muscarinic agonist, carbachol, as well as the purinergic agonist, α,β methylene ATP, and substance P did not differ between strains. Norepinephrine concentration dependently decreased and eventually abolished spontaneous myogenic contractions in WKY rats probably by stimulating β-adrenoceptors. In SHR, the response the norepinephrine was biphasic with inhibition at low concentrations (ie; β-adrenoceptors) and stimulation of contractile activity at ≥ 10μM concentrations (i.e. α-adrenoceptors). In both SHR and WKY, norepinephrine relaxed carbachol-induced contractions. In the urethra, α-adrenoceptor responses were also enhanced. These findings suggest that β-adrenoceptor is not decreased in SHR. However, postjunctional α-adrenoceptor function (α_1) is increased and prejunctional α-adrenoceptor (α_2) function decreased.

7. EFFECT OF α_1 BLOCKERS ON BLADDER FUNCTION IN SHR

Previous work has shown that systemic administration of α_1-adrenoceptor antagonists reduce urinary frequency in SHR.[12] It has been found that intrathecal administration of α_1-adrenoceptor antagonists reduce sympathetic and pudendal outflow.[19] Thus, it seemed reasonable to examine the effect of a systemic and α-adrenoceptor antagonist intrathecal doxazosin on bladder function in SHR. Both intraarterial and intrathecal decreased micturition pressure in SHR and WKY rats. However, intrathecal, as opposed to intra-arterial doxazosin, abolished unstable contractions in SHR. It appears that spinal α-adrenoceptor blockade inhibits abnormal bladder contractions in the SHR.

8. NERVE GROWTH FACTOR

NGF metabolism has been studied most thoroughly in the cardiovascular system of the SHR with comparisons made to the normotensive control strain, WKY. Tissue levels of NGF are higher in the SHR as early as 2 days of age in certain vessels and the elevated levels are maintained at least until 6 months of age.[20,21]

The increased supply of NGF from SHR viscera may be the primary cause of increased innervation in these tissues.[22] NGF mRNA levels are elevated in vessels and kidney of SHR.[21,23] These tissues are 'hyperinnervated' as shown by higher norepinephrine (NE) content and a greater density of catecholamine-containing fibers when compared to the control strain. Using tissue norepinephrine content as a measure of sympathetic innervation, the mesenteric vasculature, caudal artery, aorta, and kidneys of the SHR demonstrate elevated content as compared to the normotensive Wistar-Kyoto (WKY) control strain.[24] The development of hyper-innervation in the aorta and caudal artery paralleled the onset of hypertension implying a cause and effect.[25] This hyper-noradrenergic phenotype may be limited to certain target tissues, such as smooth muscle, because similar disturbances in NGF content are not seen in the heart of SHR.[26] The existence of a disturbed relationship between sympathetic nerve and vessel smooth muscle was further indicated by Forehand and her colleagues.[27] These investigators described increased dendritic complexity in sympathetic neurons of the hypertensive strains. Because sensory nerves are NGF-responsible, it might be anticipated that sensory hyper-innervation also develops. SHR animals also displayed enhanced excitatory junction potentials in smooth muscle which may contribute to arteriolar constriction and elevated blood pressure.[28] The increased potentials appear to be due to increased activity. Predictably a reduction of sympathetic innervation in SHR, prevents.[29]

Elevated NGF production associated with SHR vascular smooth muscle cells (VSMCs) appears to be a shared trait of SHR smooth muscle in general. Cultured bladder smooth muscle cells (BSMCs) from SHRs secrete elevated levels of NGF when compared to control WKY BSMCs. The rate of NGF secretion by cultured BSMC was nearly 20 times that of VSMC. However, there was no difference in the total RNA or mRNA for NGF between SHR and WKY BSMCs. What differed was the per cell rate of NGF secretion and amount of NGF protein synthesized per unit NGF mRNA which was higher in the SHR. Thus the SHR are apparently much more efficient at translating NGF mRNA into protein. Proliferation of vascular and detrusor smooth muscle in culture is increased in SHR. NGF production is also linked to cellular proliferation contributing to some of the increased production. Like the vasculature, it is possible that the NGF overproduction in the bladder prevents the normal apoptosis of bladder sensory and noradrenergic neurons that occurs during early development causing a relative hyper-innervation.[30]

Levels of NGF in the adult bladders were significantly higher in SHR than WKY rats per gram wet weight or mg protein. NGF is not stored. Tissue levels represent the net effect of production and uptake by nerves. However, denervation can also result in increased levels of NGF in the vasculature.[31] Since innervation is substantially increased allowing more potential uptake levels may be quite stable. Increased NGF uptake can be inferred from the relative hypertrophy of bladder DRG and MPG neurons in the SHR compared to the WKY rats. The increase in the number of bladder labeled DRG cells implies possible hyperplastic changes from reduced apoptosis.

9. WISTAR-KYOTO HYPERTENSIVE AND HYPERACTIVE RATS

One mechanistic difficulty, yet experimental opportunity, is the coexistence of hypertension with behavioral hyperactivity in the SHRs.[32,33] Behavioral hyperactivity in SHRs is not phenotypically or genotypically linked with hypertension and is a permanent fixture of the SHR genome.[34] Therefore, one is unable to determine whether SHR hyperactive voiding is attributable to traits associated with hypertension, behavioral hyperactivity or both. Wistar-Kyoto Hypertensives (WKHTs) and Wistar-Kyoto Hyperactives (WKHAs) are two new inbred strains developed through cross breeding of SHRs with WKYs.[33] One can take advantage of this breeding to investigate whether blood pressure per se can influence bladder function via cardiac, baroreceptor or vascular afferents reminiscent of other viscero-visceral reflex interactions exemplified by the ability of anal, uterine, vaginal, and other afferent input to influence efferent neural input to the bladder.

WKHAs and WKHTs void significantly more frequently but with the same total voided volume as WKY rats. However, both WKHAs and WKHTs void less frequently than SHRs. It can be inferred that traits from both the hypertensive and behaviorally hyperactive genome contribute to hyperactive voiding in SHRs. This result challenges the hypothesis that NGF-dependent changes in bladder innervation are the sole determining factor of hyperactive voiding in SHR. Excessive VSMC NGF secretion is postulated to contribute to hypertension in the SHR.[20,26,35,36] In line with this theory, both the augmented basal levels of NGF secretion and the disturbances in the pharmacological regulation of NGF output observed in cultures of SHR VSMCs co-segregate with a hypertensive phenotype.[37] The correlation between elevated NGF and hyperactive voiding was not as clear when describing the relationship in bladder smooth muscle. Basal NGF secretion from WKHA BSMCs was significantly higher than that of WKHTs.[37] Although WKHA BSMCs secreted higher basal levels of NGF than did WKHTs, α- and β- adrenergic, purinergic, peptidergic and PKA-dependent stimulation all inhibited NGF secretion in WKHA cultures.[38] In contrast, the same agents increased NGF output in WKHT cultures or had no significant effect.

In vivo, significant differences exist in bladder tissue NGF for each strain with WKY bladders exhibiting the least NGF followed by incremental increases in NGF content in WKHA, WKHT and SHR bladders.[17] Thus, although both WKHA and WKHT bladders contained more NGF than WKY bladders, WKHT bladders had significantly more NGF than WKHA bladders. These results suggest that a lack of regulatory inhibition and/or an over-sensitivity to NGF agonists in vitro may be the best indicator of bladder NGF dynamics in vivo. Moreover, augmented NGF output from bladder contributed to neurally-mediated changes in bladder function in both the hypertensive and behaviorally hyperactive strains. This finding strengthens the hypothesis that both hypertensive and behaviorally hyperactive traits are contributing to hyperactive voiding in the SHR.

SHR, WKHT and WKHA bladders contained heightened levels of NGF and would be expected to be sympathetically hyperinnervated when compared to WKY bladders. Yet, norepinephrine levels were elevated in SHR and WKHT but not WKHA bladders despite elevated NGF content in this latter strain. This result implies that the noradrenergic hyperinnervation observed in SHR bladders co-segregates with a hypertensive phenotype despite the observation that increased bladder tissue NGF occurs in both hypertensive and behaviorally hyperactive animals. When compared to WKYs, the elevations in WKHA bladder NGF were significantly less than the elevations in either SHRs or WKHTs. Perhaps the increased levels of NGF in WKHAs were not sufficient to affect the sympathetic innervation, but are elevated enough to affect sensory neurons. This observation is still under investigation. Sparse information is available concerning threshold amounts of tissue NGF necessary to induce changes in innervation.

10. DISCUSSION

Data from the SHR provides compelling evidence that an association exists between augmented smooth muscle NGF, elevated sympathetic innervation, hypertension and hyperactive voiding. NGF provides a possible link between hypertension and hyperactive voiding in animal models of hypertension. This notion is further supported by work investigating NGF output from cultured F2 BSMCs. When compared to Low BP F2 BSMCs, High BP F2 BSMCs have an elevated sensitivity to NGF-producing signaling molecules and little if any sensitivity to NGF-inhibiting agents (unpublished).

Thus, common themes exist in the three different animal model comparisons: WKY-SHR, WKHA-WKHT and Low BP F2-High BP F2. First, an apparent link between high blood pressure and hyperactive voiding exists. Second, bladder NGF production in hypertensive animals appears defective such that smooth muscle cell NGF output is not affected by inhibitory signaling pathways while excitatory signaling mechanisms are hyper-responsive. Therefore, a disturbed genetic component of the hypertensive rat genome, associated with the regulation of smooth muscle NGF output, may contribute to elevated bladder smooth muscle NGF content and ultimately neurogenic changes which potentially influence voiding frequency.

Rush has proposed in a recent review that there may be a negative feedback control system on the synthesis and/or release of NGF.[39] This idea suggests that NGF output by smooth muscle is controlled by factors secreted from sympathetic nerve terminals, such as norepinephrine. Partial support of this notion derives from experiments in which denervation of mesenteric arteries lead to decreased NGF mRNA, while antibodies against NGF caused elevated NGF levels.[39] This hypothesis predicts that the control mechanism for NGF production is significantly altered in the SHR, allowing excess NGF to be expressed.

Precisely how urinary frequency develops is unclear. It is not the result of changes in the production or concentration of urine. Cystometric data confirm a reduced threshold for micturition implying an alteration in bladder afferents. Quantification of bladder afferents has not been attempted. Similarly, NGF induced plasticity in bladder afferents or reflex pathways has been postulated to explain urinary frequency following bladder outlet.[13,40,41,42] Moreover, small bladder afferent neurons increase in size following spinal cord injury, they exhibit enhanced excitability due to a plasticity in sodium and potassium channels.[43] Altered channel properties lower high-thresholds for action potentials and cause an attenuated suppression by tetrodotoxin of Na+ currents. Furthermore, C-fiber and A(-afferents and sympathetic efferents are dependent on NGF during development (survival) and adulthood (functional maintenance).[44] NGF significantly alters the threshold for action potential generation, sensitivity of the Na+ action potential to tetrodotoxin and neural chemosensitivity.[45] Although these types of neuro-physiological changes have not been tested directly in the SHR micturition reflex pathway, enhanced nerve responsiveness has been demonstrated in cardiac and vascular reflex pathways of SHRs.[27] Whether NGF antibody can prevent urinary frequency must be tested.

It is possible that sympathetic efferent activity can influence urinary frequency. The use of 6-hydroxydopamine-sympathectomized SHR will shed insight into this.[46] Adult sympathetectomized SHR maintain hypertension suggesting other mechanisms at work. However, early destruction of sympathetic nerves can prevent the development of hypertension.[29] If heightened sympathetic efferent outflow was operative, one might expect increased bladder capacity due to heightened α-adrenoceptor function and/or detrusor hypertrophy from an increased urethral resistance. In this regard, it is undergoing that SHR tend to void at greater pressures and develop unstable contractions. Also provocative is that intrathecal α-adrenoceptor antagonists reduced unstable contractions. This observation combined with the finding that systemic α-adrenoceptor antagonists prevent urinary frequency with obstruction and the action of intrathecal α-adrenoceptor antagonists in obstructed rats suggests at least a partial spinal site of action for these agents in relieving irritative voiding.[47,48] Recent in situ hybridization data for mRNA of α_1 subtypes in the lumbar spinal cord of the rat have shown that α_{1B} and not α_{1A} are found in the dorsal horn and autonomic centers.[49] These data help explain why α_{1A} selective agents, although increasing urinary flow rates through a peripheral site of action, are much less effective at relieving irritative symptoms than non-selective α_1-adrenoceptor antagonists.

A genetic linkage of the NGF gene to high blood pressure has already been established. Kapuscinski and colleagues have demonstrated that a genetic locus in or near the

NGF gene contributes in a Mendelian dominant pattern to a significant increment in blood pressure in SHRs.[34] Furthermore, a better understanding of the connection between hypertension and voiding dysfunction may lead to more appropriate interventions for these pathologies. SHRs, in combination with WKHAs and WKHTs, may represent useful genetic models to study neurally-mediated hyperactive voiding and unravel the role of sympathetic input to the lower urinary tract.

ACKNOWLEDGMENTS

Pam Neff is acknowledged for her contributions to assays for NGF and mRNA stability. Richard McCarty is acknowledged for his work on catecholamine measurements and WKHT/WKHA work. John Spitsbergen is credited for work in glyoxylic acid staining, voiding behavior, and NGF metabolism. Worked supported by NIH grants R01DK 51364, DK 49431, DK 50164.

REFERENCES

1. Boyle, P.: Epidemiology of benign prostatic hyperplasia: risk factors and concomitance with hypertension. Bri. J. Clin. Prac., **74:** 18–22, 1994.
2. Boyle, P. and Napalkov, P.: The epidemiology of benign prostatic hyperplasia and observations on concomitant hypertension. Scand. J. Urol. Neph., **168**, 7–12, 1995.
3. Pool, J.L.: Role of the sympathetic nervous system in hypertension and benign prostatic hyperplasia. British Journal of Clinical Practice - Symposium Supplement **74:** 13–17, 1994.
4. Pool, J.L: Doxazosin: a new approach to hypertension and benign prostatic hyperplasia. Br. J. Clin. Prac., **50:** 154–163, 1996.
5. Yalla, S.V., Sullivan, M.P., Lecamwasam, H.S., DuBeau, C.E., Vickers, M.A. and Cravalho, E.G.: Correlation of the American Urological Association symptom index with obstructive and non-obstructive prostatism. J. Urol., **153:** 674–680, 1995.
6. Barry, M.J., Cockett, A.T.K., Holtgrewe, H.L., McConnell, J.D., Sihelnik, S.A. and Winfield, H.N.: Relationship of symptoms of prostatism to commonly used physiological and anatomical measures of the severity of benign prostatic hyperplasia. J. Urol., **150:** 351–358, 1993.
7. Andersson, K.-E., Lepor, H. and Wyllie, M.G.: Prostatic alpha 1-adrenoceptors and uroselectivity. [Review] [132 refs], Prostate **30:** 202–215, 1997.
8. Tong, Y.C., Hung, Y.C., Lin, S.N. and Cheng, J.T.: Alterations in urinary bladder synaptosomal neurotransmitter concentrations in two-week streptozotocin-induced diabetic rats. Neurosci. Letters, **206** (2–3): 165–168, 1996.
9. Clemow, D.B., McCarty, R., Steers, W.D. and Tuttle, J.B.: Efferent and afferent neuronal hypertrophy associated with micturition pathways in spontaneously hypertensive rats. Neurourol. Urodyn., **16:** 293–303, 1997.
10. Bachmanov, A.A.: [The characteristics of the F1 hybrids of spontaneously hypertensive and normotensive rats]. [Russian], Fiziologicheskii Zhurnal SSSR Imeni I M.: 1714–1717, 1994.
11. Clemow, D.B., Steers, W.D. and Tuttle, J.B.: personal communication 1998.
12. Kawabe, K.: Monitoring the 24-hour urinary frequency in spontaneously hypertensive, and in Wistar rats: effect of alpha-blockers. Nippon Jinzo Gakkai Shi (**9**): 1305–1309, 1996.
13. DuPont, M., Persson, K., Spitsbergen, J., Tuttle, J.B. and Steers, W.D.: The neuronal response to bladder outlet obstruction. A role for NGF., in: "Muscle, Matrix and Bladder Function.", S. A. Zderic ed., Plenum, New York, pp.41–54, 1995
14. Lundberg, J.M.:Pharmacology of co-transmission in the autonomic nervous system: integrative aspects on amines, neuropeptides, adenosine triphosphate, amino acids and nitric oxide. Pharmacological Reviews, **48 (1):** 113–178, 1996.
15. Steers, W.D.: Physiology and pharmacology of the bladder and urethra. In: Campell's Urology, edited by P Walsh, A Retik, E. D. Vaughan, and A Wein, Philadelphia:W.B. Saunders, p. 870–915, 1997.

16. Spitsbergen, J., McCarty, R., Steers, W.D. and Tuttle, J.B.: Nerve growth factor, sympathetic innervation and voiding frequency in spontaneously hypertensive rats. Soc. Neurosci. (Abstract), **21:** 227, 1995.
17. Clemow, D.B., Spitsbergen, J.M., McCarty, R, Steers, W.D. and Tuttle, J.B.: Regulation of NGF secretion in smooth muscle cells from hypertensive and hyperactive rats. Soc. Neurosci. (Abstract), **124:** 298, 1995.
18. Somogyi, G.T., Tanowitz, M. and de Groat, G.W.: Prejunctional facilitatory α_1-adrenoceptors in the rat urinary bladder. Br. J. Pharm., **114:** 1710–1716, 1995.
19. Gajewski, J., Downie, J.W. and Awad, S.A.: Experimental evidence for a central nervous system site of action in the effect of alpha-adrenergic blockers on the external urinary sphincter. J. Urol., 1**32 (2):** 403–409, 1984.
20. Donohue, S.J., Head, R.J. and Stitzel, R.E.: Elevated nerve growth factor levels in young spontaneously hypertensive rats. Hyperten. **14:** 421–426, 1989.
21. Falckh, P.H., Harkin, L.A. and Head, R.J.: Nerve growth factor mRNA content parallels altered sympathetic innervation in the spontaneously hypertensive rat. Clin. Exper. Pharm. Phys., **19:** 541–545. 1992b.
22. Head, R.J.: Hypernoradrenergic innervation: its relationship to functional and hyperplastic changes in the vasculature of the spontaneously hypertensive rat. Blood Vessels, **26:** 1–20, 1989.
23. Falckh, P.H., Harkin, L.A. and Head, R.J.: Resistance vessel gene expression of nerve growth factor is elevated in young spontaneously hypertensive rats. J. Hyperten. **10:** 913–918, 1992a.
24. Head, R.J.: Hypernoradrenergic innervation and vascular smooth muscle hyperplastic change. Blood Vessels, **28:** 173–178, 1991.
25. Donohue, S.J., Stitzel, R.E. and Head, R.J.: Time course of changes in the norepinephrine content of tissues from spontaneously hypertensive and Wistar Kyoto rats. J. Pharma. Exper. Ther. **245:** 24–31, 1988.
26. Ueyama, T., Hamada, M., Hano, T., Nishio, I., Masuyama, Y. and Furukawa, S.: Increased nerve growth factor levels in spontaneously hypertensive rats. J. Hyperten., **10:** 215–219, 1992.
27. Peruzzi, D., Hendley, E.D. and Forehand, C.J.: Hypertrophy of stellate ganglion cells in hypertensive, but not hyperactive, rats. Am. J. Phys, **261:** 22–84, 1991.
28. Brock, J.P. and Van Helden, D.F.: Enhanced excitatory junction potentials in mesenteric arteries from spontaneously hypertensive rats. Pflugers Arch **430:** 901–908, 1995.
29. Brock,J.P., Van Helden, D.F., Dosen, P. and Rush, R.A.: Prevention of high blood pressure by reducing sympathetic innervation in the spontaneously hypertensive rat. J. Autono. Ner. Sys., **61:** 97–102, 1996 .
30. Messina, A., Bertram, J.F., Bakhle, Y.S. and Bell, C.: Decreased developmental cell death in sympathetic and spinal sensory nervous systems of the Kyoto spontaneously hypertensive rat. Journal of Hypertension, **14:** 1111–1115, 1996.
31. Liu, D.T., Reid, M.T., Bridges, D.C and Rush, R.A.: Denervation, but not decentralization, reduces nerve growth factor content of the mesenteric artery. J. Neurochem., **66 (6):** 2295–2299, 1996.
32. Sutterer, J.R., Stoney, C.M. and Sanfillipo, M.: Is the hypertensive rat really hyperreactive? Hyperten., **6**: t-76, 1984.
33. Hendley, E.D., Holets, V.R., McKeon, T.W. and McCarty, R.: Two new Wistar-Kyoto rat strains in which hypertension and hyperactivity are expressed separately. Clin. Exper. Hyperten. - Part A, Theory and Practice, **13:** 939–945, 1991.
34. Kapuscinski, M., Charchar, F., Innes, B., Mitchell, G.A., Norman, T.L. and Harrap, S.B.: Nerve growth factor gene and hypertension in spontaneously hypertensive rats. J. Hypertens., **14:** 191–197, 1996.
35. Zettler, C., Head, R.J., and Rush, R.A. (1991). Chronic nerve growth factor treatment of normotensive rats. Brain Research, **538**: 251–262.
36. Zettler C., and Rush R.A. Elevated concentrations of nerve growth factor in heart and mesenteric arteries of spontaneously hypertensive rats. Brain Res **614:** 15–20, 1993.
37. Clemow, D.B., Spitsbergen, J., McCarty, R., Steers, W. D. and Tuttle J.B.: Regulation of NGF secretion in smooth muscle cells for hypertensive and hyperactive rats. Soc. Neurosci. (Abstract), 22, 1996.
38. Clemow, D.B., and Steers, W. personal communication 1998.
39. Rush, R.A., Chie, E., Liu, D., Tafreshi, A., Zettler, C. and Zhou, X.F.:Neurotrophic factors are required by mature sympathetic neurons for survival, transmission and connectivity. Clin. Exper. Pharm. Phys., **8:** 549–555, 997.
40. Steers, W.D., Ciambotti, J., Etzel, B., Erdman, S and de Groat, G.W.: Alterations in afferent pathways from the urinary bladder of the rat in response to partial urethral obstruction. J. Comp. Neur., **310:** 401–410, 1991a.
41. Steers, W.D., Kolbeck, S., Creedon, D. and Tuttle, J.B.: Nerve growth factor in the urinary bladder of the adult regulates neuronal form and function. J. Clin. Invest., **88:** 1709–1715, 1991.
42. Steers WD; Creedon DJ; Tuttle JB..: Immunity to nerve growth factor prevents afferent plasticity following urinary bladder hypertrophy. J. Urol, **155(1):** 379–85.1996.

43. Yoshimura, N., Yoshida, O. and de Groat, W.C.: Regional differences in plasticity of membrane properties of rat urinary bladder afferent neurons following spinal cord injury. J. Urol., **153:** 262a, 1995.
44. Thoenen, H. and Barde, Y.A.: Physiology of nerve growth factor. Phys. Rev., **60:** 1284–1335, 1980.
45. Aguayo, L.G., and White, G.: Effects of nerve growth factor on TTX- and capsaicin-sensitivity in adult rat sensory neurons. Brain Res., **570 (1–2):** 61–67, 1992.
46. Korsching, S., and Thoenen, H.:Treatment with 6-hydroxydopamine and colchicine decreases nerve growth factor levels in sympathetic ganglia and increases them in the corresponding target tissues. J. Neurosci., **5:** 1058–1061, 1985.
47. Steers, W.D., Albo, M. and Tuttle, J.B.:Calcium channel antagonists prevent urinary bladder growth and neuroplasticity following mechanical stress. Am. J. Phys., **266:** 1–6, 1994.
48. Ishizuka, O., Mattiasson, A., Steers, W.D. and Andersson, K.-E.: Effects of spinal alpha 1-adrenoceptor antagonism on bladder activity induced by apomorphine in conscious rats with and without bladder outlet obstruction. Neuro. Urodynam. **16:** 191–200, 1997.
49. Wada, T., Otsu, T., Hasegawa, Y., Mizuchi, A. and Ono, H.: Characterization of alpha-1 receptors in rat spinal cord. Eur. J. Pharm., **312:** 263–266, 1996.

ROLE OF Na^+-K^+-ATPase ACTIVITY IN REGULATION OF DETRUSOR CONTRACTILITY AND DIABETIC BLADDER DYSFUNCTION

Sandeep Gupta and Alan J. Wein

Division of Urology
University of Pennsylvania School of Medicine
Philadelphia, Pennsylvania 19104-4274

1. SODIUM-POTASSIUM ACTIVATED ADENOSINE TRIPHOSPHATASE

Sodium-potassium activated adenosine triphosphatase (Na^+-K^+-ATPase) is a ubiquitously expressed enzyme which utilizes the energy derived from ATP hydrolysis to transport Na^+ and K^+ across the plasma membrane. Na^+-K^+-ATPase is composed of a catalytic a (~110 kDa) subunit, and a glycosylated β subunit (35 kDa) which is required for the enzyme activity.[1] Three isoforms of the a-subunit (a_1, a_2, a_3) and two isoforms of β subunit (β_1 and β_2), encoded by a highly conserved multigene family, have been reported.[1] The expression of these genes has been shown to be hormonally regulated, and tissue and species specific.[1] Our recent studies using specific monoclonal antibodies indicate that bladder smooth muscle expresses all the a and b_1 isoforms of Na^+-K^+-ATPase (unpublished data).

2. Na^+-K^+-ATPase ACTIVITY AND REGULATION OF BLADDER SMOOTH MUSCLE CONTRACTILITY

Na^+-K^+-ATPase is electrogenic since it extrudes 3 Na^+ while bringing 2 K^+ into the cell, and by virtue of this, it plays a role in the regulation of membrane potential and the tone of vascular and nonvascular smooth muscle cells and cardiac myocytes.[2] Stimulation of Na^+-K^+-ATPase activity, similar to the activation of K^+-channels, results in membrane hyperpolarization leading inhibition of Ca^{2+}-influx via voltage operated channels, and secondarily relaxation of smooth muscle. In contrast, inhibition of Na^+-K^+-ATPase activity leads to membrane depolarization and contraction of smooth muscle cells.[2,3]

Advances in Bladder Research, edited by Baskin and Hayward.
Kluwer Academic / Plenum Publishers, New York, 1999.

2.1. Acute Effects of Na^+-K^+-ATPase Inhibitor, Ouabain, on the Resting Tone of Bladder Smooth Muscle

Ouabain, an inhibitor of Na^+-K^+-ATPase activity, has been shown to alter ion transport and electrical properties in the bladder.[4–6] Na^+-K^+-ATPase activity is an important modulator of bladder smooth muscle tone as Na^+-K^+-ATPase inhibitor, ouabain, causes acute concentration-dependent contractions in the bladder[7] (Figure 1). In other smooth muscles and cardiac myocytes, increased Na^+-Ca^{2+} exchange activity, as a result of inhibition of Na^+-K^+-ATPase activity and increased intracellular Na^+, is known to cause or enhance contraction.[2,8] However, Na^+-Ca^{2+}-exchange activity may not play a significant role in ouabain-induced contraction of bladder smooth muscle as it was completely inhibited by voltage gated Ca^{2+}-channel inhibitor, nifedipine, suggesting a prominent role of membrane depolarization.

2.2. Na^+-K^+-ATPase Activity and Agonist-Induced Contractions

Na^+-K^+-ATPase activity also plays an important role in the regulation of agonist mediated contraction of the bladder smooth muscle.[7] In contrast to the acute effects on resting tone, pretreatment of bladder smooth muscle with Na^+-K^+-ATPase inhibitor, ouabain, inhibits agonist-mediated contractions. For example, ouabain diminishes the maximum contraction induced by muscarinic receptor agonist, carbachol (Figure 2). Ouabain does not affect the affinity of carbachol for muscarinic receptors as the half maximal concentration of carbachol required for contraction is not altered in the presence of ouabain. Furthermore, ouabain inhibits non-receptor mediated contraction of bladder smooth muscle induced by depolarizing concentrations of potassium chloride (KCl). Similarly, ouabain inhibits the plateau phase of action potentials and contractions induced by histamine in ureter,[9] and acetylcholine-induced contractions in uterine smooth muscle.[8] Furthermore, inhibitors of Na^+-K^+-ATPase activity have been reported to reduce KCl, histamine and acetylcholine-induced contractions in intestinal smooth muscle.[10]

3. BLADDER DYSFUNCTION IN DIABETES

Diabetes mellitus affects ~8 million people in the United States and ~two thirds of diabetics are reported to have voiding dysfunction or diabetic cystopathy.[11–14] The most common findings include an enlarged, poorly contracting or atonic bladder with increased residual urine.[15–17] In addition, uninhibitable bladder contractions are also reported in diabetic patients with urinary symptoms. The precise mechanisms responsible for bladder dysfunction in diabetes are not known. Although it is generally thought to be related to autonomic neuropathy, pharmacological manipulations designed to enhance bladder emptying using cholinergic agents in diabetic patients have been largely ineffective.[15] In recent years, it has become increasingly apparent that smooth muscle dysfunction, in addition to any neurologic injury, plays an important role in the development of this diabetic complication.[7,18,19]

3.1. Diabetes Diminishes Bladder Smooth Muscle Contractility

Alloxan-diabetes diminishes the ability of carbachol and KCl to cause maximum contractions in rabbit bladder[7] (Figures 3 and 4). Alterations in the bladder smooth muscle

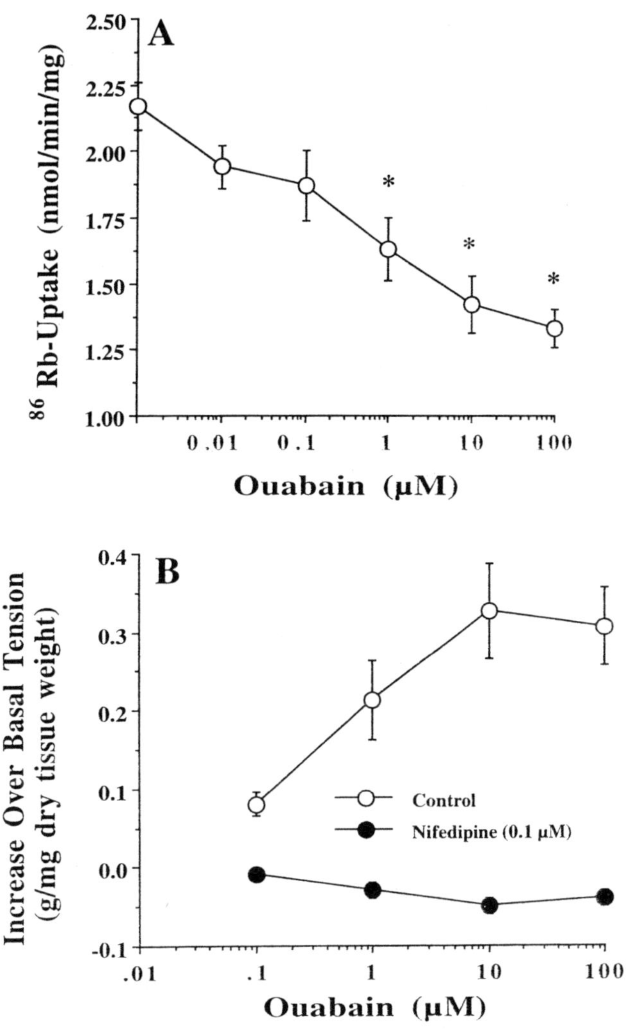

Figure 1. Effect of ouabain on ^{86}Rb-uptake (A) and bladder smooth muscle tone (B). (A) Rabbit bladder strips were incubated in Krebís bicarbonate buffer (pH 7.4, 37°C) for 2 hours and ouabain (0.01–100 μM) was added for the final 20 min of incubation. (B) Rabbit bladder strips were mounted on force transducers in the organ chambers. Ouabain dose responses were obtained in the absence and presence of 0.1 μM nifedipine after optimum isometric tension was attained as described in Methods. Results are means ± SE of 4–6 experiments. *Values significantly different from control (no ouabain) group. (Figure 1 from Gupta *et al.*, Am. J. Physiol. 1996).

contractility have previously been reported by a number of investigators in streptozotocin-induced diabetic rats and spontaneously diabetic BB rats. While some investigators demonstrated that diabetes diminished contractile responses induced by various receptor agonists, neurotransmitters, and membrane depolarization by potassium chloride,[19–21] others have shown that agonist-induced contractions of rat bladder were enhanced or unaltered in diabetes.[22–27] The discrepancy in these findings have been attributed to differences in methodology, duration of diabetes, experimental conditions and/or data normalization.[19] In bladders from streptozotocin-diabetic rats, upregulation of muscarinic receptors, increased tissue cholinesterase activity and enhanced muscarinic receptor agonist-induced contractions have also been reported.[28,29] However, these mechanisms may not be responsible for the depressed contractility of diabetic rabbit bladder, since cholinesterase-resistant carbachol was used as a muscarinic agonist, and diabetes also inhibited contractions induced by KCl (Figure 4). Thus, diabetes may affect bladder contractility via mechanisms that are downstream from agonist-receptor interaction in the signal transduction pathway.

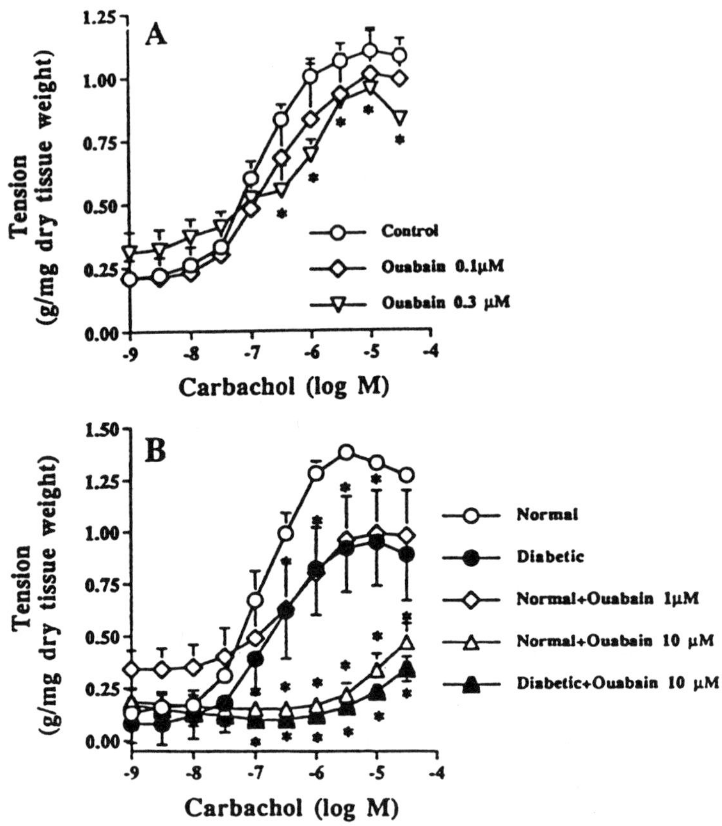

Figure 2. Effect of Na^{+}-K^{+}-ATPase inhibitor, ouabain, on carbachol-induced contraction. Bladder strips from normal and diabetic rabbits were mounted on force transducers in the organ chambers. Carbachol concentration-responses were obtained in the absence and presence of indicated ouabain concentrations (30 minute pretreatment). Results are mean ± S.E.M of 6 experiments. * Indicates values significantly different from control group. (Figure 2 from Gupta *et al.*, Am. J. Physiol. 1996).

3.2. Role of Diminished Na^{+}-K^{+}-ATPase Activity in Diabetes-Induced Bladder Dysfunction

Our results suggest that diabetes inhibits agonist induced contractions in rabbit bladder by inhibiting Na^{+}-K^{+}-ATPase activity. In support of this, diabetes caused approximately 50% reduction in bladder smooth muscle Na^{+}-K^{+}-ATPase activity as measured by ouabain-sensitive $^{86}Rb^{+}$-uptake (Figure 5). Diabetes also diminished ouabain-induced contraction of bladder smooth muscle. Furthermore, a 50% reduction in Na^{+}-K^{+}-ATPase activity in normal bladders by ouabain mimicked the effects of diabetes on bladder contractility, and the differences between normal and diabetic bladder contractility were not apparent when Na^{+}-pump activity was inhibited maximally.[7] Recently, diminished Na^{+}-K^{+}-ATPase activity has also been demonstrated to play a role in alteration of electrical properties of bladder smooth muscle from streptozotocin-diabetic rats.[30] Furthermore, diabetes diminishes Na^{+}-K^{+}-ATPase activity in a number of tissues including blood vessels, heart and peripheral nerves,[31–33] and diminished Na^{+}-pump activity is thought to be a key contributor to the development of various diabetic complications.[34]

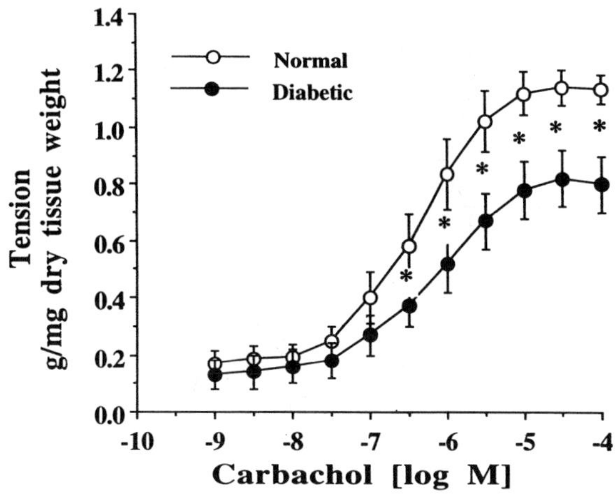

Figure 3. Effect of diabetes on carbachol-induced contraction. Bladder strips from normal and diabetic rabbits were mounted on force transducers in the organ chambers. Carbachol concentration-response was obtained as described in Methods. Results are mean ± SE of number of 9 experiments. * Indicates significant differences between normal and diabetic groups. (Figure 3 from Gupta *et al.*, Am. J. Physiol. 1996).

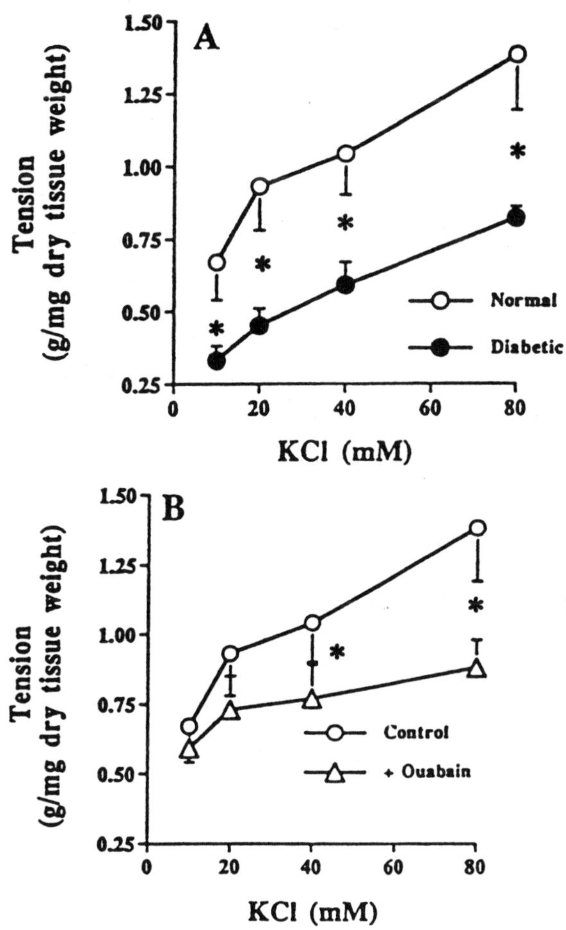

Figure 4. Effect of diabetes (A), and ouabain (B) on KCl-induced contraction. (A) Bladder strips from normal and diabetic rabbits were mounted on force transducers in the organ chambers. In normal rabbit bladder strips, or (B) ouabain (10 μM) were added after optimum isometric tension was attained. Results are mean ± SE of 6 experiments. * Indicates values significantly different from control group. (Figure 7A and 7B from Gupta *et al.*, Am. J. Physiol. 1996).

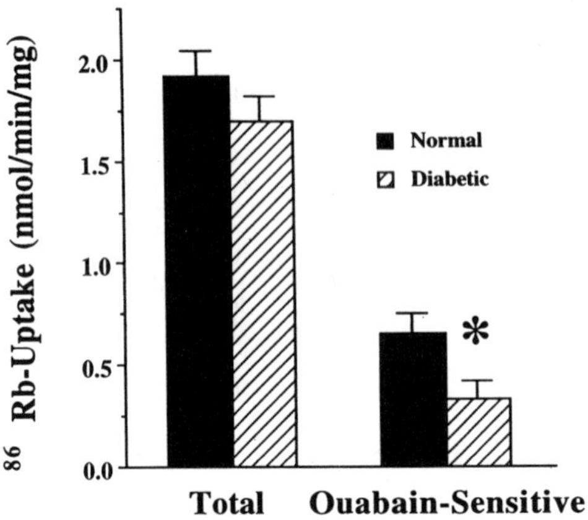

Figure 5. Effect of diabetes on ^{86}Rb-uptake. Rabbit bladder strips from normal (shaded bars) and diabetic (hashed bars) were incubated in Krebís bicarbonate buffer for 2 hours. Ouabain (100 μM) was added during the final 20 minutes of incubation total and ouabain-sensitive ^{86}Rb-uptake were determined. Results are mean ± SE of 9 experiments. * Value significantly different from normal control ($p<0.05$, Students -test). (Figure 2A from Gupta *et al.*, Am. J. Physiol. 1996).

3.3. Possible Mechanisms Underlying Diabetes-Induced Reduction of Na^+-K^+-ATPase Activity

3.3.1. The Expression of Na^+-K^+-ATPase A-Isoforms Has Been Reported to Be Altered in Diabetes. For example, a reduction in mRNA and protein levels of the a_2 and a_3 isoforms of Na^+-K^+-ATPase[35,36] and a decrease in the number of [^{3}H]-ouabain binding sites have been reported in aortas and other tissues of streptozotocin-diabetic rats.[31] In addition, an association between diabetic neuropathy and Na^+-K^+-ATPase gene polymorphism has been reported.[37]

3.3.2. Role of Protein Kinase C. Diabetes associated metabolic alterations have also been implicated in reduced Na^+-K^+-ATPase activity and in the development of diabetic complications. Particularly, increases in the activity of protein kinase C isoforms has been shown to be involved in the development of diabetic neuropathy, retinopathy and micro and macroangiopathy. Recently, an specific inhibitor of the conventional PKCbII isozyme, which requires diacylglycerol (DAG) for activation, has been shown to ameliorate diabetes associated vascular complications.[38] Furthermore, diabetes has been reported to alter PKC mediated phosphorylation of the α subunit of Na^+-pump in rat sciatic nerves in vivo.[39]

3.3.3. Role of Aldose Reductase Pathway. Diabetes and hyperglycemia also increase glucose flux through the aldose reductase (AR) pathway, leading to increases in sorbitol levels in the affected tissues,[40–42] and a number of metabolic changes as well as diminished Na^+-pump activity.[41] Inhibitors of AR have been shown to prevent, and in some cases reverse, functional and structural abnormalities as well as Na^+-K^+-ATPase activity in animal models of diabetes. Aldose reductase inhibitors also prevent acute effects of hyperglycemia on Na^+-K^+-ATPase activity and smooth muscle contractility in *in vitro* studies.[43,44] Furthermore, Luheshi and Zar[20] demonstrated that inhibitory effects of diabetes on non-cholinergic motor transmission in rat bladder was prevented by an AR inhibitor, sorbinil. Whether improvement in bladder function in the studies of Luheshi and Zar was related to improvement in Na^+-K^+-ATPase activity, which has been implicated in diabetic neuropathy, was not determined.

4. INTRACELLULAR Na^+ AND BLADDER CONTRACTILITY

The inhibition of Na^+-K^+-ATPase activity results in an increase in the intracellular Na^+-concentration. In normal bladder strips, elevation of intracellular Na^+ with the ionophore, monensin, inhibited carbachol and KCl induced contractions (Figure 6). Monensin-induced inhibition of bladder contraction was similar to that observed with ouabain. Furthermore, ouabain and monensin increased the intracellular Na^+-concentration in bladder strips as determined by ^{23}Na-NMR spectroscopy (unpublished data). This suggests that diabetes and ouabain may inhibit agonist-induced contraction of bladder smooth muscle by increasing intracellular Na^+. In further support of this notion, ouabain has been shown to increase the intracellular Na^+ concentration within minutes in cardiac myocytes leading to intracellular acidification and eventually a negative inotropic effect.[45] Akera et al. have also demonstrated that 40% inhibition of Na^+-K^+-ATPase activity increases intracellular Na^+ significantly in cardiac tissue.[46] Furthermore, diabetes has been shown to increase the intracellular Na^+-concentration in rat proximal tubules.[47]

Since both ouabain and KCl-induced contractions are dependent on activation of voltage-sensitive L-type Ca^{2+}-channels, it seems likely that Na^+-pump inhibition and subsequent increases in intracellular Na^+ alter bladder contractility by diminishing Ca^{2+} influx via these channels. In support of this, elevation of cytoplasmic Na^+ in intestinal smooth muscle is suggested to diminish contractility partially by inhibition of Ca^{2+}-influx. Furthermore, stimulation of Ca^{2+}-efflux or sequestration into organelles as a result of inhibition of Na^+-K^+-ATPase activity and resultant increase in the intracellular Na^+ concentration may contribute to the lowering of cytoplasmic Ca^{2+}, as suggested in other smooth muscles.[8]

5. LACK OF EFFECTS OF MEMBRANE DEPOLARIZATION, ELEVATION OF INTRACELLULAR Ca^{2+}, AND K_{ATP} CHANNEL BLOCKER, GLIBENCLAMIDE, ON CARBACHOL-INDUCED CONTRACTION

As stated earlier, inhibition of Na^+-K^+-ATPase activity leads to depolarization of plasma membrane and increased influx of Ca^{2+} via voltage operated channels. Thus, the possibility exists that ouabain-induced reduction in carbachol contraction is a consequence of these phenomenon. However, such was not the case as effects of ouabain on carbachol-induced contraction were not mimicked upon pretreatment of normal bladder strips with

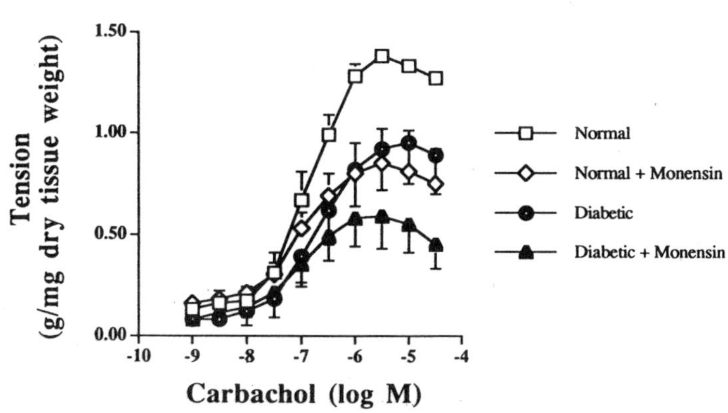

Figure 6. Effect of Na^+-ionophore monensin on carbachol-induced contraction in bladder strips from normal and diabetic rabbits. Bladder strips were pretreated with monensin (5 μM) before administering carbachol dose response. Results are mean ± SE of 4 experiments. (Figure 5 from Gupta et al., Am. J. Physiol. 1996).

KCl and/or Ca^{2+}-ionophore, A23187. Recently, it has also been suggested that carbachol causes contraction of bladder smooth muscle by inhibiting ATP activated K^+-channels (K_{ATP}), which are also thought to contribute to the maintenance of membrane potential.[48] However, K_{ATP} channel blocker, glibenclamide, did not affect carbachol-induced contraction of rabbit bladder, suggesting a lack of role for K_{ATP} channels in diabetes and ouabain-induced alteration in contractility of rabbit bladder.

6. ROLE OF Na^+-K^+-ATPase ACTIVITY IN BLADDER OVERACTIVITY DURING OUTLET OBSTRUCTION

The overactive bladder is thought to have a neurologic as well as myogenic basis. Na^+-K^+-ATPase activity may also play a critical role in the pathogenesis of detrusor instability as a result of bladder outlet obstruction. In this context, Mostwin and coworkers have demonstrated that detrusor Na^+-K^+-ATPase activity increases with partial urethral obstruction of guinea pig bladder. This increase in the membrane Na^+-K^+-ATPase activity returned to normal values following reversal of bladder outflow obstruction.[49,50] The mechanisms responsible for increased Na^+-K^+-ATPase activity after outlet obstruction are not clear. It is also not known if increased Na^+-K^+-ATPase activity plays a role in detrusor instability in humans.

7. CONCLUSIONS

Na^+-K^+-ATPase activity is an important modulator of detrusor contractility. Alterations in Na^+-K^+-ATPase activity may contribute to the development of bladder dysfunction in diabetes. Na^+-K^+-ATPase activity may also play a role in detrusor instability following bladder outlet obstruction.

REFERENCES

1. Sweadner, K.J. Isozymes of the Na+/K+-ATPase.: Biochim. Biophys. Acta. 988:185–220, 1989.
2. Blaustein M.P.: Sodium ions, calcium ions, blood pressure regulation and hypertension: a reassessment and a hypothesis. Am. J. Physiol., **232:** C165, 1977.
3. Kaplan, J.H.: Ion movements through the sodium pump. Ann. Rev. Physiol. **47:** 535, 1985.
4. Solinger, R.E., Gonzalez, C. F., Shamoo, Y. E., Wyssbrod, H. R. and W.A. Brodsky.: Effect of ouabain on ion transport mechanisms in the isolated turtle bladder. Am. J. Physiol., **215:** 249, 1968.
5. Seki, N., Karim, O. M. A. and J. L. Mostwin.: The effect of experimental urethral obstruction and its reversal on chages in passive electrical properites of detrusor muscle. J. Urol., **148:** 1957, 1992.
6. Lewis S.A.: The mammalian urinary bladder: Itís more than accomodating. N.I.P.S. **1:** 61, 1986.
7. Gupta, S., S. Yang, R.A. Cohen, R.J. Krane and I. Saenz de Tejada.: Altered Contractility Of Urinary Bladder In The Diabetic Rabbit: Relationship To Reduced Na^+-K^+-ATPase Activity. Am. J. Physiol., **271:** C2045, 1996.
8. van Breeman C., Aaronson, P. and R. Loutzenhiser.: Sodium-calcium interactions mammalian smooth muscle. Pharmacol. Rev. , **30:** 167, 1979.
9. Shuba M.F.: The effect of sodium-free and potassium-free solutions, ionic currentinhibitors and ouabain on electrophysiological properties of smooth muscle of guinea-pig ureter. J. Physiol., **264:** 837, 1977.
10. Griffin, J.D., Szaro, R. P. and J.K. Weltman.: Ouabain antagonism of smooth muscle. J. Pharmacol. Exp. Therap. , **182:** 378, 1972.
11. Buck, A.C., Reed, P. I., Siddiq, Y. K., Chisholm, G. D. and T. Russell Fraser.: Bladder dysfunction and neuropathy in diabetes. Diabetologia, **12:** 251, 1976.

12. Ellenberg, M.: Development of Urinary Bladder Dysfunction in Diabetes Mellitus. Ann. Intern. Med., **92:** 321, 1980.
13. Ioanid, C.P., Noica, N. and T. Pop.: Incidence and Diagnostic Aspects of the Bladder Disorders in Diabetics. Eur. Urol. ,**7:** 211, 1981.
14. Bradley, W.E.: Diagnosis of urinary bladder dysfunction in diabetes mellitus. Ann. Int. Med., **92:** 323, 1980.
15. Appell, R.A., Whiteside, H.V.: Diabetes and other Peripheral Neuropathies Affecting Lower Urinary Tract Function. In: Clinical Neurology, 2nd edition, Krane and Siroky, eds., 1991.
16. Kitami K.: Vesicourethral dysfunction of Diabetic Patients. Japanese J. of Urol., **82:** 1074, 1991.
17. Starer P. and L. Libow.: Cystometric evaluation of bladder dysfunction in elderly diabetic patients. Archives of Internal Medicine, **150:** 810, 1990.
18. Santiciolli, P., Gamse, R., Maggi, C. A. and A. Meli.: Cystometric changes in the early phase of streptozotocin-induced diabetes in rats: evidence for sensory changes not correlated to diabetic neuropathy. Naunyn-Schmiedebergís Arch. Pharmacol., **335:** 580, 1987
19. Longhurst, P. and J.A. Belis.: Abnormalities of rat bladder contractility in streptozotocin-induced diabetes mellitus. J. Pharmacol. Exp. Therap., **238:** 773, 1986.
20. Luheshi, G.N. and M.A. Zar.: Inhibitory effect of streptozotocin-induced diabetes on non-cholinergic motor transmissionin rat detrusor and its prevention by sorbinil. Br. J. Pharmacol., **101:** 411, 1990.
21. Moss, H.E., Lincoln, J. and G. Burnstock.: A study of bladder dysfunction during streptozotocin-induced diabetes in the rat using an in vitro whole bladder preparation. J. Urol., **138:** 1279, 1987.
22. Kolta, M.G., Wallace, L. J. and M.C. Gerald.: Streptozotocin-induced diabetes affects rat urinary bladder response to autonomic agents. Diabetes, **34:** 917, 1985.
23. Kudlacz, E.M., Gerald, M. C. and L.C. Wallace.: Effects of diabetes and diuresis on contraction and relaxation mechanisms in rat urinary bladder. Diabetes, **38:** 278, 1989.
24. Lincoln, J., Crockett, M., Haven A. J. and G. Burnstock.: Rat bladder in the early stages of strptozotocin-induced diabetes:adrenergic and cholinergic innervation. Diabetologia, **26:** 81, 1984.
25. Longhurst, P.: Urinary bladder function 6 months after the onset of diabetes in the spontaneously diabetic BB rat. J. Urol., **145:** 417, 1991.
26. Malmgren A., Andersson, P. O. and B. Uvelius.: Bladder function in rats with short- and long-term diabetes; effects of age and muscarinic blockade. J. Urol., **142:** 1608, 1989.
27. Uvelius B.: Detrusor smooth muscle in rats with alloxan-induced diabetes. J. Urol., **136:** 949, 1986.
28. Latifpour, J., Gousse, A., Kondo, S., Morita , T. and R.M. Weiss.: Effects of experimental diabetes on biochemical and functional characteristics of bladder muscarinic receptors. J. Pharmacol. Exp. Therap., **248:** 81, 1989.
29. Belis, J.A., Curley, R. M., Wagner, C. H., Murty, V. N., Winter, S. J. and T.J. Rohner.: Neurogenic function of the diabetic rat bladder: alteration by calcium channel effectors. Pharmacology, **43:** 273, 1991.
30. Hashitani H. and H. Suzuki.: Altered electrical properties of bladder smooth muscle in streptozotocin-induced diabetic rats. Br. J. Urol., **77:** 798, 1996.
31. Kjeldsen, K., Braendgaard, H., Sidenius, P., Larsen, J. S. and A. Norgaard.: Diabetes decreases Na^+-K^+-pump concentration in skeletal muscles, heart ventricular muscle, and peripheral nerves of rat. Diabetes, **36:** 842, 1987.
32. Tesfamariam, B., Gupta, S., Oates, P. J., Ruderman, N. B. and R.A. Cohen.: Reduced Na^+-K^+- ATPase activity in diabetic rabbit carotid artery: Effects of aldose reductase inhibition. Am. J. Physiol., **265:** H1189, 1993.
33. Gupta, S., Sussman, I., McArthur, C. S., Tornheim, K., Cohen, R. A. and N.B. Ruderman.: Endothelium-dependent inhibition of Na^+-K^+-ATPase activity in rabbit aorta by hyperglycemia. Possible role of endothelium-derived nitric oxide. J. Clin. Invest., **90:** 727, 1992.
34. Winegrad, A.I.: Does a common mechanism induce the diverse complications of diabetes? Diabetes. **36:** 396, 1987.
35. Ohara, T., Sussman, K. E. and B. Draznin.: Effect of diabetes on cytosolic free Ca^{2+} and Na^+-K^+-ATPase in rat aorta. Diabetes **40:** 1560, 1991.
36. Nishida K., T. Ohara, J. Johnson, J.S. Wallner, J. Wilk, N.Sherman, K. Kawakami, K.E. Sussman, and B. Draznin.: Na^+/K^+- ATPase activity and its αII subunit gene expression in rat skeletal muscle: Influence of diabetes, fasting and refeeding. Metabolism, **41:** 56, 1992.
37. Vague P., D. Dufayet, T. Coste, C. Morsicot, M.F. Jannot and D. Raccah.: Association of diabetic neuropathy with Na^+/K^+ ATPase gene polymorphism. Diabetologia, **40:** 506, 1997.
38. Ishii H. Jirousek M.R. Koya D. Takagi C. Xia P. Clermont A. Bursell SE. Kern TS. Ballas LM. Heath WF. Stramm LE. Feener EP. King GL. Amelioration of vascular dysfunctions in diabetic rats by an oral PKC beta inhibitor. Science, **272:** 728, 1996.

39. Borghini, I. Geering, K., Gjinovci, A., Wollheim, C. B. and W. F. Pralong.: In vivo phosphorylation of the Na, K-ATPase α subunit in sciatic nerves of control and diabetic rats: Effects of protein kinase modulators. Proc. Natl. Acad. Sci., USA., **91:** 6211, 1994.
40. Ruderman, N.B., S. Gupta and I. Sussman.: Hyperglycemia, Diabetes and Vascular Disease: An Overview. Editors N.B. Ruderman, Joseph Williamson and Michael Brownlee. Oxford University Press. pp 3, 1992.
41. Williamson, J.R., Kilo, C., Tilton, R.G. et al.: Hyperglycemic psuedohypoxia and diabetic complications. Diabetes, **42:** 801, 1993.
42. Greene D.A., Lattimer, S. A. and A. A. F. Sima.: Sorbitol, phosphoinositides and sodium-potassium-ATPase in the pathogenesis of diabetic complications. New Engl J Med., **316:** 599, 1987.
43. Tesfamariam B. Palacino J.J. Weisbrod RM. Cohen RA. Aldose reductase inhibition restores endothelial cell function in diabetic rabbit aorta. J. Cardiovasc. Pharmacol. **21:** 205, 1993.
44. Tesfamariam B. Gupta S. Oates P.J. Ruderman NB. Cohen RA.: Reduced $Na(^{+})$-K^{+} pump activity in diabetic rabbit carotid artery: reversal by aldose reductase inhibition. Am. J. Phys.,. **265:** H1189, 1993.
45. Kim, D., Cragoe Jr., E. J. and T.W. Smith.: Relations among sodium pump inhibition, Na^{+}-Ca^{2+} and Na^{+}-H^{+} exchange activities, and Ca^{2+}-H^{+} interaction in cultured chick heart cells. Circulation Research, **60:** 185, 1987.
46. Akera, T., Bennet, R. T., Olgaard, M. K. and T.M. Brody.: Cardiac Na,K-adenosine triphosphatase inhibition by ouabain and myocardial sodium: a computer simulation. J. Pharmacol. Exp. Therap., **199:** 287, 1976.
47. Kumar, A. K., Gupta, R. A. and A. Spitzer.: Intracellular sodium in proximal tubules of diabetic rats. Role of glucose. Kidney Int., **33:** 792, 1988.
48. Bonev, A.D. and M.T. Nelson.: Muscarinic inhibition of ATP-sensitive K^{+}-channels by protein kinase C in urinary bladder smooth muscle. Am. J. Physiol., **265:** C1723, 1993.
49. Seki, N., O.M.A. Karim and J.L. Mostwin.: Change in electrical properties of the smooth muscle membrane with bladder outflow obstruction. Am. J. Physiol., **262:** 1885, 1992.
50. Seki, N., O.M.A. Karim and J.L. Mostwin.: The effect of experimental urethral obstruction and its reversal on changes in pasive electrical properties of detrusor muscle. J. Urol. ,**148:** 1957, 1992.

24

RESTORATION OF BLADDER FUNCTION IN SPASTIC NEROPATHIC BLADDER USING SACRAL DEAFFERENTATION AND DIFFERENT TECHNIQUES OF NEUROSTIMULATION

Stefan Schumacher, Stephan Bross, Jeroen R. Scheepe, Peter Alken, and Klaus-Peter Jünemann

Neuro-Urological Laboratory
Department of Urology
Klinikum Mannheim
University of Heidelberg
Mannheim, Germany

1. ABSTRACT

Purpose. Conventional sacral anterior root stimulation (SARS) results in simultaneous activation of both the detrusor mucle and the external urethral sphincter. We evaluated the possibilities of different neurostimulation techniques to overcome stimulation induced detrusor-sphincter-dyssynergia and to achieve a physiological voiding.

Material and Methods. The literature was reviewed on different techniques of sacral anterior root stimulation of the bladder and the significance of posterior rhizotomy in patients with supraconal spinal cord injury suffering from the loss of voluntary bladder control, detrusor hyperreflexia and sphincter spasm.

Results. The achievement of selective detrusor activation would improve current sacral neurostimulation of the bladder, including the principle of "poststimulus voiding". This is possible with the application of selective neurostimulation in techniques of anodal block, high frequency block, depolarizing prepulses and cold block. Nowadays, sacral deafferentation is a standard therapy in combination with neurostimulation of the bladder because in conclusion advantages of complete rhizotomy predominate.

Conclusions. The combination of sacral anterior root stimulation and sacral deafferentation is a successful procedure for restoration of bladder function in patients with supraconal spinal cord injury. Anodal block technique and cryotechnique are excellent

Advances in Bladder Research, edited by Baskin and Hayward.
Kluwer Academic / Plenum Publishers, New York, 1999.

methods for selective bladder activation to avoid detrusor-sphincter-dyssynergia and thus improve stimulation induced voiding.

2. INTRODUCTION

Patients suffering from spinal cord injury above the second sacral spinal cord segment (S2), corresponding to a Th 12 / L 1 skeleteal spine level, develop a spastic neuropathic bladder after the initial stage of spinal shock syndrome. The neuronal activity within the reflex arches at the level of the spinal cord is pathologically increased because of abolished inhibitory influence of upper neurons on the sacral micturition center. The consequence is hyperreflexia of both bladder and urethral sphincter resulting in uninhibited contractions of the detrusor and spasticity of the pelvic floor musculature. Patient incontinence (reflex incontinence) with damage to the upper urinary tract that can lead to chronic renal insufficiency is a common outcome. Therefore, the restoration of almost physiological bladder function is of major importance in these patients, not only for quality of life but also mainly for overall prognosis and life expectancy. Anticholinergic therapy is often inadequate. Intermittent catheterization frequently becomes necessary because of inadequate bladder emptying. Nowadays, restoration of bladder function is accomplished by performing complete sacral deafferentation together with the implantation of a neurostimulator for sacral anterior root stimulation (SARS).

3. TECHNIQUE OF SACRAL ANTERIOR ROOT STIMULATION

The best clinical outcome with sacral anterior root stimulation was achieved with the Brindley[1,2,3] and Tanagho[4] neurostimulation techniques.

The electrodes are usually placed intradurally with the Brindley technique.[1] Visual identification and identification by test stimulation of the anterior and posterior roots of the sacral segments S2 to S4/S5 follows lumbosacral laminectomy and opening of the dural space. Anterior and posterior roots are separated intradurally. The anterior roots with stimulation response of urinary bladder and rectum are placed into the Brindley electrode. An extradural technique is performed if it is not possible to separate the anterior and posterior roots (e.g. severe arachnoiditis).[5] Complete sacral deafferentation is important for the management of detrusor hyperreflexia with consequent restoration of bladder storage function.[6]

Subsequent to laminectomy of L5 to S4, cuff electrodes are placed extradurally with the Tanago technique.[4,7] Deafferentation is extra[8]- or intradural.[9]

In both techniques, the stimulator implant is located in the lower abdominal wall and is connected with the cables of the electrodes. The energy transfer to the implant is via electromagnetic induction. Electrical signals are configurated in the transmitter and transferred to the reciever as high frequency impulses. Here, these impulses are transformed into electrical signals again and stimulation is carried out.

4. STIMULATION INDUCED DETRUSOR-SPHINCTER-DYSSYNERGIA

The anterior roots of the sacral segments S2 to S4/5 consist of nonmyelinated fibers to the bladder and myelinated fibers to the external urethral sphincter.[10,11] Large fibers are recruited at lower current strength than small fibers. For this reason simultaneous contractions of bladder and sphincter are observed during conventional neurostimulation of the bladder. A clinical consequence is detrusor-sphincter-dyssynergia and related complications such as increase of intravesical pressure, residual urine and recurrent urinary tract infection. To reduce detrusor-sphincter-dyssynergia a special stimulation pattern, so-called burst stimulation with a series of defined stimulation impulses and pause intervals is performed.[12] During this stimulation with interrupted pulse trains (typical 3–6s on and 6–9s off in humans) voiding is achieved between the pulse trains due to sustained high intravesical pressure (the urethral sphincter relaxes faster than the detrusor). This micturition pattern is called "poststimulus voiding".[13]

In addition, using the Tanagho stimulation technique, the infravesical outflow obstruction can be reduced by bilateral neurotomy of the pudendal nerve or sphincterotomy.[14] Furthermore selective intradural neurotomy of the large fibers of the sacral anterior roots innervating the urethral sphincter is described.[15,16]

5. SACRAL DEAFFERENTATION

In patients with supraconal spinal cord injury, detrusor hyperreflexia and sphincter spasm are major problems to manage. Anticholinergic therapy is often inadeqate. Intermittent catheterization becomes necessary for complete bladder emptying. Sacral anterior root stimulation without deafferentation of the posterior roots leads to hyperreflexia of the bladder and additional contraction of the bladder neck.[17,18] Therefore sacral posterior rhizotomy combined with sacral anterior root stimulation in the Brindley or Tanagho technique is still successfully employed for restoration of voiding function.[6,19] The rhizotomy of the posterior roots S2 to S4/5 is easier performed intradurally. In most patients, viscero-motoric spasticity and hyperreflexia of the bladder are abolished after perfoming complete deafferentation.[6] Results of this technique are complete or almost complete continence and an increase in maximum bladder capacity. Protection of the upper urinary tract is achieved by normalization of bladder compliance. Autonomic dysregulation (e.g. blood pressure crisis or headache) is abolished.

The disadvantages of deafferentation are the loss of reflexerection and reflexejaculation. On the other hand, with the neurostimulator many patients achieve an erection sufficient enough for sexual intercourse. Preoperative evaluation is possible by percutaneous test stimulation.[20] Another option for sexual intercourse is intracavernous injection therapy. Transrectal electrostimulation[21] or the implantation of a hypogastric plexus stimulator[22] is possible to obtain semen after loss of reflexejaculation. Another disadvantage of deafferentation is that the bowel or bladder can not longer be emptied with reflexive trigger mechanisms if the neurostimulator fails to function. The loss of reflexive activation of the bladder neck may lead to incontinence. In patients with incomplete supraconal lesion, rhizotomy of the posterior roots results in the complete loss of sensibility. On the other side, complete deafferentation is generally required for pain-free neurostimulation. In conclusion, the advantages of complete deafferentation predominate making this technique the present standard therapy.[23]

6. SELECTIVE NEUROSTIMULATION OF THE BLADDER

Previous experimental studies were performed to optimize stimulation technique and achieve selective activation of the bladder and low pressure voiding. Some stimulation techniques allow activation of small nonmylinated fibers without coactivation of large myelinated fibers.

In former animal studies, almost selective stimulation of the bladder without or with little accompanying activation of the urethral sphincter was achieved with the anodal block technique. This technique takes advantage of the fact that close to an anodal contact the propagation of an action potential can be blocked due to hyperpolarization of the nerve membrane. If the membrane is sufficiently hyperpolarized, action potentials cannot pass the hyperpolarized zone and are annihilated. As large fibers blocking requires a smaller stimulus than small fibers blocking, a selective block of the large fibers is possible.[24,25,26] Thus, selective activation of small fibers can be achieved by a combination of exciting both large and small fibers and by blocking, distal to the excitation point, the propagation of the induced action potentials in the large fibers. Anodal block was achieved by the application of monophasic rectangular pulses[27] or quasitrapezoidal stimulation pattern.[25,28–30] Three types of electrodes can be used to obtain selective activation of small nerve fibers: monopolar,[31,32] bipolar[33] and tripolar[25,26,30,34] electrodes. To prevent leakage of stimulus current to neighboring nerve roots, tripolar electrodes are preferred (cuff electrodes[27,29] or modified Brindley electrodes[30]). Acute animal studies showed, that anodal blocking reduced the stimulus-induced intraurethral pressure increase by more than 80%.[30,34] It has been demonstrated that the anodal block technique can also be used in humans.[35]

Alternatively, selective activation of small nerve fibers is possible using selective high-frequency block. This technique is described with the use of two electrodes (one for blocking and one for stimulation)[36] or a single tripolar electrode (driven by two different stimulators to create a pulse train with small- and high-amplitude pulses)[37,38] in peripheral nerves. Results of this technique in sacral anterior roots stimulation showed the feasibility of a normal detrusor contraction with little activation of the external urethral sphincter in acute canine experiments.[39] A small amplitude stimulus at a pulse rate of 550Hz and a larger stimulus at a pulse rate of 33Hz were utilized.

Selective stimulation of small nerve fibers can also be achieved with depolarizing prepulses. The nerve fiber membrane can be made less excitable by preceding the stimulus pulse with a subthreshold depolarizing prepulse.[40] The prepulse elicits a larger depolarization in large fibers than in small ones. Thus, the large myelinated fibers become less excitable than the small nonmyelinated fibers. The electrical stimulus following the prepulse predominately activates the small fibers.[41] Computer stimulations indicated that, with stepped prepulses, selective activation of small nerve fibers is possible.

Sacral anterior root stimulation with sinusoidal pulses (8–20Hz) is possible for selective bladder activation in a canine model.[23,42] Sinusoidal stimualtion allows selective activation of a certain range of fiber diameters as it contains only a single frequency. Using a defined frequency range stimulation results in a detrusor contraction without simultaneous contraction of the external urethral sphincter.

A new procedure (cryotechnique) has been developed for selective block of nerve fibers.[43–45] The neurophysiological blocking effect of cold on the nervous sytem is well known. Local cooling alters the membrane permeability for specific ions, such as Na^+, K^+ and Ca^{++}. The velocity at which impulses were conducted in individual axons through the cooled length of nerve slowed progressively as the temperature declined. Critical minimum conduction velocities were reached at higher temperatures in large meylinated axons rather than in small nonmyelinated ones. Peripheral myelinated fibers block at mean 7.3°C. The average of blocked nonmyelinated nerve fibers is found at 2.7°C.[46,47] The urodynamic effects of cryotechnology on sacral anterior root stimulation of the urinary bladder and the urethral sphincter were investigated in male foxhounds. Conventional neurostimulation of the spinal nerves of S2 was performed using a modified Brindley electrode. Simultaneously, the accompanying spinal nerves were cooled down in steps using a patented cryothermode. Subsequently, unilateral sacral anterior root stimulation of S2 was carried out while the ipsilateral pudendal nerves were cooled down in the same way. Cooling ranged from +25°C to a temperature initiating urethral sphincter block. During cooling of S2, a complete cold block of the urethral sphincter was recognized in all animals (100%). While the sphincter was completely blocked, the bladder was still activated.[43–45] Performing unilateral sacral anterior root stimulation of S2 and simultaneous cooling of the ipsilateral pudendal nerves, the urethral sphincter was blocked in 78%. The intravesical pressure remained unchanged before and during cold block[45] The cryoblock was completely reversible within seconds to minutes in both series.[43–45]

7. CONCLUSIONS

Sacral anterior root stimulation and sacral deafferentation are successful techniques for restoration of bladder function in patients suffering from spinal cord injury above the second sacral spinal cord segment. Although current stimulation techniques, including bladder emptying by "poststimulus voiding", are excellent procedures for patients, neurostimulation would be improved if selective activation of the bladder without simultaneous contraction of the external urethral sphincter could be achieved. This is possible by application of different selective stimulation techniques. Anodal block technique is feasible in animals and humans and also improves stimulation induced voiding. Alternatively, the cryotechnique is a new and excellent method for the selective blocking of nerve fibers. During conventional sacral anterior root stimulation of the bladder, complete and reversible block of the urethral sphincter can be achieved under preservation of bladder activity. These preliminary results must still become established in chronic animal trials and during intraoperative patient evaluation.

The future development of a modern selective bladder neurostimulator will be based on the anodal block technique or the cryotechnique.

REFERENCES

1. Brindley, G.S., Polkey, C.E. and Rushton, D.N.: Sacral anterior root stimulator for bladder control in paraplegia. Paraplegia, **20:** 365, 1982.
2. Brindley, G.S., Polkey, C.E., Rushton, D.N. and Cardozo, D.N.: Sacral anterior root stimulators for bladder control in paraplegia: The first 50 cases. J. Neurol. Neurosurg. Psychiatr., **49:** 1104, 1986.
3. Brindley, G.S. and Rushton, D.N.: Long-term follow-up of patients with sacral anterior root stimulator implants. Paraplegia, **28:** 469, 1990.
4. Tanagho, E.A. and Schmidt, R.A.: Electrical stimulation in the clinical management of the neurogenic bladder. J. Urol., **140:** 1331, 1988.
5. Sauerwein, D., Ingunza, W., Fischer, J., Madersbacher, H., Polkey, C.E., Brindley, G.S., Colombel, P. and Teddy, P.: Extradural implantation of sacral anterior root stimulators. J. Neurol. Neurosurg. Psychiatry, **53:** 681, 1990.
6. Sauerwein, D.: Die operative Behandlung der spastischen Blasenlähmung bei Querschnittlähmung: Sakrale Deafferentation (SDAF) mit der Implantation eines sakralen Vorderwurzelstimulators (SARS). Urologe A, **29:** 196, 1990.
7. Tanagho, E.A., Schmidt, R.A. and Orvis, B.R.: Neural stimulation for control of voiding dysfunction: A preliminary report of 22 patients with serious neuropathic voiding disorders. J. Urol., **142:** 340, 1989.
8. Tanagho, E.A.: Sacral root stimulation. Proceedings of the Satellite Workshop on Sacral Anterior Root Stimulation, ICS Meeting, Halifax, Nova Scotia, August 31, 1992. Neurourol. Urodyn., **12:** 505, 1993.

9. Hohenfellner, M., Thüroff, J.W., Schürmann, K., Schultz-Lampel, D., Schmidt, R.A. and Tanagho, E.A.: Sakrale Deafferentation und Implantation eines Blasenschrittmachers. Akt. Urol., **23:** I , 1992.
10. Jünemann, K.P., Lue, R.A., Schmidt, E.A. and Tanagho, E.A.: Clinical significance of sacral and pudendal nerve anatomy. J. Urol., **139:** 74, 1988.
11. Morrison, J.F.B.: Neural connections between the lower urinary tract and the spinal cord. In: The Physiology of the Lower Urinary Tract. Edited by M. Torrens and J.F.B. Morrison. New York: Springer, pp. 53, 1987.
12. Brindley, G.S.: An implant to empty the bladder or close the urethra. J. Neurol. Neurosurg. Psychiatry, **40:** 358, 1977.
13. Jonas, U., Tanagho, E.A.: Studies on the feasibility of urinary bladder evacuation by direct spinal cord stimulation. II. Poststimulus voiding: A way to overcome outflow resistance. Invest. Urol., **13:** 151, 1975.
14. Schmidt, R.A.: Applications of neurostimulation in urology. Neurourol. Urodyn., **7:** 585, 1988.
15. Hohenfellner, M., Paick, J.S., Trigo, R.F., Schmidt, R.A., Kaula, N.F., Thüroff, J.W. and Tanagho, E.A.: Site of deafferentation and electrode placement for bladder stimulation: clinical implications. J. Urol., **147:** 1665, 1992.
16. Hohenfellner, M., Trigo-Rocha, F., Paick, J.S., Gleason, C.G., Schmidt, R.A., Thüroff, J.W. and Tanagho, E.A.: Intradurale Deafferentation und superselektive ventrale Rhizotomie kombiniert mit extraduraler Stimulation: experimentelle Weiterentwicklung der sakralen Vorderwurzelstimulation. Akt. Urol., **24:** 357, 1993.
17. Schmidt, R.A., Brushini, H., Van Gool, J. and Tanagho, E.A.: Micturition and the male genitourinary response to sacral root stimulation. Invest. Urol., **17:** 125, 1979.
18. Talalla, A. and Bloom, J.W.: Sacral electrical stimulation for bladder control. In: Functional stimulation (Spinal Cord Dysfunction III). Edited by L. Illis. Oxford: Oxford University Press, pp. 206, 1992.
19. Schmidt, R.A. and Tanagho, E.A.: Neuromicturition. Extradural sacral nerve-root stimulation: guidlines for patient selection and implant techniques. World J. Urol., **9:** 114, 1991.
20. Brindley, G.S.: Future directions. Proceedings of the Satellite Workshop on Sacral Anterior Root Stimulation, ICS Meeting, Halifax, Nova Scotia, August 31, 1992. Neurourol. Urodyn., **12:** 519, 1993.
21. Brindley, G.S.: Electroejaculation: ist technique,, neurological implications and uses. J. Neurol. Neurosurg. Psychiatry, **44:** 9, 1981
22. Brindley, G.S., Sauerwein, D. and Henry, W.F.: Hypogastric plexus stimulators for obtaining semen from paraplegic men. Br. J. Urol., **64:** 72, 1989.
23. Wipfler, G., Jünemann, K.P.: Funktionelle Wiederherstellung der Blasenfunktion bei spastischer Querschnittslähmung mittels Elektrostimulation und sakraler Deafferentation. Akt. Urol., **26:** 14, 1995.
24. Accornero, N., Bini, G., Lenzi, G.L. and Manfredi, M.: Selective Activation of peripheral nerve fibre groups of different diameter by triangular shaped stimulus pulses. J. Physiol. Lond., **273:** 539, 1977.
25. Fang, Z.P. and Mortimer, J.T.: Selective activation of small motor axons by quasitrapezoidal current pulses. IEEE Trans. Biomed. Eng., **38:** 168, 1991.
26. Rijkhoff, N.J.M., Holsheimer, J., Koldewijn, E.L., Struijk, J.J., van Kerrebroeck, P.E.V., Debruyne, F.M.J. and Wijkstra, H.: Selective stimulation of sacral nerve roots for bladder control: a study by computer modelling: IEEE Trans. Biomed. Eng., 42: 413, 1994.
27. Koldewijn, E.L., Rijkhoff, N.J.M., Van Kerrebroeck, P.E.V., Debruyne, F.M.J., Wijkstra, H.: Selective stimulation of sacral nerves for bladder control in an animal model. J. Urol. 149: 357A, 1993.
28. Rijkhoff, N.J.M., Hendrikx, L.B.P.M., Debrunye, F.M.J., Wijkstra, H.: Electrical stimulation of the ventral sacral nerve roots: selective activtion in patients. Proceedings of the International Continence Society, 25th Annual Meeting, Sydeny, Australia. J. Neurourol. Urodyn. 14: 507,1995.
29. Wipfler, G., Grünewald, V., Bhadra, N., Creasy, G.H., Mortimer, J.T.: Selektive Aktivierung der Blase mittels Sakralwurzelstimulation unter Verwendung quasitrapezoidaler elektrischer Impulse beim Hund. Akt. Urol. 26: 13, (Sonderheft), 1995.
30. Wipfler, G., Scheepe, J.R., Seif, C., Gropp, A., Berle, B., Zendler, S., Jünemann, K.P., Alken, P.: Selektive sakrale Vorderwurzelstimulation mittels Anodenblocktechnik unter Verwendung einer modifizierten Brindley-Elektrode beim Hund. Akt. Urol. 28: 164, 1997.
31. Ungar, I.J., Mortimer, J.T. and Sweeny, J.D.: Generation of unidirectionally propagating action potentials using a monopolar electrode cuff. Ann. Biomed. Engin., **14:** 437, 1986.
32. Rijkhoff, N.J.M., Holsheimer, J., Debruyne, F.M.J. and Wijkstra, H.: Modeling selective activation of small myelinated nerve fibers using monopolar-point electrode. Med. Biol. Eng. Comput., **33:** 762, 1995.
33. Sweeny, J.D. and Mortimer, J.T.: An asymmetric two-electrode cuff for generation of unidirectionally propagated action potentials. IEEE Trans. Biomed. Eng., **33:** 541, 1986.

34. Rijkhoff, N.J.M., Koldewijn, E.L., van Kerrebroeck, P.E.V., Debruyne, F.M.J. and Wijkstra, H.: Acute animal studies on the use of an anodal block to reduce urethral resistance in sacral root stimulation. IEEE Trans. Rehab. Eng., **2**:92, 1994.
35. Rijkhoff, N.J.M., Hendrikx, L.B.P.M., Debruyne, F.M.J. and Wijkstra, H.: Electrical stimulation of the ventral sacral nerve roots: selective detrusor activation in patients. Neurourol. Urodyn., **14:** 506, 1995.
36. Solomonow, M.: External control of the neuromusculature system. IEEE Trans. Biomed. Eng., **31:** 752, 1984.
37. Baretta, R., Ichie, M., Hwang, S., Solomonow, M.: Method for studying muscle properties under orderly stimulated motor units with tripolar nerve cuff electrode. J. Biomed. Eng., **11:** 141, 1989.
38. Baretta, R., Ichie, M., Hwang, S., Solomonow, M.: Orderly stimulation of skeletal muscle motorunits with tripolar nerve cuff electrode. IEEE Trans. Biomed. Eng., 36: 836, 1989.
39. Zhang, T., Jiang, D.: Selective stimulation in a nerve trunk and its application in urology. Proc. 9th Annu. Int. Conf. IEEE-EMBS, Boston, USA: 1040, 1987.
40. Sassen, M. and Zimmermann, M.: Differential blocking of myelinated nerve fibers by transient depolarizing. Pflüg. Arch., **341:** 179, 1973.
41. Grill, W.M. and Mortimer, J.T.: Selective activation of distant nerve fibers. Proc. 15th Annu. Int. conf. IEEE-EMBS, San Diego, USA: 1249, 1993.
42. Jünemann, K.P., Wipfler, G., Leitzig, R., Persson-Jünemann, C., Zwick, A., Abel, K. and Alken, P.: The influence of signal sharps on selectivity and efficacy in sacral electrostimulation for emptying the canine bladder. Proc. 11th Congr. EAU, Berlin, Germany: 289, 1994.
43. Schumacher, S., Scheepe, J.R., Bross, S., Zendler, S., Seif, C., Jünemann K.P., Alken, P.: The urodynamic effects of spinal nerve cold block on the neurostimulation of external urethral sphincter and urinary bladder. Proc. 36th Annual Scientific Meeting of IMSOP, Innsbruck, Austria: 69, 1997.
44. Schumacher, S., Bross, S., Scheepe, J.R., Wipfler, G., Zendler, S., Jünemann, K.P., Alken, P.: Selective Block of urethral sphincter contractions in sacral anterior root stimulation. Proc. 2nd Annual Conference of the International Functional Electrical Stimulation Society (IFESS'97) and Neural Prostheses: Motor Prostheses V (NP'97), Burnaby, Canada: 43, 1997.
45. Schumacher, S., Seif, C., Bross, S., Jünemann, K.-P.: Urodynamische Effekte unterschiedlicher extraduraler Kryoblockformen auf die Neurostimulation der Harnblase und des urethralen Sphinkters. Urologe A, Suppl. 1/97: 57, 1997.
46. Franz, D.N. and Iggo, A.: Conduction failure in myelinated and non-myelinated axons at low temperatures. J. Physiol. (Lond)., **199:** 319, 1968.
47. Brooks, V.B.: Study of brain function by local, reversible cooling. Rev. Physiol. Biochem. Pharmacol., **95:** 1, 1983.

DISCUSSION

Nerves and Pharmacology

Q & A with William C. de Groat, Ph.D.

Saso Jezermik, Denmark.

Q: Did you study the modulation of this synaptic transmission, the one originating from the pelvic afferents? Do you think there is similar modulation in another pathway?

A: The question is, have we studied the reflexes that are activated by pelvic afferents? That is, are the ones that are coming from the bladder. Those are the ones that presumably would be activated by cold stimulation of the bladder. As far as we can tell in animals, all of the reflex mechanisms in the normal neonate and normal adult, that are driven by afferents from the bladder, go through the brain. All of them are supraspinal. And only C-spinal mechanisms emerge after the spinal cord is damaged. We don't have studies, however, on the cold-evoked reflexes. Those could be organized in different ways, but as far as tension receptor afferents, they all seem to trigger reflexes through the brain, not through the spinal cord, except after spinal cord injury.

Q: In the spinal cord injury, the pudendal nerve stimulation, would it cause inhibition of the bladder or activation/excitation?

A: We've done studies on this in cats. After spinal cord injury, pudendal afferent stimulation as electrical stimulation, or tactile stimulation of the perineum, produces very large bladder contractions. We presume that there are still inhibitory responses, but these seem to be submerged or dominated by the excitatory pathway. So, essentially, the pudendal system is capable of producing inhibition or excitation of bladder reflexes. Those effects are dependent upon age of the animal and the intactness of the nervous system. So, in young animals, or in spinal cord injured animals, the predominant effect is excitation. In animals that are mature and have an intact nervous system, the predominant effect is inhibition.

John Park, Boston, MA.

Q: In the last two years I have been studying prostaglandin regulation in the lower urinary tract, and learned something about capsaicin and sensitive C-fibers. Where do capsaicin-sensitive fibers fit in this picture of postnatal maturation involving interneurons?

A: The capsaicin-sensitive C-fibers are the ones that respond to cold stimulation, that's one. Also, capsaicin-sensitive C-fibers in the cat change their properties after spinal cord injury, and they become more excitable and able to respond to mechanical stimulation. So one of the mechanisms for the emergence of spinal reflexes is due to a change, or is mediated by a change in the afferent properties. So, not only does the central nervous system change its properties, but the peripheral nervous system, the peripheral afferents can also change their properties. I should mention that Naoki Yoshimura has found, in the rat, after spinal cord injury, that bladder afferents change their sodium channels, they change their potassium channels, they change their sensitivity to capsaicin, they change their expression of certain types of chemicals like neurofilament proteins. So many things can change after you injure or alter one part of the reflex pathway, and I think that we probably will find that the smooth muscle will change (I think that question came up yesterday), we found that bladder hypertrophy will occur after spinal cord injury, neuronal hypertrophy will occur (Bill Steers will tell us a little bit about that), and actually, synaptic transmission in the bladder from postganglionic neurons to the bladder smooth muscle can change. So I think that when we consider there are diseases that occur in one part of the system, those diseases now can influence all parts of the micturition reflex pathway.

Claire Close, Las Vegas.

Q: What do you think is stimulating the rats to void before birth?

A: We talked about that yesterday. A number of people have been interested in this. The issue is that you may simply have voiding through the urachus or voiding induced by the intrinsic properties of the bladder smooth muscle, that is that there may be a low resistance outflow and the bladder smooth muscle itself is able to trigger some voiding response. Other people, a number of years ago proposed that the peripheral plexus may mediate voiding responses in the fetus. The fourth possibility is that there may be a supraspinal pathway in the neonate, which at birth is turned off, and then turned on three weeks later. So it would be interesting to see whether the reflex that's present in the human infant, the reflex that's present in the rat after three weeks, may be there in the fetus, but after birth is turned off for a period of a few weeks to allow the mom to control voiding. So I think there are a number of different possibilities.

Q & A with Karl-Eric Andersson, M.D.

Unidentified Questioner.

Q: Have you looked at adenosine or adenine as the possible relaxation agents?

A: We have used 8-phenyl theophylline as an inhibitor of adenosine receptors. It doesn't affect the relaxation. That's the only drug we have used for inhibition of that.

William C. de Groat, Pittsburgh.

Q: On the record that you showed with urothelium removal, it seemed like the inhibition was somewhat less. You said you could still get it, but it was less. Does that imply that there may be a contribution of substances released from the urothelium?

A: We believe so.

Comment by Dr. de Groat: Because people in Pittsburgh, in my laboratory have shown that carbachol can release nitric oxide from the urothelium of the rat bladder, and probably, nitric oxide can be released by a number of different adrenergic stimuli, and may be released as well by stretch. So ATP may come out of urothelium, and nitric oxide may come out of urothelium, and there may be other unknown substances as well. And then we know that the afferent nerves in some cases are located just beneath the epithelium, so substances released from the epithelial cells may have an impact on the sensitivity of those afferent nerves, particularly the C-5 or afferents which are present there. So, you imply that maybe nitric oxide and other agents could affect afferent firing, and just as a final point, we have evidence that nitric oxide can alter calcium channels in bladder afferents. So, I think that's another potential interaction or site of action of NO in the afferent system, as well as the smooth muscle system.

A: If anything could fit with the mechanism for relaxation, it would be an endogenous release of the calcium antagonist or an calcium antagonist. But, I agree that most probably, the urothelium can contribute to this relaxation we see with carbachol, but it is not necessary for normal complete relaxation.

Penny Longhurst, Albany, New York.

Q: Have you looked at transduction mechanisms? Cyclic A or cyclic GMP?

A: We have done analysis of cyclic nucleotides. We see no changes. Maybe there are changes. You know that there are variations in the determinations of cyclic A and B, and cyclic GMP. To be completely certain that there are not small changes in this, you have to make so many experiments to have the power to detect those changes, so you have to make 30 to 50 experiments actually. But there are no dramatic changes in cyclic nucleotide levels associated with the relaxation.

Sharateg Derash, Bioscience Laboratory.

Q: I have a drug discovery standpoint question. As you pointed out earlier in your talk, the objective of drug therapy should be to reduce bladder activity without affecting the normal voiding process - if one develops a drug which directly relaxes the detrusor either by increasing cyclic A and B or cyclic GMP, wouldn't one inevitably inhibit the normal voiding process by functionally antagonizing the contraction?

A: You will do that. You will inhibit also the normal detrusor contraction. Really, there is a very high frequency of sympathetic discharge in association with bladder contraction. I think that will overcome, for example, the effect you can obtain with beta-receptor agonists or other drugs acting through cyclic nucleotides. That mechanism, the IP-3 mechanism is not very sensitive to relaxation by cyclic drugs.

Sandeep Gupta, Philadelphia.

Q: Have you considered use of sodium channel blockers in terms of inhibiting the relaxation of this unknown factor? Maybe sodium pump activation could be responsible?

A: I know that you are thinking of inhibition of sodium potassium ATPase as a possible mechanism, we haven't done those experiments.

Q: Another question is regarding the beta receptors. Do people on beta blockers, patients on beta blockers have different bladder function than normal?

A: No. Patients even with high doses of beta blockers, and I know that in the early 70s they tried to treat schizophrenia by high doses of propranolol, they give up to 5 g of propranolol to these people, and they voided without problems.

Q: So the role for beta receptors in over activity is questionable?

A: I don't think that the beta receptors have a crucial role for normal voiding.

Q & A with Osamu Nishizawa

Saso Jezermik, Denmark.

Q: How come your normal rats seem to have unstable bladders with spontaneous contractions and detrusor sphincter dyssynergia? Your data show that the normal rats also had this?

A: In the normal rats, there is an isovolumetric bladder contraction state. So we ligated at the bladder neck, so in this state, the bladder contraction developed, so this is not such an overactive state. Also, sphincter electromyographic finding seems too show worse activity, but that is not obstruction but a bladder contraction, with synchronous activity to empty bladder contents effectively. That is not a similar finding to that of the spinal rats. If we show with high-speed recording, that this is not detrusor sphincter dyssynergia.

William Steers, Charlottesville.

Q: Osamu, did you try the experiment of combining both agents? It would seem feasible that when glutamate acts, it may be acting at both receptors simultaneously or there may be an interaction. Have you thought of taking both the AMPA and NNDA antagonist together, and showing that at very low concentrations you abolish all activity?

A: We did not perform such a combination.

Q & A with Robert Levin, Ph.D.

Karl Eric Anderson, Sweden.

Q: I would like to ask you about the effects on nerves versus effects on muscle. You would expect that the nerves were affected before the smooth muscle. You showed some comparisons between nerve densities in obstructed and normal muscle. Did you consider, when you quantified that, that you had ten times bigger muscle cells? And did you quantify how many that were in relation to muscle cells? Because, just showing that you have in a certain section more synapses in the normal one than in the other, doesn't take into consideration that you have other morphological changes, that's one thing. Methodologically this is important.

A: I agree, it's extremely important. Certainly, nerve damage, we believe is the first indication of this function. The immunohistochemistry is a qualitative measure. We've used both quantitative measures of choline acetyltransferase, and that's per mg/protein or per g/tissue, and we show a marked decrease in the concentration of the enzyme which corresponds with our feeling of denervation. And that, would be of a more quantitative nature. On a functional level, it's really the response to field stimulation in relation to the response to bethanechol or KCl that we use as a functional indication of denervation, because bethanechol, and you certainly see with obstruction your first physiological change as a reduction in the response to field stimulation, with no changes in response to either KCl or bethanechol, so this would be another marker, I think for denervation. You're right, it's very difficult to get an exact quantification of nerves per unit muscle mass because of the hypertrophy of the smooth muscle.

Q: Allison Brady maintains that the denervation is patchy and that you can find perfectly normal areas in-between denervating. What is your experience in your model?

A: We've never looked for that, but I certainly wouldn't doubt that especially in the moderately decompensating bladder that it would be patchy nerves. We've never really looked at that per se, but I certainly wouldn't doubt that does exist.

Q: Do you think that Tadenan works primarily on nerves or primarily on muscles?

A: I think it has to work on nerves because the fact that it protects the response to field stimulation would indicate that. We've never looked at choline acetyltransferase with immunohistochemistry yet with an obstructed bladders in the rabbits. But I think it has to be able to work on nerves. If it didn't, then I don't think it would be very protective a function. Because if you get the denervation, then no matter what you do to the smooth muscle, you'll get marked dysfunctions in the in vivo bladder. And in vivo bladder function would probably progress in spite of protecting the membrane itself. I believe it does works on nerves, either primarily or in addition to muscle. We certainly know that the nerve membranes are more sensitive than muscle membranes to ischemic damage, or to damage due by hydrolytic enzymes.

Steve Zderic, Philadelphia.

Q: If Tadenan works, what about something just as simple as vitamin E? Maria dela Voria Popadopolous showed that if you look at brains that were rendered ischemic, iron pump activity was preserved in those animals that were pretreated with say vitamin E. Anything with a lot of double bonds in it. The other question is, then perhaps, what about using allopurinol, which also is a proposed free radical scrubbing agent? The other question would be, have you looked at the histology of these bladders after Tadenan treatment, and what do you see?

A: No we have not looked at the histology yet, only biochemistry and functional studies. The first question is about vitamin E. We don't know if Tadenan is acting as a free radical scavenger, antioxidant, or by protecting the membranes against ischemic damage. As a partial answer, Penny and I are setting up a project to actually look at vitamin E itself as a protective agent. I have a feeling what you're gonna see since this work's been published, I know there's quite a few other people with other models of obstruction that are looking at a wide variety of agents that currently are used in cardiovascular and CNS to protect against ischemic damage in the bladder. So I think over the next couple of years we're gonna get a lot of answers concerning the relevance of this hypothesis, and I should remain that this is a hypothesis at this time.

Sam Chacko, Philadelphia.

Q: If you think Tadenan has an affect on calcium activated calponin, for example, which break down a lot of the organelles or contractive proteins in the smooth muscle cells, it would be interesting to see whether Tadenan has any affect on inhibition of calponin, you can do that in a test tube. It would be interesting to do that. And also, some of the hydrolytic enzymes.

A: The real problem with Tadenan is it's only oil soluble, and it's composed of about 48 different compounds. So to be able to use this in an in vitro situation, it probably does not have relevance like you would if you had a purer compound because we don't know what the active ingredient is or how it's processed in the body. So, in cases where we've tried this, it hasn't worked at all because of the solubility problems and because of its chemical nature. So, unfortunately we cannot study Tadenan in the same way you'd be able to study a pure chemical compound in an in vitro situation, that's a problem in identifying the mechanism in the active ingredients.

Q: In your mechanism for decompensation, in response to this hypoxia and all those changes when the bladder undergoes decompensation, then is there also necrosis of smooth muscle cells? If there is, then smooth muscle is a tissue which has the ability to regenerate, then you should see a very high proliferative activity.

A: That's right. We want to look a lot more at the histological effects under different conditions, and we haven't done that yet.

Brad Kroop, Oklahoma City, Oklahoma.

Q: Follow up on Dr. Anderson's question. Have you looked at the circuit changes in association with denervation of the bladder alone and not with obstruction?

A: We believe the denervation occurs first before changes in the smooth muscle circa, just from some of the time-course experiments in mild obstruction where we can notice changes in response to field stimulation, we don't see necessarily changes in circa at that time. So we think the denervation occurs first and is a primary factor.

William C. de Groat, Pittsburgh.

Q: It would really be important to know what the neural dysfunction is, and whether its loss of axons of inputs to cells, or whether it's some other defect, for example, transmitter synthesis, storage and release. As Karl-Eric said, if you start with a muscle cell that's relatively small, and now that muscle cell becomes ten times larger but it receives the same number of synaptic contacts, you would certainly see a decline in the overall density of nerve terminals. So, if you do a quick calculation and you divide your results by the increased mass of muscle fibers, do you come up with somewhat the same distribution or density of terminals as you have here. In other words, Giorgio Gabella said that each muscle fiber in the rat bladder receives a neural input, so you could imagine that if a very large muscle fiber's still getting the same number of inputs, but there's some other defect in the way those inputs are activating the muscle.

A: We've never really looked at that, but it's a topic that has to be looked at, and that's the relationship between nerve density and hypertrophy. Cause the initial hypertrophy in a compensated bladder is largely muscle, yet the responses are the same, and the enzyme activities are the same. We haven't done choline acetyltransferase in that condition, but I would almost be willing to bet you it would be pretty much the same. But it's really a study that has to be done carefully.

Q & A with William Steers, M.D.

Unidentified Questioner.

Q: Have you looked at the effect of obstruction in these hypertensive animals?

A: Yes, we have. Katrina Pierce in Sweden did the pharmacological experiments. Looking at that, in fact, we are already at maximal rates of production, and one doesn't get any additional changes in the response in thresholds. Again, you get more changes in muscle by responses. And really, we're able to show dramatic changes. Hypertrophy of the neurons, once they were hypertrophied, was almost you might say maxed out. Urinary frequency did double.

Sharateg Derash, Bioscience Laboratory.

Q: You've shown a lot of interest in differences between the WKY and SHR. My question is, how much of these differences related to just pure hemodynamics? In other words, have you, for example, normalized the blood pressure on these animals with hypertensive drug and looked at the effects, or alternatively, have you taken a hypertensive model which is not dependent on the sympathetic system, like the renal hypertensive animal, and seen similar changes?

A: That's why we looked at the hyperactive strain rather than hypertensive, and we still saw the changes where there was no hypertension. Number two, is when you look at those due to hemodynamic changes, you would see a change in urine production which was not seen in these animals nor was the concentrating ability altered. So these animals it is not merely effect. One alternate hypothesis is, and it's quite a complex one, you might need to follow me here a minute, is visceral afferents from other organ systems could impact different organ systems. For example, if you have angina and pain from the heart, you can go into urinary retention, alter cystometrograms because of afferents coming in the cord, GI tract, etc. You have excitatory and inhibitory inputs, so you could postulate that if hypertension per se was altering any type of afferent input from the vasculature, it may feed back in the central nervous system and affect urinary frequency. And this is just feedback control through the brain stem. In fact, that's why it was so important to look at the hypertensive and hyperactive as well. So you took away at least the cardiovascular effect, and you still had the NGF upregulation and urinary frequency in those animals. So, I don't think it's merely a reflection of hemodynamics.

Unidentified Questioner.

Q: The vast majority of patients that we see with frequency, urgency are women, generally from 30 to 50. I don't know if any relationship between high blood pressures in those women, do you have any data on that? Can you suggest any?

A: Nobody has looked at that. Maybe periodic elevations. The only studies I know, are the studies for BPH. But, again, it's interesting that alpha blockers will help women as well. And intrathecal alpha blockers will help as much peripheral administration, suggesting that again, it may be the central effect of norenergic innervation as much as the peripheral that affect frequency and urgency. In fact, even in interstitial cystitis, people talk about clinical trials of giving alpha blockers. But, again, it's probably not in everybody, but there may be a certain genetic background where your response to your own system may by hyper, and that's why certain individuals with very mild obstruction have very severe symptoms. Where people, again with no obstruction can have symptoms. Again, we treat a lot of people the same way, realizing there's a whole spectrum of changes. Or other individuals that you unobstruct don't get any better. So again, it's very hard to explore the roles of the nervous system, and look at this.

William de Groat, Pittsburgh.

Q: The unstable activity that you see in the spontaneously hypertensive rat, is there evidence that that is generated by the central nervous system?

A: We've not done recordings on the nerves. The only evidence that may be true is the intrathecal alpha blockers. We've not done the alpha agonist experiments

Q: I thought Karl-Eric had a paper a couple of years ago where he measured the unstable activity with obstruction, and under those conditions came to the conclusion that that was due to intrinsic properties of the muscle, and not the central nervous system.

A: Again, we saw the changes in alpha adrenergic responses in the periphery. It just shows you can modulate that, at least centrally. I mean, it doesn't tell where the etiology is.

Q: The other point is that the sympathetic outflow to the bladder comes from L2 and L3, which is fairly close to L5 where the injections of doxazosin are made. It's very possible that when you make drug injections intrathecally, that it influences the sympathetic supply to the bladder. So, some of the changes that you see could be driven by an alteration of sympathetic outflow, and particularly in these animals where they're hyperinnervated by the adrenergic nerves.

A: Again, we haven't done the lesion experiments where you cut the chain and hypogastric nerve and shown that you abolish the effect of doxazosin which would be interesting.

John Park, Boston.

Q: You mentioned that the NGF regulation is quite complex. Is NGF something that's measurable, like in the blood that you can try to correlate the findings in the bladder?

A: NGF is measurable in the blood, although it's quite, again, complex. You have alpha macroglobulin binding. There's also debate on how you measure free versus total, and we've used a variety of methodologies to measure free NGF. But you'll see that we actually had NGF protein measurements in the bladder. But the secretion is upregulated as well in the cultured muscle. But you can get it in the blood, and I don't know if that's ever been looked at in the SHR. What is known in the SHR, is that if you exogenously give normal animals NGF, that they will get hypertensive. The other is during the first weeks of life, if you give NGF antibody to the SHR, you prevent all this. You prevent the neurgenic hyperinnervation, you prevent hypertension.

William de Groat, Pittsburgh.

Q: One of the mechanisms for inducing hyperactivity is activation of C fiber afferents, and we talked about that a little bit at this meeting. Probably, the hypertrophy of the afferents that occurs with obstruction and various models that you've used, is probably due in part to changes in the C fiber system. So, the question is, if you were to eliminate the C fiber afferent system in these animals, would that remove the hyperactivity? One way to do that, of course, is to treat with capsaicin.

A: We have done that acutely. Interestingly, we did not see very much change.

Q: Acutely probably wouldn't be very good favorable experiment.

A: We try and think of what's hyperactive, in terms of you give capsaicin intravesically, what happens to SHR versus apomorphine using a model Dr. Anderson developed in Sweden, showing that both systems, again, are up regulated equally. If you give a central stimulant to hyperactivity, apomorphine or capsaicin. So that there are capsaicin-sensitive afferents in these animals, and that you can induce hyperactivity, but we've not chronically treated, and that's actually one of the experiments we're planning, is to chronically treat these animals to see if we can abolish, as well as giving other types of NGF antagonists to abolish these effects.

Steve Zderic, Philadelphia.

Q: These bladders did not appear to undergo hypertrophy when you compared the SHRs to the wild type, and yet, with this much excess adrenergic innervation, you'd expect that the bladder base would be tighter during voiding. How can you reconcile that? Clearly, something else must be released at the same time.

A: Obviously a very fascinating topic about functional obstruction. We can't, without pressure flow studies rule out that there is a minor degree of obstruction in these animals, this is to quote Bob Levin's compensated obstruction. But even in that case, in all obstructed animals you get hypertrophy in a somewhat enlarged bladder. These animals, their bladders appear normal. So it's quite interesting. What is different about these animals? You have the hyperactivity, so experimentally, you've separated the hypertrophy from the hyperactivity, and that's why this animal's interesting. Because the only place we can show changes is innervation and nervous system, and you get the same obstruction without all the metabolic changes, ischemic changes, hypertrophy changes, etc., which in my mind raises, I mean you know that this nervous system will be very important for controlling urinary frequency. And NGF is up in both conditions.

Unidentified Questioner.

Q: I would also expect that as long as you don't have dyssynergia, when you void, basically you turn off the nor adrenergic system no matter how many receptors you have, and you have nitric oxide relaxation. But just because you have increase in alpha innervation of the urethra, you wouldn't necessarily think that that would be active during voiding.

A: Given that the bladder masses are not changing, my guess would be that their voiding pressures are normal. There's no residual urine in these animals. So again, it's an experiment of nature, the SHR animals. are useful for drug screening, and provide some insights if you upregulate noradrenergic nerves.

Saso Jezermik, Denmark.

Q: When you are talking about hypertrophy of the bladder afferents, do you have in mind the bladder afferents in the pelvic nerves or hypogastric? How does this relate to the hyperactivity of the bladder?

A: It's in both. We've not done the alkaline experiments that have been done in Chett's laboratory, where you patch clamp the DRG neurons, but that would be a very fascinating set up experiments in these animals. But when you look at L2, it's the afferents hypogastric, in the S1 it's hypogastrics in the pelvic, so they're both up, they're both again using cross-sectional areas as a measure of hypertrophy, they're both enlarged.

Q: How does this relate to the hyperactivity?

A: You can postulate that if it's similar to what happens in spinal cord injury and what may happen with obstruction, is you've lowered the threshold for firing for bladder afferents, or developed afferents that normally don't play a role, such as the C fiber afferents that Dr. DeGroat mentioned which are capsaicin sensitive and now they're active. So you can either have a different set of afferents operative or you can have normal tension afferents that actually have different channel properties. That again, is food for thought in different, new experiments.

William de Groat, Pittsburgh.

Q: Is there hyperinnervation of the blood vessels in the SHR ?

A: Yes. Same with the noradrenergic and the resistance vessels.

Q: In the bladder, how do we know that the hyperinnervation is directed at the bladder smooth muscle and not at the blood vessels in the bladder?

A: If you're basically looking at the vasculature, and you just see that the fibers are throughout the bladder not just with the blood vessels, just looking a microscope. I mean, it still could be fibers in passage, but I would expect a tremendous increase in blood vessels in the bladder. At least, when you look histologically you don't see that.

Q: It seems to me that if there is enough regulation of alpha excitatory mechanisms in the bladder muscle, then you should be able to demonstrate that with electrical field stimulation. If you were to block purinergic and cholinergic contractions, you should be left then with an adrenergically-mediated contraction that could be blocked by an alpha blocking agent. Have you tried that?

A: Those experiments have been done. I only showed you the experiments without drug electrophile stimulation, but Katrina Pierson did do that with scopolamine after alpha methyl ATP, and there was no change.

Section IV

INFECTION, IMMUNOLOGY, AND INTERSTITIAL CYSTITIS IN THE BLADDER

OVERVIEW OF INFECTION, IMMUNOLOGY, AND INTERSTITIAL CYSTITIS IN THE BLADDER

Timothy L. Ratliff

Andersen-Hebbeln Professor of Prostate Cancer Research
University of Iowa
Department of Urology
200 Hawkins Drive
Iowa City, Iowa 52242

The chapters in this section outline the basic biological principles of bladder infection and interstitial cystitis and address important issues related to inflammation and tumor therapy. The process of bacterial migration, retention and colonization are described. Many factors contribute to these processes including host-related factors, bacterial adhesins, and inflammation. The role of vaginal fluid, which may contain glycoproteins that contribute to adherence and colonization of bacteria, is discussed. This area is understudied and important issues are outlined. Further, the role of bacterial adhesins in the attachment of bacteria to mucosal surfaces is summarized. The contribution of bacterial fimbrae and non-fimbrial adhesins in attachment is delineated and the role of environment and its influence on bacterial attachment is linked to recurrent urinary tract infections. It is clear that bacterial factors alone are not sufficient to initiate an infection. Host factors such as aberrations in mucosal surfaces, induction of inflammatory mediators and the presence of secreted products all contribute in as yet ill defined mechanisms for the establishment of a bacterial infection in the bladder.

The clearance of infections is linked to inflammatory mediators. The potential contribution of nitric oxide and the induction of inhibitory cytokines in bacterial clearance are outlined. The discussion emphasizes the need to better understand the role of host factors in the elimination of infections.

Cancer biology and the potential for gene therapy are discussed. It is clear from current treatment regimens for bladder cancer that induction of inflammatory mediators, either in the form of a cellular response or through soluble cytokines, contributes to the elimination of tumor cells. Based on this established principle, the potential for biological

Advances in Bladder Research, edited by Baskin and Hayward.
Kluwer Academic / Plenum Publishers, New York, 1999.

response modifiers such as interferon as antitumor agents is presented. Such response modifiers either alone or in combination offer high potential for future therapies.

A common thread among all the topics is inflammation and immunity. This area of bladder biology is not well understood. It is clear that the bladder functions as a mucosal immune organ, but the nature of the bladder associated inflammatory responses induced by infection or the introduction of agents that activate inflammation for therapeutic purposes is not well described.

While we are gaining a better understanding of many bladder-linked processes, much work is needed to elucidate and connect the diverse yet associated disease states and therapeutic strategies.

25

PATHOPHYSIOLOGY OF BACTERIAL CYSTITIS

James A. Roberts

Tulane Regional Primate Research Center
Covington, Louisiana 70433, and
Tulane University School of Medicine
New Orleans, Louisiana

1. INTRODUCTION

Most bacterial urinary tract infections (UTI) occur by the ascending route and are due to bacteria that normally colonize the patient's gastrointestinal tract. While 80 to 90% of UTI are due to Escherichia coli [1], in young women a gram positive organism, Staphylococcus saprophyticus may cause up to 20% of the cases of acute cystitis.[2] Hospital-acquired infections associated with catheters or instrumentation, however, are more frequently due to bacterial species such as Enterobacter, Klebsiella, Proteus, or Pseudomonas. Colonization of both squamous and urothelial cells due to bacterial adhesion is necessary to initiate an ascending infection. After colonization of the perineum for various periods of time, colonization of the urethra, then bladder occurs by means of bacterial adherence. In this review I will concentrate on the pathogenesisof E. coli cystitis. Adherence by E. coli is due to P, X or type 1 fimbriae as well as non-fimbrial adhesins.[3,4] Cystitis may ensue once colonization of the bladder occurs and the bacteria remain for a time sufficient enough to reach a critical concentration necessary to cause disease. Twenty percent of females will have a urinary tract infection at some time during their lives, while twenty per cent of those who develop one infection will have a recurrence. Women may be susceptible due to genetic factors since it has been shown that those who do not have blood group antigens in their secretions (which may interfere with bacterial adhesion) more frequently have urinary tract infections.[5]

1.1. Bacterial Colonization

Urinary tract infection (UTI) does not necessarily follow gut colonization by bacteria present in the greatest number, but fecal bacteria are the source of bacteria that cause ascending UTI [6]. These bacteria that must colonize the perineum prior to causing an ascending UTI compete for their niche with normal flora of the vagina and perineal area[7].

Advances in Bladder Research, edited by Baskin and Hayward.
Kluwer Academic / Plenum Publishers, New York, 1999.

Colonization by means of bacterial adherence serves the theoretical purpose of allowing the organisms to resist currents such as urine and mucous that would wash them out of the urinary tract. However , using a recent mathematical simulation of bladder filling and emptying, it has been estimated that if bacterial doubling time was 50 minutes or less in urine, bacteria could maintain themselves within the urinary tract without adherence factors[8]. The E. coli strains studied were obtained from patients with UTI and had doubling times in urine of 50 to 60 minutes, while fecal strains not associated with UTI had much more slower doubling times. Thus the ability to grow rapidly may be almost as critical to maintain an infection as the ability to adhere. This may explain the persistence of bacterial strains found in asymptomatic bacteriuria, as these strains have no known adherence mechanisms.[9] An earlier study showed that emptying of a 500ml vessel (mimicking the human bladder) containing urine inoculated with bacteria, and refilling the vessel with new voided urine at each void did not eradicate the bacteria for 24 to 72 hours. Emptying the bladder is only one defense mechanism however, since after inoculation of the same bacteria into human volunteers a sterile urine was found by 72 hours later. Thus an additional antibacterial factor was needed for bacterial eradication.[10] After colonization of the bladder it has been shown that urothelial cells can kill bacteria,[11] and urea (a normal constituent of urine) has been shown to be bactericidal in vitro [12] as possible explanations.

1.1.1. Surface Energy Theory of Adhesion. The surface energy theory of bacterial adhesion of Derjaguin-Landau-Verwey-Overbeek (DLVO)[13,] is an attempt to explain the mechanism of bacterial adhesion, and the necessity of surface appendages for adhesion to occur in an energy-efficient manner. Net negative surface charges of both tissue cells and the bacteria, as well as diffuse ion clouds, repulse adhesion. While at 15nm there is little repulsion, as bacteria and cells come within 10 nm of each other, maximum repulsion occurs. The magnitude of both attractive as well as repulsive forces increases with the diameter of an approaching body. Thus, bacterial fimbriae or polymers, being of a much smaller diameter, allow bacterial adherence that might not otherwise occur[14] (Figure 1). Once adherent, the normal flow of body fluids such as mucous or in the case of the urinary tract, urine, will not wash away the bacteria, allowing bacterial multiplication to that critical mass necessary to produce disease.

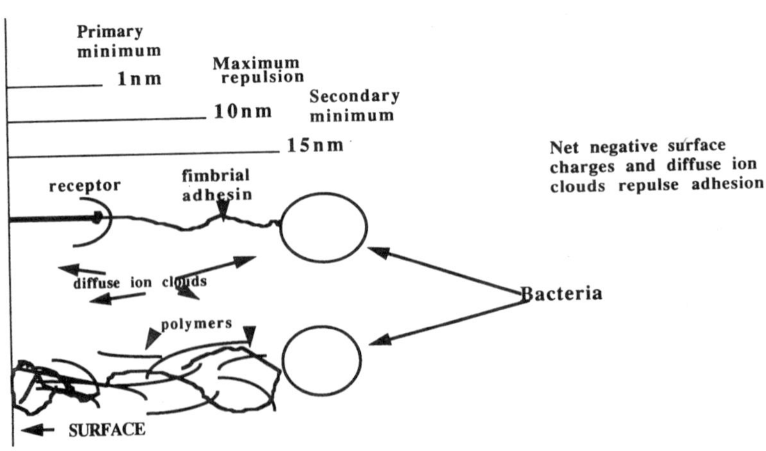

Figure 1. The surface energy theory by which bacterial adherence occurs is depicted. Note that the larger bacteria will not adhere but smaller fimbriae will because repulsive factors are dependent on their radius. Reprinted by permisssion, J. Urol., **156:** 1552, 1996.

The term adhesin designates any structure which causes bacterial adhesion to a cell or tissue. While adhesion may occur because of nonspecific characteristics such as hydrophobicity, it more frequently is due to a specific receptor-lectin interaction by means of surface hairlike appendages. These appendages have been termed either fimbriae[15]or pili[16] but carbohydrate polymers can also overcome the repulsive forces[14] The adhesins include fimbriae, adherence pedestals(similar to fimbriae), and the afimbrial adhesins such as: polymers, polysaccharides, lipoteichoic acid, and high molecular weight proteins.

In addition to E.Coli, Staphylococcus saprophyticus is an important cause of urinary tract infections among young female adults, though an uncommon cause in males.[17] Urease production appears to contribute to its pathogenicity[18], and bacterial adherence to fibronectin[19] by means of a hemagglutinin is important to its initation of infection[20].

2. FIMBRIAL ADHESINS

Fimbriae of Enterobacteriaceae that cause urinary tract infections include type 1 fimbriae, which are usually present in E. coli. Type 1 fimbriae adhere to mannose molecules of glycoproteins, thus treating the bacteria with a mannose solution prevents agglutination of red cells whose surface contains mannose(such as guinea pig cells) and the fimbriae are said to be mannose sensitive.[21] Other fimbriae have been classified as mannose resistant since mannose does not prevent hemagluttination. It was shown in 1976 that adherence to urinary tract epithelial cells by E. coli was a possible initiating step in the onset of urinary tract infection[22] as introital colonization by bacterial adherence was shown to precede urinary tract infections in women. At about the same time, the periurethral and preputial bacteria in children susceptible to urinary tract infections were found to be the same bacteria causing the infection. Bacterial adhesion was then correlated with mannose resistant agglutination of human erythrocytes in patients with non-obstructive pyelonephritis.[23] These mannose resistant adhesins include P-fimbriated E. coli, X-fimbriated E. coli (other mannose resistant strains, such as G, M, and S-fimbriated E. coli) as well as some afimbrial strains.[24] The specific urothelial receptors for P-fimbriae were found to be the P blood group antigens, thus the reason for their name.[25, 26]

2.1. Afimbrial Adhesins

The afimbrial adhesins (such as the Dr adhesin previously named type 1C) [27] are known to cause urinary tract infections and others probably will be found because of the importance of bacterial adhesion in the pathogenesis of urinary tract infections. While the non-fimbrial adhesins may be important in colonization of the perineum and urethra, it has been suggested that their effect in the urinary tract may be minimal[28]

2.1.1. P-Fimbriae. P-fimbriated E. coli strains are the most commonly associated strains causing pyelonephritis as well as approximately 20–30 % of the cases of cystitis. Several characteristics of P-fimbriae are important in understanding its role in bacterial adherence and the pathogenesis of urinary tract infections. This includes the fact that there are a multiplicity of fimbrial types and even a multiplicity of P-fimbriae with different epitopes of the tip protein[29] P-fimbriated E. coli often also have other proposed virulence factors such as bacterial hemolysin, type 1 fimbriae and aerobactin and are resistant to serum bactericidal activity.[3] A clonal theory has been advanced to explain the concomitant presence of these factors[30, 31]

2.1.2. The Tip Protein of P-Fimbriae. By means of both genetic and ultrastructural studies, P-fimbriae have been found to be heteropolymeric structures which are composed

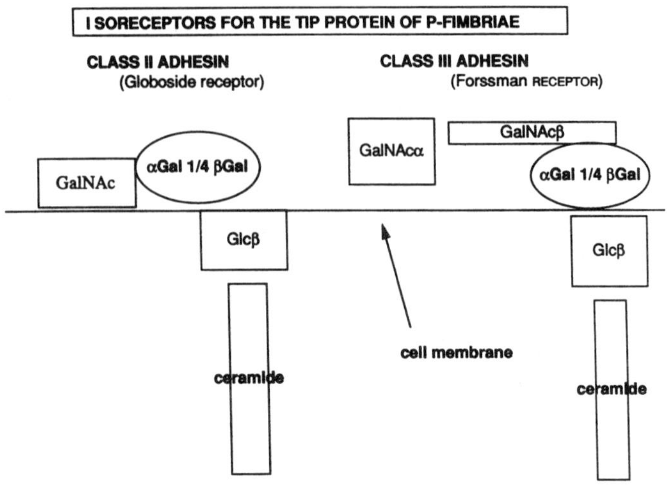

Figure 2. The architecture of the isoreceptors explains the means by which different epitopes of the tip protein of P-fimbriae adhere to different tissues, the Class 2 adhesin being associated with pyelonephritis and the Class 3 with cystitis. Reprinted by permission, J. Urol., **156:** 1552, 1996.

of a rigid stalk (the major protein) linked end to end to a flexible tip fibrillum consisting of four proteins. One of these, PapG, is the specific receptor binding adhesin located at the tip of the fimbriae[32–34]. The three classes of the G tip proteins include: Class I, which has not been associated with disease in man; the Class II tip adhesin associated with pyelonephritis; and the Class III tip adhesin associated with cystitis[35]. The isoreceptors for P-fimbriae all contain the disaccharide αGal-1—4β-Gal, but their position in the glycolipid molecule differs. The difference in adherence can be shown by looking at the surface conformation of the saccharides. The Class II adhesin, which is associated with P-fimbriated E. coli causing pyelonephritis, adheres to globoside since its architecture places the globoside disaccharide (αGal-1—4βGal) at the surface even though its terminal sugar is N-acetylgalactosamine (Gal-Nac-β) The Class III adhesin (associated with cystitis) adheres to Globo A-the Forssman antigen (Gal-Nac-α-Gal-Nac-β) since its surface orientation appears to produce steric hindrance to adherence to globoside.(Figure 2) This suggests a greater density of Globo A receptors on urothelium of the bladder and a higher density of globoside receptors on urothelium from the upper urinary tract and kidney.

2.1.3. P-Fimbriae-Nephropathogenic. A nephropathogenic strain with the Class II tip protein has been used to create a mutant without the tip protein[36]. This mutant strain, expressing P-fimbriae but lacking the PapG tip protein adhesin, was constructed by allelic replacement introducing a 1 base pair deletion early in the papG gene. This mutant with deletion of the tip adhesin was still found to be able to express morphologically normal fimbriae that however do not mediate __-Galp-1-4-__ -Galp specific attachment in vitro[37]. In cynomolgus monkeys, the wild type strain and its mutant both caused bladder infection whereas only the wild type strain could caused pyelonephritis. Since the parent strain but not its mutant adheres to renal tissue in vitro, these data underscore the critical role of microbial adherence to host tissues in infectious disease and strongly suggest that the PapG tip adhesin of P-fimbriae is essential in the pathogenesis of human kidney infection.[38] These in vivo studies of bacterial adherence by P-fimbriae of E. coli to nonhuman primate urothelial cells are only possible since monkeys have the same urothelial receptor for the fimbriae as in man.[39]

The multiple epitopes of the tip protein of P-fimbriae which adhere to different binding sites explain specific infections, such as many cases of cystitis which follow adhesion of E. coli with the Class III tip protein to human bladder cells by means of its receptor, GloboA- the Forssman glycolipid,[40]

3. TYPE 1 FIMBRIAE

Type 1 fimbriae are found in most enterobacteraciae, frequently on nonpathogenic as well as pathogenic E. coli. While in vitro adherence studies[41,42] and studies in animals[43] offer indirect evidence for the role of Type 1 fimbriae in urinary tract infections, direct evidence indicating that these structures are present in vivo are both limited and partially contradictory. Preference for adhesion of Type 1 fimbriae to mannosylated proteins leads to a number of host targets including human buccal cells, renal tubular cells, epithelial cells of the bladder, lung and intestine as well as vaginal cells.[44] It has been shown that E. coli with Type 1 fimbriae adhere to mucin coating urothelial cells. As the mucin is shed, however, adherent E. coli would be shed as well, thus creating an effective defense mechanism.[45]

3.1. Type 1 Fimbriae in Clinical Studies

In a study of cystitis in children, in this case comparing girls with urinary tract infection to normal healthy girls, it was found that bacteria colonizing the perineium as well as the bacteria causing infection had clones which were serum resistant and expressed hemolysin, aerobactin, and Type 1 fimbriae. P-fimbriae, however were not expressed by the colonizing E. coli whereas most infecting clones expressed the P-adhesin.[46] This would suggest that while Type 1 fimbriae may be important in colonization of the vagina and perineum, P-fimbriae are more important, at least in children, in the etiology of cystitis.

3.1.1. Resistance to Colonization by Type 1 Fimbriae. The binding of Type 1 fimbriated E. coli appears to be inhibited by low molecular weight molecules in urine as evidence against adherence to urothelial cells of the bladder.[47] In lower urinary tract infections Type 1 fimbriae were detected in patient's urine specimens frequently while P-fimbriated E. coli were found in only 30%. The bacteria with Type 1 fimbriae were frequently adherant to exfoliated uroepithelial cells and were less frequently free in the urine[48] as these authors had previously shown to occur by experimental studies in the mouse.[43] This study in addition showed that the expression of fimbriae in vivo do not always agree with the hemaglutination pattern when grown in vitro.

3.2. Comparison of Other Virulence Factors to Type 1 Fimbriae

A more recent study of patients with cystitis showed that the virulence factors, including Type 1 fimbriae, P. fimbriae, S. fimbriae, afimbrial adhesin, hemolysin, aerobactin, and cytotoxic necrotizing factor were more frequently detected among cystitis isolates than among fecal isolates. However, Type 1 fimbriation was as common in E. coli found in stool samples as those associated with urinary tract infection.[49] A further study of infecting strains in first time urinary tract infections used DNA colony hybridization to determine the presence of Type 1 fimbriae. Hybridization was used to rule out phenotype variation due to phase variation, which has been shown to occur in E. coli with Type 1, S, F, and P gene clusters. There were five significant associations between virulence factors and patients with first time urinary tract infection. These did not include the gene for Type 1 fimbriae.[50]

3.2.1. Molecular Biology of Type 1 Fimbriae. In a further study of the molecular epidemiology of adhesin factors among uropathogenic E. coli using operon probes , it was found that afimbrial adhesins were detected frequently among isolates from patients with cystitis. Urinary tract isolates in this study showed a low incidence of Type 1 adhesin phenotypic expression among P-positive isolates while the operon probe was usually positive for Type 1 fimbriae. P-fimbriae were significantly more frequent in patients with pyelonephritis (73%), isolates from patients with cystitis less frequently showed P. fimbriation (31%), while expression of the Type 1 adhesin was found to be similar in patients with pyelonephritis or cystitis[51].

3.2.2. Vaginal and Perineal Colonization. There is little doubt that receptors for Type 1 fimbriae are present on vaginal cells and mucous and that there is an association of adherence of E. coli to both vaginal and buccal epithelial cells in women with recurrent urinary tract infections[52]. More recent studies have shown that the Type 1 fimbriated E. coli adhere well to vaginal mucous as well[53] and this adherence is increased in patients with recurrent urinary tract infections[54]. Perhaps the most probable explanation for the association of Type 1 fimbriae with urinary tract infection is that Type 1 fimbriae adhere well to both vaginal cells and vaginal mucous. Therefore they may well be the most important factor in colonization of the vagina and perineum in women with both first and recurrent urinary tract infections. Once entering the bladder, however, it would seem that both afimbrial adhesins and P-fimbriae may well be responsible for adherence in association with cystitis. An alternative explanation for the lack of E. coli with Type 1 fimbriae in the urine of patients with cystitis may be that while these fimbriae may be important in the initiation of cystitis by adherence to urothelial cell mucins, phase variation to the non-fimbriate state occurs later in the infection. An advantage to the bacteria to be non-fimbriated is that it allows them to evade phagocytosis since the Type 1 fimbriae do adhere to the surface of phagocytes.[55] Type 1 fimbriated bacteria are important in urinary tract infections in the catheterized patient, as it has been shown that such nosocomial infections might not occur in the absence of Type 1 fimbriation [56]

4. BACTERIAL VIRULENCE FACTORS

Bacteria which grow well in urine are associated with urinary tract infections, including the Enterobacteriaceae, enterococci, Pseudomonas[57] and more recently Staphylococcus Sapropyticus. Their rigid cell wall protects such bacteria against the high osmolality of urine while porins in the cell wall allow transfer of ions between the urine and intracellular milieu of the bacteria, preventing the lysis of the cells.[58]. E. coli is the most frequent cause of UTI, thus a special pathogenicity has been assumed. Among the E. coli only several serotypes are commonly found in combination with certain bacterial virulence factors[3]

4.1. Surface Factors

Factors which predispose to urinary tract infection include the lipopolysaccharide, capsular antigens, hemoysin, as well as the adhesive factors. The capsular antigen in-

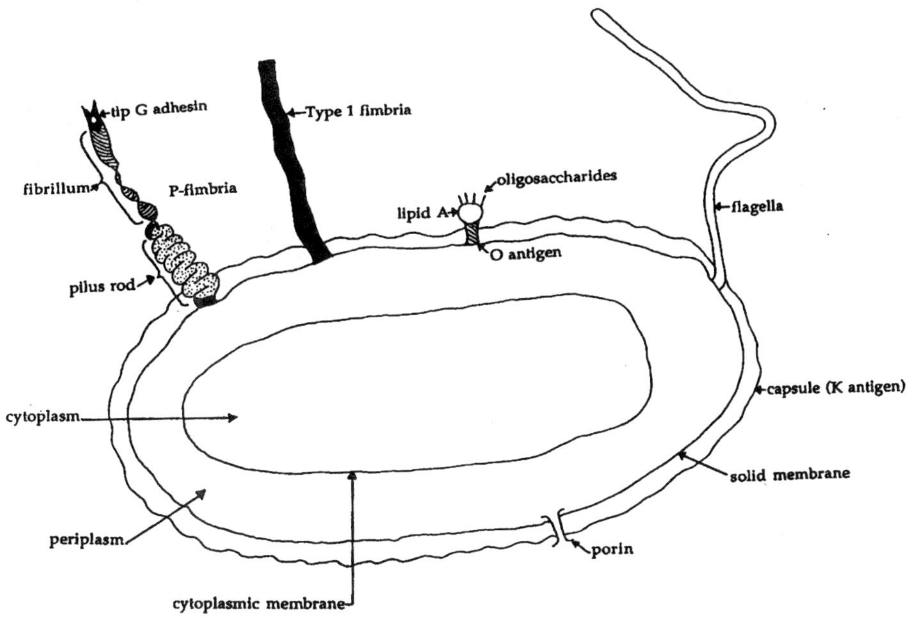

Figure 3. This cartoon depicts the anatomy of E. coli, showing its surface appendages, capsule and membranes. Reprinted by permission, Pediatr. Nephrol., **10:** 517, 1996.

creases resistance of bacteria to phagocytosis.(Figure 3) Hemolysin has been shown to kill surrounding cells, especially when bacteria are adherent to urothelial cells [59]. Uropathogenic strains appear to comprise clones of E. coli that occur more often in the urinary tract

4.1.1. Interference with Adherence. An antibacterial activity of bladder surface mucin has been postulated to be protective against bladder infection following studies in the rabbit[60]. Scanning electron microscopy of cells from humans with infection showed that the bacteria adhered to both mucin and epithelial cells, and in patients with chronic cystitis the mucin layer appeared to be thinner.[61,62] In the rat model ultrastructural studies showed that E. coli adhered to mucin on the epithelium and that this adherence then led to disruption of both the mucin layer and bladder epithelial cells[63]. It then appeared that increased adherence to sub-surface transitional cell epithelia was increased[45]. Similar studies in mice showed adherence of the bacteria first to epithelial cells with bacterial agglutination followed by adherence of the bacteria to polymorphonuclear cells by which they were phagocytosed following bladder infection[63]. In none of these studies were the presence or absence of adherence factors such as fimbriae noted.

4.2. Additional Adherence Factors

In vitro studies showed that normal human urine acted as an opsonic buffer when osmotic pressure was between 200–500 mOsm per kg, whereas over 500 mOsm or less than 200 mOsm there appeared to be a suppression of phagocytosis by polymorphonuclear cells[64]. A study of cystitis in rats showed that bacteria are first phagocytosed by epithelial cells leading to their desquamation with chemotaxis of polymorphonuclear cells to the subepithelial spaces. The leukocytes then migrated into the lumen of the bladder where phagocytosis continued[65]. Whereas type-1 fimbriae adhere to polymorphonuclear cells activating them to begin phagocytosis , the PapG tip adhesin of P-fimbriae appears to protect E. coli from neutrophil bactericidal activity and phagocytosis[66].

Whereas binding of Type-1 fimbriated E. coli to mannins of urinary mucin is thought to lead to rapid washout of such bacteria[67], it has now been found that type-1 fimbriated E. coli also bind to uroplakins 1A and 1B on urothelial apical plaques perhaps important in bladder colonization and infection[68]. Other studies in the cynomolgous monkey showed that the PapG tip adhesin of P-fimbriae of E. coli provides E. coli with a competitive edge in experimental bladder infections as when bacteria with and without the tip adhesin are innoculated into the bladder only the E.coli with the PapG tip lead to continued infection while the E.coli lacking the tip adhesin are eradicated[69].

4.3. Cellular Receptors

Following urothelial colonization an inflammatory response soon occurs with increased production of cytokines, presumably activated by endotoxin. A series of experiments with mice whose inflammatory response was normal as opposed to a strain with a deficient response to LPS suggests that a normal inflammatory response is necessary to clear a mucosal infection.[70] This poor response of LPS non- responders correlates with the lack of a cellular receptor (CD14) for LPS. [71] Once attached the LPS is believed to stimulate inflammation by means of mediators of the inflammatory response, including prostaglandins and cytokines. Both in vitro and in vivo studies have now shown that there is a rapid cytokine and inflammatory response following adherence to urothelial cells [72,73] including IL-1, IL-6 and IL8. The activation of IL-1 leads to the release of Il-6 which induces the acute phase response[74].

4.3.1. Cytokine Response. Interleukin-6 levels in the blood increase markedly within hours of bacterial infection causing the initiation of the acute phase response in the liver[75]. Bacterial lipopolysaccharide predominantly induces interleukin-1 and tumor necrosis factor production in monocytes as the monocyte has a receptor for this endotoxin (CD14) which initiates the cytokine cascade, involving endothelial cells, neutrophils, lymphocytes and fibroblasts[76]. The effect on endothelial cells increases adhesion of granulocytes to endothelial surfaces, and if granulocyte aggregation leads to capillary obstruction, can lead to ischemia. While ischemia itself may be damaging, it is during reperfusion that maximum damage appears to occur. Reperfusion injury to a great extent is due to neutrophil-mediated damage from oxygen free radicals[77]. The neutrophilic response following renal ischemia may be only a secondary response or one secondary to activation of interleukin-1[78]. The adhesion of the granulocytes to the endothelium is mediated by leukocyte adhesion molecules such as CD11a and CD18[79].

4.3.2. Cellular Response. T lymphocytes, as well as polymorphonuclear cells, secrete enzymes that degrade extracellular matrix [80]. The increase in CD4 cells in acute infection, therefore, may assist in capillary transmigration of inflammatory cells to the site of antigen expression by the bacteria.[81] While pro-inflammatory cytokines such as IL-1 and TNFα are unable to modulate endothelial permeability, they may contribute indirectly by increasing endothelial expression of adhesion molecules for polymorphonuclear leukocytes. As an example, ICAMs lead to adhesion and migration of inflammatory cells to an antigenic site where both capillary obstruction may occur and phagocytosis begins. These both lead to release of toxic forms of oxygen which may damage tissue, leading to cell death [82]. This adhesion leads to penetration of the capillary walls to reach the submucosal spaces and eventually the epithelium and urine.[83] The renal damage which occurs from bacterial infection in the monkey can also be prevented by treatment with allopurinol which stops the action of xanthine oxidase and prevents the reperfusion damage following the ischemic event[84]. Our studies of the immune response to E.coli UTI in the monkey illustrate the unique relationships between cytokines and lymphocytes in the response to bacterial infection as the inflammatory response is regulated not only by cytokine activity but by lymphocyte activation as well. The activation of Il-8 (a chemotactic cytokine) both attracts and then activates leukocytes to begin phagocytosis [73]. Cytokine activation of urothelial cells by LPS does not lead to as marked a response as that of macrophages, which not only increase cytokine activation and the inflammatory response, but also initiate the humoral response to eradicate the infection.

4.4. Immunity and Vaccines

Following spontaneous resolution of experimental cystitis in cynomologous monkeys there is both a local and systemic antibody response with serum IgG antibodies and secretory IgA antibodies in the urine. This antibody response, occuring within one week after the induction of cystitis, appears to be important in the resolution of the cystitis[85]. It has now been shown that a bladder infection with P-fimbriated E. coli in the cynomolgus monkey leads to protective immunity against E. coli bladder inoculation with P-fimbriated E. coli apparently due to a secretory IgA response in the urine[86]. Cystitis due to Type 1 fimbriated E. coli has now been prevented in the rat by vaccination with the fimH adhesin, that polypeptide associated with adherence by means of type-1 fimbriae. Further, passive systemic administration of immune sera to the fimH antigen also led to reduce bladder colonization by type-1 fimbriated E. coli[87,88]. These experiments have not been repeated with bacteria containing both type-1 and P-fimbriae, however.

5. HOST FACTORS

5.1. Bacterial Colonization

At birth the chance of an infection may be increased if the mother is fecally colonized with uropathogens, as it has been shown that vaginally-delivered infants are rapidly colonized by the bacteria which colonize the gut of the mother [89]. Indeed, four times as many urinary tract infections were found to occur in infants born of mothers who had been bacteriuric during pregnancy[90], as those found in infants born of mothers without bacteriuria during pregnancy. However, it has been shown that breast feeding in the first months of life is an advantage as it appears to protect infants against UTI[91].

5.1.1. Antibiotic Effects. Studies in the cynomolgus monkey showed that antibacterial therapy will eradicate the normal flora of the vagina, thus allowing easy colonization of the perineum and vagina with pathogenic bacteria [92]. Thus, children treated with antibiotics for upper respiratory infections which might eradicate normal flora, would make them more prone to colonization with uropathogenic bacteria and thus ascending infection [93].

5.1.2. Anatomy. It has been suggested that the incidence of urinary tract infections is greater in females because their urethra is shorter than that of males. However, uncircumcised infant males have a higher incidence of infection than do infant females in the early months of life despite the males' longer urethra, [94] thus urethral length would seem unimportant.

5.1.3. Gut Receptors for E. Coli. Since the organisms causing ascending UTI, particularly E. coli, colonize the gut, it is assumed that individuals with urinary tract infections must have receptors in the gut for these bacteria. The receptor is not known but may be the same as that which allows colonization of the gut by other Enterobacteriaceae. However, it has been shown that children who develop UTI more frequently carry the causative E. coli in their gut [95].

5.1.4. Familial Factors. There may be a familial tendency for urinary tract infection since both children 96 and women [97] with recurrent UTI lack certain blood group types associated with the non- secretor state. The non-secretor does not have the blood group antigens in their secretions such as mucous and saliva. It has been hypothesized that these antigens may sterically interfere with adherence of uropathogens. However, the most recent study in adults was not able to confirm this association, as when quantitated, the amount of either ABH or Lewis antigen on vaginal cells and in mucous did not appear to directly influence the individual susceptibility to UTI [98]. The levels of ABH and Lewis determinants did vary significantly over time in each individual as an alternative explanation for the lack of a difference between control patients and those with UTI.

6. CONCLUSIONS

As always, a combination of host and bacteria factors are important and all suchfactors must be considered to understand the pathogenesis of urinary tract infections. These include bacterial adherence by fimbrial and non-fimbrial adhesins followed by colonization of the urinary tract. Such adherence may be affected by the environment, non-secretor state, or increased receptor density in the gut and urothelial cells of the host. Colonization may not only lead to asymptomatic bacteriuria, but following mucosal adherence in symptomatic infections, by endotoxin activation of interleukins to initiate the inflammatory response. Tissue damage also occurs from toxins such as hemolysin, as well as oxygen radicals released during phagocytosis during symptomatic invasive disease.

A breakdown of host defenses may also be due to factors such as loss of normal perineal and vaginal bacterial flora, a decreased immune response, as well as the presence of foreign bodies such as catheters or instruments which may introduce bacteria into the urinary tract that would not normally cause urinary tract infections. While E. coli are the commonest cause of UTI, bacteria not usually pathogenic are more common causes of such nosocomial infections.

ACKNOWLEDGMENTS

This work was supported by USPHS grants RR00164, DK14681, and AG10292.

REFERENCES

1. Kunin, C.M. Editor : Urinary Tract Infections. 5th edition, Wms & Wilkins, Baltimore: PP6–7, 1997.
2. Wallmark, G., Arremark, I., Telander, B.: Saprophyticus saprophyticis : a frequent cause of acute urinary tract infection among female outpatients. J. Infect.Dis.,**138:** 791, 1978.

3. Johnson, J.R.: Virulence factors in Escherichia coli urinary tract infection. Clin.Microbiol . Rev., **4:** 80, 1991.
4. Roberts, J.A.: Tropism in bacterial infections:urinary tract infections. J. Urol.,156: 1552, 1996.
5. Sheinfeld, J., Schaeffer, A.J., Cardon-Cardo C., Rogatko, A., and Fair, W.R.: Association of the Lewis blood-group phenotype with recurrent urinary tract infections in women. N. Engl. Med., **320:** 773, 1989.
6. Svanborg-Eden, C., Lidin-Janson, G. and Lindberg, V.: Adhesiveness to urinary tract epithelial cells of fecal and urinary tract isolated from patients with symptomatic urinary tract infections or asymptomatic bacteriuria of varying duration. J. Urol., **122:** 185, 1979.
7. Stamey, T.A., Timothy, M., Miller, M. and Mihara, G.: Recurrent urinary infections in adult women. The role of introital Enterobacteriaceae. Calif. Med., **115:** 1,1971.
8. Gordon, D.M. and Riley, M.A.: A theoretical and experimental analysis ofbacterial growth in the bladder. Molec. Microbiol., **6:** 555, 1992.
9. Andersson, P., Egberg, I., Lidin-Jansson, G., Lincoln, K., Hull, R., Hull, S. and Svanborg, C.: Persistence of Escherichia coli bacteriuria is not determined bybacterial adherence. Infect. Immun., **59:** 2915, 1991.
10. Cox, C.E., Hinman, F. Jr.: Experiments with induced bacteriuria, vesical emptying and bacterial growth on the mechanism of bladder defense to infection. J. Urol., **86:** 739, 1961.
11. Schulte-Wisserman, H., Mannhardt, W., Schwartz J, Zepp F. and Bitter-Suermann, D.: Comparison of the antibacterial effect of uroepithelial cells from healthydonors and children with asymptomatic bacteriuria. Eur. J .Pediatr., **144:** 230, 1985.
12. Schlegel, J.U., Cuellar, J., Odell, R.M.: Bactericidal effect of urea. J. Urol., **86:** 819, 1961.
13. Rijnaarts, H.H.M., Norde, W., Bouwer, E.J., Lyklema, J. and Zehnder, A.J.B.: Reversibility and mechanism of bacterial adhesion. Colloids and Surfaces :Biointerfaces, **4:** 5,1995.
14. Jones, G.W. and Isaacson, R.E.: Proteinaceous bacterial adhesins and their receptors. CRC Crit. Rev. Microbiol., **10:** 229, 1983.
15. Duguid, J.P., Smith, I.W., Dempster, G. and Edmunds, P.N.: Nonflagellar filamentous appendages ("fimbriae") and haemagluttinating activity in bacterium coli. J. Pathol. Bacteriol., **70:** 335, 1955.
16. Brinton, C.C. Jr. : The structure, function, synthesis and genetic control of bacterial pili and a molecular mechanism for DNA and RNA transport in gram-negative bacteria. Trans. NY Acad. Sci., **27:** 1003, 1965.
17. Hovelius, B and Mardh, P.A. Staphylococcus saprophyticus as a common cause of urinary tract infection. Rev.Infect.Dis. **6:** 328, 1984.
18. Gatermann, S., John, J. and Marre, R.: Cloning and expression of Staphylococcus saprophyticus urease gene sequences in Staphylococcus canosus andcontribution of the enzyme to virulence. Infec.Immun., **57:** 2998, 1989.
19. Meyer, H.W., Wengler-Becker, U., and Gatermann, S.G.: The hemagglutinin of Staphylococcus saprophyticus Is a Major adhesin for uroepithelial cells. Infect. Immun., **64:** 3893, 1996.
20. Gatermann, S., and Meyer, H. Staphyloccus saprophyticus hemagglutinin binds fibronectin. Infect. Immun. **62:** 4556, 1994.
21. Ofek,I., Mirelman, D. and Sharon, N.: Adherence of Escherichia coli to human mucosal cells mediated by mannose receptors. Nature, **265:** 623, 1977.
22 Svanborg Edén, C., Hanson, L.Å., Jodal, U. and Lindberg, U.: Variable adherence to normal human urinary-tract epithelial cells of Escherichia coli strains associated with various forms of urinary tract infection. Lancet, **2:** 490, 1976.
23. Källenius, G. and Möllby, R.: Adhesion of Escherichia coli to human periurethral cells correlated to mannose-resistant agglutination of human erythrocytes. FEMS Microbiol. Lett., **5:** 295,1979.
24. Korhonen, T.K., Virkola, R., Westerlund, B., Holthofer, H. and Parkkinen, J.: Tissue tropism of Escherichia coli adhesins in human extraintestinal infections. Curr. Top. Microbiol. Immunol., **151:** 115, 1990.
25. Källenius, G., Möllby, R., Svenson, S.B., Winberg, J. and Hultberg, H.: Identification of a carbohydrate receptor recognized by uropathogenic Escherichia coli. Infection, **8:** 288, 1980.
26. Leffler, H. and Svanborg-Edén, C.S.: Chemical identification of a glycosphingolipid receptor for Escherichia coli attaching to human urinary tract epithelial cells and agglutinating human erythrocytes. FEMS Microbiol. Lett., **8:** 127, 1980.
27. Nowicki, B., Labigne, A., Moseley, S., Hull, R., Hull, S., and Moulds, J.: The Dr hemagglutinin, afimbrial adhesins AFA-I and AFA-III, and F1845 fimbriae of uropathogenic and diarrhea-associated Eschericha coli belong to a family ofhemagglutinins with Dr receptor recognition. Infect. Immun., **58:** 279, 1990.
28. Labigne-Rousse,l A., Lark, D., Schoolnik, G., and Falkow, S.: Cloning and expression of an afimbrial adhesin (AFA-I) responsible for P-blood-group independent mannose resistant hemagglutination from a pyelonephritic Escherichia coli strain. Infect. Immun., **46:** 251, 1984.

29. Stromberg,N., Marklund, B,-I., Lund, B., Ilver, D., Hamers, A, Gaastr, W., Karlson, KA., and Normark, S. : Host-specificity of uropathogenic Escherichiacoli depends on difference in binding specificity to Gal 1–4 Gal containing isoreceptors. EMBO J., **9:** 2001, 1990.
30. Achtman, M. and Pluschke, G.: Cl0onal anlysis of descent and virulence among selected Escherichia coli. Ann Rev. Microbiol. , **40:** 184, 1986.
31. Plos, K., Hull, S.I., Hull, R.A., Levin, B.R., Orskov, I., Orskow, F. and Svanborg-Eden, C.: Distribution of the P-associated-pilus (pap) region amongEscherichia coli from natural sources: Evidence for horizontal gene transfer.Infect. Immun., **57:** 604, 1989.
32. Kuehn, M.J., Ogg, D.J., Kihlberg, J., Slonim, L.N., Flemmer, K., Bergfors, T. and Hultgren, S.J.: Structural basis of pilus subunit recognition by the PaPD chaperone. Science, **1262:** 1234, 1993.
33. Lindberg, F., Tennent, J.M., Hultgren, S.J., Lund, B. and Normark, S.: PapD, a periplasmic transport protein in P-pilus biogenesis. J. Bacteriol., **171:** 6052, 1989.
34. Lund, B., Marklund, B.I., Stromberg, N., Lindberg, F., Karlsson, K.A. and Normark, S.: Uropathogenic Escherichia coli can express serologically identical pili of different receptor binding specificities. Molec. Microbiol., **2:** 255, 1988.
35. Hultgren, S.J., Abraham, S., Caparon, M., Falk, P., St Geme, J.W. III, and Normark, S.: Pilus and nonpilus bacterial adhesins: assembly and function in cell recognition. Cell, **73**: 887, 1993.
36. Kuehn, M., Heuser, J., Normark, S. and Hultgren, S. : P pili in uropathogenic E. coli are composite fibres with distinct fibrillar adhesive tips. Nature, **356:** 252,1992.
37. Marklund, B.I., Tennent, J., Garcia, E., Hamers, A., Baga, M., Lindberg, F., Gaastra, W. and Normark, S. : Horizonal gene transfer of the pap and prs pili operons as a mechanism for the development of tissue specific adhesive properites. Molec. Microbiol., **6:** 2225, 1992.
38. Roberts, J.A., Marklund, B.-I., Ilver, D., Haslam, D., Kaack, M.B., Baskin, G., Louis, M., Möllby, R., Winberg, J. and Normark, S.: The Gal(1–4)Gal-specific tip adhesin of Escherichia coli P-fimbriae is needed for pyelonephritis to occur in the normal urinary tract. Proc. Natl. Acad. Sci. USA, **91:** 11889 ,1994.
39. Roberts, J.A., Kaack, B., Källenius, G., Möllby, R., Winberg, J. and Svenson, S.B.: Receptors for pyelonephritogenic Escherichia coli in primates. J. Urol., **131:** 163, 1984.
40. Haslam, D.B., Boren, T., Falk, P., Ilver, D., Chou, A., Xu, Z. and Normark, S.: The amino-terminal domain of the P-pilus adhesin determines receptor specificity. Molec. Microbiol., **14:** 399, 1994.
41. Schaeffer, A.J., Jones, J.M. and Dunn, J.K.: Association of in vitro Escherichia coli adherence to vaginal and buccal epithelial cells with susceptibilitity of women to recurrent urinary-tract infections. N. Engl. J. Med., **304:** 1062, 1981.
42. Schaeffer, A.J., Chmiel, J.S., Duncan, J.L. and Falkowski, W.S.: Mannose-sensitive adherence of Escherichia coli to epithelial cells from women with recurrent urinary tract infections. J. Urol., **131:** 906, 1984.
43. Schaeffer, A.J., Schwan, W.R., Hultgren, S.J. and Duncan, J.L.: Relationship ofType 1 pilus expression in Escherichia coli to ascending urinary tract infections in mice. Infect. Immun., **55:** 373, 1987.
44. Baddour, L.M., Christensen, G.D., Simpson, W.A. and Beachey, E.H.: In: Principles and Practice of Infectious Disease. Edited by G.L. Mandell, R.G. Danglers and J.E. Bennet. New York: Churchill Livingstone, vol 2, pp. 9–22,1989.
45. Balish, M., Jensen, J. and Uehling, D.T.: Bladder mucin: a scanning electron microscopy study in experimental cystitis. J. Urol., **128:** 1060, 1982.
46. Schlager, T.A., Whittam, T.S. Hendley, J.O., Hollis, R.J., Pfaller, M.A., Wilson, R.A. and Stapelton, A.: Comparison of expression of virulence factors by Escherichia coli causing cystitis and E.coli colonizing the periurethra of healthy girls. J. Infect. Dis., **172:** 772, 1995.
47. Parkkinen, J., Virkola, R. and Korhonen, T.K:. Identification of factors in human urine that inhibit the binding of Escherichia coli adhesins. Infect. Immun., **56:** 2623, 1988.
48. Kisielius, P.V., Schwan, W.R., Amundsen, S.K., Duncan, J.L. and Schaeffer, A.J.:In vivo expression and variation of Escherichia coli Type 1 and P pili in theurine of adults with acute urinary tract infections. Infect. Immun., **57:** 1656, 1989.
49. Foxman, B., Zhang, L., Palin, K., Tallman, P. and Marrs, C.F.: Bacterial virulence characteristics of Escherichia coli isolates from first-time urinary tract infection. J. Infect. Dis., **171:** 1514, 1995.
50. Yamamoto, S., Tsukamoto, T., Terai, A., Kurazono, H., Takeda, Y. and Yoshida, O.: Distribution of virulence factors in Escherichia coli isolated from urine of cystitis patients. Microbiol. Immunol., **39:** 401, 1995.
51. Arthur, M., Johnson, C.E., Rubin, R.H., Arbeit, R.D., Campanelli, C, Kim, C., Steinbach, S., Agarwal, M., Wilkinson, R. and Goldstein, R.: Molecular epidemiology of adhesin and hemolysin virulence factors among uropathogenic Escherichia coli. Infect. Immun., **57:** 303, 1989.

52. Schaeffer, A.J., Jones, J.M. and Dunn, J.K.: Association of in vitro Escherichia coli adherence to vaginal and buccal epithelial cells with susceptibilitity of womento recurrent urinary-tract infections. N. Engl. J. Med., **304:** 1062, 1981.
53. Venegas, M.F., Navas, E.L., Gaffney, R.A., Duncan, J.L., Anderson, B.E. and Schaeffer, A.J.: Binding of Type 1-piliated Escherichia coli to vaginal mucus. Infect. Immun., **63:** 416, 1995.
54. Schaeffer, A.J., Venegas, M.F., Gaffney, R.A., Navas, E.L., Anderson, B.E., Chmiel, J.S. and Duncan, J.L.: Role of vaginal fluid in urinary tract infections (UTIs). Proc. A.U.A., **155:** 674A, 1996.
55. Tewari,R.,MacGregor, J.I.,Ikeda, T., Little, S., Hultgren, S., and Abraham, S.N. Neutrophil activation by nascent FimH subunits of type 1 fimbriae purified from the periplasm of Escherichia coli. J. Biol. Chem. **268:** 3009, 1993
56. Mobley, H.L.T., Chippendale, G.R., Tenney, J.H., Hull, R.A. and Warren, J.W.Expression of Type 1 fimbriae may be required for persistence ofEscherichia coli in the catheterized urinary tract. J. Clin. Microbiol., **25:** 2253, 1987.
57. Stamey, T.A. and Mihara, G.: Observations on the growth of urethral and vaginal bacteria in sterile urine. J. Urol., **124:** 4671, 1980.
58. Domingue, G.J. Bacterial Osmoprotection. J. Lab and Clin Med. **48:** 53, 1991.
59. A.L., Donnenberg, M.S., Cu,i X., Russell, R.G., Utsalo, S.J., Mobley, H.L.T., Warren, J.W.: Binding to and killing of human renal epithelial cells by hemolyticP-fimbriated E. coli. Kidney Int., **46:** 1083, 1994.
60. Parsons, C.L., Greeenspan, C. and Mulholland,S.G.: The primary antibacterial defense mechanism of the bladder. Invest.Urol.,**13:** 72, 1975.
61. Elliott, T.S.J., Slack, R.C.B. and Bishop, M.C.: Scanning electron microscopy of human bladder mucosa in acute and chronic urinary tract infection. Br. J. Urol.,**56:** 38, 1984.
62. Elliott, T.S.J, Reed, L., Slack, R.C.B. and Bishop, M.C.: Bacteriology and ultrastructure of the bladder in patients with urinary tract infections. J. Infection., **11:** 191, 1985.
63. Gillon, G., Small, M., Medial, O. and Aronson, M.: Sequential study of bacterial clearance in experimental cystitis. J. Med. Microbiol., **18:** 319, 1984.
64. Suzuki, Y., Fukushi, Y., Orikasa, S. and Kumagai, K.: Opsonic effect of normal and infected human urine on phagocytosis of Escherichia coli and yeasts byneutrophils. J. Urol., **127:** 356, 1982.
65. Fukushi, Y. and Orikasa, S.: The role of intravesical polymorphonuclear leukocytes in experimental cystitis. Investigative Urol., **18:** 471, 1981.
66. Tewari, R., Ikeda, T., Malaviya, R., MacGregor, J.I., Little, J.R., Hultgren S. J. and Abraham, S.N.: The PagG tip adhesin of P fimbriae protects Escherichia coli from neutrophil bactericidal activity. Infec. Immun., **62:** 5296, 1994.
67. Ofek, I., Mosek, A. and Sharon, N.: Mannose-specific adherence of Escherichia coli freshly excreted in the urine of patients with urinary tract infections, and of isolates subcultured from the infected urine. Infect.Immun., **34:** 708, 1981.
68. Wu, Z.-R., Sun, T.-T. and Medina, J.J.: In vitro binding of type 1-fimbriated Escherichia coli to uroplakins Ia and Ib: Relation to urinary tract infections.Proc. Natl. Acad. Sci. USA, **93:** 9630, 1996.
69. Winberg, J., Mollby, R., Bergstrom, J., Karlsson, R.A., Leonardsson, I., Mich, M.A., Teneberg, S., Haslam, B., Marklund, B.J. and Normark, J.: The Pap-G adhesin at the tip of P-fimbriae provides Escherichia coli with a competitive edge in experimental bladder infections of cynomolgus monkeys. J. Exp. Med., **182:** 1695, 1995.
70. Hagberg, L., Hull, R., Hull, S., Michalsek, S., McGhee, J.R., Svanborg-Eden, C.: Differences in susceptibility to gram-negative urinary tract infection between C3H/HeJ and C3H/HeN mice. Infec. Immun., **46:** 839, 1984.
71. Agace, W.W., Patarroyo, M., Svensson, M., Carlemalm, E., Svanborg, C.: Escherichia coli induces transuroepithelial neutrophil migration by an intercellular adhesion molecule-1-dependent mechanism. Infect. Immun., **63:** 4054, 1995.
72. Hedges, S., Andersson, P., Lidin-Janson, G., de Man, P. and Svanborg, C.: Interleukin-6 response to deliberate colonization of the human urinary tract with gram-negative bacteria. Infect. Immun., **59:** 421, 1991.
73. Agace, W., Hedges, S., Ceska, M. and Svanborg, C.: Interleukin-8 and the neutrophil response to mucosal gram-negative infection. J. Clin. Invest., **92:** 780, 1993.
74. Benson. M., Andreasson, A., Joda,l U., Karlsson, Å., Rydberg, J. and Svanborg, C.: Interleukin 6 response to urinary tract infection in childhood. Pediatr. Infect. Dis. J., **13:** 612, 1994.
75. Richards C., Gauldie J. and Baumann H.: Cytokine control of acute phase protein expression. Eur. Cytokine Network, **2:** 89, 1991.
76. Billiau, A. and Vandekerckhove, F.: Cytokines and their interactions with other inflammatory mediators in the pathogenesis of sepsis and septic shock. Eur. J. Clin. Invest., 21: 559, 1991.

77. Grisham, M.B., Hernandez, L.A. and Granger DN: Xanthine oxidase and neutrophil infiltration in intestinal ischemia. Am. J. Physiol., **251:** G567, 1986.
78. Dinarello, C.A.: Role of interleukin-1 in infectious diseases. Immuno. Rev., **127:** 119, 1992.
79. Patarroyo, M., Makgoba, M.W.: Leucocyte adhesion to cells in immune and inflammatory responses. Lancet, **2:** 1139, 1989.
80. Daimle, N.K. and Doyle, I.V.: Ability of human lymphocytes to adhere to vascular endothelial cells and to augment endothelial permeability to macromolecules is linked to their state of post-thymic maturation. J. Immunol., **144:** 1233, 1990.
81. Roberts, J.A., Kaack, M.B. and Martin, L.N.: Cytokine and lymphocyte activation during experimental acute pyelonephritis. Urol. Res., **23:** 33, 1995.
82. Arnaout, M.A.: Cell adhesion molecules in inflammation and thrombosis: Status and prospects. Am. J. Kidney Dis., **21:** 72, 1993.
83. Kaack, M.B., Dowling, K.J., Patterson, G.M. and Roberts, J.A.: Immunology of pyelonephritis VIII: E. coli causes granulocytic aggregation and renal ischemia. J. Urol., **136:** 1117, 1986.
84. Roberts, J.A., Kaack, M.B., Fussel, E.F. and Baskin, G.: Immunology of pyelonephritis VII: Effect of allopurinol. J. Urol., **136:** 960, 1986.
85. Hopkins, W.J., Uehling, D.T. and Balish, E.: Local and systemic antibody responses accompany spontaneous resolution of experimental cystitis in cynomolgus monkeys. Infec. Immun., **55:** 1951, 1987.
86. Soderhall, M., Normark, S., Ishikawa, K., Ishikawa, K., Karlsson, K-A., Teneberg, S., Winberg J. and Mollby, R.: Induction of protective immunity after Escherichia coli bladder infection in primates. J. Clin. Invest., **100:** 364, 1997.
87. Langermann, S., Palaszynski, S., Barnhart, M., Auguste, G., Pinkner, J.S., Burlein, J., Barren, P., Koenig, S., Leath, S., Jones, C.H., and Hultgren, S.: Prevention of mucosal Escherichia coli infection by FimH-adhesin-based systemic vaccination. Science, **276:** 607, 1997.
88. Thankavel, K., Madison, B., Ikeda, T., Malaviya, R., Sham, A.H., Arumugam P.M. and Abraham, S.N.: Localizaton of a domain in the FimH adhesin of Escherichia coli Type 1 fimbriae capable of receptor recognition and use of adomain-specific antibody to confer protection against experimental urinary tract infection. J. Clin Invest., **100:** 1123, 1997.
89. Gareau, F.E., Macke, l. D.C., Boring, J.R., Payne, F.J. and Hammett, F.L.: The acquisition of fecal flora by infants from their mothers during birth. J. Pediatr., **54:** 313, 1959.
90. Patrick, M.J.: Influence of maternal renal infection on the foetus and infant. Arch.Dis. Child., **42:** 208, 1967.
91. Piscane, A., Graziano, L., Mazzarella, G., Scapellino, B. and Zona, G. Breast feeding and urinary tract infection. J. Pediatr., **120:** 87,1992.
92. Herthelius, M., Möllby, R., Nord, C.E. and Winberg, J.: Amoxicillin promotes vaginal colonization with adhering Escherichia coli present in faeces. Pediatr. Nephrol., **3:** 443, 1989.
93. Hansson, S., Jodal, U., Lincoln, K. and Svanborg-Ede, C.: Untreated asymptomatic bacteriuria in girls: Effect of phenoxymethylpenicillin and erythromycin given for intercurrent infections. Br. Med. J., **298:** 256,1989.
94. Ginsburg, C. and McCracken, G.H., Jr.: Urinary tract infections in young infants. Pediatrics, **69:** 409, 1982.
95. Plos K, Connell H, Jodal U, Marklund B-I, Marild S, Wettergren B. and Svanborg, C.: Intestinal carriage of P fimbriated Escherichia coli and the susceptibility tourinary tract infection in young children. J. Infec. Dis., **171:** 625, 1995.
96. Lomberg, H., Hellstrom, M., Jodal, U. and Svanborg-Edén C.: Secretor state and renal scarring in girls with recurrent pyelonephritis. FEMS Microbiol. Immunol., **47:** 371, 1989.
97. Sheinfeld, J., Schaeffe,r A.J., Cordon-Cardo, C., Rogatko, A. and Fair, W.R.: Association of the Lewis blood-group phenotype with recurrent urinary tractinfections in women. N. Engl. J. Med., **320:** 773, 1989.
98. Navas E.L., Venegas, M.F., Duncan, J.L., Anderson, B.E., Kanerva, C., Chmiel, J.S., and Schaeffer AJ.: Blood group antigen expression on vaginal cells andmucus in women with and without a history of urinary tract infections. J. Urol., **152:** 345, 1994.

26

ROLE OF VAGINAL COLONIZATION IN URINARY TRACT INFECTIONS (UTIs)

Anthony J. Schaeffer, Nithya Rajan, E. Thorburn Wright, James L. Duncan, and Byron E. Anderson

Department of Urology
Northwestern University Medical School
303 East Chicago Avenue, Tarry 11-715
Chicago, Illinois 60611-3008

1. INCIDENCE AND ECONOMIC CONSEQUENCES OF URINARY TRACT INFECTIONS

Urinary tract infections (UTIs) are among the most common of urologic and bacterial diseases, ranging from asymptomatic bacteriuria to severe kidney infection and occasional renal failure.[1] Acute lower UTIs affect an estimated 10–20% of American women during their lifetime and result in over 5 million office visits to physicians, costing nearly $1 billion annually for diagnosis and treatment.[2–4] Approximately 20 percent of these women have frequent (3–6 per year) recurrences of infection following the initial episode, producing additional morbidity and time lost from work;[5] these cases of recurrent UTIs cost the health care system at least $200–350 million annually.[4–5] Additionally, 1.5 million cases of more advanced forms of UTIs, including cystitis and acute pyelonephritis, necessitate hospitalization costing over $4 billion.[1] The bacteria responsible for UTIs are normally present in the bowel, however, the factors mediating their ability to colonize, invade and damage the urinary tract are not well known. Research leading to improved prevention or treatment at the early stages of disease progression can result in significant savings to the health care system.

2. ESCHERICHIA COLI IS THE PREDOMINANT PATHOGEN CAUSING URINARY TRACT INFECTIONS IN WOMEN

Escherichia coli (E. coli) is the most prevalent uropathogen causing 75–85% of episodes of cystitis and pyelonephritis.[6,7] Other uropathogens include Staphylococcus saprophyticus, Enterococcus, Klebsiella pneumoniae, Proteus mirabilis and Pseudomonas sp.[6]

Advances in Bladder Research, edited by Baskin and Hayward.
Kluwer Academic / Plenum Publishers, New York, 1999.

Several urovirulence factors of E. coli have been identified, including pili and hemolysin.[8,9] Bacterial pili allow E. coli to adhere to mucosal surfaces and epithelial cells during colonization and infection. E. coli produce a variety of antigenically and functionally distinct pili (type 1, Pap, Prs and S pili), and some E. coli isolates produce up to four different types of pili on the same organism.[9–15] Many studies have demonstrated that type 1 and P pili are the predominant adhesins involved in UTI.[16–18] Type 1 pili, which mediate mannose-sensitive agglutination of guinea pig erythrocytes, have been identified in 76% of strains isolated from urine specimens of patients with acute lower UTIs[19] and P pili, which mediate mannose resistant agglutination of human erythrocytes, have been identified in 33% of strains from the same urine specimens. Furthermore, P pili were present in 91% of E. coli strains causing pyelonephritis, but only 19% of strains causing cystitis in young girls.[17] Therefore, research directed at the initial phases of UTIs, such as vaginal colonization, has focused on the type 1 piliated E. coli. For these reasons, our research has focused on type 1 piliated E. coli and will be extended to the type P piliated E. coli, and the vaginal surface and fluid, the first reservoir of bacterial colonization of the urinary tract.

3. VAGINAL COLONIZATION WITH E. COLI IS A PREREQUISITE FOR THE DEVELOPMENT OF ASCENDING URINARY TRACT INFECTIONS IN WOMEN

Colonization of the vaginal mucosa by uropathogens, such as E. coli, is a critical step in the pathogenesis of UTIs.[20] Urinary tract pathogens colonize in a stepwise fashion the fecal reservoir, the vaginal introitus and the urethra prior to entering the bladder.[21,22] Bacterial colonization appears to be a specific interaction influenced by: 1) bacterial properties, 2) the surface characteristics of mucosal epithelial cells and 3) fluids bathing both cell types. One essential interaction is the binding of bacteria to receptors on vaginal epithelial cells of mucosal surfaces. Numerous studies have shown that bacteria may selectively adhere to mucosal cells and that the extent to which they adhere correlates with colonization and infection.[23–25] Women with recurrent UTIs demonstrate increased adherence of bacteria in vitro to uroepithelial,[26] vaginal[27] and buccal epithelial cells[28] when compared to women who have never had an infection. This increased adherence persists despite resolution of the infection.[26] Furthermore, there appears to be a direct association between the bacterial receptivity of buccal and vaginal epithelial cells obtained from the same woman and tested with the same strain of E. coli, suggesting that differences in surface composition of mucosal epithelial cells are perhaps in part genetically determined.[28]

3.1. Changes in Vaginal Colonization Status and Susceptibility to Urinary Tract Infections Occur in Patients

Vaginal cell receptivity to bacterial attachment varies significantly from day to day in both healthy controls and patients.[29] For instance, when adherence was correlated with the days of a woman's menstrual cycle, increased numbers of bacteria adhered in the early phase and this number diminished shortly after the time of expected ovulation,[29] suggesting that hormonal fluctuations, as seen with estrogen, may modify vaginal cell receptivity and the pathogenesis of UTIs. This concept is further supported by the observation that post-menopausal estrogen-deficient women have higher vaginal pH, vaginal colonization with E. coli and rates of recurrent UTIs than pre-menopausal women.[30] Furthermore, estrogen replacement therapy has been shown to lower vaginal pH, decrease vaginal colonization with E. coli and decrease the frequency of UTIs.[30] Although the mechanism by which estrogen and pH affect attachment and colonization of the vaginal mucosa is not well understood, the association may be important.

3.2. Mucus and Fluid Secretions May Affect Bacterial Interactions with Epithelial Cells and, Thus, Vaginal Colonization Status

Biochemical factors mediating bacterial attachment to and colonization of vaginal mucosal epithelial cells include adherence factors of the bacteria, the surface receptors of the mucosal epithelial cells and the fluid of the vaginal microenvironment.[23–25] There is much information on bacterial adherence factors and some work on vaginal epithelial cell receptors, but study of the fluid which bathes the vaginal mucosa is limited.

Mucus, an extracellular secretion which is comprised of many molecules including the mucin type glycoproteins, may affect bacterial attachment and colonization by providing protection for or facilitating adherence to mucosal epithelial cells.[31] The carbohydrate structures on the mucins largely appear to determine the specificity of binding to cells, extracellular matrix, or pathogenic organisms.[32–36] The effect of mucus varies among different epithelia of the body and with different physiologic conditions such as mucus concentration or pH.[31] Many diseases can affect the production of mucus and in some cases the biochemical characteristics of the mucus are altered; in other instances the amount produced is changed, and sometimes both effects occur simultaneously.[31] Studies of adherence of bacteria to respiratory, intestinal and urinary tract mucus support this concept. For example, Pseudomonas cepacia chronically colonizes respiratory mucus in cystic fibrosis, a condition characterized by excessive mucus secretion.[37] Although P. cepacia exhibited specific, saturable binding to purified human mucin isolated from patients with cystic fibrosis, there was no preference for cystic fibrosis versus non-cystic fibrosis mucus[38] suggesting that the mucus quantity is the important determinant of P. cepacia colonization in vivo.

Physiologic changes in pH of the microenvironment may also affect mucus and, consequently, the predisposition of a host for protection from colonization by or infection with pathogens.[39–41] Wanke[39] showed that enteropathogenic E. coli strains bound to small bowel mucus in significantly greater quantities at pH 5.7 than at pH 7.4 or 8. More recent studies[42–44] demonstrated that piliated enteropathogenic E. coli or Yersinia enterocolitica that bound avidly to mucus immobilized in polystyrene wells were inhibited from binding to intestinal cells by the homologous mucus. While binding was mediated by pili, the microenvironment of the interaction, determined in part by pH, influenced the extent of inhibition and thus protection from colonization.

With regard to UTIs, mucus may protect the host by retarding access of bacteria to epithelial cells or, conversely, play a role in the pathogenesis of infection by providing a reservoir for mucus-binding bacteria, thus facilitating colonization.[32–36] Tamm-Horsfall glycoprotein of the urinary and bladder mucus inhibited binding of E. coli to transitional epithelial cells at high concentrations (>30 μg/ml) but enhanced binding at lower concentrations.[45] Taken together, these studies indicate that depending on its concentration mucus in vivo can either facilitate or inhibit bacterial attachment and colonization of mucosal surfaces. The impact of vaginal fluid glycoproteins on vaginal colonization has heretofore received little attention. Our preliminary data suggest that the binding of E. coli to vaginal fliud influences susceptibilty to vaginal colonization and subsequent UTI's.

4. BACTERIAL INTERACTIONS WITH VAGINAL FLUID

The presence of fluid on the vaginal introital cell surface may influence an individual's susceptibility to UTIs either by providing a barrier to bacterial colonization (e.g., by shielding specific epithelial receptor sites) or by providing receptor sites that increase colonization. Both of those functions may be operative in vaginal fluid, an observation consistent with the natural variation in susceptibility to UTIs in women and the day-to-day changes in the flora and biochemistry of the vaginal environment.

4.1. Type 1 Piliated E. coli Binds to Vaginal Fluid Glycoproteins

To better understand the role of vaginal fluid in the colonization of the vaginal mucosa with uropathogens and pathogenesis of UTIs, we developed a system to study the binding of a recombinant E. coli strain HB101/pWRS1–17, expressing type 1 pili, to vaginal fluid specimens collected from 28 women. Type 1 piliated E. coli were studied due to their high prevalence among uropathogens and their expression of the primary mannose binding adhesin responsible for mediating adherence to the vaginal mucosa and bladder. The details of the assay development and results are reported elsewhere.[46]

These studies represent the first documented investigation on adhesion of bacteria to vaginal fluid. Type 1 piliated E. coli bound to all of the vaginal fluid specimens. The number of bacteria bound was a function of the number of bacteria added and was dependent on the concentration of fluid specimens used to coat the wells. Furthermore, bacterial binding was mainly mannose dependent as 1% mannose added to the binding incubation mixture almost completely inhibited the adherence of bacteria to fluid. Statistical analysis of the amount of mannose-terminal residues in the fluid specimens, as determined by reactivity with anti-mannose antibody, and the amount of type 1 piliated bacteria bound to each vaginal fluid specimen showed a marginally significant correlation (Kendall's $r=0.51$, $p=0.054$), indicating that accessibility and number of mannose residues may be the primary determinants for bacterial binding.

In addition, we determined the binding capacity of E. coli to vaginal fluid obtained from 26 white women and 27 black women. The binding capacity of vaginal fluid from blacks was significantly lower than that observed for whites ($p=0.0017$). The reduced receptivity of vaginal fluid specimens from blacks represents an important demonstration of a biological UTI trait that differentiates these populations and is consistent with the previously documented lower UTI incidence among black women compared to white women.

4.2. The Binding Capacity of Vaginal Fluid from Women Colonized with E. coli Is Greater than from Noncolonized Women

Vaginal fluid specimens collected from 68 women were tested for their capacity to bind type 1 piliated E. coli. These women ranged in age from 19–88 years. The pH of collected vaginal fluid specimens ranged from 3.8–8.2. These women were categorized into four separate clinical groups, including one control group and three different patient groups, (1) Healthy control women without current colonization of the vaginal mucosa with E. coli or history of UTIs, (2) Women with a prior history of recurrent UTIs, but lacking both current colonization of the vaginal mucosa and an active UTI, (3) Women with current colonization of the vaginal mucosa, but lacking an active UTI, (4) Women with both current colonization of the vaginal mucosa and an active UTI.

As shown in Table 1, vaginal fluid collected from women colonized with E. coli had a significantly greater capacity to bind type 1 piliated E. coli than vaginal fluid from uncolonized women. Of the 68 participants, 14 were followed-up for 4–6 months from the date of vaginal fluid collection. Six out of the 14 women developed at least one UTI during this period of time. The mean RBB value (0.97) observed for the women who devel-

Table 1. Mean bacterial binding (RBB) of vaginal fluid specimens from women colonized and uncolonized with E. coli

Vaginal colonization status	N	RBB[1]	P-value
Colonized with E. coli	23	0.86 ± 0.06	0.0001
Uncolonized	45	0.54 ± 0.04	

[1]Relative Bacterial Binding values shown as the mean ± standard error.

oped a UTI was higher than that observed for the women who had no recurrence (0.61). A more detailed analysis, dividing the sample population into the four groups described above is presented in Table 2. Vaginal fluid from women exhibiting vaginal colonization at the time of an acute UTI had the highest bacterial binding capacity, and this binding was significantly higher than fluid from women without current vaginal colonization or acute UTI, whether they had no history of UTI (p=0.003) or some previous history of UTI (p=0.003). These results suggest that vaginal colonization status is associated with the bacterial binding capacity of the vaginal fluid. One could postulate that an increased binding capacity of vaginal fluid allows E. coli to establish a presence in the vaginal introitus thus mediating colonization and the development of UTI.

Our analysis also revealed significantly higher proportion of women colonized with E. coli were post-menopausal (52%) as compared to uncolonized women (22%) (p=0.026). In addition, the mean pH of vaginal fluid from colonized women (6.05) was significantly higher than from uncolonized women (5.05) (p=0.0014). These data support previous observations that post-menopausal women have both higher pH and more frequent vaginal colonization.[30]

To further define the relation between the binding capacity of vaginal fluid and colonization of the vaginal mucosa with E. coli, we used a logistic regression model. We modeled the probability of colonization as a logistic function of RBB after adjusting for pH. Our results indicate that the effect of RBB is statistically significant and independent of pH [OR=29.75, p=0.0042, 95% CI=(2.94, 298.87)], with an estimated 29.75 fold increase in odds of colonization for each RBB vaginal fluid specimen (Figure 1). Specimens were classified as colonized when their respective predicted probability of colonization was greater than or equal to 0.5. Conversely, specimens were classified as non-colonized when their predicted probability of colonization was less than 0.5. Of the 68 specimens tested, our model correctly predicted the colonization status 75% of the time. Furthermore, when comparing all possible discordant (colonized, noncolonized) pairs of specimens, this model demonstrates that the colonized specimens have a higher predicted probability of colonization 83% of the time. Such a model may prove useful for predicting probability of vaginal colonization with E. coli, and subsequent risk of recurrent UTIs, thus guiding usage of antimicrobial prophylaxis.

Table 1. Mean RBB of vaginal fluid specimens based on colonization, acute UTI, and previous history of UTI

Current vaginal colonization	Current acute UTI	Previous history of acute UTI	N	RBB[1]
no	no	no	28	0.58 ± 0.05
no	no	yes	17	0.48 ± 0.06
yes	no	yes/no	17	0.83 ± 0.07
yes	yes	yes/no	6	0.95 ± 0.13

[1]RBB values shown as the mean ± standard error.

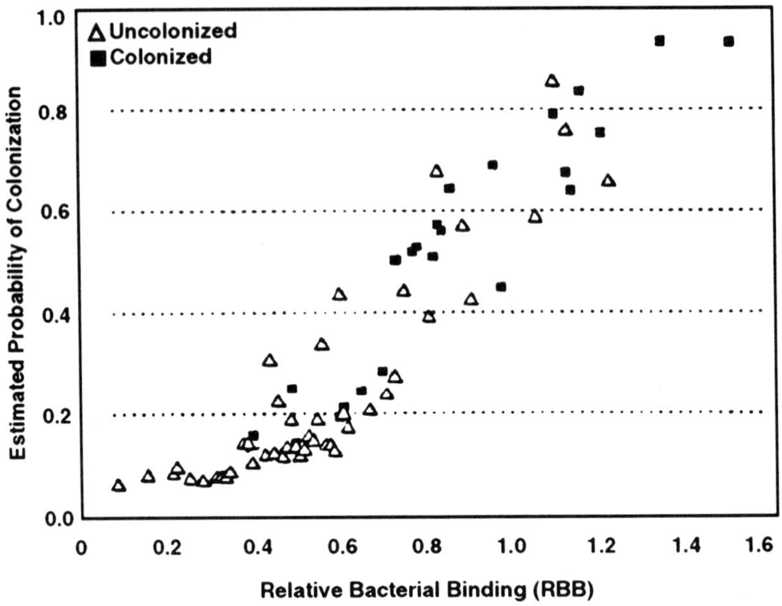

Figure 1. Variation of predicted probability of colonization with Relative Bacterial Binding (RBB).

4.3. The Binding Capacity of Vaginal Fluid Varies over Time within Individuals

We also observed that bacterial binding to vaginal fluid changes over time, with 3-fold differences in binding capacity observed between the lowest and highest specimens collected daily for five days and 17-fold differences among specimens collected weekly for 12 weeks.[47] The data clearly show temporal differences in the capacity of vaginal fluid to behave as a receptor for type 1 piliated bacterial binding. Thus, an individual may be more or less susceptible to the initial adherence of bacteria followed by colonization and UTI, depending on the concentration of glycoproteins and mannose receptors which vary from day to day.

4.4. Clinical E. coli Strains Isolated from the Vaginal Mucosa and Urine Also Bind Vaginal Fluid

Strains of E. coli were isolated from the urine and periurethral vaginal fluid of women with three or more UTI per year. The strains expressed type 1 adhesins. The positive control was E. coli HB101/pWRS1–17 and the negative control was E. coli transformed with the vector alone (E. coli HB101/pHSS22) which does not express type 1 pili.

Table 3 summarizes the adherence to pooled vaginal fluid of 4 E. coli strains isolated from the periurethral vaginal mucosa compared with the recombinant strain. Strains A and B had original colony counts of >10^5 per ml; strains C and D had original colony counts of 20 and 30 per ml respectively.

The adherence of the clinical isolates was 17–130% of control and was not correlated with guinea pig (GP) hemagglutination titer. As expected, the GP hemagglutination titer correlated with the degree of mannose inhibition which ranged from 9–64%. The re-

Table 3. Adherence of type 1-piliated Escherichia coli vaginal isolates to vaginal fluid

Strain	N	H/A Titer*	Vaginal fluid**	Relative binding	Vaginal fluid + 0.1m mannose**	% inhibition by mannose
p117^{A}	7	384 ± 57	2.52 ± 0.37	1.00	1.02 ± 0.23	60
A	5	368 ± 106	1.08 ± 0.17	0.43	0.58 ± 0.09	46
B	5	124 ± 59	1.54 ± 0.34	0.61	0.93 ± 0.19	40
p118^{B}	7	332 ± 39	2.03 ± 0.59	1.00	0.73 ± 0.07	64
C	5	66 ± 63	1.49 ± 0.45	0.73	1.35 ± 0.42	9
D	5	26 ± 6	2.64 ± 1.05	1.30	2.35 ± 0.42	11

*Guinea pig erythrocyte hemagglutination titer ± standard error of the mean.
** Number bacteria bound per well × 10^6 ± standard error of the mean.

combinant strain of E. coli that expressed type 1 pili was inhibited by 60%. This suggests that other adhesins bind to vaginal fluid in vitro and may be important for colonization in vivo. A nonpiliated recombinant stream showed a binding ratio of .09.

Table 4 summarizes the adherence to vaginal fluid of E. coli strains isolated from the urine. These strains were isolated frm the corresponding patients and showed colony counts in urine of 10–1800; thus the organisms probably represent urethral flora and not urinary bladder infection. The adherence of the clinical isolates was 20 and 54% of control and was not correlated with the GP hemagglutination titer. Mannose inhibition ranged from 9 to 46%.

To test whether a particular affinity existed between a specific vaginal fluid specimen and its corresponding bacterial isolate, 4 vaginal E. coli strains were incubated with their corresponding vaginal fluid specimens and the vaginal fluid associated with the other strains. The patients had recurrent UTIs but no UTI at the time of specimen collection. No affinity between the colonizing organism and its native vaginal fluid was observed (Table 5). Rather, receptivity of an individual's vaginal fluid was independent of the origin of the E. coli strain. For example, vaginal fluid from individual #2 was most receptive to adherence for strain p117 and 3 of the 4 difference vaginal isolates. Conversely, vaginal fluid from individual #1 was least receptive for all 5 E. coli strains.

4.5. Vaginal Fluid Affects the Binding of E. coli to Epithelial Cells

In order to assess the importance of vaginal fluid in bacteria-epithelial cell interactions, we developed an in vitro model that measures the effect of vaginal fluid on the bind-

Table 4. Adherence of type 1-piliated Escherichia coli urine isolates to vaginal fluid

Strain	N	H/A titer*	Vaginal fluid**	Relative binding	Vaginal fluid + 0.1m mannose**	% inhibition by mannose
p117^{A}	7	384 ± 57*	2.52 ± 0.37	1.00	1.02 ± 0.23	60
A	5	280 ± 86*	1.08 ± 0.17	0.46	0.67 ± 0.24	38
B	5	45 ± 28*	1.54 ± 0.34	0.54	1.03 ± 0.23	33
p118^{B}	7	332 ± 39*	2.03 ± 0.59	1.00	0.73 ± 0.07	64
C	5	32*	1.49 ± 0.45	0.46	1.06 ± 0.31	29
D	5	4*	2.64 ± 1.05	0.20	0.94 ± 0.19	64

*Guinea pig erythrocyte hemagglutination titer ± standard error of the mean.
**Number bacteria bound per well × 10^6 ± standard error of the mean.

Table 5. Adherence of type 1-piliated Escherichia coli isolate to homologous and heterologous vaginal fluid

Strain	H/A titer[a]	Adherence (vaginal fluid)[b]			
		#1	#2	#3	#4
p117	128	0.46	1.99	1.11	1.71
1	12	**0.10**	0.18	0.10	0.14
2	9	0.16	**0.31**	0.20	0.16
3	64	0.31	0.62	**0.60**	0.75
4	3	0.08	0.16	0.12	**0.14**

[a]Guinea pig erythrocyte hemagglutination titer ± standard error of the mean.
[b]Number bacteria bound per well × 10^6 ± standard error of the mean.

ing of bacteria to an epithelial cell line, A431.[48] Vaginal fluid specimens obtained from 21 women were tested. In an effort to focus on fluid-specific effects on type 1 piliated E. coli binding to epithelial cells, the epithelial cell type and bacterial strain were held constant while fluid specimens from different individuals varied according to such biochemical and clinical factors as concentration, pH during collection, and menopausal status of the subject. The data indicate that vaginal fluid has a significant effect on bacterial adherence in vitro and may either inhibit or stimulate bacterial adherence to epithelial cells at higher, physiologic, fluid concentrations. The differential effect at high concentrations was related to variations in the biochemical and clinical factors mentioned above. Increased binding of E. coli to cells in the presence of vaginal fluid correlated with increased pH of the fluid from the time of specimen collection (Kendall's r=0.33, p=0.039). Additionally, increased binding of E. coli to cells in the presence of vaginal fluid correlated with an increased capacity of vaginal fluid alone to bind E. coli (Kendall's r=0.43, p=0.006). At higher protein concentration, vaginal fluid collected from women exhibiting colonization of the vaginal mucosa with E. coli generally facilitated adherence of type 1 piliated E. coli to epithelial cells. Conversely, vaginal fluid collected from individuals not colonized with E. coli at the time of collection did not alter adherence of type 1 piliated E. coli at higher concentrations. These results, taken together, suggest that vaginal fluid may directly influence bacterial binding to cells and, in particular, the process of colonization of the vaginal mucosa. These data provide an important new biological link to critical early events of in vivo adherence.

5. CONCLUSIONS

The studies support the concept that interactions between physiologic fluid and pathogens play a specific, complex and important role in bacterial adherence to cells and in the pathogenesis of infections. The specific interaction is dependent on both the characteristics of the bacterial strain and the fluid glycoprotein that is encountered by that bacterium. The function of the glycoprotein layer depends in part on the source of the mucus and the environment in which the interaction between bacteria and mucus takes place.

Many women are susceptible to UTIs particularly when prophylactic antimicrobials are discontinued or after menopause when the vaginal pH increases and the vagina is frequently colonized by uropathogens. However, the role of vaginal fluid in UTI has not been systematically evaluated in the past. Clinical and basic research is necessary to determine how vaginal fluid influences susceptibility to UTIs and to further explain the pathophysiologic mechanisms that contribute to UTIs. Future research may result in the development of monoclonal antibodies to vaginal fluid glycoproteins identified as important for adherence and colonization. The ability to identify specific glycoproteins in vaginal fluid that play a role in vaginal colonization and subsequent infection of the urinary tract could make it possible to more accurately identify, monitor and manage patients at risk for UTIs. Additionally, this identification of specific receptor moieties involved in bacterial adherence, a first step in developing novel non-antibiotic drugs (e.g. competitive inhibitors) could enhance prevention and management of UTIs.

ACKNOWLEDGMENT

This study was supported by the NIH grant DK-42648.

REFERENCES

1. U.S. Department of Health and Human Services, Public Health Service, National Institutes of Health, The National Kidney and Urologic Diseases Advisory Board 1990 Long-Range Plan Window on the 21st Century, NIH Publication Number 90–583, March 1990.
2. Kass, E.H., Savage, W., Santamarina, B.E.: The significance of bacteriuria in preventive medicine. In: Progress in Pyelonephritis. Edited by E.H. Kass. Philadelphia: F.A. Davis Co., pp. 3–10, 1965.
3. Cypress, B.K.: Patients' reasons for visiting physicians: National Ambulatory Medical Care Survey. In: Vital and Health Statistics, Data from the National Health Survey, Series 13; no. 56 [DHHS Publ.; no. (PHS) 82–1717], Hyattsville, Maryland: National Center for Health Statistics, 1981.
4. Johnson, J.R., Stamm, W.E.: Urinary tract infections in women: Diagnosis and treatment. Ann. Int. Med., **111:** 906, 1989.
5. Stamm, W.E., Hooton, T.M., Johnson, J.R., Johnson, C., Stapleton, A., Roberts, P.L., Moseley, S.L., Fihn, S.D.: Urinary tract infections: From pathogenesis to treatment. J. Infect. Dis., **159:** 400, 1989.
6. Kunin, C.M.: An overview of urinary tract infections. In: Detection, prevention and management of urinary tract infections. Edited by C.M. Kunin. Philadelphia: Lea & Febiger, pp. 125–126, 1987.
7. Stapleton, A., Moseley, S., Stamm, W.E.: Urovirulence determinants in Escherichia coli isolates causing first-episode and recurrent cystitis in women. J. Infect. Dis., **163:** 773, 1991.
8. Eden, C.S., Hagberg, L., Hanson, L., Hull, S., Hull, R., Jodal, U., Leffler, H., Lomberg, H., Straube, E.: Bacterial adherence - a pathogenic mechanism in urinary tract infections caused by Escherichia coli. Prog. Allergy, **33:** 175, 1983.
9. Arthur, M., Johnson, C.E., Rubin, R.H., Arbeit, R.D., Campanelli, C., Kim, C., Steinbach, S., Agarwal, M., Wilkinson, R., Goldstein, R.: Molecular epidemiology of adhesin and hemolysin virulence factors among uropathogenic Escherichia coli. Infect. Immun., **57:** 303, 1989.
10. Hacker, J., Schrettenbrunner, A., Schroter, H., Duvel, H., Schmidt, G., Goebel, W.: Characterization of Escherichia coli wild-type strains by means of agglutination with antisera raised against cloned P-, S- and MS-fimbriae antigens, hemagglutination, serotyping and hemolysin production, Zbl. Bakt. Hyg., **A261:** 219, 1986.
11. Johnson, J.R.: Virulence factors in Escherichia coli urinary tract infection. Clin. Micro. Rev., **4:** 80, 1991.
12. Linstedt, R., Larson, G., Falk, P., Jodal, U., Leffler, H., Svanborg, C.: The receptor repertoire defines the host range for attaching Escherichia coli strains that recognize Globo-A. Infect. Immun., **59:** 1086, 1991.
13. Nowicki, B., Svanborg-Eden, C., Hull, R., Hull, S.: Molecular analysis and epidemiology of the Dr hemagglutinin or uropathogenic Escherichia coli. Infect. Immun., **57:** 446, 1989.
14. Stromberg, N., Nyholm, P.-G., Pascher, I., Normark, S.: Saccharide orientation at the cell surface affects glycolipid receptor function. Proc. Natl. Acad. Sci. USA, **88**: 9340, 1991.
15. Korhonen, T.K., Vaisanen-Rhen, V., Rhen, M., Pere, A., Parkkinen, J., Finne, J.: Escherichia coli fimbriae recognizing sialyl galactosides. J. Bacteriol., **159:** 762, 1984.
16. Duid, J.P., Clegg, S., Wilson, M.I.: The fimbrial and nonfimbrial haemagglutinins of Escherichia coli, J. Med. Microbiol., **12:** 213, 1979.

17. Kallenius, G., Svenson, S.B., Hultberg, H., Mollby, R., Helin, I., Cedergren, B., Winberg, J.: Occurrence of P-fimbriated Escherichia coli in urinary tract infections. Lancet. **ii:** 1369, 1982.
18. Pere, A., Nowicki, B., Saxen, H., Siitonen, A., Korhonen, T.K.: Expression of P, type-1 and type-1C fimbriae of Escherichia coli in the urine of patients with acute urinary tract infections. J. Infect. Dis., **156:** 567, 1987.
19. Kisielius, P.V., Schwan, W.R., Amundsen, S.K., Duncan, J.L., Schaeffer, A.J.: In vivo expression and variation of Escherichia coli type 1 and P pili in the urine of adults with acute urinary tract infections. Infect. Immun., **57:** 1656, 1989.
20. Neu, H.C.; Urinary tract infections. Am. J. Med., **92:** 63, 1992
21. Stamey, T.A., Timothy, M., Millar, M., Mihara, G.: Recurrent urinary infections in adult women: The role of introital enterobacteria. Calif. Med. **115:** 1, 1971.
22. Stamey, T.A., Sexton, C.C.: The role of vaginal colonization with Enterobacteriaceae in recurrent urinary infections. J. Urol., **113:** 214, 1975.
23. Smith, H.W.: Microbial surfaces in relation to pathogenicity. Bacteriol. Rev. **41:** 475, 1977.
24. Ofek, J., Beachey, E.H.: General concepts and principles of bacterial adherence in animals and men. In: Bacterial Adherence. London: Chapman and Hall, pp. 3–29, 1980.
25. Savage, D.C. In: Microbial Pathogenicity in Man and Animals. Edited by H. Smith, J.H. Pierce. London: Cambridge Press, pp. 25–57, 1972.
26. Schaeffer, A.J.: Cystitis and Pyelonephritis. In: The Biologic and Clinical Basis of Infectious Diseases. Edited by G.P. Youmans, P.Y. Paterson, H.M. Sommers. Philadelphia: W.B. Saunders Publishers, pp. 418–35, 1985.
27. Fowler, J.E., Stamey, T.A.: Studies of introital colonization in women with recurrent urinary infections. J. Urol., **117:** 472, 1977.
28. Schaeffer, A.J., Jones, J.M., Dunn, J.K.: Association of in vitro Escherichia coli adherence to vaginal and buccal epithelial cells with susceptibility to women to recurrent urinary tract infections. N. Engl. J. Med., **304:** 1062, 1981.
29. Schaeffer, A.J., Amundsen, S.K., Schmidt, L.N.: Adherence of Escherichia coli to human urinary tract epithelial cell. Infect. Immun. Dis., **24:** 753, 1979.
30. Raz, R., Stamm, W.E.: A controlled trial on intravaginal estriol in postmenopausal women with recurrent urinary tract infection. N. Engl. J. Med., **329:** 753, 1993.
31. Strous, G.J., Dekker, J.: Mucin-type glycoproteins. Crit. Rev. Bioch. Mol. Biol., **27:** 57, 1992.
32. Allen, A.: Structure of gastrointestinal fluid glycoproteins and the viscous and gel-forming properties of fluid. Br. Med. Bull., **34:** 28, 1987.
33. Carlstedt, I., Sheehan, J.K.: Structure and macromolecular properties of cervical fluid glycoproteins. In: Fluid and Related Topics. Edited by E. Chandler, N.A. Radcliffe. Cambridge: University of Cambridge. Symp. Soc. Exp. Biol. **43:** 289, 1989.
34. Thornton, D.J., Davies, J.R., Kraayenbrink, M., Richardson, P.S., Sheehan, J.K., Carlstedt, I.; Mucus glycoproteins from normal human tracheobronchial secretion. Biochem. J., **265:** 179, 1990.
35. Rose, M.C.: Characterization of human tracheobronchial mucin glycoproteins. Meth. Enzymol., **179:** 3, 1989.
36. Dekker, J., Strous, G.J.: Covalent oligomerization of rat gastric mucin occurs in the rough endoplasmic reticulum, in N-glycosylation-dependent and precedes initial O-glycosylation. J. Biol. Chem., **265:** 18116, 1990.
37. Ederer, G.M., Matsen, J.M.: Colonization and infection with Pseudomonas cepacia. J. Infect. Dis., **125:** 613, 1972.
38. Sajjan, U.S., Corey, M., Karmali, M.A., Forstner, J.F.: Binding of Pseudomonas cepacia to normal human intestinal mucin and respiratory mucin from patients with cystic fibrosis. J. Clin. Invest., **89:** 648, 1992.
39. Wanke, C.A., Cronan, S., Goss, C., Chadee, K., Guerrant, R.L.: Characterization of binding Escherichia coli strains which are enteropathogens to small-bowel mucin. Infect. Immun., **58**: 794, 1990.
40. Forstner, J.F., Wesley, A., Mantle, M., Kopelman, H., Man, D., Forstner, G.: Abnormal fluid: Nominated but not yet elected. J. Pediatr. Gastroenterol. Nutr., **3**(Suppl)**:** S67, 1994.
41. Podolsky, D.K., Fournier, D.A.: Alterations in mucosal content of colonic glycoconjugates in inflammatory bowel disease by monoclonal antibodies. Gastroenterology, **95:** 379, 1988.
42. Drumm, B., Roberton, A.M., Sherman, P.M.: Inhibition of attachment of Escherichia coli RDEC-1 to intestinal microvillus membranes by rabbit ileal fluid and mucin in vitro. Infect. Immun., **56:** 2437, 1988.
43. Mantle, M., Basaraba, L., Peacock, S.C., Gall, D.G.: Binding of Yersinia enterolitica to rabbit intestinal brush border membranes, fluid, and mucin. Infect. Immun., **57**: 3292, 1989.
44. Mantle, M., Husar, S.D.: Adhesion of Yersinia enterocolitica to purified rabbit and human intestinal mucin. Infect. Immun., **61:** 2340, 1993.

45. Duncan, J.L.: Differential effects of Tamm-Horsfall protein on adherence of Escherichia coli to transitional epithelial cells. J. Infect. Dis., **158:** 1379, 1988.
46. Venegas, M.F., Navas, E.L., Gaffney, R., Duncan, J.L., Anderson, B.E., Schaeffer, A.J.: Binding of type 1 piliated Escherichia coli to vaginal mucus. Infect. Immun. **63:** 416, 1995.
47. Gaffney, R.A., Venegas, M.F., Kanerva, C., Navas, E.L., Anderson, B.E., Duncan, J.L., Schaeffer, A.J.: Effect of vaginal fluid on adherence of type 1 piliated Escherichia coli to epithelial cells. J. Infect. Dis., **172:** 1528, 1995.
48. Schaeffer, A. J., Navas, E.L., Venegas, M.F., Anderson, B.E., Kanerva, C., Chmiel, J.S., Duncan, J.L.: Variation of blood group antigen expression on epithelial cells and mucus in secretor and nonsecretor women. J. Urol., **152:** 859, 1994.

27

HOST FACTORS IN SUSCEPTIBILITY TO URINARY TRACT INFECTIONS

Ann Stapleton

Department of Medicine
Division of Allergy and Infectious Diseases
University of Washington
Seattle, Washington 98195

1. EPIDEMIOLOGY OF URINARY TRACT INFECTIONS IN WOMEN

Acute uncomplicated urinary tract infections (UTIs) occur in millions of young women each year and the majority of these infections are caused by Escherichia coli.

The frequency of this clinical problem is illustrated by the results of a recent large study (323 person-years of observation) of UTI among women followed in a university health clinic or enrolled in a health maintenance organization.[1] This study demonstrated an annual incidence of acute cystitis of 0.5 to 0.7 episodes per person-year.[1] In addition, from previous studies, it is estimated that 20–40% of women who experience an initial urinary tract infection (UTI) go on to develop recurrent infections.[2] Since bacteriuria is more frequent among older women,[3] recurrent UTI (RUTI) may be even more common among women over age 60. An estimated 10 to 15% of women in this age group have frequent recurrences.[4]

2. KNOWN FACTORS LEADING TO AN INCREASED SUSCEPTIBILITY TO UTI

Among selected populations of pre- and post-menopausal women, certain underlying factors predictably may lead to an increased susceptibility to UTI. For example, women of any age with a history of renal stones, urinary tract structural abnormalities, underlying medical conditions causing bladder dysfunction, or whose urinary tracts have been instrumented have an increased risk of recurrent infections. Certain functional and anatomic abnormalities of the urinary tract, such as post-void residual urine, reduced blad-

Advances in Bladder Research, edited by Baskin and Hayward.
Kluwer Academic / Plenum Publishers, New York, 1999.

der tone and/or prolapse, uterine prolapse, or the alterations in the normal vaginal flora towards a uropathogen-predominant state as a result of estrogen deficiency[5,6] are more common among post-menopausal woman and appear to increase susceptibility to RUTI.

3. RECURRENT UTI IN WOMEN WITH NORMAL URINARY TRACTS

Most otherwise healthy pre-menopausal women with RUTI do not have anatomic or functional abnormalities of the urinary tract. Thus, in nearly all such women, cystography and intravenous pyelography are normal.[7,8] In addition, among post-menopausal women, the above-mentioned structural and functional factors predisposing to recurrent infections are not universally present among and thus do not account for all recurrent infections among older women. Thus, it is not clear why recurrent UTI develops in some women whose urinary tracts appear to be normal from a functional and anatomical standpoint develop recurrent UTI, while other such women are spared.

4. COLONIZATION PRECEDING UTI

Colonization of the vaginal and periurethral mucosa with the infecting bacteria appears to be a necessary prerequisite to E. coli UTI.[9] Several lines of evidence suggest that susceptibility to colonization and other aspects of the interaction between the infecting uropathogenic E. coli and women's uroepithelial cells are key to understanding the increased susceptibility to RUTI among otherwise normal young women. Women with RUTI have prolonged vaginal colonization mucosa with uropathogenic E. coli, even between times of active infection.[10] When vaginal, buccal, and voided uroepithelial cells were collected from women with a history of recurrent UTI and tested in bacterial adherence assays, three-fold more E. coli adhered to these women's cells as compared with cells from control women without a history of RUTI.[11] Thus, the urogenital mucosa of women with RUTI appears to be more prone to bacterial adherence and persistence when examined either macroscopically as an organ or on a cellular level. Taken together, these data suggest that vaginal and uroepithelial cells from women with a history of recurrent UTI may express different or greater numbers of receptors for infecting E. coli as compared with cells from women without recurrences.

5. ADHESINS IN THE PATHOGENESIS OF E. COLI UTI

Adherence to the urothelial mucosa by E. coli adhesins such as pap family of adhesins, Type 1 fimbriae, and the Dr adhesin plays a key role in the pathogenesis of uncomplicated E. coli UTI.[12] In epidemiological studies, 50–65% of strains from women with cystitis possess pap adhesins vs. 10–15% of fecal isolates from patients without UTI.[12] Thus, pap adhesins are significantly associated with strains causing acute cystitis and are present in the majority of these strains. Variations in the DNA sequences of the adhesin define three classes of pap-related adhesins: class I, exemplified by the cloned isolates containing $papG_{J96}$; class II, represented by isolates expressing $papG_{IA2}$; and class III, such as isolates expressing $prsG_{J96}$.[12,13] The pap-encoded adhesins bind to specific glycosphingolipids (GSLs) in vitro;[14] our studies have focused on studying the expression of these compounds in urogenital tissues.

6. GLYCOSPHINGOLIPIDS AS UROEPITHELIAL CELL RECEPTORS FOR E. COLI

Glycosphingolipids (GSLs) are important components of the glycocalyx surrounding mammalian cells and consist of an oligosaccharide moiety exposed on the cell surface, covalently linked to a lipid portion embedded in the outer leaflet of the plasma membrane.[15] These molecules serve as antigens and cell surface markers of differentiation and oncogenesis.[15] GSLs are involved in mammalian cell-cell interactions and they serve as eukaryotic cell adhesion sites for many pathogens and their toxins, including uropathogenic E. coli.[15]

Based on common carbohydrate core structures the GSLs are classified into families, such as the globoseries GSLs (Table 1).[16] This family of GSLs is characterized by a common internal galactose-α-1-4 galactose moiety which is the minimal receptor for uropathogenic E. coli expressing pap-encoded adhesins.[14] GSLs are synthesized by specific glycosyltransferases which sequentially link monosaccharides to specific carbohydrate core chains.[15] The expression of GSLs in a given tissue is determined by the specificities and quantities of these glycosyltransferases, the availability of substrate molecules for the enzymes, and competition among enzymes with the same substrate.[15] Some glycosyltransferases are tissue-specific and are regulated by the secretor gene (Se) or by polymorphic genes such as the ABO blood group genes.[16] As a result, GSLs on the cell surface are highly specific to a given host and play an important role both in determining tissue tropism for microbial pathogens and an individual host's susceptibility to specific infectious diseases.

7. HOST GENETIC FACTORS IN RECURRENT UTI

Several epidemiological studies have demonstrated that women with a history of recurrent UTI are three to four times more likely to be nonsecretors of ABH blood group antigens than are women without such a history.[17–19] The secretor gene, Se, encodes a glycosyltransferase which transfers a fucose residue to specific acceptor molecules, resulting in detectable blood group antigens in the secretions of people carrying the gene.[20] The Lewis genes act on the same acceptor molecules and if both Lewis and secretor genes are active in a given tissue, the Lewis blood type will reflect the Lewis antigens found in the tissues.[20] Separate fucosyltransferases are responsible for the synthesis of ABH blood group antigens on erythrocytes, as determined by the individual's ABO type.[20] Individuals who lack the secretor gene thus do not synthesize GSLs containing ABH blood groups epitopes in tissues under control of Se and instead may synthesize alternative epitopes.

Table 1. Structures of globoseries glycosphingolipids

GSL	Structure	Source
CTH	Gal-1-4 Galβ1-4 Glcβ1-1cer	human erythrocytes[21]
Globoside	GalNAcβ1-3 Gal-1-4 Galβ1-4 Glcβ1-1cer	human erythrocytes[21]
Gal globoside	Galβ1-3 GalNAcβ1-3 Gal-1-4 Galβ1-4 Glcβ1-1cer	human kidney[24]
SGG	NeuAc-2-3 Galβ1-3 GalNAcβ1-3 Gal-1-4 Galβ1-4 Glcβ1-1cer	human kidney[24]
DSGG	NeuAc-2-3 Galβ1-3 (NeuAc-2-6) GalNAcβ1-3 Gal-1-4 Galβ1-4 Glcβ1-1cer	human kidney[24]

8. GENETICALLY DETERMINED HOST CELL RECEPTORS FOR UROPATHOGENIC E. COLI

In our investigations, we integrated three areas of prior investigation by our group and others: (1) epidemiological observations regarding the association of the nonsecretor state with RUTI, (2) the propensity of women with a history of UTI to have increased degrees of vaginal E. coli colonization as compared with control subjects; and (3) laboratory observations about the adherence and bacterial binding characteristics of uropathogenic E. coli. At the onset of our investigations, it was unknown whether the secretor gene influenced the expression of and GSL in vaginal tissue nor if the synthesis of any globoseries GSL was influenced by this gene. To address these issues, we collected vaginal epithelial cells from secretors and nonsecretors and extracted the GSL fractions from vaginal epithelial cells pooled in equal quantities from women in the two groups.[21] We demonstrated that cells from nonsecretors express two unique globoseries GSLs, sialosyl galactosyl globoside (SGG) and disialosyl galactosyl globoside (DSGG).[21] These conclusions were drawn from a series of studies using several methods. First, we investigated the GSL fractions extracted from the secretors and nonsecretors, respectively, to identify GSL moieties that were capable of binding uropathogenic E. coli expressing pap-encoded adhesins.[21] We used high-performance thin layer chromatography (HPTLC) bacterial overlay assays, in which the GSLs were separated on HPTLC by size and then overlaid with a suspension of a metabolically labeled, wild-type E. coli R45[21] which expresses a Class II pap-encoded adhesin.[22] These studies showed two compounds reacting with bacteria that were found among GSLs extracted from nonsecretors' vaginal epithelial cells but not from GSLs from secretors' cells; the compounds migrated in positions corresponding with SGG and DSGG.[21] These bands were then purified from HPTLC plates and the identity of the compounds was confirmed using HPTLC immunostaining, radioimmunoassays and immunohistochemical staining using a monoclonal antibody directed against SGG.[21] These studies demonstrated for the first time that the secretor gene influences the biosynthesis of globo-series GSLs in the vaginal epithelium and suggested that genetically determined differences in receptor moieties in this tissue might explain the increased susceptibility of nonsecretors to UTI.[18,19,21,23]

In a more recent study, we demonstrated that SGG and DSGG are expressed in normal human kidney by isolating and purifying the compounds from this tissue.[24] This was accomplished through an exhaustive series of standard purification steps[25,26] to produce a neutral fraction and 0.05 M, 0.15 M, and 0.45 M ammoniun acetate fractions, the latter containing charged GSLs such gangliosides. Normal phase silica gel high-performance liquid chromatography (HPLC)[27] was then used to further separate the charged fractions. SGG and DSGG were identified and purified from these HPLC fractions by repeated cycles of HPTLC immunostaining[28,29] using specific MAbs[21,30] to identify the compounds, bacterial overlay assays[21] using uropathogenic E. coli carrying pap-encoded to identify globoseries GSLs, HPTLC in multiple solvent systems, and preparative HPTLC.[31] Figures 1A and 1B show examples of steps in this purification procedure, in which bacterial overlay assays using uropathogenic E. coli R45, expressing a pap-encoded adhesin, were performed to identify fractions containing relevant charged globoseries compounds, such as SGG and DSGG. When SGG and DSGG were purified, definitively determined the structure of purified SGG from human kidney using proton nuclear magnetic resonance spectroscopy (^{1}H-NMR), mass spectroscopy and linkage analysis.[32]

Next, we found that SGG and DSGG isolated from human kidneys bound each of 5 representative wild-type or recombinant uropathogenic E. coli expressing P fimbriae bearing pap-encoded adhesin Class I, II, or III, demonstrating that these are relevant bacterial

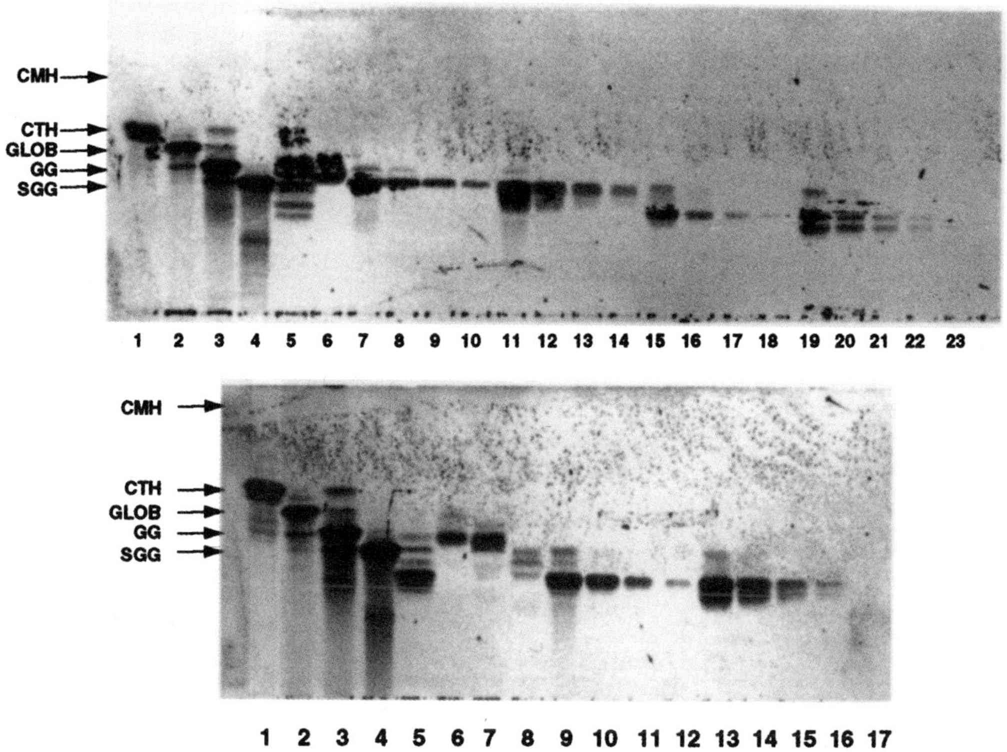

Figure 1. Purification procedures for isolating SGG and DSGG from human kidney. As described above, SGG and DSGG were purified from the upper phase GSL fractions through anion exchange chromatography followed by normal phase silica gel high-performance liquid chromatography (HPLC). To assess the relative amounts and purity of SGG or DSGG in specific HPLC fractions separated from the 0.15 M ammonium acetate and 0.45 M ammonium acetate fractions from human kidney, HPTLC bacterial overlay and immunostaining assays were used. Shown are autoradiographs from two examples of profiles of relevant HPTLC fractions using bacterial overlay assays with metabolically [^{35}S]methionine-labeled E. coli R45, which specifically binds globoseries GSLs. (A, top) 0.15 M ammonium acetate fractions from HPLC. Lane 1: ceramide trihexaosyl (CTH; Gb3) standard; Lane 2: globoside (GLOB; Gb4) standard; Lane 3: galactosyl globoside (GG) standard; Lane 4: sialosyl galactosyl globoside (SGG) standard; Lane 5: total 0.15 M ammonium acetate fraction (before HPLC); Lane 6: pooled 0.15 M ammonium acetate fractions 35–41; Lanes 7–10: pooled fractions 42–45, undiluted (Lane 7) and diluted 1:5, 1:25; and 1:125 (Lanes 8,9 and 10, respectively); Lanes 11–14: pooled fractions 46–51; undiluted (Lane 11) and diluted 1:5, 1:25; and 1: 125 (Lanes 12, 13 and 14, respectively); Lanes 15–18: pooled fractions 52–53; undiluted (Lane 15) and diluted 1:5, 1:25; and 1: 125 (Lanes 16,17 and 18, respectively); Lanes 19–22: pooled fractions 54–62; undiluted (Lane 19) and diluted 1:5, 1:25, and 1: 125 (Lanes 21 and 22, respectively); Lane 23: ceramide monohexosyl (CMH) standard. (B, bottom) 0.45 M ammonium acetate fraction from HPLC. Lane 1: ceramide trihexaosyl (CTH; Gb3) standard; Lane 2: globoside (GLOB; Gb4) standard; Lane 3: galactosyl globoside (GG) standard; Lane 4: sialosyl galactosyl globoside (SGG) standard; Lane 5: total 0.45 M ammonium acetate fraction (before HPLC); Lane 6: pooled fractions 20–42; Lane 7: pooled fractions 43–53; Lane 8: pooled fractions 54–56; Lanes 9–12: pooled fractions 57–58; undiluted (Lane 9) and diluted 1:5, 1:25, and 1:125 (Lanes 10, 11 and 12, respectively); Lanes 13–16: pooled fractions 59–64; undiluted (Lane 13) and diluted 1:5, 1:25, and 1:125 (Lanes 14, 15 and 16 , respectively); Lane 17: ceramide monohexosyl (CMH) standard.

binding entities for any of the known pap-encoded adhesins of E. coli.[24] To further investigate the potential relevance of the expression of these GSLs in the vaginal epithelium of nonsecretors, we metabolically radiolabeled the five representative E. coli isolates expressing pap-encoded adhesins and measured their relative binding affinities to serial dilutions of SGG and DSGG as well as to globotriaosylceramide (Gb_3) and globotetraosylceramide (Gb_4), two other globo-series GSLs present in urogenital tissues.[24] We found that each of the 5 representative E. coli isolates bound to SGG with the highest apparent avidity as compared with binding to DSGG, Gb_3 and Gb_4. These studies further suggest that SGG likely plays an important role in the pathogenesis of UTI and that its presence may account for the increased binding of E. coli to uroepithelial cells from nonsecretors and for their increased susceptibility to recurrent UTI. Further studies are needed to more extensively define the expression of SGG and DSGG in epithelial tissues throughout the urogenital tract. We are presently studying the GSL composition of normal human bladder epithelium, including investigating for the presence of SGG and DSGG.

9. ACQUIRED HOST FACTORS PREDISPOSING TO RUTI

Acquired factors may also affect susceptibility to UTI. The possibility that time-limited exposures or behaviors may predispose to RUTI is raised by observational studies of the natural history of UTI in young women have characterized by a temporal clustering phenomenon, in which several month periods marked by repeated infections are followed by infection free intervals.[33] Although it is not clear which specific factors might be responsible for such clusters of recurrent UTI, some of the same exposures and behaviors that increase the risk of sporadic episodes of acute uncomplicated UTI may be important. For example, multiple studies have shown that use of the diaphragm with spermicide for contraception[1,17,34–38] and recent sexual intercourse[1,17,35,38–41] greatly magnify the risk of sporadic UTI. The effect of spermicide use, with or without a diaphragm, appears to be mediated by increasing vaginal colonization with E. coli and other uropathogens,[17,36,37] perhaps through a selective microbicidal effect on the normal, protective vaginal flora such as lactobacilli.[36] Similarly, the antecedent use of antibiotics, either for treatment of a previous UTI or for other infections in the period 15 to 28 days before onset of UTI increases the proximal risk of for UTI by 2.57 to 5.83.[42] This exposure may also act by altering the indigenous urogenital flora towards a uropathogen-dominated flora.[10,43]

Many hygiene- or personal habit-related behaviors that might be postulated to affect the risk of UTI have not been thoroughly investigated or existing studies have produced conflicting results. Studies of women's voiding habits, fluid intake and the temporal sequence of micturition before and/or after intercourse are notable in this regard.[1,38,41,44,45] Behaviors such as perineal cleansing and the direction of wiping after bowel movements, menstrual products used, douching, and the use of vaginal deodorant products, have not been demonstrated to influence the risk of UTI.[38,39,41] Ongoing large epidemiological studies[46] will likely clarify whether personal habits and hygiene alter the risk of UTI.

10. SUMMARY

In summary, a variety of intrinsic and acquired factors influence the risk of RUTI in otherwise normal women, including history of prior UTIs, the woman's genetic background, and exposures to spermicides, sexual activity and antibiotics. Further studies are directed towards understanding the interplay between these factors and their relative importance among various subpopulations of women with RUTI, such as otherwise healthy pre- and post-menopausal women.

ACKNOWLEDGMENTS

This work was supported by Grants No. AI-01115 and DK-40045 from the National Institutes of Health and by Grant No. UIS DE950144 from the Women's Health Initiative, U.S. Army Medical Research and Materiel Command.

REFERENCES

1. Hooton T.M., Scholes D., Hughes J.P., et al.: A prospective study of risk factors for urinary tract infection in young women. N. Engl. J. Med., **335:** 468, 1996.
2. Mabeck C.E.: Treatment of uncomplicated urinary tract infection in nonpregnant women. Postgrad. Med. J., **48:** 69, 1972.
3. Boscia J.A., Kobasa W.D., Knight R.A., Abrutyn E., Levison M.E., Kaye D.: Epidemiology of bacteriuria in an elderly ambulatory population. Am. J. Med.;**80:** 208, 1986.
4. Romano J.M., Kaye D.: UTI in the elderly: common yet atypical. Geriatrics, 1981;**36:**113–115.
5. Parsons C.L., Schmidt J.D.: Control of recurrent lower urinary tract infections in the postmenopausal woman. J. Urol. **128:** 1224, 1982.
6. Raz R., Stamm W.E.: A controlled trial of intravaginal estrogen in postmenopausal women with recurrent urinary tract infection. N. Engl. J. Med.;**329:** 753, 1993.
7. Engle G., Schaeffer A.J., Grayhack J.T., Wendel EF.: Excretory urography, cystography, and cystoscopy in the evaluation of women with urinary tract infection. J. Urol., **123:** 190, 1979.
8. Fowler J.E., Pulaski E.T.: Excretory urography, cystography and cystoscopy in the evaluation of women with recurrent urinary tract infection. N. Engl. J. Med., **304:** 462, 1981.
9. Stamey T.A., Timothy M., Millar M., Mihara G.: Recurrent urinary tract infections in adult women. The role of introital enterobacteria. Calif. Med. **115:** 1, 1971.
10. Stamey T.A., Sexton C.C.: The role of vaginal colonization with Enterobacteriaceae in recurrent urinary tract infections. J. Urol., **113:** 214, 1975.
11. Schaeffer A.J., Jones J.M., Dunn J.K.: Association of in vitro Escherichia coli adherence to vaginal and buccal epithelial cells with susceptibility of women to recurrent urinary-tract infections. N. Engl. J. Med., **304:**1062, 1981,
12. Johnson J.R.: Virulence factors in Escherichia coli urinary tract infection. Clin. Microbiol. Rev., **4:** 80, 1991.
13. Marklund B.-I., Tennent J.M., Garcia E., et al.: Horizontal gene transfer of the Escherichia coli pap and prs operons as a mechanism for the development of tissue-specific adhesive properties. Molecular Microbiology, **6:** 2225, 1991.
14. Leffler H., Svanborg-Eden C.: Chemical identification of a glycosphingolipid receptor for Escherichia coli attaching to human urinary tract cells and agglutinating human erythrocytes. F.E.M.S. Microbiol. Lett., **8:** 127, 1980.
15. Hakomori S-I.: Glycosphingolipids in cellular interaction, differentiation, and oncogenesis. Annu. Rev. Biochem., **50:** 733, 1981.
16. Clausen H., Hakomori S-I.: ABH and related histo-blood group antigens; immunochemical differences in carrier isotypes and their distribution. Vox. Sang. **56:** 1, 1989.
17. Hooton T.M., Roberts P.L., Stamm W.E.: Effects of recent sexual activity and use of a diaphragm on the vaginal microflora. Clin. Infect. Dis.,**19:** 274, 1989.
18. Kinane D.F., Blackwell C.C., Brettle R.P., Weir D.M., Winstanley F.P., Elton R.A.: ABO blood group, secretor state and susceptibility to recurrent urinary tract infection in women. Br. Med. J., 28:7, 1982.
19. Sheinfeld J., Schaeffer A.J., Cordon-Cardo C., Rogatko A., Fair W.R.: Association of the Lewis blood-group phenotype with recurrent urinary tract infections in women. N. Engl. J. Med., **320:** 773, 1989.
20. Watkins W.M.: Biochemistry and genetics of the ABO, Lewis, and P blood group systems. In: Harris H., Hirschorn K., eds. Advances in Human Genetics 10. New York, N.Y.: Plenum Press, p1, 1980.
21. Stapleton A., Nudelman E., Clausen H., Hakomori S-I., Stamm W.E.: Binding of uropathogenic Escherichia coli R45 to glycolipids extracted from vaginal epithelial cells is dependent on the histo-blood group secretor status. J. Clin. Invest., **90:** 965, 1992.

22. Johnson J.R., Russo T.A., Brown J.J., Stapleton A.E.: papG alleles of Escherichia coli strains causing first-episode or recurrent acute cystitis in adult women. Clin. Infect. Dis., **23:** 920, 1996;
23. Lomberg H., Cedergren B., Leffler H., Nilsson B., Carlstrom A-S., Svanborg-Eden C.: Influence of blood group on the availability of receptors for attachment of uropathogenic Escherichia coli. Infect. Immun., **51:** 919, 1986.
24. Stapleton A.E., Stroud M.R., Hakomori S.I., Stamm W.E.: The globo-series glycosphingolipid sialosyl galactosyl globoside is found in urinary tract tissues and is a preferred binding receptor in vitro for uropathogenic Escherichia coli expressing pap-encoded adhesins. Infect. Immun., in press, 1998.
25. Folch J., Lees M., Sloane Stanley G.H.: A simple method for the isolation and purification of total lipides from animal tissues. J. Biol. Chem., **226:** 497, 1957.
26. Yu R.K., Ledeen R.W.: Gangliosides of human, bovine, and rabbit plasma. J. Lipid Res., **13:** 680, 1972.
27. Kannagi R., Watanabe K., Hakomori S.: Isolation and purification of glycosphingolipids by high-performance liquid chromatography. Methods Enzymol., **138:** 3, 1987.
28. Kannagi R., Nudelman E., Levery S.B., Hakomori S.: A series of human erythrocyte glycosphingolipids reacting to the monoclonal antibody directed to a developmentally regulated antigen, SSEA-1. J. Biol. Chem., **257:** 14865, 1982.
29. Magnani J.L., Smith D., Ginsburg V.: Detection of gangliosides that bind cholera toxin: direct binding of ^{125}I-labeled toxin to thin-layer chromatograms. Anal. Biochem., **109:** 399, 1980.
30. Saitoh S., Levery S.B., Salyan M.E.K., Goldberg R.I., Hakomori S.: Common tetrasaccharide epitope NeuAca2–3Galβ1–3 (NeuAca2–6) GalNAc, presented by different carrier glycosylceramides or O-linked peptides, is recognized by different antibodies and ligands having distinct specificities. J. Biol. Chem., **269:** 5644, 1994.
31. Nudelman E., Kannagi R., Hakomori S., et al.: A glycolipid antigen associated with Burkitt lymphoma defined by a monoclonal antibody. Science, (Wash DC) **220:** 509, 1983.
32. Stroud M.R., Stapleton A.E., Levery S.B.: The P-blood group related glycosphingolipid sialosylgalactosylgloboside (SGG) as a preferred ligand for uropathogenic E. coli: Isolation and structural characterization of SGG from human kidney. Submitted 1998.
33. Stamm W.E., McKevitt M., Roberts P.L., White N.J.: Natural history of recurrent urinary tract infections in women. Rev, Infect. Dis., **13:** 77, 1991.
34. Fihn S.D., Latham R.H., Roberts P., Running K., Stamm W.E.: Association between diaphragm use and urinary tract infection. J.A.M.A., **254:** 240, 1985.
35. Foxman B., Frerichs R.R.: Epidemiology of urinary tract infection: I. Diaphragm use and sexual intercourse. Am. J. Public Health., **75:** 1308, 1985
36. Hooton T.M., Hillier S., Johnson C., Roberts P.L., Stamm W.E.: Escherichia coli bacteriuria and contraceptive method. J.A.M.A., **265:** 64, 1991.
37. Hooton T.M., Fennell C., Clark A., Stamm W.E.: Nonoxynol-9: differential antibacterial activity and enhancement of bacterial adherence to vaginal epithelial cells. J. Infect. Dis., **164:** 1216, 1992,
38. Strom B.L., Collins M., West S.L., Kreisberg J., Weller S.: Sexual activity, contraceptive use, and other risk factors for symptomatic and asymptomatic bacteriuria: a case-control study. Ann. Intern. Med.,**107:** 816, 1987.
39. Leibovici L., Alpert G., Laor A., Kalter-Leibovici O., Danon Y.L.: Urinary tract infections and sexual activity in young women. Arch. Intern. Med., **147:** 345, 1987.
40. Nicolle L.E., Harding G.K.M., Preiksaitis J., Ronald A.R.: The association of urinary tract infection with sexual intercourse. J. Infect. Dis., **146:** 579, 1982.
41. Remis R.S., Gurwith M.J., Gurwith D., Hargrett-Bean N.T., Layde P.M.: Risk factors for urinary tract infection. Am. J. Epidemiol., **126:** 685, 1987.
42. Smith S., Hughes J.P., Hooton T.M., et al.: Antecedent antimicrobial use increases the risk of uncomplicated cystitis in young women. Submitted 1996.
43. Herthelius B.M., Hedstrom K-G., Mollby R., Nord C.E., Pettersson L., Winberg J.: Pathogenesis of urinary tract infections. Amoxicillin induces genital Escherichia coli colonization. Infection, **16:** 263/7, 1988.
44. Adatto K., Doebele K.G., Galland L., Granowetter L.: Behavioral factors and urinary tract infections. J.A.M.A., **241:** 2525, 1979.
45. Ervin C., Komaroff A.L., Pass T.M.: Behavioral factors and urinary tract infection [letter]. J.A.M.A., **243:** 330, 1980.
46. Hooton T.M., Scholes D., Stapleton A.E., et al.: A prospective study of asymptomatic bacteriuria in young sexually active women. Program and abstracts of the 35th annual meeting of the Infectious Diseases Society of America abstract no. 457,. 1997.

28

INDUCTION OF NITRIC OXIDE SYNTHASE WITH URINARY TRACT INFECTIONS

Marcia A. Wheeler, Shannon D. Smith, and Robert M. Weiss

Section of Urology
Yale School of Medicine
New Haven, Connecticut 06520-9041

1. WHAT AND WHY NO?

Nitric oxide (NO) is a small lipophilic gaseous molecule which is released from non-adrenergic, non-cholinergic (NANC) neurons and is synthesized in a variety of mammalian tissues including macrophages and neutrophils. NO is synthesized from L-arginine by the enzyme nitric oxide synthase (NOS) with L-citrulline being a co-product. In many systems, NO activates a cytosolic guanylyl cyclase enzyme with the resultant formation of the second messenger, cyclic GMP. Cyclic GMP, in turn, initiates a cascade of protein phosphorylations which leads to functional responses.[1]

There are three forms of NOS, historically defined as neuronal (nNOS, NOS1), cytokine inducible (iNOS, NOS2) and endothelial (eNOS, NOS3). These NOS isoforms are derived from separate genes and are regulated by and regulate, diverse signalling pathways.[2] Human inducible nitric oxide synthase, like other nitric oxide synthases requires nicotinamide-adenine-dinucleotide phosphate (NADPH), flavin-adenine-dinucleotide, flavin mononucleotide, tetrahydrobiopterin, calmodulin and heme to convert its substrate, L-arginine, to nitric oxide and L-citrulline.[3,4] nNOS has been identified as a mediator of NANC induced relaxation in the urinary tract, most notably in the urethra.[5,6,7] Cyclic GMP and NO donors relax urinary tract tissues including the bladder.[8,9] Changes in both NO and cGMP occuring during urinary tract infections (UTIs),[10,11] transplant rejection,[12] interstitial cystitis,[10,13,14] and other pathologic conditions involving the urinary tract may lead to changes in the micturition response as well as in inflammatory and anti-inflammatory responses. Characterization of the role of eNOS and iNOS in the urinary tract is now being undertaken. This review will focus on the induction and role of iNOS in the urinary tract, as we have recently reviewed the role of nitric oxide in interstitial cystitis.[15] Our recent interests have focused on the induction of iNOS, both in patients with UTIs and in the rat bladder which had been intravesically instilled with lipopolysaccharide (LPS)/protamine.[16]

Advances in Bladder Research, edited by Baskin and Hayward.
Kluwer Academic / Plenum Publishers, New York, 1999.

2. INDUCTION OF NO AND OTHER INFLAMMATORY MEDIATORS IN DISEASES OF THE URINARY TRACT

Recently, we have focused our attention on the role of inducible NOS in inflammatory and infectious processes of the urinary tract including urinary tract infections, interstitial cystitis, and renal transplant rejection. There is accumulating information that the induction of iNOS in vivo is secondary to the release of cytokines and other mediators like tumor necrosis factor (TNF) and interleukin-1 (IL-1), as pretreatment of rats with monclonal antibodies to TNF or with inhibitors of IL-1 has been shown to attenuate the induction of iNOS. IL-1 and TNFα have been shown to act alone or in combination with low doses of lipopolysaccharide (LPS) to produce a shock-like state, and to induce iNOS in vitro.[17] These cytokines along with LPS may regulate iNOS expression at more than one level. For example, INFγ not only synergizes with LPS to induce transcription of iNOS, but also stabilizes iNOS mRNA.[1] On the other hand, some anti-inflammatory cytokines, including interleukin-4 (IL-4), transforming growth factor–β (TGFβ) and interleukin-10 (IL-10) can down-regulate iNOS induction. IL-4 inhibits endotoxin stimulated release of TNF, IL-1 and prostanoids from human monocytes in vitro and inhibits the induction of iNOS in murine macrophages in vitro by interfering with iNOS transcription.[1] TGFβ reduces cytokine induced iNOS activity by inhibiting iNOS mRNA translation and by increasing iNOS protein degradation. IL-10, which is produced by human mast cells and macrophages inhibits the release of NO, perhaps by inhibiting inflammatory cytokine release.[18,19]

During infectious/inflammatory processes that occur in the bladder including bacterial infection (UTI), intravesical instillation of LPS/protamine in rats and intravesical Bacillus Calmette-Guerin (BCG) immunotherapy for carcinoma in situ, a number of chemokines, cytokines, and second messengers are elevated in the urine (Table 1). Indeed, the cytokines and chemokines, IL-1β, IL-6, IL-8, and TNFα are elevated both in the urine of patients with UTIs and in the urine of patients who are receiving BCG immunotherapy. IL-10, which may inhibit inflammatory cytokine and NO release is increased in the urine of patients receiving BCG immunotherapy and in rats after LPS/protamine intravesical instillation. The induction of inflammatory cytokines in turn causes the induction of iNOS which increases urinary cGMP levels. Cytokines most likely upregulate the enzyme cy-

Table 1. Urinary mediators that are increased or induced after the induction of infection/inflammation

UTI	LPS instillation	BCG instillation
IL-1α[20, 21]	IL-6[16]	IL-1β[28]
IL-1β[20,22]	IL-10[16]	INFγ[28,29]
IL-6[21,23,24]	NO[16]	IL-2[28, 30-32]
IL-8[25,26]	cGMP[16]	IL-2 Receptor[33]
TNFα[20]	iNOS[16]	IL-6[28, 31,34]
iNOS[11]		IL-8[28,35]
cGMP[10]		IL-10[28]
PGEs[27]		TNFα[28,31,32]
		ICAM-1[28]

Inflammatory mediators were measured in the urine of patients who had UTIs or who received BCG instillations. RT-PCR products for iNOS, Il-6 and IL-10 were measured in rat bladder and NO and cGMP were measured in rat urine after intravesical instillation of LPS.

clooxygenase 2 (COX2), the inducible form of the rate limiting enzyme involved in prostaglandin and thromboxane synthesis.[36] Elevated levels of prostaglandin E_2 are found in the urine of patients with UTIs.[27]

3. INDUCTION OF iNOS IN THE URINE OF PATIENTS WITH UTIs

Initially, we assayed NOS activity in the particulate fraction of the urine pellet of patients with UTIs and found that NOS activity was elevated compared to NOS activity in urine pellet from controls.[37] The conversion of [^{14}C]-arginine to [^{14}C]-citrulline required NADPH, was Ca^{++} independent, and was inhibited to a greater extent by L-canavanine (68±4%) than by N^G-monomethyl-L-arginine (L-NMMA) (32±3%) or NNA (N^G-nitro-L-arginine) (5±2%). These data are consistent with human infected urine containing an inducible isoform of nitric oxide synthase.[38]

Since nitric oxide can activate soluble guanylyl cyclase, we measured urinary cGMP levels in patients with UTIs and in female controls.[10] Cyclic GMP was significantly elevated (3.72 ± 0.81 μmol. cyclic GMP/g creatinine) in female patients with UTIs compared to female controls (0.82 ± 0.14 μmol cyclic GMP/ g creatinine), another indication that NOS is induced during UTIs.

We then definitively identified a source of NOS activity in human urine as the neutrophil and using molecular techniques identified iNOS as the isoform responsible for this NOS activity.[11] To characterize the cell type in urine from patients with UTIs containing iNOS, Ficoll gradient centrifugation was performed on leukocyte positive urines from patients with and without UTIs. Ficoll gradient fractionation of cell pellets from the urine of patients with UTIs shows that the majority of the NOS activity is found in neutrophil-enriched fractions. The NOS activity is 43 fold higher in neutrophil-enriched fractions obtained from patients with UTIs than from controls. When NOS activity is expressed as NADPH dependent activity, 23.5±13.8 pmol citrulline/mg protein/min. is measured in neutrophil enriched fractions obtained from patients with UTIs compared to 0.55±0.16 pmol citrulline/mg protein/min measured in neutrophil enriched fractions isolated from controls.

Inhibition of NOS activity in human urinary leukocyte enriched fractions with a series of NOS inhibitors, is consistent with the pattern seen with human recombinant iNOS[39] and with iNOS in activated rat peritoneal neutrophils.[40] The rank order of NOS-inhibitor potency on NADPH-dependent neutrophil iNOS activity (IC_{50} in brackets) is: L-NIO [3.0μM] > L-thiocitrulline [4.1μM] > L-NMMA [7.1μM] > L-canavanine [12.1μM] > aminoguanidine [22μM] > N^G-nitroarginine methyl ester (L-NAME) [116μM] = L-NNA [>200μM]. The iNOS in human neutrophils is primarily membrane associated, which is similar to the membrane partitioning of eNOS.[41] Although human neutrophil iNOS is Ca^{++} independent, it is inhibited by the calmodulin antagonists, trifluoperazine and N-(6-aminohexyl)-5-chloro-1-naphthalene-sulfonamide (W-7), with ED_{50}s of 10μM and 39μM, respectively. This suggests that calmodulin is not bound as tightly to the human neutrophil iNOS as to the rodent form. Calmodulin antagonists do not inhibit rodent iNOS activity.[42]

To molecularly identify the NOS isoform(s) in urine from patients with UTIs, using oligonucleotide primers based on human iNOS and human eNOS,[43] we performed reverse transcriptase-polymerase chain reaction (RT-PCR) on cDNA prepared from leukocyte-enriched total RNA. A 413 bp fragment is consistently amplified with human iNOS primers. DNA sequencing of the PCR product reveals 99.9% nucleotide sequence identity with its respective cloned cDNA. No primer specific RT-PCR product for iNOS is seen in samples prepared from three strains of bacteria.

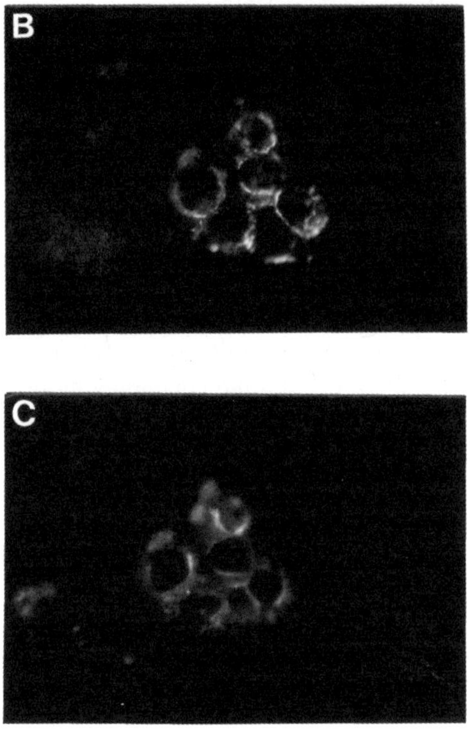

Figure 1. Immunocytochemical demonstration of iNOS in CD45-positive leukocytes. Cells labeled with CD45 (B) are visualized for iNOS(C).

Using a specific C-terminal human iNOS polyclonal antibody,[44] Western blot analysis of total lysates or 2'5' ADP-Sepharose purified samples isolated from leukocyte fractions obtained from patients with UTIs, identified iNOS protein at approximately 130 kDA, identical in M_r to the protein product seen in HEK 293 cells stably transfected with the human hepatocyte iNOS cDNA. No iNOS immunoreactivity is found in leukocyte enriched lysates, that are leukocyte esterase positive, but nitrite and culture negative or in 2'5'ADP-sepharose purified samples prepared from leukocytes obtained from uninfected patients. Using a specific eNOS monoclonal antibody and using bovine eNOS expressed in HEK 293 cells as a positive control, eNOS protein is undetectable by Western blot analysis in samples that are positive for human iNOS either by Western blotting or by RT-PCR. Neither iNOS, nNOS nor eNOS is detected in 0.8% deoxycholate extracts from three strains of bacteria which are 2'5-ADP-sepharose purified. These data, in conjunction with the Ca^{++} independence of this NOS activity demonstrates that leukocyte derived iNOS, not eNOS, is the major source of NOS activity in the urine during infection.

To determine which cell(s) contained iNOS, cells from patients with UTIs were fluorescently labelled with CD45 and sorted by flow cytometry. CD45 antigen is present on all human leukocytes. FACS sorting demonstrates that 86–93% of the total fluorescent labelling is found in the CD45 positive peak indicating that the vast majority of cells in the urine of patients with UTIs are leukocytes. CD14, a specific marker for human monocytes/macrophages, labels 6±3% of cells, indicating that the predominant population of CD45 labelled cells are neutrophils. Phase contrast microscopy confirmed that greater than 90% of these cells are neutrophils. After FACS cell sorting of CD45 labelled cells, immunocytochemistry with iNOS antibody demonstrates iNOS protein in 82.1±2.9% of the CD45 labelled cells (Figure 1). iNOS appears to be localized primarily in the perinu-

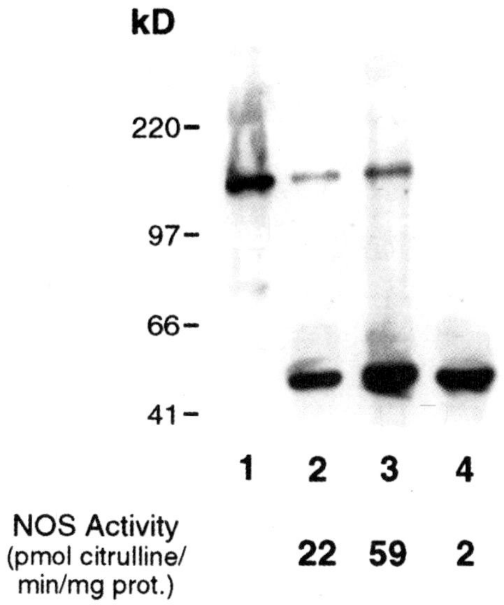

Figure 2. Immunological detection and enzyme activity of iNOS after antibiotic treatment of UTIs. Total cell lysate from urine of a patient with $>10^{-5}$ Gardnerella vaginalis at the time of diagnosis (lane 2), 3 days (lane 3) and 17 days (lane 4) after the initiation of antibiotic treatment. NOS activity was measured in membrane fractions prepared from each sample. Total lysate of HEK 293 cells stably transfected with human iNOS cDNA is shown in lane 1.

clear region of the CD45 positive neutrophils. These data provide definitive molecular proof of an iNOS in human neutrophils whose activity is increased with UTIs and provide the first evidence of the presence of a biologically active iNOS in a human inflammatory cell associated with a disease state.

It is of note that iNOS activity which is 10.7±2.9 pmol citrulline/mg protein/min. at the time of UTI diagnosis remains elevated at 23.5±8.6 pmol citrulline/mg protein/min. after 2–4 days of antibiotic therapy but decreases significantly to 1.5±0.5 pmol citrulline/mg protein/min. after 6 to 10 days of antibiotic treatment (Figure 2). NOS activity persists at a time when bacteriuria and pyuria have cleared. The induction of IL-8, a neutrophil chemoattractant which is elevated in UTIs,[25] is short lived compared to the induction of iNOS.[26] Thus, with antibiotic treatment of a UTI, iNOS activity remains elevated at a time when pyuria, and IL-8 levels are significantly decreased. The extended elevation of neutrophil iNOS during UTIs may have both anti-microbial and pro-inflammatory functions.

4. AN ANIMAL MODEL FOR THE INDUCTION OF INFLAMMATION

Human iNOS and nitric oxide have been reported to be involved in a number of infection/inflammation processes including cirrhosis,[45] asthma,[46] arthritis,[47,48] asymptomatic parasitemia in malaria,[49] ulcerative colitis,[50] cancer,[51,52] UTIs,[11,37] renal transplant rejection[12] and IC.[14] Establishing the conditions necessary for iNOS induction in human in vitro systems or characterizing NOS induction in localized systems has been fraught with difficulty.[53,54] Using an animal model of urinary tract infection provides a method to study localized inflammatory processes, since it is relatively easy to study expression of inducible enzymes and to measure induced products in the urine. Studies of relaxation-contraction responses and voiding patterns also can be analyzed. We, therefore, adapted a LPS/protamine intravesical instillation method[55] to study the induction of cytokines and iNOS in the bladder.[16]

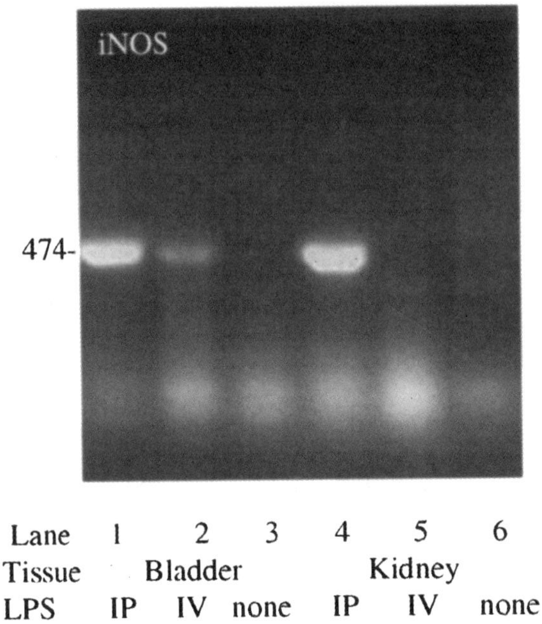

Figure 3. Induction in iNOS mRNA in bladder after intravesical instillation of protamine/LPS. mRNA was isolated from bladder (lanes 1–3) and kidney (lanes 4–6) four hours after control (protamine) instillation (lanes 3 and 6), intravesical protamine/LPS (lanes 2 and 5) and intraperitoneal LPS (lanes 1 and 4) treatment. Rodent specific iNOS primers allowed amplifications of desired cDNA.

Systemic bacterial LPS induces inflammatory responses characteristic of sepsis. LPS, the cell wall component of many bacteria including E.coli, can induce many of the host defenses required for bacterial killing.[56] LPS also can up-regulate nitric oxide and cytokine production in single cells,[3] in isolated tissues[57] and in whole animal models.[58] Intraperitoneal injection of LPS induces the expression of iNOS protein in many cells and tissues including mononuclear cells, liver, spleen, lung, kidney and bladder epithelium.[59] Instillation of protamine followed by LPS into rat bladder produces a localized inflammatory response[55] similar to that seen in urinary tract infections. We therefore compared the effect of intravesical instillation of LPS with that of intraperitoneal injection of LPS, on NOS and cytokine mRNA expression in bladder and kidney, and on urinary NO and cGMP production and NOS activity in bladder, kidney and lung. Four hours after the intravesical instillation of LPS into the bladder of female Sprague Dawley rats, neutrophils infiltrate into the bladder, and mRNA for iNOS, the pro-inflammatory cytokine, IL-6 and the anti-inflammatory cytokine, IL-10, are detected in the bladder but not in the kidney (Figures 3 and 4). Induction of iNOS protein is inferred since urinary nitrate and cGMP levels are increased four hours after intravesical instillation of LPS/protamine. When LPS is injected intraperitoneally, iNOS and IL-6 mRNA are induced both in the bladder and in the kidney. These data are consistent with the effects of intravesical instillation of LPS remaining localized to the bladder. Ca^{++} independent NOS activity increases in both particulate and soluble bladder fractions when measured four hours after intravesical instillation of LPS or four hours after LPS intraperitoneal injection (Figure 5). NOS activity was not increased in the kidney or in the lung after intravesical instillation of LPS/protamine The magnitude of these increases in iNOS activity in the bladder is not as great as when LPS is intraperitoneally injected. The localized inflammatory response produced by intravesical instillation of LPS demonstrates the importance of LPS as a mediator of the host response in UTIs and supports the use of urinary measurements of nitrate and cGMP in humans as indicative of the localized induction of iNOS in UTIs.

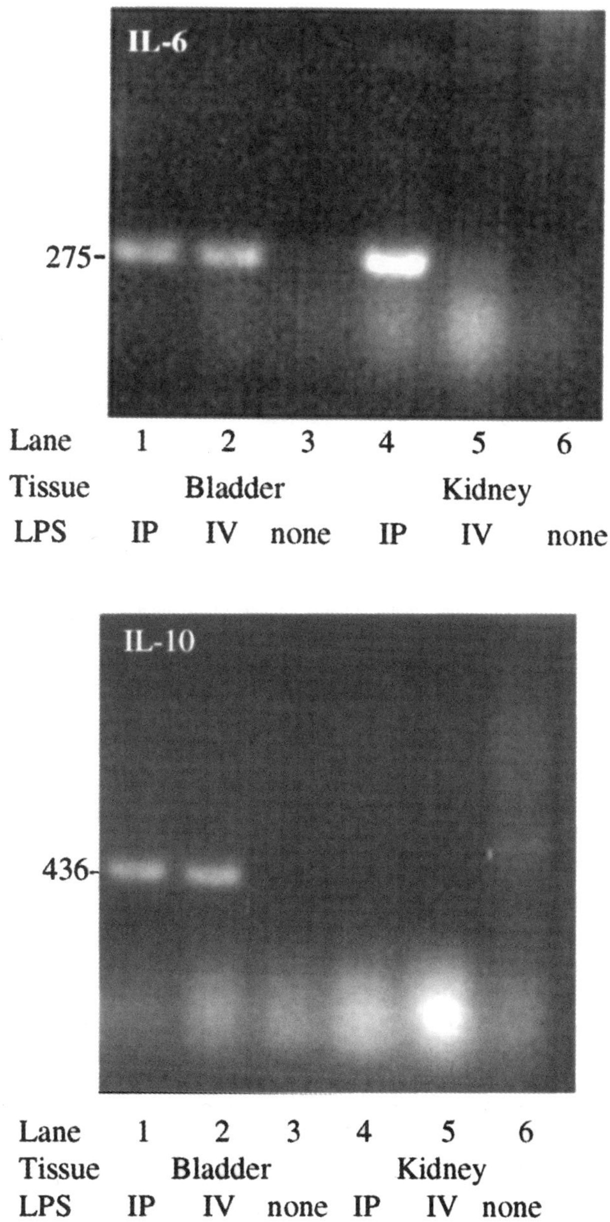

Figure 4. Induction of IL-6 (A) and IL-10 (B) in bladder after intravesical instillation of LPS. mRNA was isolated from bladder (lanes 1–3) and kidney (lanes 4–6) 4 hours after control (protamine) (lanes 3 and 6), intravesical protamine/LPS (lanes 2 and 5) or intraperitoneal LPS (lanes 1 and 4) treatment. Rodent specific IL-6 (A) or IL-10 (B) primers allowed amplifications of desired cDNA.

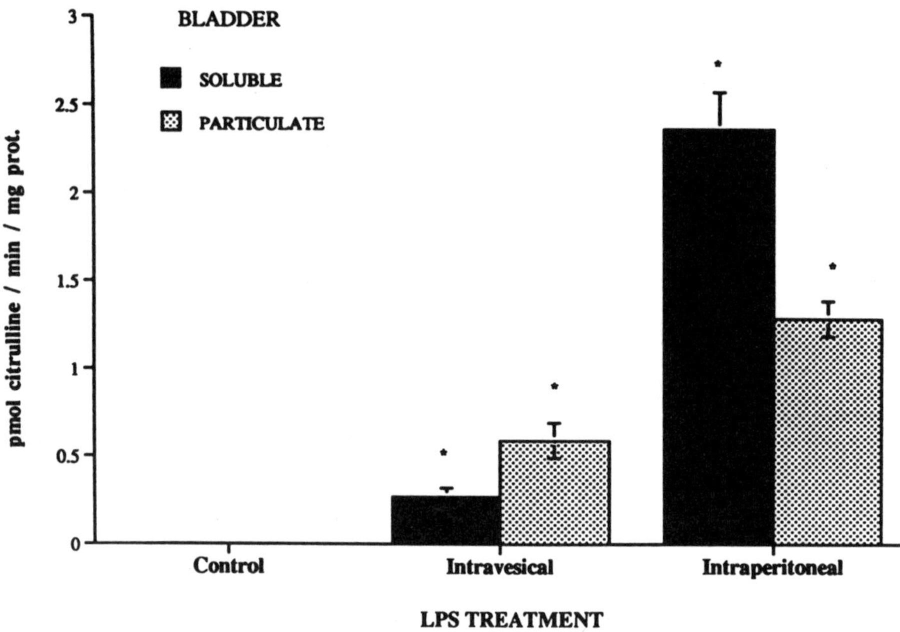

Figure 5. Intraperitoneal and intravesical LPS stimulates iNOS activity in bladder. Ca^{++}-dependent and Ca^{++}-independent, L-NMMA inhibitable NOS (iNOS) activity was measured in soluble and particulate fractions isolated from bladder 4 hours after intravesical instillation of protamine (control) or LPS/protamine and 4 hours after intraperitoneal injection of LPS, n=3 experiments done in duplicate. Data are expressed as mean plus SEM. *p>0.05 compared to control values.

5. CONCLUSIONS

During the course of a human urinary tract infection, bacteria ascend into the bladder and attach to the urothelium, causing urothelial cells to secrete pro-and anti- inflammatory cytokines and chemokines including, IL-1β, IL-6, IL-8, and TNF-α (Table 1) Smilar cytokines are elevated in the urine of patients after instillation of BCG as a form of intravesical immunotherapy for carcinoma in situ and IL-6 and IL-10 mRNA is elevated in the rat bladder after intravesical instillation of LPS/protamine. The induction of cytokines, in turn, up-regulates iNOS and most likely COX which increases urinary levels of NO, cGMP and prostaglandin E_2. We have shown that while the induction of IL-8 is short-lived in UTIs, iNOS remains elevated for several days after treatment with antibiotics. We also have shown, using a rat model, that the induction of iNOS mRNA and protein occurs rapidly, within 4 hours after intravesical instillation of LPS/protamine. We speculate that NOS may be involved in the increased urinary frequency seen in UTIs , since in the ileum[60] and in blood vessels,[61] LPS treatment reduces contractility and this reduction in contractility is reversed by treatment with NOS inhibitors. NOS also may be involved in the regulation of inflammation and in bacterial killing,[56] perhaps by reacting with superoxide to produce peroxynitrite. In an era when bacteria are becoming increasingly resistent to antibiotics, the understanding of host-defense mechanisms may enable the design of strategies to aid in the treatment of bacterial infection.

ACKNOWLEDGMENTS

Supported in part by N.I.H. Grants DK47538 and 09049.

REFERENCES

1. Xie, Q-w. and Nathan, C.: The high-output nitric oxide pathway: role and regulation. J. Leuk. Biol., **56:** 576, 1994.
2. Sessa, W.: The nitric oxide synthase family of proteins. J. Vasc. Res., **31:** 131, 1994.
3. Stuehr, D.J. and Marletta, M.A.: Mammalian nitrate biosynthesis: mouse macrophages produce nitrite and nitrate in response to *Escherichia coli* lipopolysaccharide. Proc. Natl. Acad. Sci. USA, **82:** 7738, 1990.
4. Tayeh, M.A. and Marletta, M.A.: Macrophage oxidation of L-arginine to nitric oxide, nitrite and nitrate. Tetrahydrobiopterin is required as a co-factor. J. Biol. Chem., **264:** 19654, 1989.
5. Andersson, K.-E. and Persson, K.: The L-arginine/nitric oxide pathway and non-adrenergic, non-cholinergic relaxation of the lower urinary tract. Gen. Pharmacol., **24:** 833, 1993.
6. Dokita, S., Morgan, W.R., Wheeler, M.A., Yoshida, M., Latifpour, J. and Weiss, R.M.: N^G- nitro-L-arginine inhibits non-adrenergic, non-cholinergic relaxation in rabbit urethral smooth muscle. Life Sci., **48:** 2429, 1991.
7. Dokita, S., Smith, S.D., Nishimoto, T., Wheeler, M.A. and Weiss, R.M.: Involvement of nitric oxide and cyclic GMP in rabbit urethral relaxation. Eur. J. Pharmacol., **269:** 269, 1994.
8. Wheeler, M.A., Pontari, M., Dokita, S., Nishimoto, T., Cho, Y.H., Hong, K.W. and Weiss, R.M.: Age-dependent changes in particulate and soluble guanylyl cyclase activities in urinary tract smooth muscle. Mol. and Cell. Biochem., **169:** 115, 1997.
9. Chung, B.H., Choi, S.K. and Chang, K.I.: Effects of nitric oxide on detrusor relaxation. J. Urol., **155:** 2090, 1996.
10. Smith, S.D., Wheeler, M.A., Foster, H.E., Jr. and Weiss, R.M.: Urinary nitric oxide synthase activity and cyclic GMP levels are decreased with interstitial cystitis and increased with urinary tract infections. J. Urol., **155:** 1432, 1996.
11. Wheeler, M.A., Smith, S.D., Garcia-Cardena, G., Nathan, C.F., Weiss, R.M. and Sessa, W.C.: Bacterial infection induces nitric oxide synthase in human neutrophils. J. Clin. Invest., **99:** 110, 1997.
12. Smith, S.D., Wheeler, M.A., Zhang, R., Weiss, E.D., Lorber, M.I., Sessa, W.C. and Weiss, R.M.: Rejection or infection increases nitric oxide synthase activity in urine from renal transplant patients. Kid. Internat., **50:** 2088, 1996b.
13. Smith, S.D., Wheeler, M.A., Foster, H.E., Jr. and Weiss, R.M.: Improvement in interstitial cystitis symptom scores during treatment with oral L-arginine. J.Urol., **158:** 703, 1997.
14. Wheeler, M.A., Smith, S.D., Saito, N., Foster, H.E., Jr. and Weiss, R.M.: Effect of long-term oral L-arginine on the nitric oxide synthase pathway in the urine from patients with interstitial cystitis. J. Urol. **158:** 2045, 1997.
15. Foster, H.E., Jr., Smith, S.D., Wheeler, M.A. and Weiss, R.M.: Nitric oxide and interstitial cystitis. Adv. Urol., **10:** 1, 1997.
16. Olsson, L.E., Wheeler, M.A., Sessa, W.C. and Weiss, R.M.: Bladder instillation and intraperitoneal injection of E.coli lipopolysaccharide upregulate cytokines and iNOS in rat urinary bladder. J. Pharmacol. Exp. Ther., **284:** 1203, 1998.
17. Szabo, C. and Thiemermann, C.: Regulation of the expression of the inducible isoform of nitric oxide synthase. Adv. Pharmacol., **34:** 113, 1995.
18. Becherel, P.A., Goff, L.L., Ktorza, S., Quaaz, F., Mencia-Huerta, J-M., Dugas, B., Debre, P., Mossalayi, M.D. and Arock, M.: Interleukin-10 inhibits IgE-mediated nitric oxide synthase induction and cytokine synthesis in normal human keratinocytes. Eur. J. Immunol., **25:** 2992, 1995.
19. Moore, W.J., Cunha, F.O., Lieu, F.Y. and Weiser, W.Y.: IL-10 inhibits the synthesis of migration inhibitory factor on migration inhibitory factor-mediated macrophage activation. J. Immunol., **151:** 4325, 1993.
20. Davidoff, R., Yamaguchi, R., Leach, G.E., Park, E. and Lad, P.M.: Multiple urinary cytokine levels of bacterial cystitis. J. Urol., **157:** 1980, 1997.
21. Nicolle, L.E., Brunka, J., Orr, P., Wilkins, J. and Harding, G.K.M.: Urinary immunoreactive interleukin-1α and interleukin-6 in bacteriuric institutionalized elderly subjects. J. Urol., **149:** 1049, 1993.
22. Martins, S.M., Darlin, D.J., Lad, P.M. and Zimmern, P.E.: Interleukin 1β: A clinically relevant urinary marker. J. Urol., **151:** 1198, 1994.
23. Hedges, S., Stenqvist, K., Lidin-Janson, G., Martinell, J., Sandberg, T. and Svanborg, C.: Comparison of urine and serum concentrations of interleukin-6 in women with acute pyelonephritis or asymptomatic bacteriuria. J. Infect. Dis., **166:** 653, 1992.
24. Ohta, K., Takano, N., Seno, A., Yachie, A., Miyawaki, T., Yokoyama, H., Tomosugi, N., Kato, E. and Taniguichi, N.: Detection and clinical usefulness of urinary interleukin-6 in the diseases of the kidney and the urinary tract. Clin. Nephrol., **38:** 185, 1992.

25. Agace, W.W., Hedges, S., Ceska, M. and Svanborg, C.: Interleukin-8 and the neutrophil response to mucosal gram-negative infection. J. Clin. Invest., **92:** 780, 1993.
26. Saito, N., Smith, S.D., Wheeler, M.A. and Weiss, R.M.: A comparison of IL-8 and inducible nitric oxide synthase (iNOS) in urinary tract infections and interstitial cystitis. FASEB J., **11:** A511, 1997.
27. Farkas, A., Alajem, D., Dekel, S. and Binderman, I.: Urinary prostaglandin E_2 in acute bacterial cystitis. J. Urol., **124:** 455, 1980.
28. Jackson, A.M., Alexandroff, A.B., Kelly, R.W., Skibinska, A., Esuvaranathan K., Prescott, S., Chisholm, G.D. and James, K.: Changes in urinary cytokines and soluble intercellular adhesion molecule-1 (ICAM-1) in bladder cancer patients after Bacillus Calmette-Guerin (BCG) immunotherapy. Clin. Exp. Immunol., **99:** 369, 1995.
29. Prescott, S., James, K., Hargreave, T.B. and Chisholm, G.D.: Radioimmunoassay detection of interferon-gamma in urine after intravesical BCG therapy. J. Urol., **144:** 1248, 1990.
30. Ratcliff, T.L., Haaff, E.O. and Catalona, W.J.: Interleukin-2 production during intravesical bacille Calmette-Guerin therapy for bladder cancer. Clin Immunol. Immunopathol., **40:** 375, 1986.
31. De Boer, E.C., De Jong, W.H., Steerenberg, P.A., Aarden, L.A., Teteroo, E., De Groot, E.R., Van Der Meijden, A.P.M., Vegt, P.D.J., Debruyne, F.M.J. and Ruitenberg, E.J.: Induction of urinary interleukin-1(IL-1), Il-2, Il-6 and tumor necrosis factor during intravesical immunotherapy with bacillus Calmette Guirin in superficial bladder cancer. Cancer Immunol. Immunother., **34:** 306, 1992.
32. Bohle, A., Nowe, C.H., Ulmer, A.J., Musehold, J., Gerdes, J., Holstetter, A.G. and Flad, H.D.: Elevations of cytokines interleukin-1, interleukin-2 and tumor necrosis factor in the urine of patients after intravesical bacillus Calmette-Guerin immunotherapy. J. Urol., **144:** 59, 1990.
33. Balbay, D., Ozen, H., Ozkardes, H, Barut, A., Bakkaloglu, M, Tasar, C. and Remzi, D.: Detection of urinary interleukin-2, interleukin-2 receptor and tumor necrosis factor levels in patients with superficial bladder tumors after intravesical BCG immunotherapy. Urology. **43:** 187, 1994.
34. Shamhart, D.H.J., Kurth, K.H., DeReijke, T.M. and Vleeming, R.: BCG treatment and the importance of an inflammatory response. Urol. Res., **20:** 199, 1992.
35. Bettex-Galland, M., Studer, U.E., Walz, A., Dewald, B. and Baggioloni, M. Neutrophil-activating peptide-1/interleukin-8 detection in human urine during acute bladder inflammation caused by transurethral resection of superficial cancer and bacillus Calmette-Guerin. Eur. Urol., **19:** 171, 1991.
36. Feng L., Sun, W. and Hwang, D.: Cloning two isoforms of rat cyclooxygenase: differential regulation of their expression. Arch. Biochim.Biophys., **307:** 361, 1993.
37. Smith, S.D., Wheeler, M.A. and Weiss, R.M.: Nitric oxide synthase: An endogenous source of elevated nitrite in infected urine. Kid. Internat., **45:** 586, 1994.
38. Gross, S.S., Stuehr, D.J., Aisaka, K., Jaffe, E.A., Levi, R. and Griffith, O.W.: Macrophage and endothelial cell nitric oxide synthesis: Cell-type selective inhibition by N^G-nitroarginine and N^G-methyl-arginine. Biochem. Biophys. Res. Commun., **170:** 96, 1990.
39. Nakane, M., Pollock, J.S., Klinghofer, V., Basha, F., Marsden, P.A., Hokari, A., Ogura, T., Esumi, H. and Carter, G.W.: Functional expression of three isoforms of human nitric oxide synthase in baculovirus-infected insect cells. Biochem. Biophys. Res. Comm., **206:** 511, 1995.
40. McCall, T.B., Boughton-Smith, N.K., Palmer, R.M.J., Whittle, B.J.R. and Moncada, S.: Synthesis of nitric oxide from L-arginine. Release and interaction with superoxide anion. Biochem. J., **261:** 293, 1989.
41. Pollock, J.S., F˜stermann, U., Michell, J.A., Warner, T.D., Schmidt, H.H.H.W., Nakane, M. and Murad, F.: Purification and characterization of particulate endothelium-derived relaxing factor synthase from cultured and native bovine aortic endothelial cells. Proc. Natl. Acad. Sci. USA, **88:** 10480, 1991.
42. Cho, H.J., Xie, Q.-W., Calaycay, J., Mumford, R.A., Swiderek, K.M., Lee, T.D. and Nathan, C.: Calmodulin is a subunit of nitric oxide synthase from macrophages. J. Exp. Med., **176:** 599, 1992.
43. Reiling, N., Ulmer, A.J., Durchow, M., Ernst, M., Flad, H.D. and Hauschilt, S.: Nitric oxide synthase: mRNA expression of different isoforms in human monocytes/macrophages. Eur. J. Immunol., **24:** 1941, 1994.
44. Maciejewski, J.P., Selleri, C., Sato, T., Cho, H.J., Keefer, L.K., Nathan, C.F. and Young, N.S.: Nitric oxide suppression of human hematopoiesis in vitro. Contribution to inhibitory action of interferon-gamma and tumor necrosis factor-alpha. J. Clin. Invest., **96:** 1085, 1995.
45. Laffi, G., Foshci, M., Mansi, E., Simoni, A., Mugnai, L., LaVilla, G., Barletta, G., Mannaioni, P.F. and Gentilini, P.: Increased production of nitric oxide by neutrophils and monocytes from cirrhotic patients with ascites and hyperdynamic circulation. Hepatology, **22:** 1666, 1995.
46. Persson, M.G., Zetterstrom, O., Agrenius, V., Ihre, E. and Gustafsson, L.E.: Single-breath nitric oxide measurements in asthmatic patients and smokers. Lancet, **343:** 146, 1994.

47. Farrell, A.J., Blake, D.R., Palmer, R.M.J. and Moncada, S.: Increased concentration of nitrite in synovial fluid and serum samples suggest increased nitric oxide synthesis in rheumatic diseases. Ann. Rheum Dis., **51:** 1219, 1992.
48. Grabowski, P.S, England, A.J., Dykhuizen, R., Copland, M., Benjamin, N., Reid, D.M. and Ralston, S.H.: Elevated nitric oxide production in rheumatoid arthritis detection using the fasting urinary nitrate:creatinine ratio. Arthritis and Rheumatol., **39:** 643, 1996.
49. Anstey, N.M., Weinberg, J.B., Hassanali, M.Y., Mwaikambo, E.D., Manyenga, D., Misukonis, M.A., Arnelle, D.R., Hollis, D., McDonald, M. I. and Granger, D.L.: Nitric oxide in Tanzanian children with malaria: inverse relationship between malaria severity and nitric oxide production/nitric oxide synthase type 2 expression. J. Exp. Med., **8:** 557, 1996.
50. Middleton, S.J., Shorthouse, M. and Hunter, J.O.: Increased nitric oxide synthesis in ulcerative colitis. Lancet, **341:** 465, 1993.
51. Ellie, E., Loiseau, H., Lafond, F., Arsaut, J. and Desmotes-Mainard, J.: Differential expression of inducible nitric oxide synthase mRNA in human brain tumours. NeuroReport, **7:** 294, 1995.
52. Swana, H.S., Smith, S.D., Perrotta, P., Saito, N., Wheeler, M.A. and Weiss, R.M.: Nitric oxide synthase in human transitional cell carcinoma of the bladder. J. Urol. (in press).
53. Miles, A. M., Owens, M.W., Milligan, S., Johnson, G.G., Fields, J. Z., Ing, T.S., Kottapalli, V., Keshavarzian, A., and Grisham, M.B.: Nitric oxide synthase in circulating vs. extravasated polymorphonuclear leukocytes. J. Leuk. Biol. **58:** 616, 1995.
54. Evans, T.J., Buttery, L.D.K., Carpenter, A., Springall, D.R., Polak, J.M. and Cohen, J.: Cytokine-treated human neutrophils contain inducible nitric oxide synthase that produces nitration of ingested bacteria. Proc. Natl. Acad. Sci. USA, **95:** 9553, 1996.
55. Stein, P.C., Pham, H., Ito, T. and Parsons, C.L.: Bladder injury model induced in rats by exposure to protamine sulfate followed by bacterial endotoxin. J. Urol., **155:** 1133, 1996.
56. Cross, A.: The importance of a lipopolysaccharide-initiated, cytokine-mediated host defense mechanism in mice against extra-intestinally invasive Escherichia coli. J. Clin. Invest., **96:** 676, 1995.
57. Fleming, I., Gray, G.A., Julou-Schaeffer, G., Parratt, J.R. and Stocklet, J.C.: Incubation with endotoxin activates the L-arginine pathway in vascular tissue. Biochem. Biophys. Res. Commun., **171:** 562, 1990.
58. Knowles, R.G., Merrett, M., Salter, M. and Moncada, S.: Differential induction of brain, lung and liver nitric oxide synthase by endotoxin in the rat. Biochem. J., **270:** 833, 1990.
59. Cook, H.T., Bune, A.J., Jansen, A.S., Taylor, G.M., Loi, R.K. and Catte. V.: Cellular localization of inducible nitric oxide synthase in experimental endotoxic shock in the rat. Clin. Sci., **87:** 179, 1994.
60. Weisbrodt, N.W., Pressley, T.A., Li, Y.F., Zembowicz, M.J., Higham, S.C., Zembowicz, A., Lodato, R.F. and Moody, F.G.: Decreased ileal muscle contractility and increased NOS II expression induced by lipopolysaccharide. Am.J. Physiol., **271:** G454, 1996.
61. Schneider, F.L., Bucher, B., Schott, C., Andre, A., Jules-Schaeffer, G. and Stoclet, J.-C.: Effect of bacterial lipopolysaccharide on function of rat small femoral arteries. Am. J. Physiol., **266:** H191, 1994.

29

INTERFERON ALPHA FOR THE TREATMENT OF SUPERFICIAL BLADDER CANCER

John Naitoh,[1] John Franklin,[2] Michael A. O'Donnell,[3] and Arie S. Belldegrun[1]

[1]UCLA Medical Center
Los Angeles, California 90024-1782
[2]Montefiore Medical Center
New York, New York
[3]Beth Isreal Hospital
Boston, Massachusetts

1. INTRODUCTION

The main problems that the patient with superficial transitional cell carcinoma (TCC) faces are the lifelong risk of tumor recurrence, and the risk of tumor progression and death secondary to cancer. Historically, cystectomy was the primary and only availible treatment option for patients who had recurrent bladder cancer or carcinoma in situ (CIS). However, the development of topical (intravesical) therapies for transitional cell carcinoma of the bladder has revolutionized the management of both superficial papillary tumors, and of flat carcinoma in situ. Due to the discovery of intravesical agents that have biologic activity against superficial TCC of the bladder (such as Bacillus Calmette-Guerrin (BCG) and Mitomycin), the mortality rate from bladder cancer has decreased despite a concurrent increase in the incidence of this malignancy.[1]

While chemotherapy agents such as mitomycin and doxorubicin have efficacy against superficial bladder cancer, it's also clear that BCG is the most effective intravesical agent for both the treatment of bladder tumors and the for prevention of tumor recurrence (prophylaxis).[2] Unfortunately, between 30–40% of patients have bladder tumors that are resistant to BCG therapy, while at least 5% of patients cannot tolerate treatment with BCG due to severe local or systemic side effects.[2,3] Due to these problems, efforts have been made to develop more effective and less toxic treatments for superficial bladder cancer.

Given the immunologic mechanism of action that has been proposed for BCG, cytokines such as alpha interferon (IFN-alpha) have recently been tried as intravesical agents for the treatment of superficial bladder cancer.[4] Interferon-alpha (IFN-alpha) has been par-

Advances in Bladder Research, edited by Baskin and Hayward.
Kluwer Academic / Plenum Publishers, New York, 1999.

ticularly promising given that it has both an immunostimulatory function as well a direct antitumor action.[5,6] While the studies to date that have looked at the use of IFN-alpha are limited, it has been shown that this cytokine has biologic activity against bladder cancer, both when it was used as a single agent, or when it was used in combination with other, more traditional intravesicular agents. This review will discuss the rationale behind intravesicular IFN-alpha therapy, will list the clinical studies that have been performed to date that used IFN-alpha, and will address the future directions that will have to be explored before IFN-alpha can be uniformly accepted as a first line single agent for superficial bladder cancer, or as part of a combination regimen for the treatment or prophylaxis of transitional cell carcinoma. There will also be a discussion regarding the use of the other types of interferon (gamma and beta) for the management of superficial TCC, but it must be remembered that the data regarding the use of these alternate biologic response modifiers is extremely limited.

1.1. Superficial Bladder Cancer: Defining the Problem

Bladder cancer is the second most common urologic malignancy in the United States. Approximately 52, 900 new cases of transitional cell carcinoma were diagnosed last year, with 11,700 deaths expected to occur from this disease during 1997.[7] Over seventy percent of these tumors were superficial or limited to the lamina propria, with the majority of these lesions therefore amenable to treatment by transurethral resection (TUR) alone. Unfortunately, TCC of the bladder is a lifelong disease process, with the majority of patients eventually having a recurrence of their malignancy.[8] Furthermore, while the majority of patients who have invasive bladder cancer were found to have muscle invasive disease at the time of initial diagnosis, between 2 to 45% percent of the bladder tumors that are diagnosed initially as superficial will eventually progress to either invade the muscle wall of the bladder or become metastatic.[9]

While transurethral resection or fulguration represent the primary surgical therapies for superficial bladder carcinoma, in certain circumstances treatment via surgical resection alone will be inadequate. In patients who present with multifocal, superficial TCC, or who have a history of rapidly recurrent TCC, there is a high risk for relapse following TUR alone and these patients should have intravesical therapy to decrease their risk of recurrence.[1,10] In addition, patients who have high grade tumors, diffuse CIS, lamina propria invasion (T1), or tumor involvement of the prostatic urethra, the risk of progression to muscle invasive carcinoma following TUR alone can be as high as 80%.[11] In these patients, intravesical therapy should also be considered since it has been shown that it can decrease the risk that an invasive or metastatic cancer will develop.[8,12]

While the goals of intravesical chemotherapy are to decrease the recurrence rates of the tumor in comparison to TUR alone, decrease the risk of cancer progression, treat residual tumor not eradicated by TUR/fulguration, and improve overall survival, it has been difficult to quantify the efficacy of intravesical treatment due to the variations in the patient populations that have been studied, and due to the confusing nomenclature that has been used to define outcome. The majority of superficial bladder cancer clinical trials have enrolled patients that had widely varying risks of tumor recurrence and progression. Due to the differences in the recurrence and progression rates that are seen in patients who have low grade versus high grade tumors, sessile lesions versus diffuse CIS, solitary versus multiple lesions, tumors that have failed 1 or more courses of intravesical therapy, and completely versus incompletely resected lesions, a review of the literature must account for any differences in patient selection. The apparent differences in treatment efficacy that were seen in some of the early clinical studies were probably more influenced by patient selection than the type of intravesical treatment that was used. In reviewing the literature, it also must be remembered that intravesical agents will have differing degrees of activity depending on what type of tumor is being treated, and which end point of therapy is being used.

Adding to the confusion is the fact that most of the studies that looked at intravesical therapy used different end points to define success or failure: the goals of therapy have historically been the eradication of residual cancer, the prevention of recurrence, and the prevention of tumor progression. Following complete endoscopic tumor resection, intravesical therapy is defined as prophylactic if it is intended to prevent tumor recurrence. In contrast, when a residual papillary lesion or residual CIS is left following endoscopic treatment, tumor responses are characterized as a complete response (CR), a partial response (PR), stable disease (SD), or progressive disease (PD), depending on the changes that are observed in the bladder based on endoscopy, cytology, and biopsy following treatment. These definitions of response are important to remember so that an accurate comparison of outcomes between different studies can be made. Additionally, while intravesical agents are useful if they can decrease recurrence rates (which therefore decreases the treatment burden that the patient must undergo involving mutiple TUR procedures under anesthesia), decreasing the rate of tumor progression to a higher stage or grade is perhaps an even more important endpoint if the treatment is to have an impact on the morbidity and mortality rates of bladder cancer.

Besides the difficulties that exist in comparing the results of different studies that involved the use of different patient populations, and involved the use of different end points for success, problems also exist due to the haphazard dosing schedules and dosages of the agents that have been used in clinical trials. Furthermore, distinctions must be made between induction therapy versus maintenance therapy, especially since some maintenance regimens may look more like a series of multiple induction cycles due to their intensity. This differentiation is important due to the invasiveness, cost, and potential toxicity of the intravesical agent used, and since the ideal dose and treatment regimen for most intravesical agents remain poorly defined. Finally, since bladder cancer is a lifelong process, duration of follow-up in these studies is another important parameter that has to be examined before deciding if a treatment is effective.

Because all of these parameters come into play when one is comparing the outcomes of different intravesical therapy trials for superficial bladder cancer, it is inherently difficult to compare the efficacy of the different intravesical agents. These pitfalls are most evident when reviewing the reports of the efficacy of interferon therapy for superficial bladder cancer. It's our belief that the limitations of all these studies can only be resolved by using randomized, prospective clinical trials.[13] However, despite all of the limitations that exist in the medical literature, it's also apparent that it can be concluded from prior studies that the interferons, especially IFN-alpha, do have some biologic activity in the prophylaxis and treatment of superficial bladder cancer. This review will now discuss the underlying rationale for interferon (mainly IFN-alpha) in the management of bladder cancer, and will look at the clinical trials that suggested that IFN-alpha has a role for treating patients who have superficial TCC. The studies will be divided into those that used IFN-alpha as monotherapy and those that used it as a part of a combination regimen. This paper will also look at the use of IFN-alpha as either a prophylactic agent or as an agent that can be used to treat residual tumor/CIS. Finally, this review will summarize the results of the clinical trials to date that demonstrated the ability of IFN-alpha to either decrease the recurrence rate or progression rate of bladder cancer.

1.2. IFN-Alpha: Mechanism of Action for Superficial Bladder Cancer

Intravesical therapy has been well accepted due to its relatively low toxicity, its ability decrease the risk of tumor recurrence, its ability to reduce the incidence of disease progression, and its ability eradicate residual cancer cells or CIS.[1] To date, Bacillus Calmette-Guerin (BCG) appears to be the most effective intravesical agent for the prophylaxis and treatment of superficial bladder cancer.[1,8,14] Bacillus Calmette-Guerin is believed to function as a biologic response modifier that is able to produce a local and systemic immune reaction against the tumor. Although many theories exist about how BCG works, the actual anti-tumor mechanism is still unknown and it is likely to be complex. [15]

Unfortunately, the therapeutic efficacy of BCG is limited, and the intravesical administration of BCG is also associated with significant toxicity.[3] Furthermore, approximately 5% of patients are unable to tolerate BCG therapy due to local or systemic toxicity, while up to 90% of patients will complain of bladder irritation.[15] Due to these limitations of BCG, a number of other agents have also been tried for intravesical use. Given the apparent ability of BCG to activate the immune system to fight superficial TCC, a significant amount of attention has been focused on the intravesical administration of other biologic response modifiers such as IFN-alpha.

The interferons (IFNs) were first discovered by Isaacs and Lindenmann in 1957, and are classified as either interferon-alpha (IFN-α), -beta (IFN-β), or -gamma (IFN-γ) according to their cellular origin and mode of induction.[16,17,18] Interferons are proteins of approximately 146–166 amino acids: they were first discovered during the study of the immune system's response to viral infection. [18,19] Interferons activate their target cells by binding to specific cell surface receptors, with the subsequent activation of intracellular signaling proteins that induce the expression of specific interferon response genes in the cell.[18]

It has been shown that interferons can directly activate natural killer cells and cytotoxic-T-cells, and that they can induce the expression of the major histocompatibility complex (MHC) class I and class II antigens on the surface of cells.[20,21] It has also been shown that interferons can increase the expression of tumor antigens on the surface of bladder cancer cells, which should further sensitize them to killing by the immune system.[22–24] Thus, through these multiple actions, it is believed that interferons are able to improve the immune system's ability to recognize and destroy any abnormal cells that have foreign antigens on their surface (such as virally infected cells or tumor cells).

However, besides the local immune effect that is seen in the wall of the bladder, systemic immune modulation also appears to occur following intravesical interferon therapy. A potential effect of topical interferon therapy is to reverse the systemic immune suppression that has been associated with bladder cancer.[25] It has been shown that systemic changes in peripheral blood mononuclear cell (PBMC) function can be seen up to 6 months after discontinuation of the local treatment; increased PBMC production of IFN gamma and decreased IL-4 production were both detected following topical IFN-alpha therapy, along with increased cytotoxic activity of peripheral blood T-lymphocytes and NK cells.[26] This activation of the systemic immune system was also described by Alvarez-Mon, who showed that there was an increase in the blastogenic response of PBMCs following intravesical interferon therapy. Interestingly, a similar response was not seen in the patients whose tumors did not respond to the treatment, suggesting that the systemic immune system plays a significant role in determining the local response of the bladder tumor to the treatment.[25]

Besides the activation of the local and systemic immune system that occurs with IFN therapy, it has been discovered that the interferons also have direct antitumor activity due to their ability to inhibit cell proliferation, their ability to inhibit angiogenesis, and due to their interaction with growth factors, cell surface receptors, and cytokines such as tumor necrosis factor.[17,18] For example, it has been shown that interferon-alpha have a direct antiproliferative effect on bladder cancer cells that are grown in vitro: in this case, interferon has a direct cytostatic effect on cells due to its ability to lengthen the cell cycle.[6,27]

Thus, given the ability of interferon alpha to increase the susceptibility of the bladder tumor to killing by the immune system, as well as its ability to directly inhibit cellular proliferation, interferon alpha has significant potential as an anti-cancer agent. Furthermore, given its immunomodulatory function, it has been hypothesized that interferon should have therapeutic synergy with other immune system modulators such as BCG.[5] And finally, given its direct antiproliferative effects on tumor growth, synergy with chemotherapeutic agents such as mitomycin can also be expected.[28]

1.3. Clinical Trials Using Interferon

Due to the multiple immunomodulatory and antiproliferative effects that interferon has in experimental models of bladder cancer, it has entered clinical use for the management of superficial bladder cancer. While many different types of interferon have been used, involving administration by intravesical, subcutaneous and systemic routes, the majority of the clinical trials have focused on the intravesical administration of interferon alpha (IFN-alpha) for the management of superficial disease.

Thus far, multiple treatment schedules and doses have been used to deliver IFN-alpha to the cancerous cells. The heterogeneity of the dosing regimes and the different patient populations that were entered into these trials has made the interpretation of the results difficult. However, despite the limitations in the medical literature, there is some evidence to show that the intravesical route is the most effective route for dosing interferon, that IFN-alpha has activity in both the treatment and prophylaxis of bladder cancer, and that IFN-alpha is effective either as a single agent or in combination with another agent. Unfortunately, at this time there is not enough data in the literature that defines an optimal dose or dosing schedule, or defines which patients can best benefit from IFN-alpha.

1.4. Intramuscular and Intralesional Interferon Therapy

Intramuscular administration of IFN-alpha has demonstrated only modest response rates, with only 33% of patients having partial responses and no complete responses seen in 18 patients.[29] Two subsequent studies that investigated the intralesional injection of interferon-alpha demonstrated that it had somewhat greater efficacy than IM administration.

In a small, preliminary study that was performed by Ikic et al., 8 patients with papillary TCC underwent intralesional injection of 2 million units (MU) of human leukocyte interferon alpha for 21 days.[30] With a mean follow up of 26 months, they reported a complete response rate of 75%. Toxicity was mild and was limited to flu like symptoms.

More recently, the effectiveness of intralesional injection of IFN-alpha was evaluated by DiStasi and associates, who treated 15 patients with 3 MU IFN-alpha weekly for 4 weeks.[31] At 4 weeks following the last injection, all patients were restaged by endoscopic examination and TUR of the injection site. Responses were observed in 7 patients (1 CR, 6 PR), with responses limited to patients that had tumors $\leq$ 20 mm. The treatment was well tolerated, with mild flu-like symptoms the most common side effect.

With intralesional injection, although increased levels of serum IFN-alpha were seen following IFN injection, no differences in the phenotype of peripheral lymphocytes were identified before, during, or after IFN-alpha administration. Similarly, no significant changes were seen in the degree of inflammatory cell infiltration following treatment when TUR specimens of injected patients were compared to untreated controls. These findings suggested that there are direct antitumor effects of recombinant IFN-alpha that are mediated by a non-immunologic mechanism.

2. INTRAVESICAL INTERFERON THERAPY

2.1. Interferon Monotherapy for the Eradication of Residual Cancer or CIS

In evaluating the ability of an intravesical agent to eradicate residual malignant disease, consideration must be made as to the definition of residual cancer following TUR, since leaving bulky unresected tumors has a different implication than leaving behind microscopic (or endoscopically undetectable) disease. In general, it has been shown that the tumor burden has a direct relationship to the efficacy of the therapy, with resection or fulguration of all gross disease preferable in order to optimize the efficacy of intravesical therapy.[1] In this regard, intravesical fluorescent dyes can certainly be helpful to facilitate complete endoscopic tumor identification and destruction.[32]

Interferon's ability to eradicate CIS or residual tumor following TUR has been evaluated in a number of prospective studies (see Table 3).[33–36] In one of the earliest studies of intravesical therapy, Oliver et al. treated 16 patients who had recurrent papillary TCC or CIS with 50 million units (MU) of lymphoblastoid interferon weekly for 8 weeks. Six of eight patients who had papillary tumors (75%) had a response to therapy (3 CRs and 3 PRs), but no patient with CIS had any kind of response.[34]

In the Cooperative Northern California Oncology Group Trial, Torti et al. treated 35 patients with escalating, intravesically administered doses of recombinant IFN-alpha-2b.[35] Thirty-five patients in this study received 8, weekly IFN-alpha doses that ranged from 50 to 1,000 MU. In this study, 16 patients had papillary TCC: 15 of these patients had a history of multiple prior papillary tumor recurrences (range 1–10), and 13 of these patients (81%) had failed prior courses of intravesical therapy that used other agents. Out of these 16 patients, 4 (25%) achieved CRs for a mean duration of 17.5 months. No partial responses were observed.

This study also treated 19 patients who had CIS (2 of these patients also had coexisting papillary lesions). Ten of these 19 patients (53%) had failed prior intravesical chemotherapy, and 15 (79%) had a history of prior CIS or papillary bladder cancer, with a mean of 3.0 (range 1 to 10) recurrences of CIS prior to the initiation of the interferon therapy. Six of these 19 patients with CIS (32%) had complete responses as determined by negative urinary cytology and bladder biopsy. Out of these 6 responders, 3 recurred at 10, 14, and 17 months, while the other 3 patients remained disease free without any further intravesical therapy at 18, 32, and 37 months. While this study was not designed to determine the dose that had optimal therapeutic efficacy, the maximum tolerated dose of interferon that could be administered intravesically was not reached. Although responses were seen at all dose levels administered, 4/10 of the total responders (40%) were seen at the low dose level (40–100 MU), and another 4/10 (40%) of the CRs were observed at the intermediate dose level (100 to 300 MU). Overall the toxicity rates were low at all dose levels, and the treatments were well tolerated.

In a subsequent multi-institutional study reported by Glashan et al. 87 patients who had CIS were prospectively randomized to receive either low dose or high dose (10 MU or

100 MU) intravesical recombinant IFN-alpha.[36] The dosing schedule involved weekly IFN-alpha for 12 weeks followed by monthly prophylactic instillations for a maximum of 1 year. Complete responses were observed in 20/47 (43%) of the high dose patients, and 2/38 (5%) of the low dose patients. Six of the 9 patients who had failed prior intravesical BCG therapy responded to the high dose IFN-alpha therapy (2 CRs, 4 PRs). Additionally, another 23% of patients that received high dose therapy achieved a partial response (as defined by persistent positive urinary cytology results with no endoscopic or biopsy proven evidence of carcinoma in situ). The responses proved to be durable with this maintenance regimen of interferon, with disease free durations that were greater than 12 months in 10 of the 20 patients who received the high dose IFN-alpha regimen. The most frequent side effect that was encountered involved mild to moderate flu-like symptoms, which were reported in 3/38 (8%) of the low dose and 8/47 (14%) of the high dose patients. No patient in this study had their therapy terminated due to treatment-related adverse experiences.

Taken in combination, these results indicate that intravesically administered IFN-alpha alone has significant antitumor activity in patients that have incompletely resected papillary TCC or CIS. Toxicity was mild and was not related to the dosage that was used, with some evidence that doses of at least 50 to 100 MU are required to see a beneficial effect. Furthermore, intravesical IFN-alpha was able to induce responses in a substantial portion of patients that had failed treatment with other intravesical agents, including BCG.

In comparison to the experience with IFN-alpha intravesical therapy, the clinical experience using IFN-β is much more limited. In one study, Niijima and colleagues treated 51 patients intravesically with 3–36 MU IFN -β, and noted that responses seemed to be correlated with the frequency of the instillations that were given.[37] In the 10 patients who received between 3–36 MU of IFN -β on days 1–3 every week for 2 weeks, no responses were seen. In the 32 patients who were treated with 3–36 MU IFN -β daily for 10–20 days, there were 8 (25%) partial responses. Finally, in the 9 patients who were treated with divided doses of 18 MU IFN-alpha-β twice a day for 10–20 days, partial responses were seen in 6 (67%). Problematically, no complete responses were seen using this agent. Besides IFN-alpha and IFN-beta, there is currently only one series that used recombinant IFN-γ for the treatment of bladder cancer. Out of the 13 patients who received twice weekly treatments of 1 mg IFN-γ, only 1 complete response and 1 partial response was observed.[38]

2.2. Interferon Monotherapy for the Prevention of Bladder Tumor Recurrence

The efficacy of intravesical interferon to prevent the recurrence and progression of superficial bladder tumors following complete TUR has been evaluated in several studies (see Table 1).[39–43] In a non randomized study by Kostakopoulos et al., 30 patients with newly diagnosed papillary superficial bladder cancers were treated with 8 weekly bladder instillations of 10 MU of IFN-alpha following complete tumor resection.[41] The patients who were treated in this study were largely at high risk for recurrence or progression: 8 of the patients had multifocal disease at presentation, tumor stage was T1 in 23 (77%) patients, and 1 patient had a papillary tumor concurrent with diffuse CIS. Following the course of intravesical interferon, patients in this study were monitored for recurrence for between 12–28 months. Out of the 30 patients who were treated, 19 (63%) remained free of tumor recurrence at the time the paper was written. However, tumor recurrences did develop in 11 (37%) patients, with 4 of these patients progressing to develop a muscle invasive carcinoma. While these patients reportedly did better than historical controls, the lack of an actual control group made the conclusions of the study hard to interpret.

Table 1. Interferon monotherapy for the prevention of tumor recurrence: non-randomized studies

Study	Agent	Dosing (weeks)	Median F/Up (months)	Recurrence rate	Mean time to recurrence (months)	Comments
Kostakopoulos	IFN-alpha	10MU × 8	12–18	37%	11.2	13 % progression rate: mostly high risk patients
Bartoletti	IFN-alpha	54MU × 2 (10 Tx)	23.8	21%	9.4	Mostly low risk patients

In contrast, Bartoletti et al. looked at the impact of intravesical IFN-alpha prophylaxis in a lower risk population.[42] In this study, the vast majority of the patients had solitary low grade (G1) lesions, and only one patient in the study group had received prior intravesical chemotherapy. The 20 patients who were treated in this study received 54 MU of IFN-alpha intravesically for 5 days during 2 consecutive weeks. The treatment regimen was well tolerated, with only 3 (15%) patients complaining of bladder irritation immediately following the treatment. With a median follow-up of 23 months, 15 of 19 (79%) evaluable patients remained tumor free, with only four (21%) patients relapsing at a median of 9.4 months after treatment. Unfortunately, conclusions about the efficacy of this regimen were again difficult to reach due to the lack of a control group.

To better answer the question of efficacy, several randomized, comparative trials have been performed to compare the ability of IFN-alpha and other intravesical agents to prevent tumor recurrence (Table 4). In one multi-center study, 287 patients with primary Ta to T1, grade 2 superficial bladder cancer were randomized to receive either intravesical IFN-alpha or Mitomycin-C (MMC) following complete TUR.[44] Patients received weekly instillations of either 50 MU of IFN-alpha, or 40 mg of MMC for a total of 8 weeks. Exclusion criteria included a prior history of bladder cancer, tumor involving the prostatic urethra, a positive cytology after TUR, or any evidence of concurrent carcinoma in situ. The treatment groups were well matched for patient age, gender, number of primary tumors, maximum tumor size, tumor stage, tumor grade, and compliance to treatment. Overall, MMC appeared to be superior to IFN-alpha in preventing tumor recurrence, with tumor recurrences occurring in 52 of the 141 (36%) patients who were treated in the MMC group, compared to 70 of 146 (48%) who were treated in the IFN-alpha group. Local toxicities including hematuria and stranguria were more common in the MMC treated patients, while systemic toxicities (fever, flu symptoms) were more common in the IFN-alpha treated group. Other randomized studies have also been performed that compared IFN-alpha to other chemotherapy agents, and all demonstrated the inferiority of IFN as a prophylactic agent.[45,46]

Similarly, studies have been performed that proved that IFN-alpha is inferior to BCG as a prophylactic agent (Table 2).[47–49] In a randomized, prospective study by Kalbe et al., 78 patients were treated with either 12 monthly instillations of IFN-alpha (10 MU) or BCG (120 mg).[47] After a median follow up of 24.5 months, the BCG group had a significantly lower recurrence rate than the IFN alpha group (16% versus 60%), although the treatment related side effects were more common and more severe in the BCG treated group. Jiminez-Cruz and associates had similar conclusions following their prospective study of 122 patients with T1, grade 1–3 tumors who were randomized to receive either BCG or IFN-alpha.[49]

Recently, one of the most compelling studies that looked at the utility of IFN-alpha as a prophylactic agent was reported by Portillo et al.[50] Using a double blind study design

Table 2. Interferon monotherapy: prevention of recurrence. Randomized studies of BCG versus IFN-alpha

Study	Agent	Dosing	Median F/up	Recurrence rate	Mean time to recurrence	Comments
Kalbe	IFN-alpha vs BCG	10 MU X 1 year	25 months	60%	8 months	BCG more effective
		120 mg X 1 year	24 months	16%	15 months	
Donovan	TUR Alone		24 months	23%		BCG more effective
	TUR+IFN-alpha		24 months	18%		
	TUR + BCG		24 months	13%		
Portillo	TUR	—	43 months	51.2%	9.6 months	randomized, double blind study design: no advantage to IFN-alpha
	TUR + IFN-alpha	60MU × 12 Weeks + 1 tYear Maintinence Tx	43 months	53.8%	17 months	

(with sterile water instilled in the bladders of the control group patients), 90 patients with stage T1, grade 2–3 TCC were prospectively randomized to receive weekly intravesical treatments of either 60 MU of IFN-alpha or water for 12 weeks, followed by monthly instillations for 1 year. After an average of 43 months of follow up, no statistically significant difference was seen between the two groups in terms of tumor recurrence rate, tumor progression rates, or toxicity of the treatment. However, despite these apparent negative results, there were trends towards a decreased recurrence rate after the first year (28.2% versus 35.8%), an increased average time to tumor recurrence (17 months versus 9.6 months), and decreased tumor progression by either stage or grade (14.2% versus 35%). While this is the only randomized study that effectively compared TUR and IFN-alpha versus TUR alone, it was unfortunately statistically underpowered to definitively prove or disprove the efficacy of interferon treatment, and it focused on a group of patients (stage T1) that have historically responded poorly to intravesical therapy.[51]

Thus, while the ability of intravesical IFN-alpha to prevent tumor recurrences is suggested by the data, it has yet to be definitively proven in a randomized, prospective manner. Additionally, while it is clear that IFN-alpha is inferior to both standard chemotherapeutic agents (mitomycin, ethoglucid) and to BCG, it is also clear that IFN-alpha treatment is well tolerated with few side effects.[50]

Table 3. Interferon monotherapy for the treatment of residual tumors or CIS

Study	Agent	Dosing	Response rate	Median duration of recurrence	Comments
Oliver	IFN-alpha	50 MU × 8 weeks	0	—	—
Torti	IFN-alpha	50–1,000 MU × 8 weeks	32% CR	21+ Months	50% failed prior intravesical Tx: response not dose dependent
Glashan	IFN-alpha	10 MU × 1 year	5% CR	12+ Months	*22% BCG failures had CR
	IFN-alpha	100 MU × 1 year	43% CR/11% PR		

Table 4. Interferon as part of a combination regimen for tumor prophylaxis

Study	Agent	Dosing	Median follow-up	Recurrence rate	Comments
Engelmann	IFN-alpha	10 MU	6.2 months	18%	Patients Enrolled were pTa-pT1, Grade 1-3:
	IFN-alpha + MMC	10 MU/20 mg	6.2 months	0%	Combination treatment more effective
	MMC	20 mg	6.2 months	22%	
Ferrari	IFN-alpha	50 MU × 6 Weeks	19 months	24%	Slight advantage to combination tx.
	IFN-alpha + Epirubcin	50MU/80 mg × 6 Weeks	19 months	16%	
Pavone-Macaluso	IFN + epirub	5 MU/30 Mg	24 months	86%	*Separating the Dose of the Two Agents by 24 hours improved efficacy
	IFN + epirub	5 MU/30 Mg*	24 months	56%	
	IFN + epirub	10 MU/30 Mg	24 months	55%	
	IFN + epirub	10 MU/30 Mg*	24 months	43%	

2.3. Interferon in Combination with Intravesical Chemotherapy

Although it is now clear from the clinical data that standard intravesical chemotherapy agents and BCG are superior to IFN-alpha for the treatment of superficial bladder cancer, IFN-alpha may still have a role as part of a multidrug, intravesical instillation regimen. Intravesical agents such as mitomycin or epirubicin work by inhibiting cell proliferation through the inhibition of DNA synthesis. Given the direct, anti-proliferative effect that IFN-alpha has on bladder cancer cell lines in vitro, a synergistic effect between IFN-alpha and chemotherapeutic agents should be seen given the different mechanisms of action that these agents use to slow cellular proliferation.

In vitro studies of murine and human bladder cancer cell lines have shown that combinations of IFN-alpha with doxorubicin and mitomycin in fact do have a synergistic ability to inhibit the proliferation of the tumor cells.46 Subsequent clinical trials in humans have confirmed that this in vitro synergy is also seen in vivo (Table 4). In a study reported by Engelmann et al., the ability of low level (10 MU) IFN-alpha to prevent tumor recurrence was compared to either mitomycin alone (20 mg) or to a combination of IFN-alpha and mitomycin.[52] Sixty seven patients with recurrent Ta or T1 (grade 1 to 3) tumors were randomized to one of the 3 treatment arms. After 6 months of follow-up, recurrence rates of 18% and 22% were seen in the treatment groups that used IFN-alpha or MMC alone, while no recurrences were seen in the group that underwent combination therapy. While this data was compelling, the follow-up interval in this study was too short to make any definite conclusions.

A similar trial that compared the efficacy of IFN-alpha alone (50 MU) versus IFN-alpha plus epirubicin (80 mg) also showed that there was a therapeutic advantage with the combination of the two agents.[53] In this study, IFN-alpha monotherapy resulted in a 24% recurrence rate at 19 months, compared to the 16% rate that was seen following combination therapy.

2.4. Interferon in Combination with Intravesical BCG

Besides using combinations of IFN-alpha with chemotherapy, combinations using IFN-alpha with other biologic response modifiers (such as BCG) may also have synergistic effects. Although the exact mechanisms by which BCG and IFN-alpha activate the immune system are not completely defined, the antiproliferative and immunomodulatory effects of these two agents appear to be complementary.[5,54] It has been shown that intravesical BCG results in the infiltration of inflammatory cells into the bladder wall, and increases the ratio of CD4 to CD8 T cells.[1] In contrast, IFN-alpha increases expression of MHC and tumor antigens on the surface of tumor cells, activates cytotoxic T cells and NK cells, increases the susceptibility of tumor cells to attack by CTLs, and directly inhibits the proliferation of tumor cells.[6,20,23,27] Thus, given the ability of BCG to attract lymphocytes to the bladder wall, and given IFN-alpha's ability to activate these inflammatory cells, while also increasing the susceptibility of the tumor cells to killing by the immune system, synergy between BCG and IFN-alpha should be seen with combination therapy.

Besides synergy, here are other reasons that make combination therapy using BCG and IFN-alpha appealing. Because BCG and IFN-alpha are biocompatible, these agents can be combined and instilled intravesically as a single procedure.[5,55] Additionally, given the dose related toxicity that is seen with BCG, any dose reductions of BCG that can be accomplished using combination therapy with IFN-alpha could result in decreasing the toxicity of treatment regimen, increase the numbers of patients who can tolerate the full course of intravesical therapy, and potentially improve the efficacy of immunotherapy to eliminate superficial bladder cancer (Table 6).[56]

Recently, the results of two clinical trials were reported, suggesting that combination therapy with BCG and IFN-alpha was more effective than either agent alone (Table 5). In an early pilot study, combination therapy was used to treat 12 patients who had either residual papillary tumors or CIS.[57] In this protocol, low dose BCG (60 mg) was combined with escalating doses of IFN alpha (10, 30, 60, and 100 MU): the treatments were administered as 6 weekly instillations of both agents together. Initially after the treatment, all of

Table 5. Interferon as part of a combination regimen with BCG for treatment of CIS or residual cancer

Study	Agent	Dosing	Median follow-up	Recurrence or response rate	Comments
Stricker	IFN-alpha + BCG	10–100 MU/60 mg BCG × 6 weeks	12 months	25% recurrence rate	Pt did better than expect- ed: no actual control group for comparison.
O'Donnell (CIS)	IFN-alpha + BCG	50 MU/80 mg BCG × 8 Weeks	3–7 months	75% CR	All of these patients failed prior courses of BCG or IFN alone
O'Donnell (recurrent, papillary TCC)	IFN-alpha + BCG	50 MU/80 mg BCG × 8 Weeks	8	75% CR	Most tumors in this study were Grade 2-3, Rapidly Recurrent or Multifocal

Table 6. Review of toxicities of BCG and interferon

Toxicity	Interferon	Low Dose BCG	High Dose BCG
Dysuria	42% (1.4% severe)	27% (5% severe)	90%
Hematuria	10% (1.4% severe)	2% severe	32%
Fever	10% (1.4% severe)	17% (2% severe)	42%
Malaise/flu like symptoms	8-14%	not listed	33%

the patients were classified as complete responders (8 patients with CIS, 4 patients with multifocal Ta tumors). At twelve months post-treatment, there has been no progression to muscle invasion, although 3 patients did have tumor recurrence. While these results were better than expected, the strength of this study was limited by the small number of patients, the lack of a control group, and the relatively short duration of follow-up.

Based on this earlier study, a prospective, open label trial using 50 MU of IFN-alpha in combination with either full or reduced doses (1/3 strength) of BCG was recently initiated by O'Donnell and colleagues.[58] The population that was treated in this study all had a high risk for relapse, with approximately 80% of the study population having CIS that was refractory to prior courses of BCG. In a preliminary report, these investigators described the outcomes of first five patients that were treated. With a follow-up of 6 months or more, no serious adverse events were noted. Furthermore, 4 of the patients (all of whom had BCG refractory lesions) had complete responses to the combination treatment, and 1 patient had a complete resolution of CIS following combination therapy, but had a recurrent papillary bladder tumor 6 months later.

Thus, while the results of combination therapy using IFN-alpha and BCG appear to be promising, more studies will be needed before such this approach can be put into widespread application. To achieve this end, a double-blind, prospective, randomized study for patients with aggressive superficial bladder cancer (Ta G3, T1, or CIS) was recently started.[59] This study will compare the efficacy and toxicity of full dose BCG alone to reduced dose BCG (1/3 strength) plus IFN-alpha (50 MU). The trial will involve a standard 6 week induction course of BCG, followed by a maintenance regimen (3 weekly instillations every 3 months) of either full dose BCG alone or reduced dose BCG (1/3 strength) plus IFN-alpha (50 MU). In a second part to this study, non-responders to the induction course of BCG will be randomized to receive either a second induction cycle of BCG, or to an induction cycle that will use reduced doses of BCG in combination with IFN-alpha.

2.5. Future Directions

While it has been shown that intravesical IFN-alpha has moderate antitumor effects, the overall role that interferons and IFN-alpha will have in the management of superficial bladder cancer has yet to be defined. While there is strong evidence to show that other intravesical agents such as mitomycin and BCG have greater efficacy in both eradicating residual tumors in the bladder, and in preventing tumor recurrences, there is evidence that IFN-alpha can benefit some patients that have tumors that are resistant to BCG or to other intravesical agents.

However, before IFN-alpha becomes acceptable as a primary agent for the management of superficial bladder cancer, many clinical issues need to be resolved. While the minimum effective dose appears to be between 50 to 100 MU per instillation, it's unknown if further dose escalations will improve efficacy or cause an increase in toxicity. Additionally, the frequency of intravesical administration during the induction cycle, and the optimal duration for the maintenance course of IFN-alpha are unknown. Furthermore, some of the other basic questions that need to be addressed in future randomized trials using IFN-alpha include the identification of patients who will best benefit from IFN-alpha therapy, the determination of the optimal timing of combination therapy (Which agent is given first? Should they be given concurrently?), determining the durability of the response to IFN-alpha, measuring the effect of IFN-alpha therapy on tumor progression and on overall disease specific survival, and characterizing the underlying mechanism of action by which the anti-tumor effects of IFN-alpha are mediated.

3. CONCLUSIONS

The management of superficial bladder cancer continues to evolve. Although intravesical BCG has been established as the gold standard in the treatment of this disease, there are a number of limitations to intravesical BCG therapy. Recurrence rates of 20% to 40% with BCG therapy are observed, and the severity of intravesical BCG toxicity limits therapy in some patients. The search for alternative therapies has included the evaluation of various cytokines including the interferons. Interferon, particularly IFN-alpha, has been shown to have some efficacy against superficial bladder cancer.

When compared to BCG or to intravesical chemotherapeutic agents, intravesical IFN-alpha appears to be less effective. However, there are several factors that support a role for intravesical IFN-alpha in the treatment of superficial bladder cancer, either as a single agent or in combination with mitomycin, epirubicin, or BCG. Interferon-alpha alone can salvage some patients who failed to respond to intravesical BCG, with combination therapy using IFN-alpha producing occcasional dramatic responses in patients that were refractory to BCG alone or to intravesical chemotherapy alone. Furthermore, there is hope that a regimen of reduced-dose-BCG combined with IFN-alpha will have a degree of efficacy that is comparable to full dose BCG, but with a marked reduction in the side effects of the treatment. Although the initial results of intracavitary interferon are encouraging, further evaluation of the efficacy of IFN-alpha is needed, along with further definition of the optimal dosing schedule for this potent and nontoxic anticancer agent.

REFERENCES

1. Nseyo UO, Lamm DL: Immunotherapy of Bladder Cancer. Seminars in Surgical Oncology 13:342–349, 1997.
2. Brosman SA: Bacillus Calmette-Guerin Immunotherapy. Urol Clin N.A. 19:557–564, 1992.
3. Lamm DL. Complications of Bacillus Calmette-Guerin Immunotherapy. Urol Clin N.A. 19:565–572, 1992.
4. Sargent ER, Williams RD: Immunotherapeutic Alternatives in Superficial Bladder Cancer. Urol Clin N.A. 19:581–589, 1992.
5. Zhang Y, Khoo HE, Esuvranthan K: Effects of Bacillus Calmette-Guerrin and Interferon alpha on Human Bladder Cancer in vitro. Int J Cancer 71: 851–57, 1997.
6. Borden EC, Groveman DS, Nasu T, Reznikoff C, and Bryan GT: Antiproliferative Activities of Interferons Against Human Bladder Carcinoma Cell Lines in vitro. J Urol 132:800–803, 1984.
7. Parker SL, Tong T, Bolden S, Wingo PA: Cancer statistics, 1996. CA Cancer J Clin 1996;46:5–27.
8. Herr HW, Schwalb DM, Zhang Z-F, Sogani PC, Fair WR, Whitmore WF, Jr, and Oettgen HF. Intravesical Bacillus Calmette-Guerin Therapy prevents tumor progression and death from superficial bladder cancer: Ten-year follow-up of a prospective randomized trial. J Clin Oncol., 13:1404–1408, 1995.
9. Heney NM: Natural History of Superficial Bladder Cancer.Urol Clin N.A. 19:429–433, 1992.

10. Sarosdy MF: Principles of Intravesical Chemotherapy and Immunotherapy. Urol Clin N.A. 19:509–519, 1992.
11. Utz DC, Hanash KA, Farrow GM: The Plight of the Patient with Carcinoma in situ of the Bladder. J Urol 103:160–4, 1970.
12. Herr HW, Laudone VP, Badalament RA: Bacillus Calmette-Guerin Therapy Alters the Progression of Superficial Cancer. J Clin Oncol 6:1450–55, 1988.
13. Franklin J, and Benson MC: New techniques in management and treatment of superficial bladder cancer. In Neal DE (ed.): Tumors in Urology. London, Springer-Verlag, pp 65–78, 1994.
14. Lamm DL: BCG in Perspective: Advances in the treatment of superficial bladder cancer. Eur Urol., 27(Suppl):2–8, 1995.
15. Berry DL, Blumenstein BA, Magyary DL: Local Toxicity Patterns Associated with Intravesical bacillus Calmette-Guerrin: A Southwest Oncology Group Study. Int J Urol 3:98–101, 1995.
16. Isaacs A, and Lindenmann J: Virus inference: I. The interferon. Proc R Soc Lond (Biol), 147: 258–267, 1957
17. Kurzrock R, Gutterman JU, and Talpaz M: Interferon-a,b,g: basic principles and preclinical studies. DeVita VT, Jr. et al. (Eds): Biologic Therapy of Cancer, p 247. Philadelphia, JB Linnincott Co., 1991.
18. Baron S, Tyring SK, Fleischmann R, Coppenhaver DH, Niesel DW, Kimpel GR, Stanton J, Hughes TK: The Interferons. Mechanism of Action and Clinical Applications. JAMA 266:1365–1381, 1991.
19. Kurzrock R, Gutterman JU, and Talpaz M: Interferon-a,b,g: basic principles and preclinical studies. DeVita VT, Jr. et al. (Eds): Biologic Therapy of Cancer, p 247. Philadelphia, JB Linnincott Co., 1991.
20. Hawkyard S, James K, Prescott S, Jackson AM, Ritchie AWS, Smyth JF, Chisolm GD: The Effects of Recombinant Human Interferon-Gamma on a Panel of Uman Bladder Cancer Cell Lines. J Urol 145:1078–81, 1991.
21. Molto L, Alvarez-Mon M, Carballido J, Manzano L, Guillen C, Prieto A, Olivier C, and Rodriguez-Zapata M: Intracavitary prophylactic treatment with interferon alpha 2b of patients with superficial bladder cancer is associated with a systemic T-cell activation. Br J Cancer, 70: 1247, 1994. 22) Patard J, Muscatelli-Groux B, Saint F, Popov Z, Maille P, Abbou C, Chopin D, et al. Evaluation of local immune response after intravesical bacille Calmette-Guerin treatemnt for superficial bladder cancer. Br. J. Urol. 78:709–714, 1996.
23. Tzai TS, Lin SN: Interferon alpha can alter the Lytic Susceptibility of Murine Bladder Transitional Cell Carcinoma (MBT-2) by Their Original Poor Specific Cytotoxic Tumor Infiltrating Lymphocytes. J Urol 146: 523/7, 1992.
24. Jackson AM, Alexandrov AB, Prescott S, James K, Chisolm GD: Expression of Adhesion Molecules by Bladder Cancer Cells: Modulation by Interferon-Gamma and Tumor Necrosis Factor. J Urol 148:1583, 1992.
25. Alverez-Mon M, Molto LM, Manzano L, Olivier C, and Carballido JA: Immunomodulatory effect of interferon-a2b on natural killer cells and T lymphocytes from patients with transitional cell carcinoma of the bladder. Anti-Cancer Drugs, 3(Suppl): 5, 1992.
26. Molto L, Carballido J, Manzano L, Reyes E, Oliver C, Alvarez-Mon, M: Prophylactic Intracavitary Treatment with Interferon alpha Increases Interferon gamma production by Peripheral Blood Mononuclear Cells in Patients with Superficial Transitional Cell Carcinoma of the Bladder. Brit J Cancer 75:1849–53, 1997.
27. Fuchsberger N, Kubes M, Kontsek P, Borecky L, Hornak M, Godal A, Silvanova, Svec J: In vitro antiproliferative effect of interferon alpha in solid tumors: a potential predictive test. Neoplasma, 40: 293–6, 1993.
28. Boccardo F, Cannata D, Rubagotti A, Guameri D, Decensi A, Canobbio L Curotto A, Martorana G, Pegoraro C, Selvaggi F, Salvia G, Comeri G, Bano A, Borella T, and Giuliani L: Prophylaxis of superficial bladder cancer with mitomycin or interferon alpha-2b: Results of a multicentric Italian study. J Clin Onc., 12: 7, 1994.
29. Scorticatti C, de la Pena N, Bellora O, et al.: Systemic IFN-alpha treatment of multiple bladder papilloma grade I or II patients: pilot study. J Interferon Res., 2: 339–43, 1987.
30. Ikic D, Nola P, Maricic Z, et al.: Application of human leukocyte interferon in patients with urinary bladder papillomatosis, breast cancer, and melanoma. Lancet 1: 1022, 1981.
31. Di Stasi SM, Virgili G, Vespasiani G, Porena M, and Micali F: Intralesional alpha interferon therapy in papillary superficial transitional cell carcinoma of the bladder: A pilot study. B J Urol., 71: 422, 1993.
32. Koenig F, McGovern FJ: Fluorescence Detection of Bladder Carcinoma. Urology 50:778–9, 1997. 33) Ikic D, Nola P, Maricic Z, et al.: Application of human leukocyte interferon in patients with urinary bladder papillomatosis, breast cancer, and melanoma. Lancet 1: 1022, 1981.
34. Oliver R, Waxman J, Kwok H, et al.: Alpha lymphoblastoid interferon for non-invasive bladder cancer. Br J Cancer 53:432, 1986.

35. Torti FM, Shortliffe LD, Williams RD, Pitts WC, Kempson RL, Ross JC, Palmer J, Meyers F, Farrari M, Hannigan J, Spiegel R, McWhirter K, and Freiha F: Alpha-interferon in superficial bladder cancer: A Northern California Oncology Group Study. J Clin Onc., 6: 476, 1988.
36. Glashan RW: A randomized controlled study of intravesical a-2b-interferon in carcinoma in situ of the bladder. J Urol., 144: 658, 1990. 37) Niijima T: Intravesical treatment of bladder cancer with recombinant human interferon-b. Cancer Immunol Immunoth., 30: 81, 1989.
38. Geboers ADH, Van Bergen TNLM, and Ousterlinck W: Gamma-interferon in the therapeutic and prophylactic management of superficial bladder cancer: J Urol 137:276A, abstract 691, 1987.
39. Geboers ADH, Van Bergen TNLM, and Ousterlinck W: Gamma-interferon in the therapeutic and prophylactic management of superficial bladder cancer: J Urol 137:276A, abstract 691, 1987.
40. Migliari R, El Demiry M, Muscas G, and Usai E: Intravesical instillation of beta-interferon in the treatment of bladder cancer. Brit J Urol., 70: 169, 1992.
41. Kostakopoulos A, Deliveliotis Ch, Mavromanolakis E, Aravnatinos G, and Dimopoulos MA: Intravesical interferon alpha-2b administration in the treatment of superficial bladder cancer. Eur Urol., 18: 201, 1990.
42. Bartoletti R, Massimini G, Criscuolo D, and Rizzo M: Interferon alpha 2a in superficial bladder cancer prophylaxis: Toleration and long-term follow-up. A phase I-II study. Anti-Cancer Res., 11: 2167, 1991.
43. Boccardo F, Cannata D, Rubagotti A, Guameri D, Decensi A, Canobbio L Curotto A, Martorana G, Pegoraro C, Selvaggi F, Salvia G, Comeri G, Bano A, Borella T, and Giuliani L: Prophylaxis of superficial bladder cancer with mitomycin or interferon alpha-2b: Results of a multicentric Italian study. J Clin Onc., 12: 7, 1994.
44. Boccardo F, Cannata D, Rubagotti A, Guameri D, Decensi A, Canobbio L Curotto A, Martorana G, Pegoraro C, Selvaggi F, Salvia G, Comeri G, Bano A, Borella T, and Giuliani L: Prophylaxis of superficial bladder cancer with mitomycin or interferon alpha-2b: Results of a multicentric Italian study. J Clin Onc., 12: 7, 1994.
45. Hoetl W, Hasun R, Albrecht W, Marberger M: How effective is topical alpha-2b interferon in preventing recurrence of superficial bladder cancer? Brit J Urol., 68: 495, 1991.
46. Okamoto E, Kinne RKH, Sokeland J: Interferons modify in vitro proliferation of human bladder transitional cell carcinoma in the presence of doxorubicin and mitomycin C: J Urol 156:1492–95, 1996.
47. Kalbe T, Beer M, Mendoza E, Ikinger U, Link M, Reichert HE, Frangenheim T, Klein E, and Fabricius PG: BCG vs interferon A for prevention of recurrence of superficial bladder cancer. A prospective randomized study. Urologe Ausgabe a, 33: 133–7, 1994.
48. Donovan MG, Grainger R, Hegarty JH, et al.: A prospective, randomized study of intravesical BCG and interferon in superficial bladder cancer. Presented as a poster at the 4th European Conf on Clinical Oncology and Cancer Nursing, Madrid 1987.
49. Jiminez-Cruz JF, Vera-Donoso CD, Leiva O, Pamplona M, Rioja-Sanz LA, Martinez-Lasierra MM, Flores N, Unda M: Intravesical Immunoprophylaxis in Recurrent Superficial Bladder Cancer (Stage T1): Multicenter Trial Comparing BCG and Interferon-Alpha. Urology 50:529–535, 1997.
50. Portillo J, Martin J, Hernandez R, Correas M, Gutierrez J, Valle JD, Roca A, Vega A, Villanueva A, Gutierrez R: Results at 43 Months' Follow-up of a Double Blind, Randomized, Prospective Clinical Trial Using Intravesical Interferon Alpha 2b in the Prophylaxis of Stage pT1 Trasitional Cell Carcinoma of the Bladder. Urology 49:187–190, 1997.
51. Lamm DL: Results at 43 Months' Follow-up of a Double Blind, Randomized, Prospective Clinical Trial Using Intravesical Interferon Alpha 2b in the Prophylaxis of Stage pT1 Trasitional Cell Carcinoma of the Bladder (Editorial). Urology 49:190, 1997.
52. Engelmann U, Knopf H, Graff J, and the Project Group Bochum-interferon and superficial bladder cancer: Interferon-a2b instillation prophylaxis in superficial bladder cancer—a prospective, controlled three-arm trial. Anti-Cancer Drugs, 3: 33, 1992.
53. Ferrari P, Castagnetti G, Pollastri CA, Ferrari G, Tavoni F, Grassi D: Chemoimmunotherapy for prophylaxis of recurrence in superficial bladder cancer: Interferon-a2b with epirubicin. Anti-Cancer Drugs, 3: 25, 1992.
54. Gomella LG, Keeley FX, Jr., McCue PA, and Lattime EC: A comparison of the local immune response to intravesical alpha-interferon and Bacillus Calmette-Guerin (BCG) in patients with superficial bladder cancer. J Urol., 151: 473A, 1994.
55. Downs T, Szilvasi, O'Donnel M: Pharmacological Biocompatibility between intravesical preparations of BCG and Interferon-Alfa 2B. J Urol 158:2311–2315, 1997.
56. Pagano F, Bassi P, Milani C, Meneghini A, Maruzzi D, Garbeglio A: A Low Dose Bacillus Calmette-Guerin Regimen in Superficial Bladder Cancer Therapy: Is It Effective? J Urol 146:32–35, 1991.

57. Stricker P, Pryor K, Nicholson T, Goldstein D, Golovsky D, Ferguson R, Nash P, Ehsman S, Rumma J, Mammen G, Penny R Bacillus Calmette-Guerin plus intravesical interferon alpha-2β in patients with superficial bladder cancer. Urology 48:957–961, 1996.
58. O'Donnell MA, Chen X, Luo Y, DeWolf WC: Experimental and Clinical Evidence of Enhancement of BCG Efficacy by Adding Interferon-Alpha. J Urol 157:1502A, 1997.
59. O'Donnell MA: Personal Communication, 1998.

DISCUSSION

Infection, Immunology, and Interstitial Cystitis in the Bladder

Q & A with James A. Roberts, M.D.

Richard Grady, Seattle.

Q: I noticed that your non tumorigenic cell lines were not expressing any accessory molecules such as V7. Do you think that normal urothelium can act as an antigen-presenting cell or is that really more limited to the tumor lines?

A: We have developed primary cultures from murine bladder, and we see the same markers expressed, and they are capable of presenting antigen. There were a couple of papers presented, one by Mike O'Donnel and one by Ralph Butyen at the AUA a couple of years ago, where they did some immunohistochemical studies, at least Ralph Butyen's reported the presence of FAS antigen on epithelial cells. So, yes, I think that they do express this in vivo. We haven't actually looked at V7 yet, though.

Unidentified Questioner.

Q: One of the questions that I have as a pediatric urologist, concerns the common problems of dysfunctional voiding. What I tell the parents is that you're child's running around with a tank more than half empty most of the time. What are your biases about residual urine in cases like that, so that there's not an anatomic blockage but rather just simply a failure to empty on a more often than not basis?

A: There is no evidence, either clinical or experimental, to show that residual urine predisposes to infection. There is, however, evidence that residual urine allows the bacteria to grow. So the two should be carefully separated. It does not initiate disease to have a residual urine.

Anthony J. Schaeffer, Chicago.

Q: With regards to phase variation, does this occur with p-fimbriated bacteria as it does with type 1, and might that kind of concept play a role in the ascent of the bacteria from the bladder to the kidney? In other words, we have patients with cystitis, and why then do some of them develop pyelonephritis?

A: P-fimbria do have phase variation as do type 1. Exactly how that is helpful to the bacteria, I don't think has been shown in the studies that I know of.

Comment: The second question I have is with regards to the concept of colonization, vaccination, and prevention. When we treat women or girls with urinary tract infections, we prevent the normal antibody response that might prevent recurrence. I don't think there's much evidence that children who have recurrent urinary tract infections have an immune deficiency, but rather simply increased receptors and decreased immune response because we treat them and prevent continued bacterial stimulation, and that's the reason for thinking of a vaccine.

David Thomas, Leeds, UK.

Q When we talk about urinary infection, we think of it as being synonymous with bacterial infection, which is clearly the most important etiological factor, but I wonder if there's any information on whether other infective agents (viruses, Mycoplasma, and so on) cause urinary infection? Is there a role for other infective agents?

A: As far as we know, viruses do not cause disease in the urinary tract. There certainly is a viruria, when you have a viral infection of almost any kind, but there is no viral disease in the bladder, in the urinary tract, except from cytomegalovirus, and that virus, of course, causes disease only in severe immunodeficiency. The only other time that I've seen an affect on the bladder is with herpes zoster, and herpes zoster can cause the same kind of appearance in the bladder as it shows you on the flank in the case of shingles, it's the site of the nerves. This does cause sufficient urinary tract infections for the patients who present. That's the only one that I know of that causes disease from a virus.

Richard Grady, Seattle.

Q: I was wondering, with the girls who are partially treated and don't have adequate antibody response, whether anyone has looked at whether they have a cell-mediated immune response to these bacteria afterwards?

A: We had tried to sort that out years ago, and we found that urinary tract infection does not have any cell-mediated response which is important in eradicating the infection. It's the humoral response that's important. There is very little cell mediating.

Comment: Just a comment on that. It is an interesting question because you do get isotype switching of your antibodies, so you must get a, T-cell dependent response, so you get a T-cell activation, but it doesn't appear to be necessary for elimination of the bacteria.

Q & A with Ann Stapleton, M.D.

Anthony J. Schaeffer, Chicago.

Q: I wondered if you have some thoughts about the possibility about what might modulate susceptibility of nonsecretors, i.e., even the worst-case patient may have three or four infections per year, but essentially be healthy the rest of the time. Do you think that some of the receptors you've identified might be expressed differentially at different times in a given patient, perhaps even influenced by their estrogen cycles and/or say pre-, post-menopausal replacement therapy, those kinds of things that clinically seem to influence susceptibility?

A: That's a very interesting question, which we are trying to address in a study in which we're looking at cells from a lot of people whose genetic background, as well as their cell cycle (these are younger women who are cycling regularly) we know where they'll be, and we can by inference assume their estrogen status. Maybe Linda will tell us if that's true or not - we'll have a general idea of where they stand with respect to their hormonal state, so hopefully in a couple of years when we've done about a million cell-sorter assays we'll have a better idea of that. There is a little bit of in vitro data on other stored hormones that the glycosyltransferase activity in various tissues can be affected by treatment with these hormones, so this is an area that I would very much like to get into, and I am getting into it, but I think it's a very young field in terms of being able to take paradigms from other areas. The other question is, I think sometimes some of these behavioral factors that are a little more squeezy, probably do come into play. Another area that I'm interested in and I'm pursuing as well, is whether or not the actual inflammation associated with infection can modulate the expression of glycolipids. There are some other paradigms in other diseases in which the actual cytokines that are elaborated, for example, the virucytotoxin story, some of the cytokines elaborated actually upregulate the virutoxin receptors in endothelial cells. These are actually the same family of glycolipids, and this has been shown to be specifically dependent upon the increase in activity of a certain glycosyltransferase, so a definite explanation for why this is happening. So there's probably all these kinds of different factors coming into play. It's not as simple as we had hoped it was. That bacterial factor was pretty much the only thing that determines susceptibility.

M'liss Hudson, St. Louis.

Q: I realize that most of your work has concentrated on the nonsecretors, but I wondered if you'd ever looked at the women who are secretors? We know in bladder cancer that if you lose your ability to secrete blood group antigens on the cell surface, that that's a sort of negative prognostic factor. I'd wondered if you'd looked at the secretors serially over time, whether there is any bacterial influence or another influence that caused some of these women to lose their ability to secrete blood group antigens on the cell surface, and whether this might or might not be an explanation for the recurrent infections in that small group of secretor-positive women who still have the infections?

A: You make a very interesting connection there between the bladder cancer findings. We may be able to get the answer to that from our study of cells from a lot of different women, some of which are serial samples. I should point out that what we think is happening with bladder cancer, why you can't detect blood group antigens anymore, say with antibody, it is thought to be dedifferentiation. I didn't show you the data, but we've looked at T-24s and J-82s, and they don't make complex carbohydrates on glycosphingolipids like we see that bind the bacteria with higher avidity. So, if people did down regulate the transferases that make these longer sugar molecules, they might not actually protect themselves. So, we don't really know whether in either group of patients that happens.

Comment: True, although chronic infection is supposed to be a risk factor, at least for some forms of bladder cancer. Usually, it's more likely to be squamous cell, but some of them will absolutely develop papillary or transitional cell tumors.

Barry Kogan, Albany, New York.

Q: Just a question to see if I understood this right. You're suggesting these glycolipids, whether you're a secretor or nonsecretor, cover the surface of transitional cell epithelia as well?

A: We assume they do, because pretty much all mammalian cells have them. They're probably in patches because in cells that have been studied, there're specialized patches that have certain structures.

Q: So not every epithelial cell might have these?

A: Well, they should, from what we understand of mammalian cells. We've stained cultures, and we don't see anything that's negative.

Q: Okay. So my real question then relates to yesterday's presentations on uroplakins which also bind to the cell membranes. If these are all bound to the cell membranes and the uroplakins are bound to all the cell membranes, how do they fit in? Henry Sun's pictures showed sheets of these uroplakins. I mean, where are these glycolipids anatomically in that particular model system, or in that envisionment of how the bladder cell looks?

A: Maybe we should ask Dr. Sun, to make a contribution to that question as well. I think it's not really well understood. The cells that I showed you should be sort of more basal intermediate in their phenotype, essentially, since they were not differentiated. We've done some work with differentiating cells with calf serum. We don't see any differences in fairly large number of cells and numbers of bugs per cell. But that's a kind of a crude phenotype, so we don't really know whether or not. The organisms that I showed you do not express any p-fimbria under the conditions that we grow them for the assays. That's in part because if you grow organisms to express p-fimbriae, they don't usually express hypo infimbriae just because of the way they regulate. So, we might not pick up on type 1 mediated adherence under those conditions using those organisms. In terms of models that might have been made of it, I don't know of any models that have been made to look at this. As a general rule, the glycolipid in relation to other intrinsic proteins would be surrounding it so that the sugar parts theoretically come up, they're actually much smaller than a protein molecule, would come up surrounding them in patches. What I'm referring to has not been looked at in urinary epithelium to my knowledge. It's in other cell types, not even epithelia, where people like Dr. Hakamori, who is a mammalian cell-cell person, really more has looked. There's a specialization where on some sort of very small, very, very small level, there is a certain protein surrounded by what they call patches of glycolipids of a certain family or structure. So this could certainly happen, and I don't know that we have ruled that out or in by the kinds of beautiful images that Dr. Sun has shown.

Q & A with Anthony J. Schaeffer, M.D.

James Roberts, Tulane.

Q: You did eradicate the lactobacilli from the mucin used in your assays?

A: Yes, that's a good point Jim. All these vaginal fluids were filtered and freed from all bacteria, Lactobacillaceae, as well as pathogens.

Q: You previously showed that the menstrual cycle is associated with a difference in adherence. Does this also correlate with pH in the menstrual cycle?

A: Yes. As Ann has shown, and others, the vaginal pH is high postmenopausally, and when you give hormone replacement therapy, the pH drops to the premenopausal, as we call it, the healthy pH. Sort of mimicking that slide where I showed you that it is possible to adjust the pH, and so that might be an in vitro correlate of what would happen if you administered hormones to a patient who is postmenopausal. As Ann is showing, these women have a favorable response by reducing their numbers of recurrent urinary tract infections significantly.

Q: Can you show this effect too on the cycling female?

A: We haven't done that. We don't have enough data to correlate that. We showed before, as you know, that vaginal cell receptivity fluctuates and tends to be higher in the first two weeks as compared to the last two weeks of the cycle. The data from the vaginal fluid seems to suggest that the same things are happening, that it at least correlates with hormone status. We're in the process now, this is the third year of collecting vaginal fluid and cells from women who are on the estrogen replacement therapy study that the NIH is running, so we have a large number of blinded samples of cells and fluids that we've collected. After four or five years, we should be able to determine what the hormonal status of the patient has vis a vis the ability of her cells or her fluid to bind E. Coli.

Unidentified Questioner.

Q: Did you filter out the immunoglobulins as well?

A: No, we did not. When we asked the question, could immunoglobulins be binding the E. coli, we found out that they were not the three putative receptors that we have identified. Immunoglobulins are presumably still present in the vaginal fluid, as they would be naturally.

Q: Because, actually, probably nonspecifically, and also specifically, if you add some of the bacteria back to the vaginal fluid, you could actually absorb out the immunoglobulins, so that they may coat bacteria in some nonspecific or specific way.

A: Right.

Ann Stapleton, Seattle, WA.

Q: Have you characterized the vaginal fluid, say in an individual over time, or comparing individuals with respect to even something like protein concentration? Do you know what is varying between people or within through the cycle?

A: The vaginal fluid concentration of these glycoproteins varies. The problem we have in determining what the true variation is, that the way we collect it by simply swabbing the surface, and is not reproducible. You're collecting more or less quantities of vaginal fluid, but if you accept the fact that we over time run hundreds of samples, that would be relatively common among patients. There does appear to be a large difference in the amounts of glycoprotein that we recover, suggesting that there could be more production of this vaginal fluid at different times in the cycle which would correlate with my clinical bias that there are different amounts of fluid present.

Q: Do you think any other proteins are from sloughed cells?

A: Yes. We've ruled out the urine, which would have been our first guess. I believe this is a transudate and/or a sort of surrogate for the vaginal cell surface. As you know, we looked at the expression of secretory proteins on the surface and in the fluid, and saw a good correlation there, so I suspect this is the same thing that when we swab the surface, we're collecting some of the glycoproteins that are exposed on the surface of the cells, as well as some of those that may be naturally in suspension. So, I would expect that they would be very closely related.

Q & A with Michael Kiarash

Joel Hutchinson, Philadelphia.

Q: Could you tell me how do things stand with BCG or interferon alpha used in BCG failures?

A: That is actually one of the questions that is needs to be addressed and hopefully there will be a randomized study looking at two things:

1. The comparison of interferon versus interferon plus BCG.
2. And then just as you pointed, patients who have already failed BCG therapy whether or not there will be a high response to a combination of interferon alpha and BCG. So your question is very appropriate and unfortunately the study is just not done yet.

Ginsburg, Los Angeles.

Q: I know interferon therapy can be quite toxic given systemically. Can you comment on what toxicity is seen with intravesical therapy?

A: Certainly. It has turned out that the side-effects noted with intravesical interferon alpha, even at high doses, has been very, very limited. And typically what is seen is just cystitis. Some patients can develop flu-like symptoms, but again, the incidence is extremely low--more like 5%-10%. Some studies have gone us as high as 30%, but it is typically a very low incidence.

Jenny Southgate, Leeds, UK.

Q: Can I just check with you the tables that you showed as the difference in the interferon studies. Were they all interferon alpha?

A: No, some were interferon beta, some were interferon gamma, but most of the studies have been done with interferon alpha--the majority of them I should say.

Q: Because I think it is important to appreciate that there are different biological effects of the different interferon, so they should not all be clumped together.

A: Exactly. And that is one of the difficulties that we have is that there is just not enough data out there with enough patient sample sizes and follow-up to really render any deductions whatsoever. Especially for the interferon beta and gamma studies are even fewer with smaller sample sizes. So again, it is very difficult to test them. And based on in vitro studies, I think that interferon beta and gamma may actually be more efficacious in treating these patients, yet most the studies have been done with interferon alpha.

Section V

CANCER OF THE BLADDER

OVERVIEW OF CANCER OF THE BLADDER

S. W. Hayward

Departments of Anatomy and Urology
University of California
San Francisco, California 94143-0540

Carcinoma of the urinary bladder is a major source of both morbidity and mortality. Bladder cancer is the fifth most common solid tumor in the United States. The 1998 figures show approximately 54,400 new cases of bladder cancer and 12,500 deaths due to the disease. Although the incidence of bladder cancer has been rising steadily throughout this century, probably mirroring behavioral changes and aging in the population, the mortality rates have remained relatively constant. This likely reflects gradual improvements in detection and treatment methods.

The vast majority of bladder tumors in the western world are transitional cell carcinomas.[1] In other parts of the world squamous cell carcinomas prevail, due to infection by the parasite Schistosoma haematobium. The natural history of transitional cell carcinoma is complex, involving multiple stages. Benign (normal urothelium, atypia and dysplasia) and malignant (papillary [Ta], flat carcinoma in situ [Tis], and invasive [T1-T4]) tumors are commonly multifocal. It is important to note that papillary carcinomas tend to be low grade with a high incidence of recurrence but a low rate of progression. In contrast to Ta tumors, Tis is generally a high-grade disease with a high rate of progression to invasive disease. Tis is typified by loss of the superficial umbrella cells and increased pleiomorphism of the nuclei in the remaining urothelial layers.[2] Some stromal changes, such as increased vascularity in the vicinity of the tumor, are also reported to be associated with these early stage tumors. If left untreated bladder tumors can invade through the muscularis mucosa and lamina propria into the surrounding bladder smooth muscle wall.[2,3] They can then progress to metastatic disease.

Bladder cancer was one of the first diseases to be linked to environmental carcinogens with the description in 1895 by Rehn of a high disease incidence in workers in the aniline dye industry.[4] This was followed by numerous studies through the early part of this century which firmly established the link between bladder cancer and environmental carcinogen exposure, particularly in workers in the dye, rubber and leather industries (for review see [5,6]). The link between cigarette smoking and bladder cancer was first established by Lillienfield et al. in 1956[7] and has since been confirmed by numerous other studies (see [5,6]). Most case

Advances in Bladder Research, edited by Baskin and Hayward.
Kluwer Academic / Plenum Publishers, New York, 1999.

controlled studies have reported a relative risk of approximately 2 for cigarette smokers as compared to non-smokers, with steadily higher risks observed with increasing cigarette consumption.[5] The exact mechanism by which cigarette smoke induces bladder cancer is unclear, partly because of the complex mixture of known and potential carcinogens present in tobacco smoke. Smokers are exposed to a large number of compounds which are known to be bladder carcinogens in experimental models. Aromatic amines such as 2-naphthylamine and 4-aminobiphenyl are leading candidates as specific etiologic agents.[8] 4-aminobiphenyl DNA adducts are found in exfoliated bladder cells from smokers. The presence of these adducts is consistent with mutation of the p53 gene which is commonly found in bladder cancer (for review see [1]).

Bladder carcinogenesis is clearly a multistage process. Early work by Hicks and coworkers established the idea that multiple doses of carcinogen were synergistic in producing tumors.[9] They also demonstrated that urothelium which had been "initiated" by exposure to a single (non-tumorigenic) dose of a strong carcinogen could be promoted to form a tumor by chronic exposure to low levels of compounds which were not themselves tumorigenic.[10,11] These studies were followed by extensive genetic analyses of bladder tumors, which have demonstrated a range of anomalies. These include non-random changes in chromosomes 1, 5, 7, 9, 11 and 17, activation of the ras and erbB-2 oncogenes, and mutations to the p53 and retinoblastoma (Rb) tumor suppresser genes. Mutations in the p53 and Rb genes are common in bladder cancer, occurring in approximately 50% and 33% of tumors, respectively.[1,12] These genetic changes have led to the proposal of various models for bladder cancer progression based on specific genetic damage (see [1,12,13]). Such models are entirely consistent with the earlier initiation and promotion models of Hicks et al.

Traditionally much of the research effort devoted to cancer has been concentrated on changes to the genetic material of the epithelial cells which form carcinomas. Concomitant with the epithelial genetic changes, there are associated stromal changes which support and promote the development of the nascent tumor. Currently the most widely discussed of these is the increased vascularization which occurs around tumors and which is required for their growth. It is clear that there is increased vascularization around growing bladder tumors and that this is related to disease prognosis. It is also clear that there are a multitude of other changes occurring in the stroma surrounding growing tumors, including inappropriate expression of extracellular matrix modifying enzymes such as matrix metalloproteinases and decreases in the levels of tissue inhibitors of metalloproteinases. Regulation of these systems will be discussed by Dr. M'Liss Hudson and the topic which will also be visited, though from a somewhat different angle in a chapter written by Dr. Catherine Booth. There is growing evidence that both genetic and epigenetic factors in the carcinoma and in the surrounding stromal cells play significant roles in the development of cancers. Thus, it is appropriate that this session begins with a talk by Dr. Peter Jones. Dr. Jones will be discussing the role of DNA methylation in permanently inhibiting the expression of specific genes and describing a new technique to genomically screen for such methylation. Dr. Frederick Waldman will discuss gains and losses of specific genes in bladder cancer as detected by cytometric analysis and, I am sure, will remind us that individual tumors are heterogeneous entities. Dr. Monica Liebert will describe the large range of differentiation markers expressed by urothelial cells providing us with a valuable resource with which to continue investigation into this field of genetic and phenotypic variability in tumor cell morphology and progression. Dr. Robert Hurst will discuss the complex interrelationships between gene expression and cellular and tissue phenotype. His talk should serve to remind us all that tumor biology is far more complex than simply the reflection of the expression of one or a group of genes, and includes factors such as environmental exposure to carcinogenic agents and the genetic background of the affected individua. In addition to the work presented in this Oncology session a number of other speakers at this meeting have also presented data relevant to bladder cancer. I would therefore call particular attention to the work presented by Drs. Resnikoff and Southgate in other sections.

REFERENCES

1. Reznikoff, C.A., Belair, C.D., Yeager, T.R., Savelieva, E., Blelloch, R.H., Puthenveettil, J.A. and Cuthill, S.: A molecular genetic model of human bladder cancer pathogenesis. Seminars in Oncology, **23:** 571,1996.
2. Tannenbaum, M., Romas, N.A. and Droller, M.J.: The pathobiology of early urothelial cancer. In: Diagnosis and Management of Genitourinary Cancer. Edited by Skinner, D.G. and Lieskovsky, G. Philadelphia: W.B. Saunders Co, chapt., 55, 1988.
3. Melicow, M.M.: Tumors of the bladder: A multifaceted problem. J. Urol., **112:** 467,1974.
4. Rehn, L.: Blasengeschwultse bei fuschsin-arbeitern. Arch Clin Chir, **50:** 588,1895.
5. Ross, R.K., Paganini-Hill, A. and Henderson, B.E.: Epidemiology of bladder cancer. In: Diagnosis and Management of Genitourinary Cancer. Edited by Skinner, D.G. and Lieskovsky, G. Philadelphia: W.B. Saunders Co, chapt., 23, 1988.
6. Ross, R.K., Jones, P.A. and Yu, M.C.: Bladder cancer epidemiology and pathogenesis. Seminars in oncology, **23:** 536,1996.
7. Lillienfield, A.M., Levin, M.L. and Moore, G.E.: The association of smoking with cancer of the urinary bladder in humans. Arch. Intern. Med., **98:** 129,1956.
8. Patrianakos, C. and Hoffman, D.: On the analysis of aromatic amines in cigarette smoke. J. Anal. Toxicol., **3:** 150,1979.
9. Hicks, R.M. and Wakefield, J.S.J.: Rapid induction of bladder cancer in rats with N-methyl-N-nitrosourea. I. Histology. Chem.-Biolo. Interactions, **5:** 139,1972.
10. Hicks, R.M.: Initiation and promotion in the transitional epithelium of the rat bladder. Br. J. Cancer, **41:** 504,1980.
11. Hicks, R.M. and Chowaniec, J.: The importance of synergy between weak carcinogens in the induction of bladder cancer in experimental animals and humans. Cancer Res, **37:** 2943,1977.
12. Knowles, M.A.: Molecular genetics of bladder cancer. Brit. J. Urol., **75 Suppl1:** 57,1995.
13. Cordon-Cardo, C., Dalbagni, G., Sarkis, A.S. and Reuter, V.E.: Genetic alterations associated with bladder cancer. In: Important Advances in Oncology. Edited by DeVita, V., Hellman, S. and Rosenberg, S.A. Philadelphia: J.B. Lippencott, chapt., 1994.

30

UROKINASE (u-PA) AND THE u-PA RECEPTOR

Modulation of in Vitro Invasiveness of Human Bladder Cancer Cell Lines

M'Liss A. Hudson and L. Michelle McReynold

Division of Urologic Surgery
Washington University School of Medicine
4960 Children's Place
St. Louis, Missouri 63110

1. ABSTRACT

1.1. Background

Urokinase (u-PA) and the u-PA receptor (u-PAR) influence tumor invasion and metastasis.

1.2. Purpose

The purpose of this study is to determine whether u-PAR display in 4 human bladder cancer cell lines (RT4, 253J, EJ and T24) can be modulated with substances including phorbol 12-myristate 13-acetate, interferon gamma, epidermal growth factor, and transforming growth factor beta and to correlate changes with u-PAR expression with the ability of these cells to invade artificial basement membrane (Matrigel).

1.3. Methods

u-PAR display was determined using flow cytometry and immunohistochemical staining with an anti-u-PAR monoclonal antibody (Mab3936). Matrigel invasion chamber assays were used to assess the invasive capacity of the cell lines.

Advances in Bladder Research, edited by Baskin and Hayward.
Kluwer Academic / Plenum Publishers, New York, 1999.

1.4. Results

The T24, EJ and 253J cells expressed the u-PAR while the RT4 cells did not. The EJ cells (expressing the highest u-PA antigen levels and the u-PAR) invaded Matrigel. The T24 cells, which expressed the u-PAR but do not produce u-PA, invaded Matrigel only when pretreated with high molecular weight urokinase (HMW u-PA). HMW u-PA pretreatment of EJ and 253J cells also enhanced their invasive potential. Blocking u-PA/u-PAR attachment with an anti-u-PAR monoclonal antibody (Mab3936) inhibited invasion. Modulation of u-PAR expression on cell lines displaying the u-PAR directly affected in vitro invasiveness of the cell lines. The RT4 cells which lack the u-PAR were not invasive under conditions tested. Thus, bladder cancer cell lines require both u-PA and the u-PAR to invade Matrigel and modulation of u-PAR display directly affects their in vitro invasive capacity.

1.5. Conclusions

This study shows that bladder tumor cells produce u-PA and express the u-PAR and require both for in vitro invasion to occur. When bladder cancer cells express the u-PAR, invasiveness can be enhanced by exogenous u-PA and inhibited by anti-u-PAR antibodies. Modulation of u-PAR expression on the cell surface of bladder cancer cells can also affect their ability to invade Matrigel.

1.6. Implications

Histologically similar bladder tumors show differences in their propensity to invade locally or metastasize. Intrinsic differences in the tumor cells such as the production of u-PA antigen and u-PA receptor on the cell surface and extrinsic differences in the tumor cell environment such as substances influencing u-PAR display or antibodies blocking the u-PAR may affect the biological potential of human bladder cancers and offer one explanation for the aggressive or indolent tumor behavior observed in individual patients.

2. BACKGROUND

Components of the plasminogen activator system provide important mechanisms for implantation, local tumor invasion, and metastasis in many forms of cancer.[1] Plasminogen activators are thought to degrade extracellular matrix protein and basement membrane and allow local tumor invasion and access to the intravascular system for metastasis. Urokinase (u-PA) is the most common form of plasminogen activator associated with malignancies.[1,2] The u-PAR is hypothesized to focus and prolong effective plasminogen activation enhancing basement membrane destruction and tumor invasion while preventing inactivation by inhibitors.[1,3,4] u-PA must bind to the u-PAR on the surface of cells for invasion (degradation of extracellular matrix and basement membrane) to occur in other tumor systems.[5] Many tumors have been shown to express the u-PAR[6–10] and its expression has been correlated with cell migration and tumor invasion.[11–14]

Early reports have related components of the plasminogen activator system to bladder cancer.[15–20] High u-PA antigen levels in bladder tumor extracts have correlated directly with advancing bladder tumor stage and grade [17–19,21] and an increased u-PA antigen level has been suggested as a poor prognostic factor in bladder cancer patients.[21] A direct correlation between u-PAR expression and increasing bladder tumor grade and stage has been observed.[21] We have previously characterized the different factors of the plasminogen activator system (u-PA, t-PA, PAI-1, and

PAI-2 antigen levels and u-PAR display) produced by 4 human transitional cell carcinoma cell lines and not other components of tumor extracts.[22] u-PA antigen production and u-PAR expression together appeared to be required for in vitro invasion to occur.[22]

In other systems, modulation of u-PAR expression on the cell surface has been effected by cytokines and growth factors. Phorbol 12-myristate 13-acetate (PMA), gamma interferon (IFN gamma), transforming growth factor beta (TGF beta), and epidermal growth factor (EGF) have all been shown to increase u-PAR expression on vascular smooth muscle cells.[23–25] The effects of PMA, INF gamma, TGF beta, and EGF on u-PAR expression by human bladder tumor cells has not yet been studied, but was investigated herein. The purpose of this study is to determine whether substances including PMA,IFN gamma,TGF beta, and EGF affect u-PAR expression in four human bladder cancer cell lines (RT4, 253J, EJ and T24) and to correlate modulation of u-PAR display on the cell surface of these cell lines with invasion of artificial basement membrane (Matrigel).

3. METHODS

3.1. Tumor Cell Lines

The RT4 cell line was developed from a primary transitional cell carcinoma of the bladder with both papillary and solid components demonstrating superficial muscle invasion.[26] The T24 cell line was developed from a primary grade 3 transitional cell carcinoma of the bladder, depth of invasion unknown.[27] The 253J cell line was developed from a retroperitoneal metastasis from a patient with multiple transitional cell tumors of the urinary tract.[28] The EJ cell line was developed from an invasive, poorly differientiated transitional cell carcinoma of the bladder.[29,30] The RT4, 253J and T24 cell lines were obtained from Dr. Timothy L. Ratliff at Washington University. The EJ cell line was obtained from Dr. Kevin Scanlon at City of Hope Hospital. The HT-1080 cell line, derived from a human fibrosarcoma (ATCC CCL121, American Type Culture Collection, Rockville, MD, USA) is known to express the u-PAR [8,31] and was used as a positive control in the studies below. 3T3 cells (ATCC CRL 6442, American Type Tissue Collection, Rockville, MD, USA) is known to lack the u-PAR[6] and was used as a negative control in the studies below.

The RT4 and T24 cell lines were grown in MEM medium supplemented with 10% fetal calf serum, 1% L-glutamine and 0.05% penicillin/streptomycin. The 253J cell line was grown in DMEM medium supplemented with insulin 5 μg/ml, transferrin 5 μg/ml, thyroxine 0.0136 μg/ml, biotin 0.244 μg/ml, adenine 25 μg/ml, 10% fetal calf serum, and 0.05% penicillin/streptomycin. The EJ cell line was grown in RPMI-1640 medium supplemented with 10% fetal calf serum and 0.05% penicillin/streptomycin. The HT-1080 cells were grown in Dulbecco's modified Eagle's medium supplemented with 10% fetal calf serum and 0.05% penicillin/streptomycin. The 3T3 cells were grown in Dulbecco's MEM supplemented with 2 mM glutamine, 10% fetal calf serum, and 0.05% penicillin/streptomycin. Medium was replaced every 3 days and the cells were subcultured at 80% confluency.

3.2. Flow Cytometry for u-PA Receptors

u-PARs were detected using the commercially available monoclonal antibody Mab3936 (American Diagnostica, Greenwich, CT) directed against the u-PA receptor. Mab3936 has been shown to bind both unoccupied and occupied u-PA receptors on the surface of tumor cells.[32] Single cell associated fluorescence was quantified by flow cytometry (FACS) with the FACscan flow cytometer (low power argon laser excitation at 488 nm).

Table 1. Detection of the u-PAR using a flow cytometry technique with the Mab3936 directed against the u-PAR

Cell lines	T24	RT4	253J	EJ	3T3	HT1080
Trial 1						
2° Ab + 7AAD	3.52 ± 1.97	2.88 ± 1.49	2.67 ± 1.34	2.29 ± 0.70	4.26 ± 1.59	2.51 ± 1.47
Anti-u-PAR + 2° Ab	18.4 ± 10.9	3.17 ± 2.31	31.9 ± 17.7	26.3 ± 18.1	2.7 ± 1.63	24.7 ± 11.3
Trial 2						
2° Ab + 7AAD	2.29 ± 0.70	3.43 ± 1.55	2.13 ± 1.01	2.21 ± 0.79	2.46 ± 2.38	3.28 ± 1.67
Anti-u-PAR + 2° Ab	19.8 ± 17.0	3.18 ± 4.56	21.0 ± 20.1	23.0 ± 11.2	2.56 ± 1.05	23.9 ± 21.6
Trial 3						
2° + 7AAD	6.20 ± 2.75	8.00 ± 3.39	6.67 ± 3.13	5.56 ± 2.97	4.22 ± 1.36	3.28 ± 1.67
Anti-u-PAR + 2° Ab	25.8 ± 12.0	8.29 ± 4.10	23.0 ± 11.4	17.3 ± 11.1	4.07 ± 1.5	26.7 ± 18.1

7-aminoactinomycin (7AAD) was used to exclude non-viable cells. 2° Ab = secondary antibody = FITC-anti-mouse IgG
Anti-u-PAR = Mab3936 (American Diagnostica, Greenwich, CT.)
HT1080 cell line is known to express the u-PAR and was used as a positive control while the 3T3 cells which lack the u-PAR served as a negative control.
Results expressed as the mean channel fluorescence ± the standard deviation.
The Mann-Whitney U test was used to compare the background fluorescence (2° + 7AAD) to the fluorescence with the anti-u-PAR antibody + 2° antibody for each cell line. The EJ, 253J and T24 cells and the positive control (HT1080 cells) all showed a significant increase in florescence (all p values ≤ 0.01 for each of the 4 cell lines) indicating the presence of the u-PAR on the cell surface of these cell lines. No significant change in fluorescence was noted for the RT4 cells or the negative control (3T3 cells) ($p > 0.8$ for each, respectively.)

10^6 living cells after the manner of Chucholowski et al.[32] were reacted with Mab3936 (2 μg/ml) in RPMI-1640–0.1% bovine serum albumin (BSA) at 22°C for 30 minutes. Cells were then washed twice with RPMI-1640–0.1% BSA and resuspended in the same containing FITC-antimouse IgG (4 μg/ml) at 22°C for 30 minutes. Cells were extensively washed with RPMI-1640–0.1% BSA and then resuspended in 2.5 ml of the same. 7-aminoactinomycin (7AAD) was used to exclude nonviable cells. Cell-associated fluorescence was determined by flow cytometry (FACScan). Autofluorescence of the cells and non-specific binding of the FITC-antibody in the absence of Mab3936 was subtracted from the total fluorescence for each determination. HT1080 cells which are known to express the u-PAR[8,31] were used as a positive control while 3T3 cells were used as the negative control.[6] Table 1 shows the mean channel fluorescence ± the standard deviation expressed by each cell line. Each experiment was run with 20,000 events per sample. Three independent experiments were run and similar results were obtained in each experiment.

For the modulation studies, 5×10^5 cells were pretreated with following agents for the following periods of time. For the PMA studies, cells were exposed for 4 hours to phorbol 12-myristate 13-acetate (PMA) (Sigma P8139, Sigma Chemical Co., St. Louis, MO) at concentrations of 0, 5, 50, 500 nM. For the EGF studies, cells were exposed to varying concentrations (0, 10, 100, 200 ng/ml) of epidermal growth factor (EGF) (Sigma E-4127, Sigma Chemical Co., St. Louis, MO) for 6 days. For the interferon gamma studies, cells were exposed for 2 days to recombinant human interferon gamma (interferon gamma) (Sigma I-3265, Sigma Chemical Co., St. Louis, MO) at concentrations of 0, 100, 500, 1000 iu/ml. For the TGF beta studies, transforming growth factor beta (TGF beta) (Sigma T1910, Sigma Chemical Co., St. Louis, MO) was used at concentrations of 0, 1, 2, and 5 ng/ml for 6 days. Table 2 shows the mean channel fluorescence ± the standard deviation expressed by cell lines exposed to varying concentrations of PMA, IFN gamma, TBG beta, and EGF for various time periods. Again, each experiment was run with 20,000 events per sample. Three independent experiments were run and similar results were obtained in each experiment.

Table 2. Effect of substances used to pretreat cells on u-PAR display

Cells/substance	HT1080	3T3	T24	253J	EJ	RT4
PMA	–	NC	–	+	–	NC
IFN gamma	+	NC	+++	NC	NA	NC
TGF beta	NC	NC	–	+	NA	NC
EGF	+	NC	+	NC	NA	NC

NC = no change in u-PAR display.
+ = significantly enhanced u-PAR display.
+++ = markedly enhanced u-PAR display.
– = significantly inhibited u-PAR display.
The positive or negative effect described for each substance refers to the concentration of substance that provided the most marked effect and usually referred to the highest concentration of that substance to which the cells were exposed.
HT1080 cell line is known to express the u-PAR and was used as a positive control while the 3T3 cells which lack the u-PAR served as a negative control.
The Mann-Whitney U test was used to compare the background fluorescence (2° + 7AAD) to the fluorescence with the anti-u-PAR antibody + 2° antibody for each cell line after exposure to varying concentrations of the substances listed.

3.3. Immunohistochemical Staining for u-PA Receptor

Cell pellets were prepared from the cell lines grown in tissue culture. Cells were removed from the flasks using 10 mM EDTA, washed twice in PBS and resuspended in serum-free MEM medium. Cell pellets were obtained by centrifugation at 1500 rpm for 8 minutes, then the cell pellets were fixed in 10% formalin for 24 hours and embedded in paraffin. Cell pellets from the HT1080 (positive control) and 3T3 (negative control) cell lines were similarly prepared. Immunohistochemical staining of such cell pellets for u-PAR expression has been previously described.[7]

Immunohistochemical studies were performed on 5 μm sections from formalin-fixed, paraffin embedded cell pellets, mounted on acid-cleaned glass slides, and heated for 1 hour at 55°C. Slides were dewaxed and dehydrated and then immersed for 30 minutes in absolute methanol containing 0.6% (v/v) hydrogen peroxide, to quench endogenous peroxidase activity. After rehydration through graded ethanols and distilled water, sections were immersed in 10 mM citrate buffer (10 mM citrate monohydrate in distilled water, pH 6) in plastic Coplin jars and subjected to microwave irradiation (600 W, 2450 MHz) for 4 minutes. After cooling to room temperature, slides were removed to PBS, pH 7.4, for 10 minutes. Sections were overlaid with Mab3936 at 10 μg/ml (American Diagnostica, Greenwich, CT) for 18 hours in moisture chambers at 4°C. The avidin-biotin-peroxidase complex (ABC) method was performed (Vectastain ABC Elite Mouse Universal Kit; Vector Laboratories, Burlingame, CA) 33. Chromogenic development resulted from immersion of slides in 3,3'-diaminobenzidine tetrahydrochloride [Sigma Chemicals, St. Louis, MO; 0.5 mg/ml, containing 0.003% (v/v) hydrogen peroxide] for ≤10 minutes. The reaction was enhanced by brief exposure to 0.125% (v/v) osmium tetroxide. Slides were counterstained in Harris' hematoxylin and coverslipped with a synthetic mounting medium.

3.4. Matrigel Invasion Assays

Biocoat® Matrigel® Invasion Chambers (Collaborative Research, Becton Dickinson Labware, Bedford, MA) were used to assess the invasive potential of the bladder cancer cell lines expressing the u-PAR or lacking the u-PAR in vitro. 24 well plate size plates (Becton Dickinson Labware, Bedford, MA) were used. In each assay, the upper and lower chambers were separated by PET filters with 8-μm pores coated with Matrigel basement membrane matrix (125 μg/cm^2). Matrigel basement membrane matrix is a solubilized basement membrane preparation extracted from Engelbreth-Holm-Swarm mouse sarcoma. It contains laminin, collagen type IV, heparin sulfate proteoglycan, entactin and growth factors, including TGF beta, basic FGF and others. The chemoattractant placed in the lower chamber was 3% human serum in RPMI-1640.

10^6 cells harvested with 10 mM EDTA/trypsin and then washed twice in 0.1% bovine serum albumin (BSA)/DMEM. Either 10^4 cells were suspended in 200 μl of 0.1% BSA/DMEM or 5 × 10^4 cells harvested with 10 mM EDTA in PBS were suspended in 25μl of warm, serum-free MEM and placed in the upper compartment. Some cells were pre-treated in the presence or absence of high molecular weight urokinase (HMW u-PA) (0 iu, 2500 iu, or 5000 iu) (American Diagnostica, Greenwich,CT) for 1 hour prior to placement in the invasion chambers. Mouse monoclonal antibody anti-protein kinase C (Sigma Chemical Co., St. Louis, MO) was used as the control antibody. Cells were incubated in the invasion chambers for 36, 48, or 72 hours in a sealed chamber at 37°C at 5% $C0_2$. Non-invasive cells were removed from the upper surface of the membrane with a cotton swab before staining with Diff-Quik stain and microscopically counting invasive cells on the underside of the membrane. Cells were counted at 250X magnification under oil immersion with a Leitz phase contrast microscope and 100 fields were counted.

For the inhibition studies, the Matrigel invasion chambers and cells were prepared as above with the following modifications. 5 × 10^4 cells were pretreated with varying concentrations of Mab3936 (0 μg/ml, 2.5 μg/ml, 5 μg/ml, 10 μg/ml, 15 μg/ml, or 20 μg/ml) or a control antibody (mouse IgG2a, kappa [UPC 10] at the same concentrations) (Sigma Chemical Co., St. Louis, MO) for 15 minutes prior to placement in the upper chambers. Mab 3936 or control antibody were added to the upper chambers at 3 hour invervals for 24 hours. Additional antibody was added during the invasion assay to replace that potentially internalized by the cells and to block new u-PARs being expressed on the cell surface during the incubation period. After the last addition of antibody, the chambers were incubated another 12 hours so that the total incubation time was 36 hours. Membranes were removed, stained and invading cells counted as above.

For the modulation studies, 5 × 10^5 cells were pretreated with following agents for the following periods of time. For the PMA studies, cells were exposed for 4 hours to phorbol 12-myristate 13-acetate (PMA) (Sigma P8139, Sigma Chemical Co., St. Louis, MO) at concentrations of 0, 5, 50, 500 nM. For the EGF studies, cells were exposed to varying concentrations (0, 10, 100, 200 ng/ml) of epidermal growth factor (EGF) (Sigma E-4127, Sigma Chemical Co., St. Louis, MO) for 6 days. For the interferon gamma studies, cells were exposed for 2 days to recombinant human interferon gamma (IFN gamma) (Sigma I-3265, Sigma Chemical Co., St. Louis, MO) at concentrations of 0, 100, 500, 1000 iu/ml. For the TGF beta studies, transforming growth factor beta (TGF beta) (Sigma T1910, Sigma Chemical Co., St. Louis, MO) was used at concentrations of 0, 1, 2, and 5 ng/ml for 6 days. Cells were incubated in the invasion chambers for 36 hours. Membranes were removed, stained and counted as above.

HT1080 cells served as the positive control and 3T3 cells as the negative control for these studies.

3.5. Statistical Analysis

Nonparametric tests were used to evaluate the data in the experiments described above. The Mann-Whitney U test was used for pairwise comparisons while the Kruskal-Wallis test was used to compare data with more than 2 groups.

4. RESULTS

4.1. Flow Cytometric Detection of the u-PAR

The T24, EJ and 253J cell lines appeared to express the u-PAR while the RT4 cell line did not. Mean channel fluorescence ± the standard deviation are shown in Table 1.

Table 2 shows the effects of PMA, IFN gamma, TGF beta and EGF on the bladder cancer cell lines and positive and negative controls. PMA enhanced u-PAR display on the 253J cell line, but appeared to decrease u-PAR display on the HT1080 (positive control) and T24 cell lines. IFN gamma exposure markedly enhanced u-PAR display on the T24 cells as well as the HT1080 (positive control) cells. TGF beta exposure showed mixed effects on u-PAR display. The most consistent response was observed after exposure to EGF. All cell lines expressing the u-PAR showed an increase in u-PAR display after EGF pretreatment. None of these substances induced u-PAR display in the cells not naturally expressing the u-PAR (RT4 and 3T3 cells).

4.2. Immunohistochemical Staining for u-PAR

When anti-u-PAR antibodies such as the Mab3936 have been used to detect the u-PAR on formalin-fixed, paraffin-embedded tissue sections or cell pellets, staining is observed to be primarily cytoplasmic and/or associated with the plasma membrane 7,10,32. The T24, 253J, and EJ cell pellets showed cytoplasmic staining by Mab3936. The brown color in the cytoplasmic extensions of the T24 cell pellet in Figure 1A is a representative illustration of u-PAR immunoreactivity by these cells. Cells vary in the amount of associated cytoplasm and some "naked nuclei" (those without associated cytoplasm) may occur in processing thus a brown color is not associated with all of the T24 cells. The absence of brown color in all of the cells of the RT4 cell pellet in Figure 1B indicates these cells are not immunoreactive for the u-PAR.

4.3. Matrigel Invasion Chamber Assays

The 253J and EJ cell lines demonstrated the capacity to invade the artificial basement membrane (Matrigel) when compared to the 3T3 cells (low invasive potential, negative control). They did not have the same ivasive potential as the highly invasive HT1080 cells used as the positive control (Table 3 shows three separate trials). The RT4 and T24 cell lines were initially noninvasive.

Pretreatment of the cell lines with high molecular weight urokinase (HMW u-PA) to occupy the u-PAR enhanced the ability of all cells expressing the u-PAR (253J, EJ, and T24) to invade Matrigel, but did not affect the invasive potential of cells lacking the u-PAR (RT4 and 3T3). In particular the pretreatment of the T24 cells (which express the u-PAR, but produce only minimal amounts of u-PA) with high molecular weight urokinase (HMW u-PA) converted these cells from a non-invasive to an invasive cell line in vitro (Table 3).

Pretreatment of the cells with the anti-u-PAR antibody Mab3936 (which is known to bind to both the occupied and unoccupied receptor) blocked invasion of Matrigel by the EJ cell line (the most naturally invasive of the bladder cell lines tested) expressing the u-PAR and producing u-PA (Table 4). HMW u-PA pretreated T24 cells which express the u-PAR and which were invasive only when pretreated with u-PA were also inhibited from invading Matrigel by the anti-u-PAR antibody Mab3936.

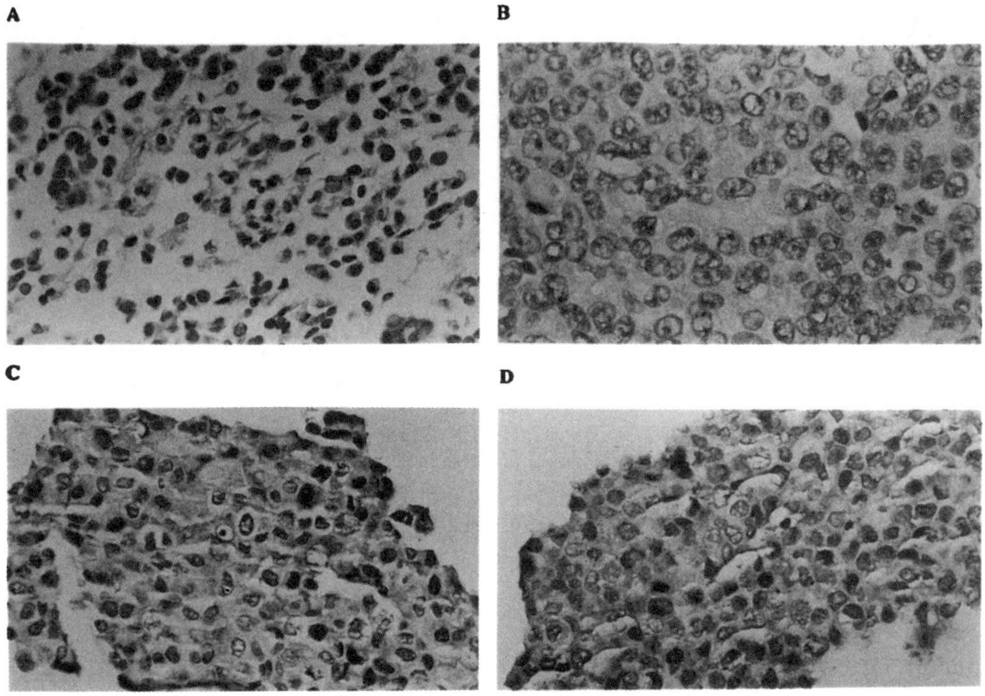

Figure 1. Immunoreactivity for the antibody directed against the u-PAR is seen as brown staining of the cytoplasmic extensions of the (A) T24 cells at 300X magnification Similar staining was noted in the cell pellets from the EJ and 253J cell lines (C and D). The (B) RT4 cell line at 400X maginification does not demonstrate any cytoplasmic staining and thus is not immunoreactive for the antibody directed against the u-PAR.

Increased u-PAR display on the cell surface was shown to correlate directly with an increased ability of cells to invade Matrigel (Table 5). None of the substances tested induced invasion of Matrigel by noninvasive cells.

5. DISCUSSION

u-PA has been the predominant PA found in association with urothelial malignancies. We previously confirmed this finding in 3 of the 4 cell lines (RT4, EJ and 253J) tested.[22] u-PA production alone did not predict tumor invasiveness, although the tumor cell lines producing the highest levels of u-PA were found to have the highest in vitro invasive potential.[22]

Previous studies have confirmed that u-PA must bind to the cell surface u-PAR for tumor cell invasion and metastasis to occur 1.[34,5,35,36] The u-PAR binds pro-u-PA which is then converted to its active form. It strongly enhances the production of surface-bound plasmin which is protected from inactivation by alpha $_2$-antiplasmin.[35] The u-PAR has been found at foci of tumor invasion [34,36] and has been suggested as a poor prognostic factor in several tumor types.[34,37] Both melanoma and osteosarcoma cell lines expressing the u-PA and the u-PAR have been shown to be invasive.[11,12] u-PA and the u-PAR facilitate migration of nonmalignant cells as well.[13,14]

In vitro studies have shown that colon carcinoma cell lines expressing the u-PAR but without u-PA production could not initially breakdown extracellular matrix.[38] When these

Table 3. Number of cells invading a Matrigel artificial basement membrane with or without pretreatment with high molecular weight urokinase (HMW-u-PA)

Conc. of HMW urokinase	0 iu/ml	2500 iu/ml	5000 iu/ml
HT1080			
Trial 1	2282	3783	5232
Trial 2	2554	6160	11774
Trial 3	806	3247	7405
3T3			
Trial 1	25	4	3
Trial 2	3	0	8
Trial 3	1	1	2
253J			
Trial 1	58	101	104
Trial 2	10	30	210
Trial 3	64	370	385
T24			
Trial 1	18	341	1984
Trial 2	88	621	2518
Trial 3	93	360	1268
EJ			
Trial 1	1128	1672	3414
Trial 2	1084	1296	3581
Trial 3	1668	2618	2654
RT4			
Trial 1	6	3	1
Trial 2	0	1	7
Trial 3	1	1	0

Conc. HMW urokinase = concentration of high molecular weight urokinase.
Matrigel invasion chambers were run in duplicate and the average number of invading cells recorded for each of 3 trials.
The Kruskal-Wallis Test was used to compare the increase in number of invading cells seen with addition of exogenous u-PA. A significant increase in number of cells able to invade Matrigel was noted with u-PA pretreatment of the EJ cells ($p < 0.04$), the T24 cells ($p < 0.03$), and the positive control (HT1080 cells) ($p < 0.04$). The 253J cells showed a trend toward, but not a significant increase ($p = 0.11$) in the ability to invade Matrigel with u-PA pretreatment.

cells were transfected with u-PA cDNA and were able to produce u-PA, they developed the capacity to invade artificial basement membrane in vitro.[38] Our previous work showed the T24 cells express the u-PAR but secretes minimal amounts of u-PA.[22] Our studies have shown the T24 cell line alone is noninvasive in Matrigel invasion assays but develops an invasive capacity when pretreated with HMW u-PA. In colon cancer, the source of u-PA production may be stromal fibroblast cells, not the carcinoma cell;[36] our data suggest a similar possibility in bladder cancer. Thus, a bladder cancer cell (like the T24) may be noninvasive until u-PA becomes available from another cell, whereupon it then exhibits an invasive phenotype. The RT4 cell line, despite secretion of detectable levels of u-PA, appeared to be noninvasive in our study[22] and those of other investigators[39] because it lacks the efficient proteolysis that expression of the u-PAR may confer. The 253J and EJ cell lines which both secrete u-PA and express the u-PAR are invasive in vitro. These data suggest that an invasive bladder tumor requires both cells expressing the u-PAR and a source of u-PA if the plasminogen-plasmin system is responsible for tumor invasion. Pretreatment of the cell lines expressing the u-PAR with HMW u-PA enhanced their invasive potential

Table 4. Inhibition of Matrigel Invasion by EJ or u-PA pretreated T24 cells pretreated with anti-u-PAR antibody (Mab3936)

Conc. of Ab	0 Ab	2.5 μg	5 μg	10 μg	15 μg	20 μg
EJ cells						
Anti-u-PAR Ab	79.5 ± 3.0	52.5 ± 6.3	31 ± 1.2	42 ± 2.1	16 ± 3.7	16 ± 1.7
Control Ab	79.5 ± 0.8	67.5 ± 3.1	59 ± 1.1	55 ± 3.7	58 ± 4.2	47 ± 4.3
u-PA/T24 cells						
anti-u-PAR Ab	103 ± 6.7	78.8 ± 4.8	54.6 ± 8.3	28.3 ± 5.9	24.5 ± 3.1	11.2 ± 2.2
Control Ab	101 ± 5.5	102 ± 7.7	105 ± 6.1	99.5 ± 4.2	95.8 ± 2.8	97.7 ± 6.4

Conc. of Ab = concentration of antibody in 1 ml of fluid volume.
anti-u-PAR Ab = Mab3936.
Control Ab = mouse IgG2a, kappa [UPC 10].
u-PA/T24 cells = T24 cells pretreated with 2500iu HMW u-PA.

in Matrigel invasion chambers. Cell lines lacking the u-PAR did not developed the ability to invade Matrigel when pretreated with HMW u-PA. Thus, neither production of u-PA alone, nor the presence of the u-PAR without a source of u-PA were sufficient to allow tumor invasion in vitro.

Furthermore, blocking the binding of u-PA to the u-PAR has been shown to inhibit invasion and metastasis.[40–43] Our studies also suggest that blocking attachment of u-PA to the u-PAR prevents tumor invasion by bladder tumor cells in vitro.

Induction of u-PAR expression has been observed in other tumor systems by exogenous substances cytokines and growth factors. Pretreatment of endothelial cells with PMA has been shown in vitro to induce increased u-PAR mRNA.[24] PMA has also been shown to induce u-PAR expression in human myeloid leukemic cells 44. In the current study, PMA has noted to enhance u-PAR display by one cell line (253J), but seemed to inhibit u-PAR expression by two bladder cancer cell lines (EJ and T24) as well as the HT1080 cells. Heterogeneity of u-PAR display on U937 cells after PMA exposure has been reported.[45] This alteration of u-PAR distribution following PMA treatment was thought to be related to changes in glycosylation and ligand affinity of the purified u-PA receptor. Thus the anti-u-PAR antibody used in our studies to detect u-PAR expression may had a different binding affinity to the u-PA receptor after PMA exposure of the cells which may not translate into an actual decrease in the expression of the u-PAR on the cell surface.

Table 5a. Number of cells invading a Matrigel artificial basement membrane after pretreatment of cells with varying concentrations of PMA for 4 hours

Cell lines	HT1080	3T3	T24	EJ	253J
PMA 0 nM	704	4	16	645	41
PMA 5 nM	519	0	166	61	96
PMA 50 nM	122	2	131	53	562
PMA 500 nM	100	1	24	55	666

Matrigel invasion assays were run in duplicate and the average number of cells invading was recorded.
Cells were exposed to PMA at the concentrations shown for 4 hours prior to placement in the Matrigel invasion chambers.
The Kruskal-Wallis Test was used to compare the change in number of invading cells seen with pretreatment of cells with PMA. PMA pretreatment of EJ and the positive control (HT1080) cells significantly reduced the number of cells invading Matrigel ($p < 0.05$, respectively for each cell line). A significant increase in invasive capacity was noted for the 253J cells pretreated with PMA, while a mixed effect was noted in the T24 cells. Low dose PMA (5nM) enhanced T24 cell invasion ($p < 0.05$) while higher doses (50 or 500 nM) did not.

Table 5b. Number of cells invading a Matrigel artificial basement membrane after pretreatment of cells with varying concentrations of interferon gamma for 2 days

Cell lines	HT1080	3T3	T24	EJ	253J
0 iu/ml INF	17,102	1	621	27,796	179
100 iu/ml IFN	12,202	13	568	9,770	194
500 iu/ml IFN	10,158	2	1,850	7,742	424
1000 iu/ml IFN	6,076	9	4413	1,119	1451

Matrigel invasion assays were run in duplicate and the average number of cells invading recorded.
Cells were exposed to interferon gamma for 2 days prior to placement in the Matrigel invasion chambers.
The Kruskal-Wallis Test was used to compare the change in number of invading cells seen with pretreatment of cells with IFN gamma. IFN gamma pretreatment enhanced the invasiveness of T24 cells ($p < 0.05$ at a concentration of 500 iu/ml and 1000 iu/ml) and 253J cells ($p < 0.05$ at a concentration of 1000 iu/ml). IFN gamma inhibited in vitro invasiveness of the EJ and positive control (HT1080) cells at all concentrations tested.

Table 5c. Number of cells invading a Matrigel artificial basement membrane after pretreatment of the cells with varying concentrations of TGF beta

Cell lines	HT1080	3T3	T24	253J
0 ng/ml TGF beta	1109	4	266	156
1 ng/ml TGF beta	1081	0	86	299
2 ng/ml TGF beta	994	11	77	387
5 ng/ml TGF beta	1245	6	32	318

Matrigel invasion assays were run in duplicate and the average number of invading cells recorded.
Cells were exposed to TGF beta for prior to placement in the Matrigel invasion chambers.
The Kruskal-Wallis Test was used to compare the change in number of invading cells seen with pretreatment of cells with TGF beta. TGF beta pretreatment of 253J cells enhanced in vitro invasion of these cells ($p < 0.05$) at the 2 ng/ml and 5 ng/ml concentrations. TGF beta pretreatment of the T24 cells inhibited the invasiveness of these cells at a concentration of 5 ng/ml ($p < 0.05$). Pretreatment of the other cell lines did not affect their ability to invade (HT1080) or be noninvasive (3T3 or RT4 cells) on Matrigel.

Table 5d. Number of cells invading a Matrigel artificial basement membrane after exposure of cells to varying concentrations of EGF for 7 days

Cell lines	HT1080	3T3	T24	253J
0 ng/ml EGF	1025	40	106	133
10 ng/ml EGF	1808	19	232	614
100 ng/ml EGF	4600	16	289	588
200 ng/ml EGF	8135	37	353	396

Matrigel invasion assays were run in duplicate and the average number of invading cells recorded.
Cells were exposed to EGF for 7 days prior to placement in the Matrigel invasion chambers.
The Kruskal-Wallis Test was used to compare the change in number of invading cells seen with pretreatment of cells with EGF. EGF pretreatment of the T24 and positive control (HT1080) cells significantly enhanced the invasive potential of these cells in vitro ($p < 0.05$ for all concentrations tested). EGF pretreatment of the 253J cells inhibited invasion of Matrigel ($p < 0.05$ at concentration of 200 ng/ml). EGF pretreatment did not affect the negative control (3T3) cells.

Interferon gamma has been shown to upregulate u-PAR expression in U937 cells.[25] A marked increase in u-PAR display was observed in the T24 bladder cancer cells after exposure to interferon gamma as well as in the positive control (HT1080) cells. This upregulation of u-PAR display was not observed in all cells expressing the u-PAR, however.

Both EGF, TGFβ and PMA have been observed to increase u-PAR expression on cultured smooth muscle cells of bovine aorta.[23] Our studies confirm that EGF pretreatment of bladder cancer cells expressing the u-PAR enhanced u-PAR display. TGF beta showed enhancement of u-PAR display on one bladder cancer cell line (T24) and the positive control (HT1080) cells, but decreased u-PAR display on another bladder cancer cell line (253J). Again this may be attributable to a change in the binding of the anti-uPAR andtibody to the u-PAR after TGF beta exposure.

A direct correlation between increased u-PAR expression on bladder tumor cell surfaces and the ability of bladder tumor cells to invade Matrigel was observed after modulation of u-PAR expression by PMA, interferonγ, TGF β and EGF.

6. CONCLUSIONS

Bladder tumor cells produce u-PA, express the u-PAR, and require both for in vitro invasion by bladder cancer cells to occur. Bladder tumor cell invasiveness can be enhanced by pretreatment with exogenous u-PA to cells expressing the u-PAR and inhibited by anti-u-PAR antibodies. u-PAR expression on the cell surface of bladder tumor cells can be modulated by environmental substances which also affect the in vitro invasive potential of these cells. Different bladder cancer cell lines will react differently to environmental exposures regarding u-PAR display and changes in in vitro invasiveness. Histologically similar bladder tumors often show marked differences in their propensity to invade locally or metastasize. Modulation of u-PAR expression by substances in the tumor enviroment as well as differences in the ability of bladder cancer cells to display the u-PAR may affect the invasive and metastatic potential of human bladder cancers and offer one explanation for the aggressive or indolent tumor behavior observed in different bladder tumor patients. Elucidation of the mechanisms including the plasminogen activator system through which bladder tumor invasion and metastasis occurs could lead to novel therapies for the treatment of this disease. Further in vivo work is needed to confirm that u-PAR expression and u-PA antigen production together are important for tumor invasion and/or metastasis to occur.

ACKNOWLEDGMENTS

Supported by USPHS Grant #CA54277 from the National Cancer Institute. We would also like to thank Dr. Paul Swanson and Prosperidad Amargo for help with the immunohistochemical staining of the cell pellets and Dr. Peter Humphrey for assistance in preparing the photographs of the immunohistochemically stained cell pellets. We also like to thank Dr. Debra Smith for her help with the statistical analysis.

REFERENCES

1. Conese M., Blasi F.: The urokinase/urokinase-receptor system and cancer invasion. Baillieres Clin. Haematol., **8:** 365, 1995.

2. Cajot J.F., Kruitfof E.K., Schleuning W.D., Sordat B., Bachmann F.: Plasminogen activators, plasminogen activator inhibitors and procoagulant analyzed in twenty human tumor cell lines. Int. J. Cancer; **38:** 719, 1986.
3. Nykjaer A., Kjoller L., Cohen R.L., Lawrence D.A., Garni-Wagner B.A., Todd R.F. 3rd, et al.: Regions involved in binding of urokinase-type-1 inhibitor complex and pro-urokinase to endocytic alpha 2-macroglobulin receptor/low density lipoprotein receptor-related protein. Evidence that the urokinase receptor protects pro-urokinase against binding to the endocytic receptor. J. Biol. Chem., **269:** 25668, 1994.;
4. Blasi F.: Urokinase and urokinase receptor: a paracrine/autocrine system regulating cell migration and invasiveness. Bioessays, **15:** 105, 1993.
5. Mignotti P., Rifkin D.B.: Biology and biochemistry of proteinase tumor invasion. Physiol. Rev. **73:** 161, 1993.
6. Mizukami I.F., Garni-Wagner B.A., DeAngelo L.M., Liebert M., Flint A., Lawrence D.A., et al.: Immunologic detection of the cellular receptor for the urokinase plasminogen activator. Clin. Immunol. Immunopathol., **71:** 96, 1994.
7. Bianchi E., Cohen R.I., Thor A.T., Todd R.F. 3rd, Mizukami I.F., Lawrence D.A., et al.: The urokinase receptor is expressed in invasive breast cancer but not in normal breast tissue. Cancer Res., **54:** 861, 1994.
8. Carriero M.V., Franco P., Del Vecchio S., Massa O., Botti G., D'Ainto G., et al.: Tissue distribution of soluble and receptor-bound urokinase in human breast cancer using a panel of monoclonal antibodies. Cancer Res., **54:** 5445, 1994.
9. de Vries T.J., Quax P.H., Denijn M., Verrijp K.N., Verheijen J.H., Verspaget H.W., et al.: Plasminogen activators, their inhibitors, and urokinase receptor emerge in late stages of melanocytic tumor progression. Am. J. Pathol., **144:** 70, 1994.
10. Pyke C., Ralfkiaer E., Ronne E., Hoyer-Hansen G., Kirkeby L., Dano K.: Immunohistochemical detection of the receptor for urokinase plasminogen activator in human colon cancer. Histopathology, **24:** 131, 1994.
11. Stahl A., Mueller B.M.: Binding of urokinase to its receptor promotes migration and invasion of human melanoma cells in vitro.: Cancer Res., **54:** 3066, 1994.
12. Kariko K., Kuo A., Boyd D., Okada S.S., Cines D.B., Barnathan E.S.: Overexpression of a urokinase receptor increases matrix invasion without altering cell migration in a human osteosarcoma cell line. Cancer Res., **53:** 3109, 1993.
13. Gyetko M.R., Todd R.F. 3rd, Wilkinson C.C., Sitrin R.G.: The urokinase receptor is required for human monocyte chemotaxis in vitro. J. Clin. Invest. **93:** 1380, 1994.
14. Pepper M.A., Sappino A.P., Stocklin R., Montesano R., Orci L., Vassalli, J.D.: Upregulation of urokinase receptor expression on migrating endothelial cells. J Cell. Biol., **122:** 673, 1993.
15. Dubeau L., Jones P.A., Rideout W.M. 3rd, Laug W.E.: Differential regulation of plasminogen activators by epidermal growth factor in normal and neoplastic human urothelium. Cancer Res., **48:** 5552, 1988.
16. Brunner G., Pohl J., Erkell L.J., Radler-Pohl A., Schirrmacher V.: Induction of urokinase activity and malignant phenotype in bladder carcinoma cells after transfection of the activated Ha-ras oncogene. J. Cancer Res. Clin. Oncol., **115:** 139, 1989.
17. Hasui Y., Suzumiya J., Marutsuka K., Sumiyoshi A., Hashida S., Ishikawa E.: Comparative study of plasminogen activators in cancers and normal mucosae of human urinary bladder. Cancer Res., **49:** 1067, 1989.
18. Hasui Y., Marutsuka K., Suzumiya J., Kitada S., Osada Y., Sumiyoshi A.: The content of urokinse-type plasminogen activator antigen as a prognostic factor in urinary bladder cancer. Int. J. Cancer, **50:** 871, 1992.
19. Hasui Y., Marutsuka K., Nishi S., Kitada S., Osada Y., Sumiyoshi A.: The content of urokinase-type plasminogen activator and tumor recurrence in superficial bladder cancer. J Urol., **151:**16, 1994.
20. Hisazumi H., Andersson L., Collins V.P.: Fibrinolytic activity of in vitro cultivated human bladder cancer cell lines. Urol. Res., **5:** 133, 1977.
21. Dickinson A.J., Savage P.B., Newcomb P.V., Lodge R., Sibley G.N.: The expression of urokinse, its receptor and plasminogen activator inhibitor-1 in bladder cancer [abstract]. J. Urol., **153:** 406A, 1995.
22. Hudson M.A., McReynolds L.M.: Urokinase and the urokinase receptor: association with in vitro invasiveness of human bladder cancer cell lines. J. Natl. Cancer Instit., **89:** 708, 1997.
23. Reuning U., Bang Nu.: Regulation of the urokinase-type plasminogen activator receptor on vascular smooth muscle cells is under the control of thrombin and other mitogens. Arterioscler. Thromb., (AZ1) **12:** 1161, 1992.
24. Matsumoto H., Ueshima S., Fukao H., Mitsui Y., Matsuo O.: Identification of urokinase-type plasminogen activator receptor in human endothelial cells and its modulation by phorbol myristate acetate. Cell Struct. Funct., **20:** 429, 1995.

25. Sitrin R.G., Todd R.F. 3rd, Mizukami I.F., Gross T.J., Shollenberger S.B., Gyetko, M.R: Cytokine-specific regulation of urokinase receptor (CD87) expression by U937 mononuclear phagocytes.: Blood, **84:** 1268, 1994.
26. Rigby C.C., Franks L.M.: A human tissue culture cell line from a transitional cell tumour of the urinary bladder: growth, chromosome pattern and ultrastructure. Br. J. Cancer, **24:** 746, 1970.
27. Bubenik J., Baresova M., Viklicky V., Jakoubkova J., Sainerova H., Donner J.: Established cell line of urinary bldder carcinoma (T24) containing tumor-specific antigen. Int ,J. Cancer., **11:** 765, 1973.
28. Elliott A.Y., Cleveland P., Cervenka J., Castro A.E., Stein N., Hakala, T.R., et al.: Characterization of a cell line from human transitonal cell cancer of the urinary tract. J. Natl. Cancer Inst., **53:** 1341, 1975.
29. Evans D.R., Irwin R.J., Havre P.A., Bouchard J.G., Kato T., Prout G.R., Jr.: The activity of the pyrimidine biosynthetic pathway in MGH-U1 transtional carcninoma cells grown in tissue culture. J. Urol., **127:** 712, 1977.
30. Hastings R.J., Franks L.M.: Cellular heterogeneity in a tissue culture cell line derived from a human bladder carcinoma. Br. J. Cancer **47:** 233, 1983.
31. Chucholowski N., Schmitt M.,Rettenberger.: Flow cytometric analysis of the urokinse receptor (uPAR) on tumor cells by fluorescent uPA-ligand or monoclonal antibody#3936. Fibrinolysis, **6:** 95, 1992.
32. Blasi F., Stoppelli M.P., Cubellis M.V.: The receptor for urokinase-plasminogen activator. J. Cell Biochem., **32:** 179, 1986.
33. Hsu S.M., Raine L., Fanger H.: Use of avidin-biotin-peroxide complex (ABC) in immunoperoxidase techniques: a comparison between ABC and unlabelled antibody (PAP) procedures. J. Histochem. Cytochem., **29:** 577, 1981.
34. Pyke C., Kristensen P., Ralfkiaer E., Grondahl-Hansen J., Eriksen J., Blasi F., et al.: Urokinase-type plasminogen activator is expressed in stromal cells and its receptor in cancer cells at invasive foci in human colon adenocarcinomas. Am. J. Pathol., **138:** 1059, 1991.
35. Ellis V., Pyke C., Eriksen J., Solberg H., Dano K.: The urokinase receptor: involvement in cell surface proteolysis and cancer invasion. Ann. N.Y. Acad. Sci**., 667:** 13, 1992.
36. Dano K., Grondahl-Hansen J., Eriksen J., Nielsen B.S., Romer J., Pyke C.: The receptor for urokinase plasminogen activator:stromal cell involvement in extracellular proteolysis during cancer invasion. In: Bond JS, Barrett AJ, editors. Proteolysis and protein turnover. London:Portland Press,. p239, 1993.
37. Pedersen H., Brunner N., Francis D., Osterlind K., Ronne E., Hansen H.H., et al.: Prognostic impact of urokinase, urokinase receptor, and type 1 plasminogen activator inhibitor in squamous and large cell lung cancer tissue. Cancer Res., **54:** 4671, 1994.
38. Reiter L.S., Kruithof E.K., Cajot J., Sordat B.: The role of the urokinase receptor in extracellular matrix degradation by HT29 human colon carinoma cells. Int .J. Cancer, **53:** 444, 1993.
39. Redwood S.M., Liu B.C., Weiss R.E., Hodge D.E., Droller M.J.: Abrogation of the invasion of human bladder tumor cells by using protease inhibitor(s). Cancer, **69 :**, 1212, 1992.
40. Ossowski L., Russo-Payne H., Wilson E.L.: Inhibition of urokinase-type plasminogen activator by antibodies: the effect on dissemination of a human tumor in the nude mouse. Cancer Res., **51:** 274, 1991.
41. Ossowski L.: Invasion of connective tissue by human carcinoma cell lines: requirment for urokinase, urokinse receptor, and interstitial collagenase. Cancer Res., **52:** 6754, 1992.
42. Hearing V.J., Law L.W., Conti A., Apella E., Blasi F.: Modulation of metastatic potential by cell surface urokinase of murine cells. Cancer Res., **48:** 1270, 1988.
43. Crowley C.W., Cohen R.L., Lucas B.K., Liu G., Shuman M.A., Levinson A.D.: Prevention of metastasis by inhibition of the urokinase receptor. Proc. Natl. Acad. Sci. U.S.A., **90:** 5021, 1993.
44. Schuster W.A., Medcalf R.L., Kruithof E.K.: Retinoic acid potentiates phorbol ester-mediated induction of urokinase and plasminogen activator inhibitor type 2 in human myeloid leukemic cell lines. Endocrinology, **133:** 1724, 1993.
45. Hansen S.H., Behrendt N., Dano K., Kristensen P.: Localization of urokinase-type plasminogen activator receptor on U937 cells: phorbol ester PMA induces heterogeneity. Exp. Cell. Res., **187:** 255, 1990.

THE ROLE OF MATRIX METALLOPROTEINASES IN AN IN VITRO MODEL OF BLADDER TUMOR INVASION

Catherine Booth, Patricia Harnden, Peter J. Selby, Ludwik K. Trejdosiewicz, and Jennifer Southgate

Imperial Cancer Research Fund Cancer Medicine Research Unit
St James's University Hospital
Leeds, UK

1. INTRODUCTION

Invasive TCC typically progresses via local advance into the superficial and deep muscle layers of the bladder wall, through the perivesical fat and thence to adjacent organs, including the regional (pelvic) lymph nodes. Hematogenous spread to more distant sites is a frequent feature of subsequently fatal disease. An understanding of the genetic and epigenetic factors governing the invasive and metastatic cell phenotype is critical to the development of effective treatments[1].

Tumor invasion is a multifactorial process in which proteolysis is believed to play a central role[2]. In addition to weakening the tissue matrix, proteolysis has also been implicated in other processes which promote tumor cell spread, including the release and activation of matrix-bound cytokines such as bFGF[3], angiogenesis[4] and the disruption of homotypic binding between epithelial cells by cleavage of the extracellular portion of E-cadherin to release soluble E-cadherin fragments[5].

We have established an in vitro reconstituted organ culture model for studying the interactions of normal and neoplastic human urothelial cells with the stroma[6,7]. Whereas normal human urothelial (NHU) cells seeded from monolayer cultures onto a de-epithelialized stroma reconstituted a homeostatic stratified transitional neo-epithelium, the ability of TCC cell lines to reform a urothelium was related to the grade of tumor from which they were derived. Two of the cell lines, RT112 and EJ, showed invasive patterns of behavior, with RT112 cells intravasating the vessels of the sub-epithelial capillary bed and EJ cells showing aggressive invasion through the stroma[6]. The aim of this study was to determine if differences in invasive properties of the cell lines in the in vitro model could be

Advances in Bladder Research, edited by Baskin and Hayward.
Kluwer Academic / Plenum Publishers, New York, 1999.

due to differences in expression or activity of the matrix metalloproteinases, MMP2 and MMP9, and their inhibitors, TIMP1 and TIMP2.

2. METHODS

NHU cells were isolated and grown in monolayer culture, as previously described[8]. Also studied were established TCC-derived cell lines, RT4, RT112 and EJ which were well-, moderately and poorly-differentiated in vitro, respectively, reflecting the grades of tumor from which the cells were originally derived[9,10]. Reconstituted tissues were prepared by seeding NHU or TCC cell lines onto the basement membrane of de-epithelialized human urinary tract stromal tissues. The organoids were placed epithelial side up on permeable membranes, which were maintained at an air-liquid interface for 2 weeks, as previously described[6,7].

In order to determine whether NHU and TCC cell lines in monolayer culture were secreting MMP2 and MMP9 conditioned serum-free culture media were collected from cells grown in adherent tissue culture and analyzed by gelatin zymography. MMP and TIMP expression was also investigated by immunoblotting lysates prepared from the cell lines using monoclonal antibodies to MMP2, MMP9 (Calbiochem-Novabiochem (UK) Ltd., Nottingham), TIMP1 and TIMP2 (Chemicon International Ltd., Harrow, UK). The localization of MMP9 was studied by immunoperoxidase labeling of paraffin wax-embedded reconstituted tissue organoids using an MMP9-specific antibody (Chemicon) following microwave antigen retrieval in 10mM citrate buffer pH 6.0.

3. RESULTS

Several species of gelatinase were detected on zymograms on the basis of molecular size and co-migration with proteinase standards (Figure 1). The dominant components were identified as latent and activated MMP2 (72,000 M_r and 62,000 M_r, respectively) and latent and activated MMP9 (92,000 M_r and 82,000 M_r, respectively), although several other gelatinases were also present. Fetal bovine serum was found to contain very high amounts of latent gelatinases, thereby illustrating the importance of serum-free medium when analyzing proteinase secretion by cells in culture.

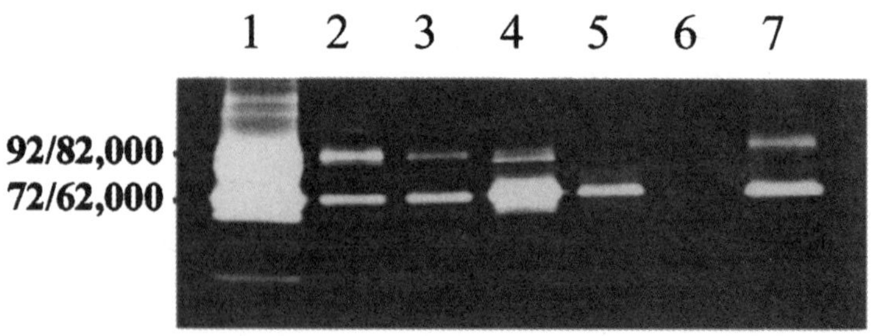

Figure 1. Zymography of concentrated conditioned serum-free tissue culture supernatants from NHU (lane 2), RT4 (lane 3), RT112 (lane 4) and EJ (lane 5) cell cultures. The MMP2 and MMP9 proteinase standards (lane 1) produced doublet bands of 92,000 M_r / 82,000 M_r (MMP9) and 72,000 M_r / 62,000 M_r (MMP2). An additional band of approximately 40,000 M_r represents an additional activated form of MMP2. Serum free KSFM in which NHU cells are grown contained no gelatinase activity (lane 6). FBS contained high levels latent MMP-2 and -9 (lane 7).

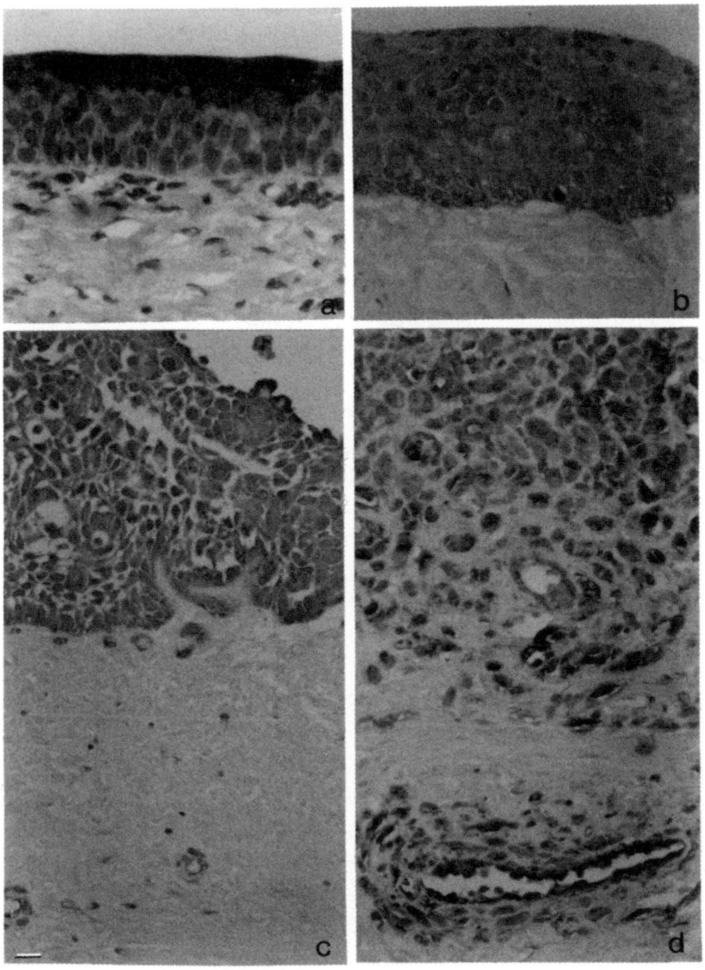

Figure 2. Indirect immunoperoxidase labeling of paraffin sections showing expression of MMP9 in organoids reconstituted with NHU (A), RT4 (B), RT112 (C) and EJ (D) cells after 4 weeks in organ culture. Scale bar ≡ 25 μm.

Zymograms showed that proteins with gelatinase activity were secreted into the culture medium by NHU and all three TCC cell lines grown in monolayer culture. The 72,000 M_r latent form of MMP2 was secreted by both normal and TCC cell lines, with RT112 cells showing highest activity. The latent 92,000 M_r form of MMP9 was secreted by NHU, RT4 and RT112 cell lines. NHU and RT112 cells additionally secreted the active 82,000 M_r form of MMP9, which was the predominant MMP9 species in RT112 cells (Figure 1). No MMP9 activity was detected in EJ cell supernatants. Immunoblotting of cell lysates demonstrated that MMP2 and MMP9 were produced by NHU and all three TCC cell lines, including EJ cells. TIMP1 and TIMP2 were detected in all cell lines, with the exception that EJ cells were negative for TIMP2.

In organoids reconstituted with NHU cells, MMP9 was expressed weakly in the basal and the majority of the intermediate cell layers and intensely in late intermediate and superficial cells (Figure 2A). In organoids constituted with RT4 and RT112 cells, MMP9 was expressed throughout the surface neo-epithelia (Figures 2B-C). MMP9 expression was retained by RT112 cells which had infiltrated the sub-epithelial capillary bed (Figure 2C). In EJ organoids, where no surface epithelium formed, all EJ cells were MMP9-positive, but highest expression was seen in EJ cells located within blood vessels (Figure 2D).

4. DISCUSSION

Proteases serve important physiological roles in extracellular matrix turnover and cell migration during inflammation, wound healing and other tissue remodeling processes[11]. The regulation of protease activity involves activation cascades, specific inhibitory molecules and the targeting of proteolytic activity by binding and activation at cell surface receptors[12]. The multiple levels of regulation imply that protease activity must be tightly controlled in normal physiological processes and provides ample opportunity for dysregulation to occur in pathological conditions such as neoplasia.

Metalloproteinases (MMPs) are a family of zinc-dependent proteases which have been implicated in tumor progression. In particular, MMP2 and MMP9 have been found to be elevated in many cancers, including bladder carcinoma[13]. A correlation has also been found between the degree of MMP2 and MMP9 activation and tumor grade using zymography of bladder tumor extracts[13]. The tissue inhibitors of metalloproteinases (TIMPs) are a family of four members which form tight-binding 1:1 inactive complexes with the MMPs[14]. TIMP2 can be secreted as a specific complex with latent MMP2[2] and similarly, TIMP1 can bind and be secreted as a complex with the latent form of MMP9[15]. Although TIMP1 and TIMP2 do not bind other latent matrix metalloproteinases, they can inhibit all MMPs following activation.

Although NHU cells have the capacity to produce MMPs in active form, NHU cells did not invade the basement membrane when seeded onto a stroma. RT4 cells, which showed a similar proteolytic repertoire to NHU cells, were also non-invasive. This suggests that in noninvasive cells, proteolytic activity is maintained under strict regulation and emphasizes the importance of inhibitory mechanisms, such as the expression of inhibitory proteins. RT112 and EJ cells both displayed invasive characteristics in the reconstituted organoids. RT112 cells specifically invaded via intravasation into the subepithelial capillary bed of the basement membrane. This implies a mechanism which is not dependent on matrix degradation per se. The only cells to show evidence of matrix-degrading activity in situ were EJ cells, which degraded the epithelial basement membrane and traversed through the stroma[6]. Although EJ cells appeared not to secrete MMP9 in monolayer culture, they did express MMP9 and showed evidence of secretion around invading cells within the stroma of the reconstituted organoids. This implies that the secretion of MMP9 is dependent upon interactions between EJ cells and the stroma, perhaps mediated by integrins. This would result in high specific proteolytic activity in the immediate vicinity of the tumour cell. Furthermore, EJ cells were the only cells which did not express TIMP2. In conclusion we suggest that MMP9 activity and absence of TIMP2 expression may be determining factors in the ability of EJ cells to invade the stroma.

ACKNOWLEDGMENTS

This work was supported by the Association for International Cancer Research (AICR), UK and the Imperial Cancer Research Fund (ICRF), UK.

REFERENCES

1. Harnden P., Parkinson M.C. Transitional cell carcinoma of the bladder: diagnosis and prognosis. Current Diagnostic Pathology, 3: 109, 1996.
2. Stetler-Stevenson W.G., Liotta L.A., Kleiner D.E., Jr.: Extracellular matrix 6: role of matrix metalloproteinases in tumor invasion and metastasis. F.A.S.E.B. J. **7:** 1434, 1993.
3. Saksela O., Rifkin D.B.: Release of basic fibroblast growth factor-heparan sulfate complexes from endothelial cells by plasminogen activator-mediated proteolytic activity. J. Cell Biol., **110:** 767, 1990.
4. Moses M.A.: The regulation of neovascularization of matrix metalloproteinases and their inhibitors. Stem Cells, **15:** 180, 1997.
5. Katayama M., Hirai S., Kamihagi K., Nakagawa K., Yasumoto M., Kato I.: Soluble E-cadherin fragments increased in circulation of cancer patients. B. J. Cancer, **69:** 580, 1994;
6. Booth C., Harnden P., Trejdosiewicz L.K., Scriven S., Selby P.J., Southgate J.: Stromal and vascular invasion in an human in vitro bladder cancer model. Lab. Invest., **76:** 843, 1997.
7. Scriven S.D., Booth C., Thomas D.F., Trejdosiewicz L.K., Southgate J.: Reconstitution of human urothelium from monolayer cultures. J. Urol., **158:** 1147, 1997.
8. Southgate J., Hutton K.A., Thomas D.F., Trejdosiewicz L.K.: Normal human urothelial cells in vitro: proliferation and induction of stratification. Lab. Invest., **71:** 583, 1994.
9. Masters J.R., Hepburn P.J., Walker L., Highman W.J., Trejdosiewicz L.K., Povey S., Parkar M., Hill B.T., Riddle P.R., Franks L.M.: Tissue culture model of transitional cell carcinoma: characterization of twenty-two human urothelial cell lines. Cancer Res., **46:** 3630, 1986.
10. Ahlering T.E., Dubeau L., Jones P.A.: A new in vivo model to study invasion and metastasis of human bladder carcinoma. Cancer Res., **47:** 6660, 1987.
11. Salo T., Makela M., Kylmaniemi M., Autio-Harmainen H., Larjava H.: Expression of matrix metalloproteinase-2 and -9 during early human wound healing. Lab. Invest., **70:** 176, 1994.
12. Schmitt M., Wilhelm O., Janicke F., Magdolen V., Reuning U., Ohi H., Moniwa N., Kobayashi H., Weidle U., Graeff H.: Urokinase-type plasminogen activator (uPA) and its receptor (CD87): a new target in tumor invasion and metastasis. J. Obstet. Gynaecol., **21:** 151, 1995.
13. Davies B., Waxman J., Wasan H., Abel P., Williams G., Krausz T., Neal D., Thomas D., Hanby A., Balkwill F.: Levels of matrix metalloproteases in bladder cancer correlate with tumor grade and invasion. Cancer Res., **53:** 5365, 1993.
14. Goldberg G.I., Marmer B.L., Grant G.A., Eisen A.Z., Wilhelm S., He C.S.: Human 72-kilodalton type IV collagenase forms a complex with a tissue inhibitor of metalloproteases designated TIMP-2. Proc. Natl. Acad. Sci. US A., **86:** 8207, 1989.
15. Wilhelm S.M., Collier I.E., Marmer B.L., Eisen A.Z., Grant G.A., Goldberg G.I.: SV40-transformed human lung fibroblasts secrete a 92-kDa type IV collagenase which is identical to that secreted by normal human macrophages. J. Biol. Chem., **264:** 17213, 1989.

DNA METHYLATION IN DEVELOPMENT OF BLADDER CANCER

Peter A. Jones

Department of Biochemistry and Molecular Biology
University of Southern California/Norris Comprehensive Cancer Center and Hospital
1441 Eastlake Avenue
Room 8315, Mail Stop #83
Los Angeles, California 90033-0800

1. INTRODUCTION

In this chapter, the role of DNA methylation in the generation of human bladder cancer will be reviewed. As is known, there are several different kinds of bladder cancer including: papillary tumors, carcinoma in situ (CIS), invasive cancers and squamous cancers. Fortunately several tumor suppressor genes which are important in the genesis of these different diseases are now known. For example, there seems to be a gene on chromosome 9q which is probably involved in the genesis of papillary tumors and the p53 gene, as well as p16, p19 genes play important roles (see chapter by Reznikoff). Tumor suppressor genes have different penetrance in these different kinds of bladder cancers. For example, a p53 mutation in a papillary tumor is virtually never seen. The mechanisms by which tumor suppressors are inactivated are interesting and we are familiar with mutations, loss of heterozygosity and homozygous deletions as being major routes to their disablement. Presently, the concept of DNA methylation as being a new mechanism that people are realizing could play an important role in switching genes off in cancer development, will be discussed. The mechanisms by which a given suppressor is inactivated tends to vary depending upon that suppressor. For example, the p53 is very often disabled by point mutations whereas the p16 and p19 genes are rarely found to be mutated in bladder cancer. In most cases, the genes are completely missing and mutations seem to be a much more rare event. So, the type of disease, the tumor suppressor, and the mechanism for inactivation are important in understanding the molecular basis.

Advances in Bladder Research, edited by Baskin and Hayward.
Kluwer Academic / Plenum Publishers, New York, 1999.

2. DNA METHYLATION

DNA methylation which used to be an esoteric field, is gaining a lot more scientific interest because of its potential to permanently, yet potentially reversibly, inactivate genes. Almost all of the methylation found in human DNA occurs at the dinucleotide sequence CpG where the cytosine becomes modified to 5-methylcytosine. CpG sites can either be methylated or not, and the methylation occurs at the 5 position of the cytosine ring after DNA has been replicated. Organisms like Drosophila which can methylate their DNA, have lots of CpG sites, and none of them are methylated because Drosophila lacks the enzyme that methylates DNA. On the other hand, organisms such as ourselves, which do methylate their DNAs, show a strong suppression of the frequency of occurrence of the CpG site and most of these sites are missing. In fact, 80% of them are missing because it turns out that methylating these sites is very mutagenic, and most of them have been mutated during the course of evolution[1]. The distribution of CpG sites in our DNA is very asymmetric. CpG sites occur infrequently in DNA, but there are stretches of DNA where they are clustered in regions of about a kilobase in length, which are called CpG islands. These are the most important regions since when CpG islands occur in the promoters of genes they can switch a gene off, if they become methylated. We know that this happens in development. For example, the promoters of the genes which are on the inactive X chromosome all become methylated, and that keeps the gene off in female mammals. It is a very stable mechanism to switch a gene off, because once itís happened, itís very difficult to reverse it. In fact, methylation in these promoter regions can be inherited and keep that gene off for the rest of the life of that individual who might live to be 100 years old with a lot of cell division. [2]

3. CPG ISLANDS

CpG islands are a very interesting part of DNA, because they are always associated with genes, and sometimes although not always, are in promoters. CpG islands are intriguing because they are always unmethylated in the germline and in somatic cells, except for genes on an inactive X chromosome and for imprinted genes. What makes them particularly interesting from the point of view of cancer research, is that unlike the situation in normal somatic cells, CpG islands are very frequently methylated in cancer. So, we start out with a lack of methylation of CpG islands in normal cells, whereas in cancer cells, there are profound differences in the distribution of CpG methylation. The reason why this is important is because methylation has the potential to silence genes in a very stable fashion. So, the main idea in the field is that a growth regulatory gene such as p16, which has an unmethylated island within its promoter, could be disabled by progressive methylation of its promoter, resulting in down regulation of the gene. [3] What makes this kind of inactivation different, is that unlike the situation where a gene becomes mutated, or when a gene becomes lost, here what you have is a perfectly good tumor suppressor gene, which is dormant within the cell. It is still there, it is a perfectly good copy of the gene, it is just not being used.

4. THE P19/P16 LOCUS

The p19/p16 locus will be focussed upon because, evidence that this situation does actually happen within this region containing two genes on top of each other, can be provided. There is no methylation at all in this region in either sperm or white blood cells.

Exon 2 of the p16 gene which is the coding portion of the gene, is also a small CpG island in addition to the CpG island in the promoter, so there are two CpG islands in tandem here. Up until recently it has been difficult to assess the methylation status of a given CpG site within DNA, because when genes are cloned, the methylation pattern information is lost, making it very difficult to tell whether a given locus is methylated or not. However, a new technique has been developed in Australia in which DNA is treated with bisulfite, so that the methylation status of any site within a particular region of DNA can be assessed. This technique has been modified here in a procedure called Ms-SnuPE to measure the methylation status of three sites, in the p16 gene promoter in transitional cell and squamous cell carcinoma in an individual with bladder cancer.[4] No methylation was found in the transitional cell carcinoma, however, there was methylation in the squamous carcinoma within the promoter region of the gene. Thus, the p16 gene has become inactivated by DNA methylation in the squamous carcinoma component, but not in the transitional cell carcinoma. It has also been shown directly in test tube experiments that methylation of this region switches the gene off. It is therefore known that the methylation of these sites in the promoter of the gene is capable of extinguishing its expression.

5. T24 CELLS

To probe in more detail how this methylation is relevant to the expression of p16, we moved into a tissue-culture model system using T24 cells. The whole of this region is completely methylated in these cells and the gene is completely dormant. Nowadays, drugs can be used to reverse the methylation change which has occurred within a promoter in an attempt to switch the gene back on again. 5-Azacytidine, which we found some years ago to be incorporated into the DNA of dividing cells, is a powerful methylation inhibitor because when the enzyme encounters this fraudulent molecule in DNA, it binds very tightly and forms a suicide complex with the base in DNA. The cell becomes depleted of methyltransferase, so the methylation pattern is no longer copied. T24 cells were exposed for 24 hrs to 5-azacytidine, then the drug removed and cells recovered which are expressing the p16 gene. So this gene which has been lying dormant within the T24 cells for probably more than 30 years can be resurrected into action by briefly treating it with the methylation inhibitor. The gene still has no mutations in it, and it has been shown in other experiments that it seems to be capable of suppressing the growth of those cells in which it is reactivated. [5]

To try to understand the relationship between the methylation within the promoter of the gene and expression of p16, a more detailed examination was performed. Not only the promoter of the gene is a CpG island, but also exon 2, which is part of the coding region, is a CpG island. Christina Bender, has measured the methylation status not only of the promoter, but also of exon 2 and the result was surprising. If clones of cells which are strong expressors of the gene are used, the promoter region is relatively unmethylated, whereas, exon 2 is almost completely methylated. The reason why this is important, is there is a misconception in the field that methylation of a CpG island equals no expression. In this case the gene is being expressed but the location of the CpG island is what is the critical issue. If the promoter of the gene is methylated, then the gene is turned off. But if a CpG island in a coding region is methylated, it does not prevent transcription.

6. TUMOR SUPPRESSOR GENES

Tumor suppressor genes are normally found through familial studies looking for mutations or genetic loci which are important for the particular disease. Alternatively, loss of heterozygosity studies in tumors, where we look for losses of chromosomal material, can identify tumor suppressors. However, tumor suppressor genes become inactivated through preferential pathways, so p16 for example, is mostly lost by homozygous deletion even though there is some evidence that this gene can also become inactivated by promoter methylation. The point is that you only see what youíre looking for. Therefore looking for genes which become methylated in cancers may be a way of identifying new tumor suppressors.

Looking for changes in methylation and working backwards to find new tumor suppressor genes may therefore be a relevant way to look at cancer development. A technique has been developed in our laboratory by Mark Gonzalgo and Giangning Liang, which is methylation-sensitive called arbitrary primed PCR. It is a genome scanning mechanism to look for DNA methylation changes. [6] Short primers are used which are very CG rich, and under low stringency conditions PCR is used to get a series of bands which tend to be CpG rich. This method was first developed by Michael McClelland. It has modified by treating DNA with bacterial enzymes which either cut or doními cut the DNA, depending upon whether it is methylated. The primers sit down arbitrarily on the DNA, and if a given site is unmethylated a specific pattern results which can be identified. By comparing two lanes in this genome scanning approach, methylation changes can be identified, and then a new suppressor gene may be located. This method works quite well and has been applied to bladder cancers, the interesting bands cut out, sequenced and identified.

The extent of methylation changes which occur in human tumors, is quite dramatic. Analysis of colon, bladder, and prostate tumors for 45 of these arbitrary bands shows that there are lots of changes. Interestingly, in some patients, there is little hypermethylation whereas other tumors have many changes suggesting that there may be two separate methylation pathways to cancer development.

7. ENDOTHELIN-B RECEPTOR GENE

Some of the bands which are frequently altered in many different cancers have been isolated, and sequenced to find out what they are. The first one hit was exciting because it turned out to be the human endothelin-B receptor gene. Nelsonís group at Hopkins had just published a paper in Cancer Research, saying that this gene becomes frequently methylated in prostate cancer. In their case, they started with a gene and looked for the methylation change whereas this laboratory started with a methylation change and found the gene. Why would the endothelin-B receptor be important in cancer? It turns out that endothelins can act as growth factors, and their signal is transmitted through the endothelin-A receptor to cause cell proliferation. The endothelin-B receptor is actually a dummy receptor which binds the ligand, but doesnít signal cells to divide. Having found the gene, a methylation assay was designed to look in patientsí tumors to see exactly what is the relationship between methylation of this promoter and expression of the endothelin-B receptor. An analysis of two cases showed that normal urothelium expressed the receptor gene; whereas, the tumor does not. The tumor is extensively methylated at key sites within the promoter. This is exciting because it suggests that a set of new tumor suppressor genes which are relevant to bladder cancer progression and development may soon be discovered.

8. PAX-6

Whilst doing this screen, about 20 different anonymous bands of DNA were found, most of which are not in the database, but most of which are CpG islands. Since CpG islands are almost always associated with genes, the advantage of the approach is that if anything is found, there is a good chance there is a gene. One of the genes found, by Carol Salem in the laboratory, was very surprising because it was exon 5 of a gene which is called Pax-6. Pax-6 is a well-known gene important in the development of the eye, and it has been suggested that Pax-6 and the other Pax genes, which are all homeobox-related genes might be capable of acting as oncogenes. What made it particularly surprising was that the fragment pulled out of the screen was actually in the exon, that is in the transcribed region of the gene, not in the promoter. In fact the increased methylation on exon 5 of Pax-6, is actually correlated with over expression of Pax-6 in bladder cancer. This seems to be counter intuitive since methylation is mostly assumed to be correlated with even repression. Whether Pax-6 is an oncogene in bladder cancer is now being investigated.

9. DISCUSSION

In summary this kind of scanning approach may be useful in finding new genes which are important for cancer. CpG islands are not methylated in normal cells unless theyíre on the inactive X chromosome or on imprinted genes. Our results show that when methylation occurs in the promoter of a gene, this equals extinction of the gene. Evidence was provided with two genes; there was no expression on the p16 gene, when the promoter was methylated and when this promoter is demethylated, expression occurrs. The endothelin-B gene receptor, is also methylated in the promoter in bladder cancer. The new finding is that with Pax-6 exon 5 and with the exon 2 region of p16, methylation activity correlates with expression. Methylation screens therefore give two opportunities at once. First of all, genes can be picked out because the CpG islands can be found and, methylation of fragments like the endothelin-B receptor promoter, might be associated with down regulation. Some of the changes may however be associated with up regulation. It will be interesting to come up with a much more complete picture on how different genes become switched off and switched on during bladder carcinogenesis by changing the way different epigenetic defects are located.

REFERENCES

1. Yang, A.S., Jones, P.A., and Shibata, A.: The mutational burden of 5-methylcytosine. In: Epigenetic Mechanisms of Gene Regulation, Cold Spring Harbor Laboratory Press, New York, 1996, pp 77–94.
2. Gonzalgo, M.L., and Jones, P.A.: Mutagenic and epigenetic effects of DNA methylation. Mutat. Res., **386:** 107, 1997.
3. Gonzalgo, M.L., Hayashida, T., Bender, C.M., Pao, M.M., Tsai, Y.C., Gonzales, F.A., Nguyen, H.D., Nguyen, T-D.T., and Jones, P.A.: The role of DNA methylation in expression of the p19/p16 locus in human bladder cancer cell lines. Cancer Res., **58:** 1245, 1998.
4. Gonzalgo, M.L., and Jones, P.A.: Rapid quantitation of methylation differences at specific sites using methylation-sensitive single nucleotide primer extension (Ms-SNuPE). Nucleic Acids Res., **25:** 2529, 1997.
5. Bender, C.M., Pao, M.M., and Jones, P.A.: Inhibition of DNA methylation by 5-aza-2í-deoxycytidine suppresses the growth of human tumor cell lines. Cancer Res., **58:** 95, 1998.
6. Gonzalgo, M.L., Liang, G., Spruck III, C.H., Zingg, J-M., Rideout III, W.M., and Jones, P.A.: Identification and characterization of differentially methylated regions of genomic DNA by methylation-sensitive arbitrarily primed PCR. Cancer Res., **57:** 594–599, 1997.

33

CELLULAR PROLIFERATION AND CELL-CELL CYCLE REGULATORY PROTEINS AS PROGNOSTIC MARKERS FOR TRANSITIONAL CELL CARCINOMA OF THE BLADDER

Gary D. Grossfeld,[1,3] Michael Muscheck,[1,2] John P. Stein,[4] Karen Chew,[2,3] Richard J. Cote,[5] David A. Ginsberg,[4] Frederic M. Waldman,[2,3] and Peter R. Carroll[1,3]

[1]Department of Urology
[2]Department of Laboratory Medicine
University of California San Francisco
San Francisco, California 94143
[3]The Program in Urologic Oncology, and
UCSF/Mt. Zion Cancer Center
San Francisco, California 94115
[4]Department of Urology
[5]Department of Pathology
University of Southern California
Los Angeles, California

1. INTRODUCTION

Normal cellular proliferation is a tightly regulated process. The two most important events of the eukaryotic cell cycle are DNA synthesis (replication of chromosomes) and mitosis (segregation of chromosomes into dividing cells),which occur during the S-phase and the M-phase of the cell cycle, respectively.[1] The remainder of the cell cycle consists of two gap phases (G1 and G2) which occur prior to S-phase (G1) and prior to M-phase (G2) (see Figure 1). As growth quiescent cells enter and progress through the cell cycle, a number of "checkpoints" are encountered that are positioned at transition points between the different cell cycle states.[2] These checkpoints are positioned at the initiation and completion of DNA replication (S-phase) and at the initiation and completion of cell division (M-phase). The most extensively studied of these checkpoints occurs late in G1, at the G1 to S transition, when the cell commits itself to another round of DNA replication. [2] Bypass

Advances in Bladder Research, edited by Baskin and Hayward.
Kluwer Academic / Plenum Publishers, New York, 1999.

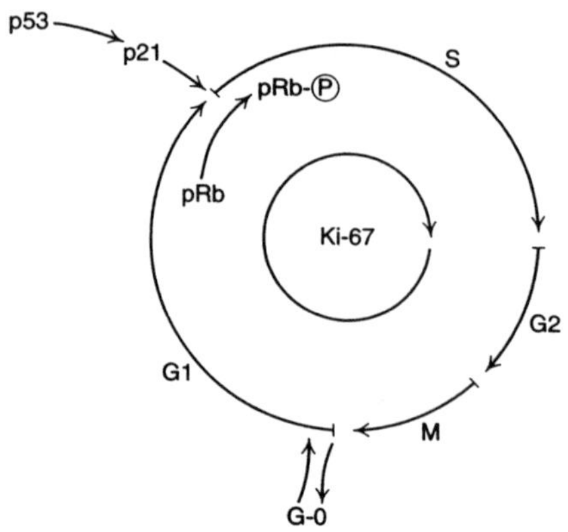

Figure 1. Schematic representation of the eukaryotic cell cycle.

of this important cell cycle checkpoint may allow cells to replicate damaged DNA thereby resulting in the accumulation of genetic mutations that may eventually contribute to a neoplastic phenotype.

Progression through this late G1 checkpoint is predominately controlled by protein complexes that are composed of cyclins and cyclin-dependent kinases, the most important of which appear to be cyclin D and E and CDK 2 and 4[2, 3]. These complexes control progression through the cell cycle by phosphorylating key proteins that are involved in cell cycle transition points, including the protein encoded by the retinoblastoma gene (pRB). Both negative and positive regulators of these protein complexes exist. It has been hypothesized that aberrant expression of these regulators, leading to a loss of cell cycle control, may represent an early step in cancer initiation and ultimately, progression.

Neoplastic disease has been characterized as a proliferative disorder in which there is uncontrolled cellular growth. [3] Consequently, neoplastic cells must acquire the ability to bypass normal cellular control mechanisms as they progress from a normal to a malignant state. Understanding the mechanisms by which malignant cells acquire this ability to bypass the normal regulatory mechanisms may provide insight into which tumors will have the greatest malignant potential. In addition, this understanding should provide a basis for designing therapeutic strategies that will be successful in retarding unregulated cellular proliferation.

Recent studies have examined the expression of several genes (and their protein products) which are thought to be involved in cell cycle regulation in order to determine their prognostic significance in a variety of human malignancies, including bladder cancer. The purpose of the current study is two-fold; (1) to review evidence to date regarding the prognostic significance of cellular proliferation and expression of cell cycle regulatory proteins in transitional cell carcinoma of the bladder , and (2) to present recent data describing the interactions between these prognostic markers in patients with superficial and muscle invasive bladder cancer.

2. RETINOBLASTOMA TUMOR SUPPRESSOR GENE

The Retinoblastoma (Rb) gene was the first tumor suppressor gene identified. The Rb gene includes 27 exons that span over 190 kilobases on chromosome 13 (mapping to band 13q14) and encodes for a 110 kDalton nuclear phosphoprotein.[4, 5] Although initially found to be deleted in patients with inherited retinoblastoma, altered Rb gene expression has been reported in a variety of other human tumors, including transitional cell carcinoma of the bladder.[4, 6–8] In its physiologic active hypophosphorylated form, pRb inhibits cell-cycle progression at the G1 to S checkpoint by binding to a number of cellular proteins including the transcription factor E2F.[9, 10] The overall amount of pRb expression does not change significantly during progression of the cell cycle.[3] However, pRb is phosphorylated in a cell cycle dependent manner, and it is the phosphorylation state of pRb that determines its function. For example, pRb exists predominately in its hypophosphorylated form during the G1 phase of the cell cycle, thereby inhibiting cell cycle progression and resulting in a tumor suppressor effect. After phosphorylation, pRb no longer inhibits cell cycle progression and cells are allowed to progress into early S-phase (Figure 1).

Recent evidence suggests that inactivation of the Rb gene may be an important step in bladder cancer progression. Monoclonal and polyclonal antibodies have been discovered which allow the reliable immunohistochemical evaluation of pRb expression.[11] Using both immunohistochemical techniques and molecular analysis, several groups have demonstrated that the proportion of bladder tumors demonstrating Rb alterations increases with increasing tumor grade and tumor stage.[12, 13] These results suggest that the loss of pRb expression may be an important prognostic factor for patients with transitional cell carcinoma of the bladder. Presti and colleagues analyzed 34 bladder tumors of various grades and stages and found that pRb alteration was significantly more likely in invasive versus superficial bladder tumors.[8] Cordon-Cardo et al. subsequently evaluated pRB expression in 48 bladder tumors obtained at the time of radical cystectomy or transurethral resection and found that pRb alterations were present in nearly 30% of these unselected tumors.[4] Five year survival rates were significantly lower in those patients with altered Rb protein expression when compared to those patients who maintained normal Rb protein expression ($p < 0.001$).[4] This difference in survival was also seen when patients with muscle invasive disease were analyzed separately ($p < 0.001$). Logothetis and associates also evaluated pRb expression in 43 patients with locally advanced bladder cancer and reported that Rb protein expression was altered in 37% of these patients.[7] Rb protein expression was found to be independent of other well established prognostic variables for patients with bladder cancer, including clinical stage and the rate of cellular proliferation. Similar to what was reported by Cordon-Cardo et al., those patients who had lost Rb protein expression had a shorter overall survival than patients who maintained Rb expression.[7] Based on the aforementioned data, pRb expression appears to be an important prognostic factor in patients with invasive bladder cancer.

3. p53 TUMOR SUPPRESSOR GENE

The p53 gene, located on chromosome 17 (mapping to band 17p13), encodes for a 53-kDalton nuclear protein. Mutation of the p53 gene is the most commonly described genetic abnormality in human cancers.[14] p53 is involved in several critical cellular processes including inhibition of cell growth, cellular response to genetic damage and entrance into apoptotic pathways.[15] With respect to carcinogenesis, p53 functions as a tumor suppressor gene through its role in cell cycle regulation.[16] When cellular DNA is damaged, the level of p53 protein increases causing cell cycle arrest. This allows for DNA repair, preventing the propagation of DNA mutations. Alterations in the p53 gene result in the production of an abnormal (and usually dysfunctional) protein product with a prolonged half-life compared to the wild-type protein. Thus, this abnormal protein accumulates in the cell nucleus and can be detected by immunohistochemical techniques. Previous studies have demonstrated that p53 nuclear reactivity, as determined by immunohistochemical methods, correlates well with gene mutations as detected by DNA sequence analysis.[17, 18]

Several recent studies have demonstrated that p53 alteration may be an important prognostic indicator for bladder cancer progression.[19–25] Sidransky and colleagues initially reported p53 alterations in 11 of 18 (61%) patients with invasive bladder carcinoma using polymerase chain reaction (PCR) techniques.[24] Interestingly, these authors were also able to detect the same p53 mutations in exfoliated cells isolated from preoperative urine specimens in 3 patients with known p53 mutations. Subsequent studies have demonstrated significant associations between p53 nuclear reactivity and bladder tumor grade, stage, disease progression and overall survival. Wright et al. examined p53 expression in 82 patients with bladder cancer.[25] p53 immunoreactivity was detected in 54% of these patients, and it was significantly associated with tumor grade, tumor stage and c-erbB-2 expression. Esrig and colleagues evaluated p53 nuclear reactivity in 243 patients with invasive bladder cancer who were uniformly treated by radical cystectomy.[20] Patients with altered p53 (p53 nuclear reactivity) were found to have a significantly higher risk of disease recurrence, and a significantly lower overall survival compared to those patients with wild-type (unaltered) p53. This association was independent of tumor grade, pathologic tumor stage, and lymph node status, and it was strongest in those patients with pathologically organ confined tumors (P1, P2, P3a, lymph node negative). Furthermore, in a multivariable analysis, nuclear accumulation of p53 was found to be the only independent predictor of disease progression in a model that included histologic grade, pathologic stage and lymph node status.

Lipponen and associates also evaluated p53 immunoreactivity in 212 patients with primary bladder cancer.[22] Higher tumor grade and tumor progression were significantly associated with p53 alteration, thereby confirming that p53 nuclear reactivity was a significant adverse prognostic factor in patients with muscle invasive bladder cancer. Two separate studies have evaluated p53 expression in patients with minimally invasive, stage T1 bladder cancer.[21, 23] Sarkis and associates examined p53 expression in 43 patients with stage T1 bladder cancer who had not received adjuvant therapy.[23] In this study, p53 nuclear reactivity was not associated with patient age, sex, tumor grade, or the presence of associated carcinoma-in-situ or vascular invasion. However, p53 nuclear reactivity was significantly associated with disease progression. These authors reported that patients with stage T1 bladder tumors which demonstrated p53 nuclear reactivity (altered p53) had a significantly higher likelihood of disease progression (20.5% per year) than those patients with stage T1 tumors which demonstrated wild-type p53 (2.5% per year) ($p < 0.001$). This finding has recently been supported by Hermann et al. who reported that the presence of p53 nuclear reactivity was significantly associated with a lower probability of survival in patients in patients with stage T1 transitional cell carcinoma of the bladder who were under 75 years of age (relative risk ratio of 1.8 when compared to similar patients with wild-type p53) ($p < 0.05$).[21]

Based on a substantial amount of evidence that has been obtained from a number of independent studies, p53 immunoreactivity appears to be a reliable and consistent prognostic marker for bladder cancer progression. p53 nuclear reactivity, as determined by immunohistochemical techniques, is generally seen in high grade and high stage bladder tumors. Furthermore, p53 nuclear reactivity appears to be associated with a increased likelihood of bladder cancer progression and cancer-specific death.

4. COMBINATION OF Rb AND p53 TUMOR SUPPRESSOR GENES

Two studies have examined the prognostic significance of combining Rb protein expression with p53 nuclear reactivity in patients with transitional cell carcinoma of the bladder.[26, 27] Preliminary data from these studies suggest that bladder tumors which exhibit alterations in both p53 and Rb have a worse prognosis than bladder tumors with normal expression of both prognostic markers. Tumors containing only one alteration (either a p53 or an Rb alteration) behave in an intermediate fashion. These data suggest that both p53 and Rb are important prognostic markers in bladder cancer, and that these two proteins may act in an independent, yet synergistic manner in this disease.

5. p21 TUMOR SUPPRESSOR GENE

The studies summarized above provide convincing evidence that p53 nuclear reactivity, as detected by immunohistochemical methods, is a significant predictor of bladder cancer progression. However, these studies also demonstrate that not all p53-altered bladder tumors recur or progress. p53 has several important functions, one of which is regulation of the cell cycle. [3, 18] The inhibitory effect of p53 on the cell cycle appears to be mediated through its regulation of $p21^{WAF1/CIP1}$ expression. [3] Alterations in p53 lead to loss of p21 expression, which subsequently results in unregulated cellular growth. This is thought to be one of the mechanisms by which p53 alterations may influence tumor progression. However, recent evidence demonstrates that p21 expression may also be mediated through pathways which are independent of p53.[28–30] This important evidence suggests that despite p53 alteration, p21 expression may be preserved thereby allowing for the maintenance of cell cycle control. Consequently, it is possible that p21 expression may provide important prognostic information (in addition to that provided by p53) for patients diagnosed with bladder cancer. Specifically, the presence or absence of p21 expression in bladder tumors that demonstrate p53 alteration could be an important factor in predicting clinical behavior. This is based on the hypothesis that p21 expression, through p53-independent pathways, may abrogate the deleterious effects of p53 alterations by resulting in the maintenance of cell cycle control.

To test this hypothesis, Stein and colleagues evaluated the immunohistochemical expression of p21 in bladder tumors from 242 patients with invasive bladder cancer who were uniformly treated by radical cystectomy [31]. Of these 242 tumors, 102 (42%) had p53 alterations, while 156 (64%) maintained p21 expression (p21 positive). This study demonstrated that loss of p21 expression was associated with a significantly higher probability of disease-recurrence ($p < 0.0001$) and a significantly lower probability of overall survival ($p < 0.0001$) for patients with all stages of bladder cancer. This association was strongest for those patients with organ-confined (Pa, Pis, P1, P2, P3a, lymph node negative) disease ($p < 0.00001$). Although p21 expression was significantly associated with p53 status, a multivariable analysis demonstrated that p21 expression was an independent predictor of both disease recurrence and overall survival when considered along with tumor grade, tumor stage, lymph node status and p53 nuclear reactivity.

In this study, Stein and colleagues also analyzed the association between p21 expression, p53 status, disease recurrence and overall survival.[31] Patients with p53 altered tumors who maintained normal p21 expression exhibited similar rates of disease recurrence and overall survival as those patients with p53 wild-type tumors. Conversely, patients with p53 altered tumors who lost p21 expression demonstrated significantly higher rates of disease recurrence, and significantly worse overall survival than those patients with either p53 wild-type tumors or p53 altered tumors which maintained normal p21 expression. These important findings support the hypothesis that preservation of p21 expression through p53 independent pathways could result in the maintenance of normal cell cycle control. p53 alteration combined with loss of p21 expression identifies a high risk bladder cancer phenotype that may benefit from early, aggressive treatment.

6. MEASUREMENT OF CELLULAR PROLIFERATION USING Ki-67 LABELING INDEX

One of the defining characteristics of malignancy is uncontrolled cellular growth. The fraction of proliferating cells (growth fraction) may help to define the biologic potential of a tumor and thus, act as an important prognostic feature for any malignancy. In this regard, several studies have examined the relationship between cellular proliferation and tumor grade, stage and disease progression to determine if proliferative activity can be used as a prognostic marker for patients with bladder cancer.

Cellular proliferation can be determined by a number of techniques including the measurement of mitotic count, silver-stained nucleolar organizer regions (AgNORs), Ki-67 immunoreactivity, immunohistochemical expression of the proliferating cell nuclear antigen (PCNA) or tissue labeling with bromodeoxyuridine (BrdUrd)[32]. The most promising marker of cellular proliferation appears to be Ki-67 labeling index. Ki-67 expression can be determined on routinely processed formalin-fixed, paraffin-embedded tissue specimens using immunohistochemical techniques. Increased expression of this nuclear antigen indicates a high level of proliferative activity in tumor cells, and thus appears to indicate a more aggressive biologic phenotype with an increased propensity for tumor progression and metastasis.[33]

Ki-67 is a monoclonal antibody that recognizes a nuclear protein which may form a part of the DNA replicase complex.[34, 35] Ki-67 is an excellent marker of cellular proliferation as it is expressed during all phases of the cell cycle except G0.[36] Since the initial discovery of this antibody in 1983, it has been evaluated in a variety of human malignancies,[37] including bladder cancer.[32, 34, 35, 38–41] Two studies have confirmed that Ki-67 labeling index (LI) correlates well with BrdUrd LI in transitional cell carcinoma of the bladder. [32, 41] These studies also demonstrated that Ki-67 LI is associated with bladder tumor grade, [32, 41] stage and tumor size.[41]

Okamura and associates determined the proliferative fraction of 55 bladder tumors using Ki-67 immunostaining and found a highly significant association between increasing Ki-67 LI and tumor grade and stage. [35] Other studies have not only confirmed this relationship, but they have also demonstrated a significant association between proliferative index (as determined by Ki-67 nuclear reactivity) and tumor recurrence.[32, 34, 38–41] Mulder and colleagues reported that bladder tumor progression occurred only when the Ki-67 LI was greater than 10%.[40] Fontana et al. found a significant correlation between tumor recurrence and the location of tumor cell proliferation.[38] These authors reported no recurrences in 19 patients with stage Ta bladder tumors in whom cellular proliferation was located in the vegetative (superficial) portion of the tumor. However, recurrences were noted in 16 of 17 stage Ta patients (94%) in whom the proliferative index was present near the lamina propria ($p < 0.001$). King et al. reported that disease-free survival following cystectomy was significantly higher for those patients with a Ki-67 LI less than 20% compared to patients with a tumor Ki-67 LI greater than 20%.[39]

These studies suggest that the proliferative index of a bladder tumor, as determined by Ki-67 nuclear reactivity, can provide important prognostic information in addition to that obtained from histologic grade and tumor stage. Because Ki-67 LI can be easily determined from routinely processed tissue, the measurement of cellular proliferation may prove to be a useful prognostic marker for a wide variety of tumors.

7. ASSOCIATION BETWEEN CELLULAR PROLIFERATION AND CELL CYCLE REGULATORY PROTEINS IN TRANSITIONAL CELL CARCINOMA OF THE BLADDER

The relationship between different cell cycle regulatory genes in bladder cancer tissue specimens, and their relationship to tumor cell proliferation is currently uncertain. This is because most studies to date that have examined cellular proliferation or the expression of cell cycle regulatory proteins in bladder cancer have evaluated these prognostic markers individually. Consequently, several candidate genes have been identified as putative mediators of bladder cancer initiation and/or progression, but their relative importance in the regulatory pathways involved in bladder cancer progression remains unknown. By evaluating the expression of these different prognostic markers in the same bladder tumor specimen, and determining the relationships between these regulatory genes in patients with bladder cancer, it may become possible to identify the in vivo mechanisms by which bladder tumors originate and progress.

In order to determine if the interactions that occur between cell cycle regulatory proteins in vitro also occur in vivo in patients with bladder cancer, we evaluated the associations between cellular proliferation (Ki-67 labeling index), p53, Rb and p21 in 97 patients with transitional cell carcinoma of the bladder. The population under consideration consisted of patients who were diagnosed with, and treated for stage Ta-T4 transitional cell carcinoma of the bladder at the University of California, San Francisco. Patient characteristics, according to tumor stage, are summarized in Table 1.

Bladder tumors for immunohistochemical analysis were obtained either at the time of transurethral resection or at the time of radical cystectomy. Hematoxylin and eosin stained tissue sections were examined to confirm the presence of bladder cancer in each case, and appropriate tissue sections were selected for immunohistochemical analysis. Routinely processed formalin-fixed, paraffin-embedded tissue blocks were cut into 5μm sections and mounted on positively charged slides. Tissue sections were deparaffinized with xylene, and then rehydrated with graded alcohol solutions. Endogenous peroxidase was quenched in 3% hydrogen peroxide/phosphate buffered saline (PBS) for 5 minutes. Slides were then washed with PBS and water. Sections for Ki-67 immunostaining were trypsinized (0.25% concentration) at 37°C for 15 minutes. Antigen retrieval was then per-

Table 1. Patient characteristics according to tumor stage for 97 patients with transitional cell carcinoma of the bladder

Tumor stage	No. of patients	Mean Age*	Male:Female
Ta	35	68.4	26:9
T1	27	68.1	21:6
T2-T4	35	70.9	25:10

* p=0.59.

formed for all specimens by heating the slides for 10 minutes in a General Electric microwave oven at the highest power setting (900 W) in 10 mM citrate buffer at pH 6.0. Following antigen retrieval, slides were cooled for 30 minutes. Blocking of non-specific binding was carried out in 1.5% normal horse serum in PBS for 30 minutes at room temperature.

Tissue sections were then incubated with the appropriate primary antibody overnight at 4°C. Antibody dilutions in PBS were those recommended by the manufacturer: monoclonal MIB-1 (Ki-67) at 1:200 (Immunotech Inc; Westbrook, ME), pAb1801 (p53) at 1:1000 (Genosys Biotechnologies; Cambridge, England), Pharmingen #14001A (Rb) at 1:1000 (Pharmingen, San Diego, CA) and WAF1 (Ab1), clone EA10 (p21) at 1:250 (Calbiochem, Cambridge, MA). After washing with PBS, tissue sections were incubated with a biotinylated secondary antibody after which reactivity was visualized using streptavidin-HRP (for Ki-67 and p53) (Zymed Laboratories; South San Francisco, CA) or the avidin-biotin-peroxidase system (for p21 and Rb) (Vector Laboratories; Burlingame, CA) using diamino-benzidine as the chromagen. Slides stained for p53 expression were counterstained with 1% methyl green, while slides stained for Ki-67 expression, p21 expression and Rb expression were counterstained with hematoxylin. Tissue sections were then dehydrated in graded alcohols, cleared in xylene and cover slipped.

Only areas of bladder tumor were included for analysis; areas of benign transitional epithelium were not considered for immunohistochemical evaluation. Ki-67 labeling index (LI) was calculated by determining the percentage of cells demonstrating nuclear reactivity after counting at least 1000 tumor cells at 400X magnification. Fields with the highest degree of nuclear reactivity were chosen for evaluation. For p21, p53 and Rb, the percentage of tumor cells staining positive was estimated to the nearest decile. Specimens were also scored as p53 positive or Rb positive based upon a pre-determined cut-point of ≥10% positive nuclei.

The chi squared test was used to evaluate the association between p53 expression and tumor stage, Rb expression and tumor stage, and p53 expression and Rb expression (when p53 and Rb were graded as positive or negative). Analysis of variance was used to evaluate the association between the various labeling indexes and tumor stage. The paired t-test was used to compare p53 and Rb expression (graded as positive or negative) with Ki-67 LI and p21 LI. Logistic regression analysis was used to evaluate the association between Ki-67 LI and p21 LI. All p-values reported are from 2 sided tests.

Of the 97 bladder tumors evaluated, 46 (47%) demonstrated p53 nuclear reactivity in at least 10% of cell nuclei (p53 altered), while 86 (89%) demonstrated Rb nuclear reactivity in at least 10% of cell nuclei. As summarized in Table 2, all four of the immunohistochemical markers examined were significantly associated with bladder tumor stage (Table 2). Although p53 expression (graded as positive or negative) did not significantly correlate with tumor stage, p53 LI significantly increased with increasing depth of tumor cell invasion. Both Rb expression (graded as positive or negative) and Rb LI were inversely associated with tumor stage. As the depth of tumor cell invasion increased, the Rb LI significantly decreased and significantly fewer tumors were graded as Rb positive. Tumor proliferative index (as measured by Ki-67 LI) and p21 LI were also significantly associated with bladder tumor stage. While Ki-67 LI significantly increased as tumor stage increased, p21 LI was inversely correlated with the depth of tumor cell invasion (Table 2).

To further evaluate the interactions between p53, Rb, p21 and cellular proliferation, we analyzed the associations between these various immunohistochemical markers in this patient population with bladder cancer. We found that cellular proliferative index (as measured by Ki-67) was significantly associated with p53 expression (positive or nega-

Table 2. Association between tumor stage and p53, Rb, p21 and Ki-67 LI

Marker	Stage Ta	Stage T1	Stage T2-4	p-value
p53 expression (percent cases positive)	34%	48%	60%	0.09
p53 LI	12.3%	18.3%	45.3%	0.0001
Rb expression (percent cases positive)	97%	93%	77%	0.02
Rb LI	75.6%	65.5%	58.5%	0.02
Ki-67 LI	24.7%	44.4%	62.8%	0.0001
p21 LI	62.6%	54.8%	38.3%	0.0002

tive) and p21 LI. Cellular proliferation was significantly higher in p53 altered tumors (p53 positive) than in p53 wild-type tumors (p=0.002). Furthermore, proliferative index significantly increased as p21 LI decreased (p=0.0001). Proliferative index was not associated with Rb status (p=0.2).

Although p21 LI was higher in p53 wild-type tumors (p53 negative) than in p53 altered tumors (p53 positive), it was surprising to find that this difference did not reach statistical significance in our cohort of patients (p=0.08). However, p21 LI was significantly associated with Rb status (p=0.005). p21 LI was significantly higher in Rb positive than in Rb negative tumors. Finally, we found no association between p53 expression and Rb expression (p=0.07).

Our results demonstrate that the cell cycle regulatory proteins, p53 and p21, are significantly associated with cellular proliferation in patients with transitional cell carcinoma of the bladder. p53 expression and Rb expression appear to be independent in these patients, while the level of p21 expression was significantly associated with Rb and not p53. Our inability to demonstrate a correlation between p53 and p21 in these tumors may reflect either a small sample size or the maintenance of p21 expression through p53 independent pathways.

8. CONCLUSIONS

Tumor heterogeneity is the hallmark of transitional cell carcinoma of the bladder. Conventional histopathologic parameters, such as tumor grade and stage, are unable to reliably predict the behavior of most bladder tumors. Newer prognostic markers, that are better able to identify a bladder tumor's true malignant potential, are urgently needed. With a better understanding of cell cycle regulation, as well as the ability to reliably and easily measure the proliferative index of a particular tumor, it may soon be possible to stratify patients into risk groups based upon a panel of immunohistochemical markers of disease progression. The markers that appear most promising in this regard include p53 nuclear reactivity, Rb expression, p21 labeling index and Ki-67 labeling index. Previous studies have demonstrated that each of these markers is independently associated with bladder tumor grade, tumor stage and/or tumor progression. In the current study, we confirmed these associations and evaluated the interactions between these markers in patients diagnosed with, and treated for bladder cancer. The ultimate goal is not only to use these immunohistochemical markers of bladder cancer progression when making treatment decisions in the clinic, but also to use this understanding of cell cycle regulation to design rational treatment strategies aimed at decreasing tumor cell proliferation and prolonging survival.

REFERENCES

1. Nurse, P.: Ordering S phase and M phase in the cell cycle. Cell, **79:** 547,1994.
2. Hunter, T. and Pines, J.: Cyclins and cancer. II: Cyclin D and CDK inhibitors come of age. Cell, **79:** 573,1994.
3. Cordon-Cardo, C.: Mutations of cell cycle regulators. Biological and clinical implications for human neoplasia. Am. J. Pathol., **147:** 545,1995.
4. Cordon-Cardo, C., Wartinger, D., Petrylak, D., Dalbagni, G., Fair, W.R., Fuks, Z. and Reuter, V.E.: Altered expression of the retinoblastoma gene product: prognostic indicator in bladder cancer. J. Natl. Cancer Inst., **84:** 1251,1992.
5. Fung, Y.K., Murphree, A.L., T'Ang, A., Qian, J., Hinrichs, S.H. and Benedict, W.F.: Structural evidence for the authenticity of the human retinoblastoma gene. Science, **236:** 1657,1987.
6. Cairns, P., Proctor, A.J. and Knowles, M.A.: Loss of heterozygosity at the RB locus is frequent and correlates with muscle invasion in bladder carcinoma. Oncogene, **6:** 2305,1991.
7. Logothetis, C.J., Xu, H.J., Ro, J.Y., Hu, S.X., Sahin, A., Ordonez, N. and Benedict, W.F.: Altered expression of retinoblastoma protein and known prognostic variables in locally advanced bladder cancer. J. Natl. Cancer Inst., **84:** 1256,1992.
8. Presti, J.C., Jr., Reuter, V.E., Galan, T., Fair, W.R. and Cordon-Cardo, C.: Molecular genetic alterations in superficial and locally advanced human bladder cancer. Cancer Res., **51:** 5405,1991.
9. Bagchi, S., Weinmann, R. and Raychaudhuri, P.: The retinoblastoma protein copurifies with E2F-I, an E1A-regulated inhibitor of the transcription factor E2F. Cell, **65:** 1063,1991.
10. Wang, J.Y., Knudsen, E.S. and Welch, P.J.: The retinoblastoma tumor suppressor protein. Adv Cancer Res., **64:** 25,1994.
11. Geradts, J., Hu, S.X., Lincoln, C.E., Benedict, W.F. and Xu, H.J.: Aberrant RB gene expression in routinely processed, archival tumor tissues determined by three different anti-RB antibodies. Int. J. Cancer, **58:** 161,1994.
12. Ishikawa, J., Xu, H.J., Hu, S.X., Yandell, D.W., Maeda, S., Kamidono, S., Benedict, W.F. and Takahashi, R.: Inactivation of the retinoblastoma gene in human bladder and renal cell carcinomas. Cancer Res., **51:** 5736,1991.
13. Xu, H.J., Cairns, P., Hu, S.X., Knowles, M.A. and Benedict, W.F.: Loss of RB protein expression in primary bladder cancer correlates with loss of heterozygosity at the RB locus and tumor progression. Int. J. Cancer, **53:** 781,1993.
14. Hollstein, M., Sidransky, D., Vogelstein, B. and Harris, C.C.: p53 mutations in human cancers. Science, **253:** 49,1991.
15. Levine, A.J.: p53, the cellular gatekeeper for growth and division. Cell, **88:** 323,1997.
16. Lane, D.P.: Cancer. p53, guardian of the genome. Nature, **358:** 15,1992.
17. Esrig, D., Spruck, C.H.d., Nichols, P.W., Chaiwun, B., Steven, K., Groshen, S., Chen, S.C., Skinner, D.G., Jones, P.A. and Cote, R.J.: p53 nuclear protein accumulation correlates with mutations in the p53 gene, tumor grade, and stage in bladder cancer. Am. J. Pathol., **143:** 1389,1993.
18. Dalbagni, G., Cordon-Cardo, C., Reuter, V. and Fair, W.R.: Tumor suppressor gene alterations in bladder carcinoma. Translational correlates to clinical practice. Surg. Oncol. Clin. N. Am., **4:** 231,1995.
19. Cordon-Cardo, C., Dalbagni, G., Saez, G.T., Oliva, M.R., Zhang, Z.F., Rosai, J., Reuter, V.E. and Pellicer, A.: p53 mutations in human bladder cancer: genotypic versus phenotypic patterns. Int. J. Cancer, **56:** 347,1994.
20. Esrig, D., Elmajian, D., Groshen, S., Freeman, J.A., Stein, J.P., Chen, S.C., Nichols, P.W., Skinner, D.G., Jones, P.A. and Cote, R.J.: Accumulation of nuclear p53 and tumor progression in bladder cancer. N. Engl. J. Med, **331:** 1259,1994.
21. Hermann, G.G., T., H. and Steven, K.: The influence of the level of lamina propria invasion and the prevalence of p53 nuclear accumulation on survival in stage T1 transitional cell bladder cancer. J. Urol., **159:** 91,1998.
22. Lipponen, P.K.: Over-expression of p53 nuclear oncoprotein in transitional-cell bladder cancer and its prognostic value. Int. J. Cancer, **53:** 365,1993.
23. Sarkis, A.S., Dalbagni, G., Cordon-Cardo, C., Zhang, Z.F., Sheinfeld, J., Fair, W.R., Herr, H.W. and Reuter, V.E.: Nuclear overexpression of p53 protein in transitional cell bladder carcinoma: a marker for disease progression. J. Natl. Cancer Inst., **85:** 53,1993.
24. Sidransky, D., Von Eschenbach, A., Tsai, Y.C., Jones, P., Summerhayes, I., Marshall, F., Paul, M., Green, P., Hamilton, S.R., Frost, P. and et al.: Identification of p53 gene mutations in bladder cancers and urine samples. Science, **252:** 706,1991.

25. Wright, C., Mellon, K., Johnston, P., Lane, D.P., Harris, A.L., Horne, C.H. and Neal, D.E.: Expression of mutant p53, c-erbB-2 and the epidermal growth factor receptor in transitional cell carcinoma of the human urinary bladder. Br. J. Cancer, **63:** 967,1991.
26. Esrig, D., Shi, S.R., Bochner, B., Nichols, P.W., Chen, S.C., Yang, C., Jones, P., Skinner, D.G. and Cote, R.J.: Prognostic importance of p53 and Rb alterations in transitional cell carcinoma of the bladder. J Urol, **153:** 362,1995.
27. Lerner, S., Linn, D., Chakraborty, S., Trong, L., Xu, H.-J., Wheeler, T.M., Kattan, M.W. and Benedict, W.F.: Correlation of p53 and retinoblastoma protein expression with established pathologic prognostic features in radical cystoprostatectomy specimens. J. Urol., **153:** 363,1995.
28. Bond, J.A., Blaydes, J.P., Rowson, J., Haughton, M.F., Smith, J.R., Wynford-Thomas, D. and Wyllie, F.S.: Mutant p53 rescues human diploid cells from senescence without inhibiting the induction of SDI1/WAF1. Cancer Res., **55:** 2404,1995.
29. Michieli, P., Chedid, M., Lin, D., Pierce, J.H., Mercer, W.E. and Givol, D.: Induction of WAF1/CIP1 by a p53-independent pathway. Cancer Res., **54:** 3391,1994.
30. Parker, S.B., Eichele, G., Zhang, P., Rawls, A., Sands, A.T., Bradley, A., Olson, E.N., Harper, J.W. and Elledge, S.J.: p53-independent expression of p21Cip1 in muscle and other terminally differentiating cells. Science, **267:** 1024,1995.
31. Stein, J.P., Ginsberg, D.A., Grossfeld, G.D., Esrig, D., Freeman, J.A., Dickinson, M.G., Groshen, S., Taylor, C.R., Skinner, D.G. and Cote, R.J.: The effect of $p21^{WAF1/CIP1}$ expression on tumor progression in bladder cancer. J. Natl. Cancer Inst., in press.
32. Cohen, M.B., Waldman, F.M., Carroll, P.R., Kerschmann, R., Chew, K. and Mayall, B.H.: Comparison of five histopathologic methods to assess cellular proliferation in transitional cell carcinoma of the urinary bladder. Hum. Pathol., **24:** 772,1993.
33. Fradet, Y. and Cordon-Cardo, C.: Critical appraisal of tumor markers in bladder cancer. Semin. Urol., **11:** 145,1993.
34. Bush, C., Price, P., Norton, J., Parkins, C.S., Bailey, M.J., Boyd, J., Jones, C.R., A'Hern, R.P. and Horwich, A.: Proliferation in human bladder carcinoma measured by Ki-67 antibody labelling: its potential clinical importance. Br. J. Cancer, **64:** 357,1991.
35. Okamura, K., Miyake, K., Koshikawa, T. and Asai, J.: Growth fractions of transitional cell carcinomas of the bladder defined by the monoclonal antibody Ki-67. J. Urol., **144:** 875,1990.
36. Gerdes, J., Becker, M.H., Key, G. and Cattoretti, G.: Immunohistological detection of tumour growth fraction (Ki-67 antigen) in formalin-fixed and routinely processed tissues. J. Pathol., **168:** 85,1992.
37. Brown, D.C. and Gatter, K.C.: Monoclonal antibody Ki-67: its use in histopathology. Histopathology, **17:** 489,1990.
38. Fontana, D., Bellina, M., Gubetta, L., Fasolis, G., Rolle, L., Scoffone, C., Porpiglia, F., Colombo, M., Tarabuzzi, R. and Leonardo, E.: Monoclonal antibody Ki-67 in the study of the proliferative activity of bladder carcinoma. J. Urol., **148:** 1149,1992.
39. King, E.D., Matteson, J., Jacobs, S.C. and Kyprianou, N.: Incidence of apoptosis, cell proliferation and bcl-2 expression in transitional cell carcinoma of the bladder: association with tumor progression. J. Urol., **155:** 316,1996.
40. Mulder, A.H., Van Hootegem, J.C., Sylvester, R., ten Kate, F.J., Kurth, K.H., Ooms, E.C. and Van der Kwast, T.H.: Prognostic factors in bladder carcinoma: histologic parameters and expression of a cell cycle-related nuclear antigen (Ki-67). J. Pathol., **166:** 37,1992.
41. Tsujihashi, H., Nakanishi, A., Matsuda, H., Uejima, S. and Kurita, T.: Cell proliferation of human bladder tumors determined by BrdUrd and Ki-67 immunostaining. J. Urol., **145:** 846,1991.

34

UROTHELIAL DIFFERENTIATION AND BLADDER CANCER

Monica Liebert, Daniel Gebhardt, Christopher Wood, I-Ling Lee Chen, Jeffrey Ellard, Daniel Amancio, and H. Barton Grossman

Departments of Urology and Cell Biology
The University of Texas-M.D. Anderson Cancer Center
Houston, Texas 77098

1. INTRODUCTION

Cancer is the endpoint of a multi-step process of carcinogenesis.[1–5] This process involves multiple genetic and epigenetic alterations and cellular selections.[2–4,6] These changes result in cell populations with unregulated growth and ultimately, the ability to invade and metastasize. Many of these cancer cell characteristics may reflect a loss of the regulation present during the normal differentiation process.[7]

Bladder cancer will occur in over 54,000 new cases in the US in 1998, with approximately 12,500 deaths due to bladder cancer.[8] Bladder cancer presents to the clinician in two forms: approximately 70% of patients have a low stage, non-invasive cancer (Ta and carcinoma in situ, TIS). The other 30% of bladder cancer patients present with invasive cancer (T1, minimally invasive, through T4, invading though the bladder wall[9]). Some (up to 80%) patients with low stage bladder cancer will have recurrence of their cancer and a small number of patients with low stage bladder cancer will have their tumors progress and become invasive.[9]

Since patients with superficial low stage bladder cancer are at high risk for recurrence, this population is a target for chemoprevention. Treatment with chemopreventive agents may induce apoptosis of transformed cells and/or increase the differentiation of cells. Consequently, characterization of differentiation in the target organ may provide essential biomarkers (surrogate endpoints) for monitoring of therapy. This approach has been highly successful in cancers of the head and neck and oral cancers, where the well-characterized squamous differentiation markers are well used.[10] Unfortunately, the molecular changes associated with urothelial differentiation are still poorly understood, although actively under investigation by several groups. This paper will summarize some of the current knowledge.

Advances in Bladder Research, edited by Baskin and Hayward.
Kluwer Academic / Plenum Publishers, New York, 1999.

Urothelial cells, the cells that line the urinary tract from the renal pelvis through the urinary bladder to the proximal urethra, are the cells of origin for the vast majority of bladder cancers. The urothelium is a transitional epithelium, normally 3–7 layers deep, with the least differentiated cells at the basal layer near the lamina propria, and the most differentiated cells at the lumenal surface. The most differentiated cells are a large, multinucleate,specialized cell called the umbrella cell.[11] The normal orderly cellular and nuclear architecture of the urothelium is altered in bladder cancer. Increased cell number and mitoses, loss of tissue polarity and demonstration of nuclear pleomorphisms serve as histologic signposts for grading of bladder cancer.[12] A number of molecular and antigenic changes are associated with these histologic alterations, including cytoskeletal changes, changes in adhesion molecules, and alterations in cell surface markers and growth factor receptors, as will be described in more detail in the following sections.

2. MARKERS OF PROLIFERATION

Normal urothelial cells are fairly long-lived epithelial cells and show little mititotic activity under most conditions.[13] When mitoses occur, they are limited to the basal layer, which contain the least differentiated cells. Increasing numbers of mitotic figures are characteristic of bladder cancers; furthermore, in cancers, cell division is no longer limited to the basal layer but is observed throughout the tissue, which is often increased in cell number as well.[12] Immunologic markers such as PCNA and Ki67 that identify dividing cells also label more cells throughout the cell layers in bladder cancers.[14,15]

3. CYTOSKELETAL MARKERS OF UROTHELIAL DIFFERENTIATION

Since alterations in cell shape and motility are hallmarks of cancer cells,[16] it is not surprising that the cellular structures that maintain cell shape, the cytoskeleton, is altered in bladder cancer. Two main components of the cytoskeleton, actins and cytokeratins, have been intensely investigated.

3.1. Actin

Actin is one of the major contractile proteins of the cytoskeleton, and exists in two forms: G-actin, a globular non-polymerized form, and F-actin, a filamentous polyermized form.[17] Actin is linked to the surface of the cell through linking proteins associated with focal adhesions and, at the cell surface, integrins, especially members of the β1integrin family.[18] In differentiated normal urothelial cells, the amount of F-actin is relatively high compared to the amount of G-actin; in cancer cells, the relationship is reversed.[19,20] Furthermore, a shift to lower G-actin expression may be useful in monitoring chemoprevention trials.[19]

Gelsolin is one of the actin "capping" proteins that modulate assembly of F-actin networks or bundles.[17] Gelsolin expression is reduced or absent in bladder cancers.[21] In the same study, Tanaka and his coworkers showed that transfection of gelsolin into a human bladder cancer cell line with low native gelsolin expression resulted in slowed growth, induced morphologic changes, and reduced tumorigenicity.[21]

3.2. Cytokeratins

Cytokeratins are members of the intermediate filament family of the cytoskeleton. Cytokeratins are thought to provide a cellular scaffolding to maintain cellular shape and integrity. Furthermore, cytokeratins are linked to cell contacts through interactions with desmosomes, cadherins and hemidesmosomes.[18]

Cytokeratins are a large family of cytoskeletal proteins, and expression of specific types is characteristic of the tissue and differentiation state of the cell. In urothelium, cytokeratin 4 is typically expressed in more basal or immature cells, while cytokeratins 7,8, 13, 18 and 19 are expressed throughout the urothelium.[22] Expression of cytokeratins 8, 18 and 19 appears to persist in bladder cancers.[22]

Another cytokeratin, cytokeratin 20, has been recently identified. In normal urothelim, cytokertain 20 expression is limited to the surface of the most superficial cells (umbrella cells), while in bladder cancers, cytokeratin 20 is frequently found throughout the lesion in both superficial and invasive cancers.[23] Southgate and her coworkers observed that recurrent superficial bladder cancers had a greater dysregulation of cytokeratin 20 expression (e.g., expression throughout the tumor) while primary tumors tended to have a more normal cytokeratin 20 expression limited to the cell surface.[24] She also observed abberant cytokeratin 20 expression in urothelial dysplasia.[25]

4. ADHERENCE MARKERS

Cells maintain their relationship to one another within tissues through a number of cellular adherence structures which link cells to each other and to the extracelluar matrix.[26] These structures include desmosomes, hemidesmosomes, integrins, and cadherins.

4.1. Desmosomes

Desmosomes are part of the cell-cell adherence network, and are linked within the cell through the cytokeratin network.[18] In normal urothelial cells and tissues, desmosomes are found at the cell-cell contacts.[27–29] In cultured rat cell lines and in bladder tumors, reduced expression of desmosomes and desmosomal proteins is associated with greater tumorigenic potential and progression.[30,31]

To demonstrate alterations in desmosomal structure in bladder cancers, we used immunostaining with antibodies to desmoplakin I+II, proteins of the desmosome, followed by secondary fluorescence detection to demonstrate desmosomal expression in a human bladder cancer cell line, UM-UC-9. As shown in Figure 1A, this cancer cell line shows patchy and abberant expression of desmoplakin I+II, while normal differentiated urothelial cells (Figure 1B) show normal orderly expression at cell contacts. Many other bladder cancer cells show little or no desmoplakin I+II expression (unpublished; data not shown).

4.2. E-Cadherin

Cadherins are a family of cell-cell adherence molecules. In urothelial cells, E-cadherin has been most extensively studied, although other members of the cadherin family may also be expressed. E-cadherin is normally found at cell junctions both in differentiated cultured urothelial cells[27] and in tissues.[32] However, E-cadherin expression is lost in

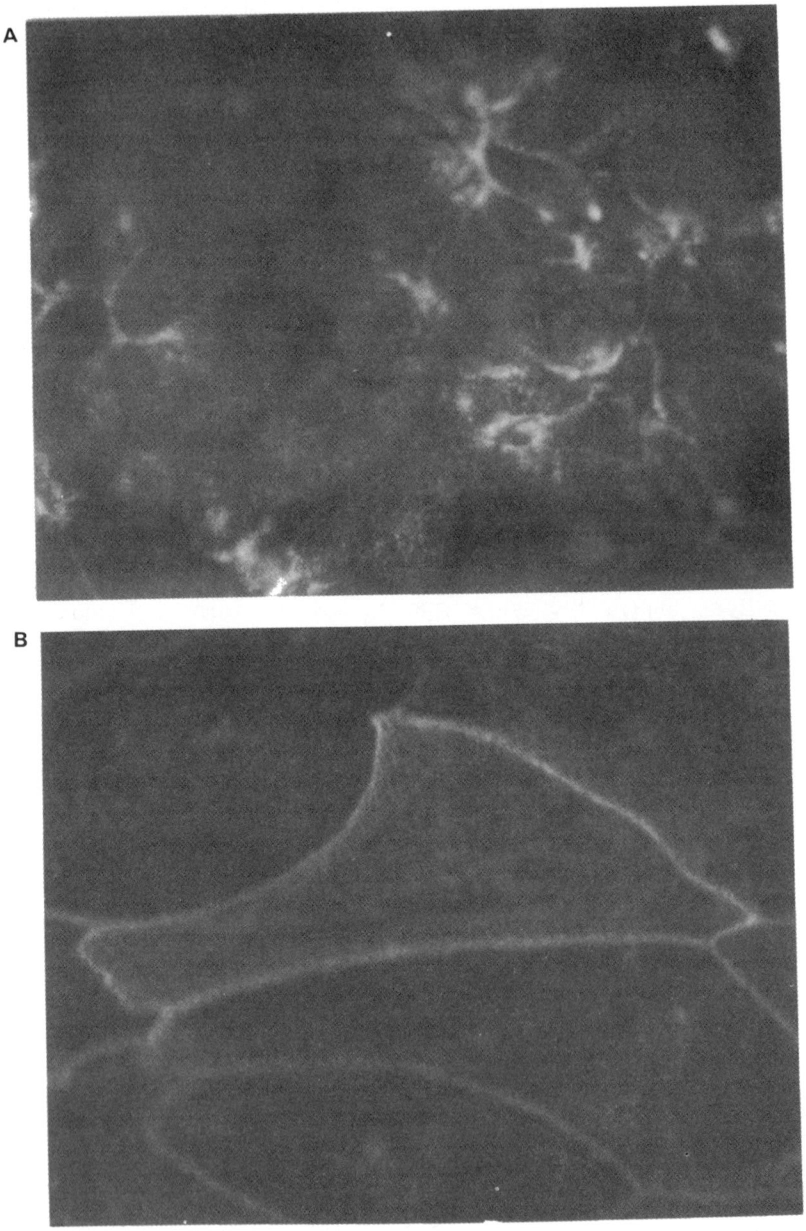

Figure 1A, B.

C

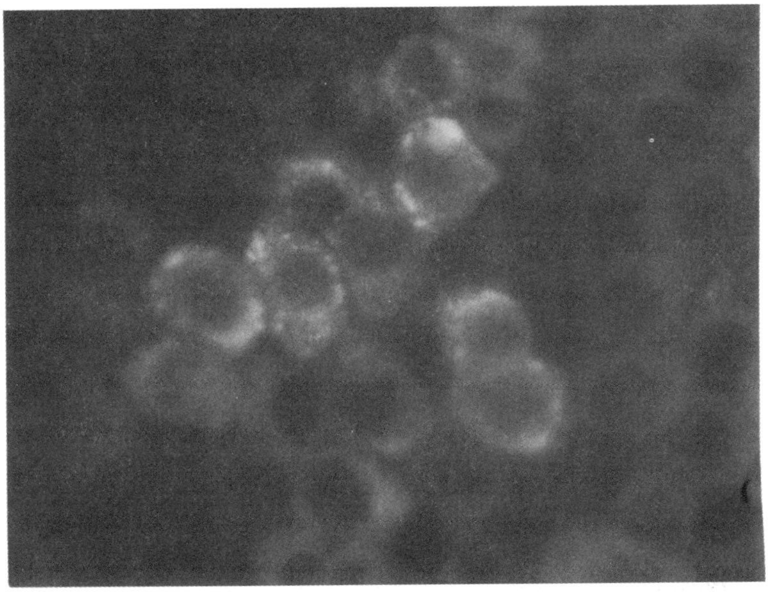

Figure 1C.

many bladder cancers [32], and this loss is a poor prognostic sign. In other studies, re-expression of E-cadherin in cells is associated with reduced tumorigenicity, indicating that E-cadherin may function as a tumor suppressor [33, 34].

Again, to demonstrate the abberant expression of E-cadherin, we immunostained human bladder cancer UM-UC-9 cells with antibodies to E-cadherin followed by fluorescent antibody detection. As shown in Figure 1C, only cytoplasmic expression of E-cadherin is observed, while a normal expression would be similar to the desmoplakin I+II staining[27] seen in Figure 1B.

4.3. Integrins

Integrins are a large family of heterodimeric receptors for extracellular matrix proteins. Each integrin is composed of an alpha and beta dimer and interacts with specific extracellular matrix components. For example, the $\alpha5\beta1$integrin interacts with the fibronectin molecule [35].

In normal urothelium, several integrin types are observed, including $\alpha2\beta1$, $\alpha3\beta1$and $\alpha6\beta4$[36–38]. The $\alpha2\beta1$ and $\alpha3\beta1$ integrins are found in the basal and intermediate layers in normal urothelial tissues, but $\alpha2\beta1$ expression is lost in superficial bladder cancers, while $\alpha3\beta1$ expression appears to be retained by the majority of tumors[36,37]. Studies of other cell types suggest that $\alpha2\beta1$ and $\alpha3\beta1$ are important in maintaining cell-cell contact and tissue integrity[39,40].

In contrast, $\alpha6\beta4$ integrin is normally found only at the basolateral surface of the basal cells in contact with the lamina propria[37,38], and in both superficial and most invasive cancers, $\alpha6\beta4$ no longer has a polarized expressed but is found throughout the tumor[38].

4.4. Hemidesmosomes

Hemidesmosomes are cellular structures associated with anchoring of epithelial cells to the extracellular matrix, and loss of these structures are associated with a number of blistering diseases of the skin. The α6β4integrin is associated with the hemidesmosome in normal cells. While the α6β4 integrin is overexpressed in many bladder cancers, the normal association with collagen VII, a component of the hemidesmosome, is lost, with increasing loss of collagen VII expression.[38] These data suggest that the hemidesmosome structure is lost early during development of bladder cancers.

5. CELL SURFACE MARKERS

Alterations in cell surface markers have long been recognized in cancers. These changes may be epigenetic or may reflect the genetic alterations occuring in the cancer cells. Among these, three will be discussed: blood group antigens, uroplakins and the epidermal growth factor receptor.

5.1. Blood Group Antigens

Blood group antigens are the result of glycosylation of proteins and lipids on the surface of cells. A, B, O [H] expression of human tissues, including urothelial cells, is consistent with the blood type and secretor status of the individual. Thus, A,B, O [H] antigens are normally found throughout the urothelium.[41–44] In type A or B patients, Limas found an increase in the H antigen (which is the biochemical precursor[43]), and, in further elegant studies, Orntoft and coworkers confirmed that the glycosyltransferases responsible for generation of the subsequent A or B antigens are lost in on bladder cancer tissues.[44] Further, another blood group antigen, the LeX antigen, is normally found only on the most differentiated umbrella cells in the normal bladder; however, in bladder cancers, LeX is frequently found throughout the cell layers.[41]

5.2. Uroplakins

The most differentiated urothelial cells, the specialized umbrella cells, have a unique feature called the asymmetric unit membrane, that is thought to function during stretching of the bladder.[45] These structures include a series of proteins called uroplakins.[46] In some bladder cancers, expression of uroplakins is observed, especially at cell membranes, but not restricted to the surface.[47] In many cases this expression is focal, but in some cancers the expression is observed throughout the tumor.

5.3. Epidermal Growth Factor Receptor

Expression of the epidermal growth factor receptor (EGFr) is restricted to the basal cells in the normal urothelium.[48] The epidermal growth factor receptor (EGFr) is overexpressed in bladder cancers although the gene is not amplified.[48–51] The overexpression of EGFr correlated with earlier recurrence and shorter patient survival.[50] Overexpression of EGFr is also found in apparently normal areas of the bladder distant from the tumor, frequently described as a "field effect"[20,48]. This expression may also be useful in monitoring chemoprevention trials, since EGFr expression in bladder washings or urine cytology reflects not only expression in tumor but also in cells shed from the distant, apparently normal areas. Furthermore, since EGFr functions in signal transduction, EGFr may be a therapeutic target in bladder cancers.

Table 1. Characteristics of undifferentiated and differentiated cultured normal human urothelial cells in our in vitro model

Marker	Expression in undifferentiated urothelial cells	Expression in differentiated urothelial cells
Blood group A, B	low	high
MUC-1 mucin	low	high
Desmosomes	cytoplasmic	cell junctions
E-cadherin	cytoplasmic	cell junctions
Tight junctions	cytoplasmic	cell junctions
Epithelial resistance	low	high

After references 27,59.

6. nm23 AND DIFFERENTIAL DISPLAY REVERSE TRANSCRIPTASE POLYMERASE CHAIN REACTION

We have used an in vitro model of normal human urothelial cells that can be induced to differentiate by addition of a high calcium, serum containing mediumto study gene expression during urothelial differentiation.[27] Some of the characteristics of undifferentiated and differentiated normal human urothelial cells in this model are described in Table 1.

We have used the technique of differential display reverse transcriptase polymerase chain reaction (ddRT-PCR[52,53]) to analyze gene expression in differentiated compared to undifferentiated normal human urothelial cells. We have successfully used this technique in the past to identifiy the expression of mal in differentiated urothelial cells,[27] and have also used a similar technique to evaluate differentiation in prostate cancer cells (unpublished, data not shown). Briefly, total RNA is isolated from the target cells, reverse transcribed and amplified by PCR using relatively nonspecific short primers (10–12mers) in the presence of a radioactive nucleotide. The goal is to amplify a subset of the total RNA expressed in the cells. The resulting products are separated side-by-side on a sequencing gel, and bands revealed by autoradiography. Bands present in one RNA but lacking in another may represent genes expressed in one cell but not another.

Using this technique, we identified that the nm32-H1 gene is expressed during prostatic differentiation. The nm23 gene was originally identified because of its reduced expression in metastatic cell lines; subsequent work supported its potential role as a metastasis suppressor in various xenograft models and in some human cancers.[54] Evidence also is emerging that nm23 expression may be associated with differentiation during embryogenesis.[54]

We also evaluated the expression of nm23-H1 in urothelial differentiation. As shown in Figure 2A, nm23-H1 is found in both undifferentiated and differentiated urothelial cell RNA by northern blotting, but the differentiated cells express significantly more. GADPH serves as the loading control for this experiment. Immunostaining, shown in Figure 2B, using rabbit antibodies reactive with both nm23-H1 and H2 protein, results in staining of all urothelial cells in this normal specimen. In contrast, as shown in Figure 3, northern blotting of a panel of human bladder tumor cell line RNA indicates that all tumor cells express approximately equal amounts of nm23H1 mRNA. Similar results have been obtained by immunostaining of human bladder cancer tissues in our laboratory (unpublished, data not shown) and by other investigators.[56] Thus it is unlikely that nm23 expression is prognostic in bladder cancer. However, one caveat is that nm23 may be a gene family[54,55] and a specific member of this family may be more informative than the family as a whole.

A

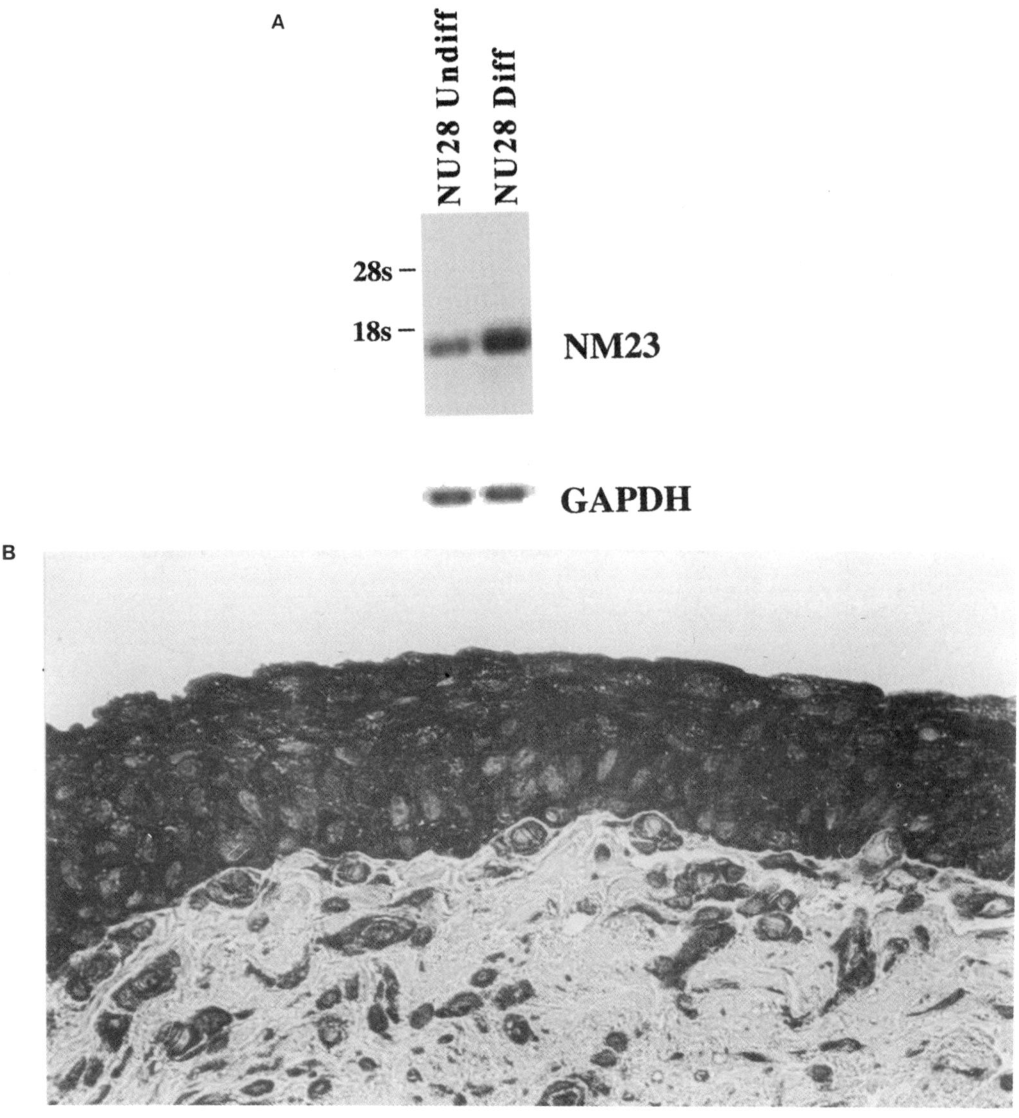

Figure 2. A,B

7. CONCLUSIONS

In this review, we have presented the current knowledge about urothelial differentiation and potential applications in bladder cancer, and, in particular, in monitoring of chemoprevention treatments. A summary is shown in Table 2.

As seen in Table 2, alterations of normal urothelial marker expression in bladder cancer can take several forms. Frequently, a marker expressed in normal urothelial cells is lost with development of cancer (e.g., gelsolin). It is also common for a marker expressed in differentiated normal urothelial cells to be disregulated and inappropriately expressed, in amount and in location, in bladder cancer (e.g., cytokeratin 20). Markers associated

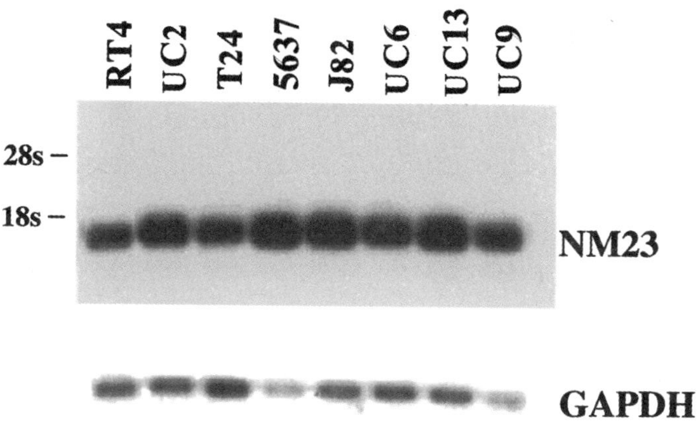

Figure 3.

Table 2. Urothelial differentiation markers and their alterations in bladder cancer

Marker	Normal expression	Expression in differentiated or undifferentiated cells	Expression in bladder cancers	References
PCNA, Ki67	basal (minimal)	undifferentiated	increased, nonbasal	14,15
G-actin	low	undifferentiated	increased	19,20
F-actin	high	differentiated	reduced	19,20
Gelsolin			reduced or lost	21
Cytokeratin 20	umbrella	differentiated	high, throughout	23,24
Desmosomes	throughout (cell junctions)	differentiated	lost	30
E-cadherin	throughout (cell junctions)	differentiated	lost	32
$\alpha 2\beta 1$ integrin	basal and intermediate (cell junctions)	both	lost	36
$\alpha 6\beta 4$ integrin	basal	undifferentiated	increased, nonbasal	38
Hemidesmosomes	basal	undifferentiated	lost	38
Blood group A, B, O	intermediate and luminal	differentiated	lost	41-44
Blood group LeX	umbrella	differentiated	increased, altered location	41
Uroplakins	umbrella	differentiated	focal	47
Epidermal growth factor receptor	basal	undifferentiated	increased, nonbasal	20, 50,51
nm23	throughout	differentiated	throughout	56, this report

with undifferentiated urothelial cells may be overexpressed in bladder cancers (e.g., $\alpha 6\beta 4$I integrin), or lost (e.g., hemidesmosomes).

Although the field is progressing, studies of urothelial differentiation lag behind the well developed and established investigations in hematologic cells and keratinocytes. With the development of new in vivo and in vitro models and further understanding of urothelial gene expression, the field should expand rapidly in the next few years to meet the challenges of chemoprevention and contribute to the new fields of differentiation and gene therapy.[57,58]

ACKNOWLEDGMENTS

Supported by grants CN-154 and RPG-96-044-03-MGO from the American Cancer Society

REFERENCES

1. Fearon E.R., Vogelstein B.: A genetic model for colrectal tumorigenesis. Cell, **61:** 759, 1990.
2. Hollingsworth R.E., Lee W-H.: Tumor suppressor genes: New prospects for cancer research. J.N.C.I., **83:** 91, 1991.
3. Knudson A.G.: Hereditary cancer, oncogenes and antioncogenes. Cancer Res., **45:** 1437, 1985.
4. Nowell P.C.: How many human cancer genes? J.N.C.I. **83:** 1061, 1991.
5. Sporn M.B.: Carcinogenesis and cancer: Different perspectives on the same disease. Cancer Res., **51:**6215, 1991.
6. Weinberg R.A.: Oncogenes, Antioncogenes, and the molecular bases of multistep carcinogenesis. Cancer Res., **49:** 3713, 1989.
7. Harris H.: The role of differentiation in the suppression of malignancy. J. Cell Sci., **97:** 5, 1990.
8. Landis S.H., Murray T., Bolden S., Wingo P.: Cancer statistics, 1998. CA: A Cancer Journal for Clinicians, **48:** 6, 1998.
9. Raghavan D., Shipley W.U., Garnick M.B., Russell P.J., Richie J.P.: Biology and management of bladder cancer. New Engl. J. Med., **322:** 1129, 1990.
10. Lotan R.: Squamous cell differentiation markers in normal, premalignant, and malignant epithelium: Effects of retinoids. J Cell Biochem.,**17F:** 167, 1993.
11. Monis B. and Dorfman H.D. : Some histochemical observations on transitional epithelium of man. J Histochem Cytochem Aug., **15 (8):**475, 1967.
12. Bane B.L., Hemstreet G.P., Rao J.Y.: Pathology and staging of bladder cancer. Semin. Oncol., **23:** 546, 1996.
13. Martin B.F. Cell replacement and differentiation in transitional epithelium: a histological and autoradiographic study of the guinea-pig bladder and ureter. J Anat., **112:** 433, 1972.
14. Okamura K., Miyake K., Koshikawa T., Asai J.: Growth fraction of transitional cell carcinomas of the bladder defined by the monoclonal antibody Ki-67. J Urol., **144:** 875, 1990.
15. Lipponen P.K., Eskelinine M.J.: Cell proliferation of transtional cell bladder tumors determined by PCNA/cyclin immunostaining and its prognostic value. Br J Cancer., **66:** 171, 1992.
16. Alberts B., Bray D., Lewis J., Roff M., Roberts K., Watson J.D.: Molecular Biology of the Cell. Garland Publishing, NY, NY , p 1127, 1989.
17. Pollard T.D., Cooper J.A.: Actin and actin-binding proteins. A critical evaluation of mechanisms and functions. Ann Rev Biochem **55:** 987, 1986.
18. Alberts B., Bray D., Lewis J., Roff M., Roberts K., Watson J.D.: Molecular Biology of the Cell. Garland Publishing, NY, NY, p 791, 1989.
19. Hemstreet G.P., Bane B.L., Liebert M., Grossman H.B., Waliszeski P., Bonner R.B., Hurst R.E., Rao J.: G-actin as a risk factor and modulatable endpoint for cancer chemoprevention trials. J. Cell Biochem., **25:** 197, 1996.
20. Rao J.Y., Hemstreet G.P., Hurst R.E., Bonner R.B., Jones P.L., Min K.W., Fradet Y.: Alterations in phenotypic biochemical markers in bladder epithelium during tumorigenesis. P.N.A.S., 90: 8287, 1993.

21. Tanaka M., Mullauer L., Ogiso Y., Fujita H., Moriya S., Furuuchi K., Harabayashi T., Shinohara N., Koyanagi T., Kuzumaki N.: Gelsolin: A candidate for suppressor of human bladder cancer. Cancer Res., **55:** 3228, 1995.
22. Moll R., Achtstatter T., Becht E., Balcarova-Stander J., Ittensohn M., Franke W.W.: Cytokeratins in normal and malignant transitional epithelium. Maintenance of expression of urothelial differentiation features in transitional cell carcinomas and bladder carcinoma cell culture lines.Am. J. Pathol., **132:**123, 1988.
23. Moll R., Lowe A., Laufer J., Franke W.W.: Cytokeratin 20 in human carcinomas. A new histodiagnostic marker detected by monoclonal antibodies. Am. J. Pathol., **140:** 427, 1992.
24. Harnden P., Allam A., Joyce A.D., Patel A., Selby P., Southgate J.: Cytokeratin 20 expression by non-invasive transitional cell carcinomas: Potential for distinguishing recurrent from non-recurrent disease. Histopathology, **27:** 169, 1995.
25. Harden P., Southgate J., Joyce A.D., Eardley I.: Cytokeratin 20 as an objective marker of urothelial dysplasia. Br. J. Urol., **78:** 870, 1996.
26. Gumbiner B.M.: Cell adhesion: The molecular basis of tissue architecture and morphogenesis. Cell, **84:** 345, 1996.
27. Liebert M., Hubbel A., Chung M., Wedemeyer G., Lomax M.I., Hegeman A., Yuan T.Y., Brozovich M., Wheelock M.J., Grossman H.B.: Expression of mal is associated with urothelial differentiation in vitro: Identification by differential display reverse transcriptase polymerase chain reaction. Differentiation, **61:** 177, 1997.
28. Green K.J., Parry D.A.D., Steinert P.M., Virata M.L.A., Wagner RM, Angst BD, Nilles LA: Structure of the human desmoplakins. J. Biol. Chem., **265:** 2603, 1990.
29. Moll R., Cowin P., Kapprell H-P., Franke W.W.: Biology of Disease. Desmosomal Proteins: New markers for identification and classification of tumors. Lab. Invest. **54:** 4, 1986.
30. Alroy J., Pauli B.U., Weinstein R.S.: Correlation between numbers of desmosomes and the aggressiveness of transitional cell carcinoma in human urinary bladder. Cancer, **47:** 104, 1981.
31. Green K.J., Stappenbeck T.S., Noguchi S., Oyasu R., Nilles L.A.: Desmoplakin expression and distribution in cultured rat bladder epithelial cells of varying tumorigenic potential. Exp. Cell Res., **193:** 134, 1991.
32. Bringuier P.P., Umbas R., Schaafsma H.E., Karthaus H.F.M., Debruyne F.M.J., Schalken J.A.: Decreased E-cadherin immunoreactivity correlates with poor survival in patients with bladder tumors. Cancer Res., **53:** 3241, 1993.
33. Frixen U.H., Behrens J., Sachs M., Eberle G., Voss B., Warda A., Lochner D., Birchmeier W.F.: E-cadherin-mediated cell-cell adhesion prevents invasiveness of human carcinoma cells. J. Cell Biol., **113:** 173, 1991.
34. Behrens J., Mareel M.M., Van Roy F.M., Birchmeier W.: Dissecting tumor cell invasion: Epithelial cells acquire invasive properties after the loss of uvomorulin-mediated cell-cell adhesion. J. Cell Biol., **108:** 2435, 1989.
35. Hynes R.O.: Integrins: Versatility, modulation and signalling in cell adhesion. Cell, **69:**11, 1992.
36. Liebert M., Washington R., Stein J., Wedemeyer G., Grossman H.B.: Expression of the VLA $\beta1$ integrin family in bladder cancer. Amer. J. Pathol., **144:** 1016, 1994.
37. Southgate J., Trejdosiewicz L.K., Hutton K.A., Kennedy W.: Expression and in vitro regulation of integrins by normal human urothelial cells. Cell Adhes. Comm., **3:** 231, 1995.
38. Liebert M., Washington R., Wedemeyer G., Carey T.E., Grossman H.B.: Loss of co-localization of $\alpha6\beta4$integrin and collagen VII in bladder cancer. Am. J. Pathol., **144:** 787, 1994.
39. Wheelock M.J., Jensen P.J.: Regulation of keratinocytes intercellular junction organization and epidermal morphogenesis by E-cadherin. J. Cell Biol., **117:** 415, 1992.
40. Carter W.G., Wayner E.A., Bouchard T.S., Kaur P.: The role of integrins $\alpha2\beta1$ and $\alpha3\beta1$ in cell-cell and cell-substrate adhesion of human epidermal cells. J. Cell Biol., **110:** 1387, 1990.
41. Sheinfeld J., Reuter V.E., Sarkis A.S., Cordon-Cardo C.: Blood group antigens in normal and neoplastic urothelium. J. Cell Biochem., **16I:** 50, 1992.
42. Summers J.L., Coon J.S., Ward R.M., Falor W.H., Miller A.W., Weinstein R.S.: Prognosis in carcinoma of the urinary bladder based upon tissue blood group ABH and Thomsen-Friedenreich antigen status and karyotope of the initial tumor. Cancer Res., **43:** 934, 1983.
43. Limas C., Lange P.: Altered reactivity for ABH antigens in transitional cell carcinomas of the urinary bladder. Cancer,. **46:** 1366, 1980.
44. Orntoft T.F., Wolf H., Watkins W.M.: Activity of human blood group ABO, Se, H, Le and X gene-encoded glycosyltransferaces in normal and malignant urothelium. Cancer Res., **48:** 4427, 1988.
45. Koss L.G.: The asymmetric unit membranes of the epithelium of the urinary bladder of the rat. Lab. Invest., **21:** 154, 1969.

46. Wu X.R., Manabe M., Yu J., Sun T.T.: Large scale purification and immunolocalization of bovine uroplakins I, II, and III. Molecular markers of urothelial differentiation. J. Biol. Chem., **265:** 19170, 1990.
47. Moll R., Wu X-R., Lin J-H., Sun T-T.: Uroplakins, specific membrane proteins of urothelial umbrella cells, as histological markers or metastatic transitional cell carcinomas. Am. J. Pathol., **147:** 1383, 1995.
48. Messing E.M., Reznikoff C.A.: Epidermal growth factor and its receptor: Markers of--and targets for--chemoprevention of bladder cancer. J. Cell Biochem. Supp .,**16I:** 56, 1992.
49. Wood D.P., Fair W.R., Chaganti R.S.K.: Evaluation of epidermal growth factor receptor DNA amplification and mRNA expression in bladder cancer. J. Urol., **147:** 274, 1992.
50. Neal D.E., Sharpes L., Smith K., Fennelly J.A., Hall R.R., Harris A.L.: The epidermal growth factor receptor and the prognosis of bladder cancer. Cancer, **65:** 1619, 1990.
51. Berger M.S., Greenfield C., Gullick W.J., Haley J., Downward J:, Neal D.E., Harris A.L., Waerfield M.D. Evaluation of epidermal growth factor receptors in bladder tumors. Br. J. Cancer.., **56:** 533, 1987.
52. Wright C., Mellon K., Johnston P., Lane D.P., Harris A.L., Horne C.H.W., Neal D.E.: Expression of mutant p53, c-erb-B2 and the epidermal growth factor receptor in transitional cell carcinoma of the human urinary bladder. Br. J. Cancer., **63:** 967, 1991.
53. Liang P., Pardee A.B.: Differential Display of eukaryotic messenger RNA by means of the polymerase chain reaction. Science, **257:** 967, 1992.
54. Liang P., Averboukh L., Pardee A.B.: Distribution and cloning of eukaryotic mRNAs by means of differential display; refinements and optimization. Nuc. Acid Res., 21: 3269, 1993.
55. MacDonald N.J., De La Rosa A., Steeg P.S.: The potential roles of nm23 in cancer metastasis and cellular differentiation. Eur. J. Cancer., **31A:** 1096, 1995.
56. Martinez R., Venturelli D., Perrotti D., Veronese K.L., Kastury K., Druck T., Huebner K, Calabretta B: Gene structure, promoter activity, chromosomal location of the DR-nm23 gene, a related member of the nm23 gene family. Cancer Res., **57:** 1180, 1997.
57. Shiina H., Igawa I., Urakami S., Shirakara H., Ishibe T.: Immunohistochemical analysis of nm23 protein in transitional cell carcinoma of the bladder. Br. J. Urol., **76:** 708, 1995.
58. Hong W.K., Lippman S.M.: Cancer chemoprevention. J. Natl. Cancer Inst. Monog., **17:** 49, 1995.
59. Scott R.E.: Differentiation, differentiation/gene therapy and cancer. Pharmacol. Ther.,. **37:** 51. 1997.
60. Liebert M., Wedemeyer G., Chang J.H.C., Stein J.A., McKeever P.E., Carey T.E., Flint A., Steplewski Z., Buchsbaum D.J., Wahl R..L., Grossman H.B.: Comparison of antigen expression on normal urothelial cells in tissue section and tissue culture. J. Urol., **144:** 1288, 1990.

35

COMPLEXITY, RETINOID-RESPONSIVE GENE NETWORKS, AND BLADDER CARCINOGENESIS

Robert E. Hurst,[1,2,3] Przemyslaw Waliszewski,[1] Miroslawa Waliszewska,[1] Rebecca B. Bonner,[1] Doris M. Benbrook,[2,4] Arindam Dar,[2,4] and George P. Hemstreet, III [1,3,5]

[1]Department of Urology
[2]Department of Biochemistry and Molecular Biology
[3]Department of Environmental Health
[4]Department of Obstetrics and Gynecology
[5]Department of Pathology
University of Oklahoma Health Sciences Center
P.O. Box 26901, Oklahoma City, Oklahoma 73190

1. ABSTRACT

Carcinogenesis involves inactivation or subversion of the normal controls of proliferation, differentiation, and apoptosis. However, these controls are robust, redundant, and interlinked at the gene expression levels, regulation of mRNA lifetimes, transcription, and recycling of proteins. One of the central systems of control of proliferation, differentiation and apoptosis is retinoid signaling. The hRARαnuclear receptor occupies a central position with respect to induction of gene transcription in that when bound to appropriate retinoid ligands, its homodimers and heterodimers with hRXRαregulate the transcription of a number of retinoid-responsive genes. These include genes in other signaling pathways, so that the whole forms a complex network. In this study we showed that simple, cause-effect interpretations in terms of hRARαgene transcription being the central regulatory event would not describe the retinoid-responsive gene network.

A set of cultured bladder-derived cells representing different stages of bladder tumorigenesis formed a model system. It consisted of 2 immortalized bladder cell lines (HUC-BC and HUC-PC), one squamous cell carcinoma cell line (SCaBER), one papilloma line (RT4), and 4 transitional cell carcinomas (TCC-Sup, 5637, T24, J82) of varying stages and grades. This set of cells were used to model the range of behaviors of bladder cancers. Relative gene expression before (constitutive) and after treatment with 10 μM all-trans- retinoic acid (aTRA) was measured for androgen and estrogen receptor; a set of

Advances in Bladder Research, edited by Baskin and Hayward.
Kluwer Academic / Plenum Publishers, New York, 1999.

genes involved with retinoid metabolism and action, hRAR α nd β, hRXR αand β CRBP, CRABP I and II; and for signaling genes that are known to be sensitive to retinoic acid, EGFR, cytokine MK, ICAM I and transglutaminase. The phenotype for inhibition of proliferation and for apoptotic response to both aTRA and the synthetic retinoid 4-HPR was determined. Transfection with a CAT-containing plasmid containing an aTRA-sensitive promoter was used to determine if the common retinoic acid responsive element (RARE)-dependent pathway for retinoid regulation of gene expression was active. Each of the genes selected is known from previous studies to react to aTRA in a certain way, either by up- or down-regulation of the message and protein.

A complex data set not readily interpretable by simple cause and effect was observed. While all cell lines expressed high levels of the mRNAs for hRXRαand β that were not altered by treatment with exogenous aTRA, constitutive and stimulated responses of the other genes varied widely among the cell lines. For example, CRABP I was not expressed by J82, T24, 5637 and RT4, but was expressed at low levels that did not change in SCaBER and at moderate levels that decreased, increased, or decreased sharply in HUC-BC, TCC-Sup and HUC-PC, respectively. The expression of hRARα, which governs the expression of many retinoid-sensitive genes, was expressed at moderate to high levels in all cell lines, but in some it was sharply upregulated (TCC-Sup, HUC-PC and J82), remained constant (5637 and HUC-BC), or was down-regulated (SCaBER, T24 and RT4). The phenotypes for inhibition of proliferation showed no obvious relationship to the expression of any single gene, but cell lines that were inhibited by aTRA (HUC-BC and TCC-Sup) were not sensitive to 4-HPR, and vice versa. One line (RT4) was insensitive to either retinoid. Transfection showed very little retinoid-stimulated transfection of the CAT reporter gene with RT4 or HUC-PC. About 2-fold enhancement transactivation was observed with SCaBER, HUC-BC, J82 and T24 cells and 3–8 fold with 5637, TCC-Sup cells. In HUC-BC, a G to T point mutation was found at position 606 of the hRARα gene. This mutation would substitute tyrosine for asparagine in a highly conserved domain.

These data indicate that retinoid signaling is probably a frequent target of inactivation in bladder carcinogenesis. However, it seems this inactivation is a random process, but the strongly interlinked nature of signaling pathways can lead to unexpected manifestations. The phenotype of a cell is probably determined more by a neural net type of calculation involving numerous inputs (risk factors) than it is by strict cause and effect, i.e. inactivation of of a particular gene may lead to widespread alterations in the activities of genes linked to it. A novel perspective based upon complexity for the investigation of carcinogenesis is presented.

2. INTRODUCTION

In recent years there has been a veritable explosion of knowledge concerning genetic changes that are associated with the process of carcinogenesis. This knowledge has been applied to develop mechanistic models of carcinogenesis, the archetype of which is the ordered model of Vogelstein and Kinzler for the development of colon cancer.[1] In this model, specific genetic mutations are associated with the morphologic progression from normal epithelium to carcinoma. Mutation in APC is first and followed by K-Ras, DCC, and p53. While this model is intrinsically satisfying and follows a system of cause and effect, it is not clear whether this model describes most colon cancers. For example, as discussed by Shackney and Shankey,[2] only 10% of colon cancers showed all four of the mutational events described by Kinzler and Vogelstein, and only 40% showed any three. It is not clear, as Richmond Prehn pointed out, whether ìthe hypothesis that cancer is usually the result of genomic mutations may be wrong. We should at least examine the alternative hypothesis, for which there is considerable evidence, that mutations do not commonly beget cancer, but rather that cancer phenotypes result from confused or aberrant patterns of normal-gene expression; the abnormal patterns are postulated to result from epigenetic mechanisms rather than from mutations[3].

The model is both ordered and deterministic, in the sense that if these errors occur, then cancer must result. However, most natural systems are more complex than deterministic. Determinate systems are special cases in which either the system is isolated or the components do not interact strongly at the level of organization being studied. Under such conditions, individual variables can be isolated and the effects of their alterations studied. In real systems, the components interact and the system becomes nonlinear and at least partially indeterminate. Predictions become statistical, and the systemís properties are dependent upon system history and environment. An excellent example is from the work of Bissell,[4] in which the extracellular matrix and its receptors, particularly the ratio of expression of cell-matrix and cell-cell adherence receptors, dictate the phenotype of mammary epithelial cells, and thus in their model system the tissue phenotype is dominant over the cellular genotype. In this chapter we review some of the mechanistic models of carcinogenesis and then re-examine the process in terms of a model based upon complexity and the interactions among exogenous environment, endogenous environment, cell proliferation, differentiation, and death and genetics. These points will be illustrated with retinoid signaling in bladder cancer.

2.1. Bladder Cancer: Mechanisms of Carcinogenesis

Bladder cancer affected an estimated 50,500 Americans and 59,200 Europeans in 1995 and 11,200 and 27,380 respectively died of the disease.[5,6] The main symptoms that lead to diagnosis are hematuria or dysuria, but approximately half of high-grade lesions have metastasized by the time symptoms of hematuria and dysuria appear.[7] Although 85% of bladder cancers are low-stage papillary tumors at initial diagnosis in the U.S., 15% of these eventually progress to dangerous high-grade disease with greatly increased risk of patient mortality, and most will recur.[8]

The normal bladder, or any other solid organ, represents a complex ecosystem of interacting epithelial and stromal cells whose growth is highly regulated, and the progressive subversion of proliferation, apoptotic and differentiation controls[9–14] leads to emergence of cells with tumorigenic phenotypes. Some of these altered phenotypes result from genotypic changes, from altered differentiation arising from a changed cytokine and stromal environment[15,16] and others from epigenetic effects driven by exogenous agents.[3,17–20] Bladder cancer seems to typically develop by a process of "field disease," frequently involving widespread histopathologic changes such as carcinoma in situ or phenotypic changes in morphologically normal urothelium[21] and carrying increased risk for progression and recurrence.[8,22,23] Clonal expansion occurs early such that the progeny of one initiated stem cell may take over large areas of bladder uroepithelium,[24] and then further promotion or additional genetic instability may lead to emergence of additional tumors.[21] That tumors with different p53 mutations can emerge during recurrence has been shown,[25] suggesting that the community of initiated cells can progress independently. The potential for recurrence seemingly is directly related to the number of partially transformed cells remaining after therapy and to the continuance of epigenetic events promoting further carcinogenesis.[26]

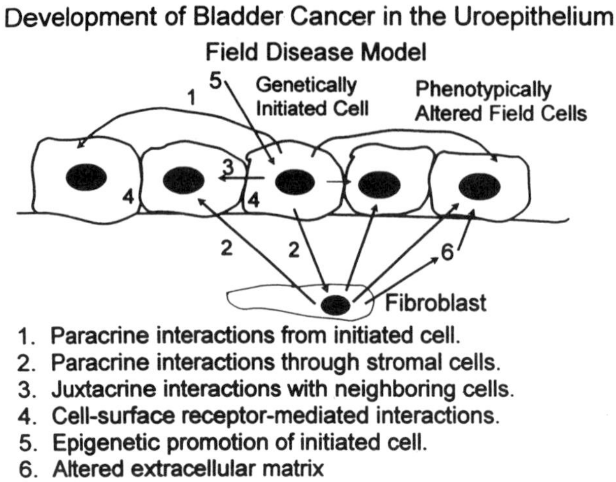

Figure 1. Illustration of biochemical communication processes that modulate cellular differentiation, proliferation, and apoptosis. These factors are significant in establishment of phenotype and in chemoprevention.

Bladder cancer appears to develop along two main tracks; a deeply invasive, high-grade form involving p53 mutations that rapidly becomes life-threatening,[25, 27] and a much less dangerous low-grade form.[22, 25, 27, 28] Although low-grade tumors are usually cured readily, by simple resection if detected early, or by BCG therapy in the case of multiple tumors, preventing recurrence and progression is pressing because approximately 15% of patients with these tumors progress to dangerous disease.[8]

Although recent research has tended to emphasize the role of mutational events in carcinogenesis, it is not clear to what extent they are the cause or the result of the process of tumorigenesis.[3] What is becoming more evident is that cancer is a disease that affects an entire cell population and may represent more complex events than genetic mutation alone, including the rates of proliferation[29] and apoptosis,[30] which themselves reflect a balance among extracellular matrix and various endocrine, paracrine, autocrine and juxtacrine means of cellular communication.[15, 16, 31] Some of the biochemical communications are illustrated in Figure 1. Certainly, a tissue has multiple and redundant controls on cellular proliferation and differentiation; thus there may be many ways for a cell population to subvert the normal controls. However, while there is undoubtedly a multiplicity of malignant genotypes, it is likely there are far fewer malignant phenotypes, As a simple example, receptor-mediated signaling can be subverted by inactivating the kinase itself, the extracellular receptor, or the ras protein that modulates it. These facts are relevant both to detection of bladder cancer and to chemoprevention. For detection, they suggest that identifying phenotypes may be much more sensitive than identifying genotypes, each of which may represent only a small fraction of total cancers. For cancer prevention, they suggest that agents that alter cell signaling may have multiple and complex actions at the tissue level in preventing further progression toward a tumorigenic phenotype.

2.2. Retinoids and Bladder Cancer

Agents that prevent further cancer development are termed chemopreventive agents. Among the best-studied chemopreventive agents are retinoic acid and its derivatives, which were earlier shown to modify expression of several genes important for cellular proliferation and differentiation.[32, 33] That these compounds regulated the pathways that are subverted in carcinogenesis provided a background for chemoprevention, a relatively new approach to cancer control, in which cellular differentiation, proliferation, or induction of apoptotic death are manipulated in transformed cells, initiated cells progressing toward malignancy, and adjacent normal cells comprising the epithelial and stromal field.[34] Retinoids therefore modulate the signaling pathways illustrated in Figure 1.

The actions of retinoids in cells are complex and involve a number of different chemical forms of retinoids, nuclear receptors that alter expression of a number of genes involved with cellular signaling, and direct effects on cytoplasmic enzymes. All-trans-retinoic acid, actively synthesized in various tissues as a result of enzymatic transformation of vitamin A and β–carotene[35, 36] was found to be one of the most biologically active retinoids in exerting an antitumorigenic effect in a variety of transformed epithelial cells.[33, 34] Retinol is bound in the cytoplasm by a high-affinity binding protein, CRBP, while all-trans-retinoic acid and its derivatives are bound by CRABP I and CRABP II[37–40] These latter two proteins seem to modulate the activity of aTRA, at least in part, by inhibiting interaction with the nuclear receptors. The bound retinoids also are subject to metabolic transformation to more polar compounds, and it is not clear the extent that these metabolites are biologically active.[41, 42] Although many details of retinoid action in altering gene expression or inducing differentiation and apoptosis are becoming clear, the interplay between nuclear receptors and carrier protein-bound retinoids acting in the cytoplasm is complex.[43]

The signaling pathway involves binding to specific high-affinity nuclear receptors.[32,44–47] The retinoid receptors are classified into the RAR and RXR subfamilies, each manifesting α,β, and γsubtypes.[32, 45–49] The highest affinity ligands for the RAR and RXR subfamilies are aTRAand 9-cis retinoic acids, respectively.[50] The RAR and RXR proteins form heterodimers that bind much more efficiently to DNA retinoid-responsive elements than do RAR-RAR homodimers.[51,52] The ligand-receptor complexes bind to retinoid-responsive elements (RARE) located in the 5' flanking regions of gene promoters and modulate gene transcription in a ligand- and dose-dependent manner by acting as transcriptional activators.[45,53] Expression of the retinoid-responsive genes is believed to determine cellular sensitivity to retinoids.[32] The RXR receptors also form homodimers with other receptors in the steroid-retinoid superfamily, a step which has been proposed to be necessary for transactivation of the other receptor.[54] In addition, retinoids can also interact with the transcription factor AP-1[55] and inhibit its activity. When this occurs, retinoids act as transrepressors and serve to inhibit cell proliferation

One of the complexities of retinoid research is that many retinoids have multiple actions. In recent years, a number of more receptor-selective derivatives have been synthesized.[54, 56, 57] These more selective retinoids seem to show lower toxicity and produce more limited biological responses than the broad-spectrum parent compounds. In HL-60 cells, activation of the RARα and RXRα receptors are reported specifically to induce differentiation and apoptosis respectively.[58] The synthetic retinoid N-(4-hydroxyphenyl)retinamide (4-HPR) was synthesized a number of years ago during studies seeking retinoid derivatives that were less toxic that aTRA[59] and is reported to be specific in inducing apoptosis. In neural crest tumors, 4-HPR was cytostatic at an order of magnitude lower concentration that was aTRA, and, unlike aTRA did not induce differentiation but did induce apoptosis.[60] In leukemic cells, 4-HPR also induced apoptosis, but apparently not differentiation, by mechanisms that seemed to act independently of p53 and bcl-2.[61] However, in spite of this alleged specificity, careful examination showed that pathways were complex and redundant and that the ìselectiveî retinoids often had unexpected actions. Moreover, refractoriness was shown to be complex and occur even when retinoid signaling appeared to be functional.[62] Retinoids also are capable of blocking tumor promoter-induced transformation, apparently by blocking AP-1 action rather than by acting through RARE.[56, 63, 64] In addition, the synthetic retinoid 4-HPR also functions through receptor-independent mechanisms.[65]

Retinoids have been used successfully in a number of animal experimental carcinogenesis experiments to inhibit the onset of tumor. The retinoid 4-HPR has shown strong activity in chemoprevention of bladder cancer in mice and rats[66–68] with lower toxicity than is seen with aTRA. In human clinical trials, however, retinoids have been disappointing, though yielding positive results.[69–71] The animal and human studies appear to be in conflict except that in animals the retinoid was administered before appearance of the first tumor, while in human clinical trials, the drug was administered after appearance of the first tumor. Thus, in the human clinical trials, it is likely that more malignant phenotypes were present that were not inhibited or killed by the retinoid. Inclusion of a few subjects who were resistant to the drug would quickly destroy the statistical significance in response of drug and placebo groups unless the sample size were very large. In support of this hypothesis we present data showing that a number of bladder cancer and premalignant cell lines were resistant to aTRAand 4-HPR. Moreover, the finding that time to first recurrence was not altered by etretinate[70] or megavitamin[72] treatment, but that subsequent recurrences were inhibited also supports the hypothesis that more advanced phenotypes may also be retinoid-resistant.

3. EXPERIMENTAL

3.1. Cell Culture, Retinoid Stimulation, and RNA Isolation

Human embryonal cell line FHS 738BL, human urinary bladder papilloma-derived cell line, RT4, and human urinary bladder carcinoma-derived cell lines HT 1376, J82, SCaBER, T24, TCC-Sup, and 5637 were obtained from the American Type Culture Collection. SV40-immortalized human uroepithelial cell lines HUC-BC and HUC-PC were obtained from Dr. C. Reznikoff.[73]

Initially, we attempted to perform the these studies in serum-free medium, but only J82 cells grew well. Range-finding experiments with 10nM, 100 nM, 1μM, and 10 μ M were performed, which showed that inhibition was only significant at 10 μ M. Hence, 2,000 cells of each line were seeded into three 96-well plates with 10% fetal bovine serum-containing medium and allowed to attach for 24 hrs. Thirty wells in each plate were controls without added retinoid, two were blanks, and the remaining 64 were treated with all-trans retinoic or 13-cis retinoic acid (Sigma, St. Louis, MO) with daily changes of medium to ensure a rougly constant concentration of retinoid. The media recommended by the American Type Culture Collection (ATCC) were used; MEM for TCC-Sup, J-82 and SCaBER, McCoyís 5A for T-24, RPMI for 5637, and Hamís F-10 for HUC-PC, HUC-BC and RT-4. The concentrations of all-trans retinoic acid and vitamin A in the FBS used to supplement the medium were checked by HPLC[42] and were below the detection limit of 10 nM. Retinoids were kept at -20°C, and dilutions were kept on ice in foil-wrapped tubes. To minimize photolysis, lights in the room and hood were switched off when retinoids were pipetted. After 6 days the number of cells was measured by the 3-(4,5-dimethylthiazol-2-yl)-2,5-diphenyltetrazolium bromide (MTT) assay.[74] Because the data were found to be normally distributed by the Shapiro-Wilk test, the means of growth relative to control were compared by the Student t-test.

To isolate total RNA for gene expression studies, cells were cultured with or without 10μM all-trans retinoic acid for 48 hrs in Phenol Red-free medium,[75] washed three times with cold PBS, and treated with 4M guanidinium thiocyanate according to Chomczynski and Sacchi.[76] The RNA quality was checked by electrophoresis.

3.2. Transactivation of CAT Gene under RARE Control and CAT Assay

The pBLCAT2 reporter plasmid containing the chloramphenicol acetyl transferase (CAT) gene linked to a thymidine kinase (tk) promoter on the 5' end to allow for a low basal activity and containing the SV40 polyadenylation signal attached to the 3' end of the CAT gene to allow stabilization of the message was used as a control for the basal level of activity. The same plasmid containing the retinoic acid response element (RARE) from the RARβ gene (RARE-tk-CAT) was used to test for functionality of retinoic acid signaling. The plasmids were transfected by the calcium phosphate method.[77] CAT activity was measured with a CAT ELISA kit (Boehringer-Mannheim, Indianapolis, IN) after stimulation with 10 μM all-trans retinoic acid in comparison to the control.

3.3. Ligand-Binding Domain Probes

The EcoRI fragments of hRARα were prepared from the full-coding sequence cDNA, generously provided by Magnus Pfahl to one of us (DMB). The RARβ EcoRI/BamHI fragment was isolated from the full-length cDNA.[49] Labeling was performed with the Pharmacia random priming kit.

3.4. RT-PCR

Two μg total RNA was digested with DNase I to eliminate artifacts from genomic sequences and processed to first strand cDNA by reverse transcription using random primers (Boehringer-Mannheim, Indianapolis, IN). Conditions for RT-PCR for each cDNA were optimized, but most conditions used thirty-two cycles of PCR at 96°C for 1 min, 52°C for 1 min and 72° C for 2.5 min. The primers concentration was 150 ng/μl. The PCR products (25 μl/lane) were separated on 1.5% agarose gel. Identities of the bands were confirmed by: (a) a single band of the expected size was observed, (b) restriction enzyme digestion to yield the expected size products and (c) in the case of hRARα and hRARβ Southern blot analysis with independent probes. With the exception of the primers for EGFR, which was purchased from Stratagene, the other primers were synthesized at the local Oklahoma Center for Molecular Medicine core facility. The sequences and the size of the expected products are listed in Table 1.

3.5. Relative RT-PCR

To compare constitutive levels of expression of retinoid-responsive genes with levels found subsequent to stimulation, and to compare the levels of expression of the cell lines among each other, the technique of Relative RT-PCR was used. This technique normalizes the amounts of cDNA to the concentration of a common housekeeping gene, in this case β-actin, which is not affected by retinoid treatment.[78] The cDNA samples were amplified for β-actin using conditions optimized for its amplification. Amplification for 21 cycles produced bands that were within the linear range of the PCR procedure, and under such conditions the band densities were proportional to the relative concentrations of β-actin cDNA. The band densities were used to normalize the amount of cDNA in each

Table 1. Sequences of primers made locally for amplification of specific retinoid-responsive genes

Gene	Sense Primer (5'-3')	Antisense Primer	Size (bp)
HRARα	TTGTCTGTCAGGACAAGTCC	CCACTTCAAAGCACTTCTGC	188
HRARβ	Same as hRARα sense primer	TCGTCTGCCAGGACAAATCA	189
HRXRα	CAGGCAAGCACTATGGAGTGT	TGGCCAGGCACTTCTGGTA	165
hRXRα	CTTTGTGCAATCTGCGGGGAC	TGGCCAGGCACTTCTGATATC	192
CRBP I	GTCGACTTCACTGGGTACTGGA	TTGAATACTTGCTTGCAGACCACA	441
CRABP I	CGGCACCTGGAAGATGCG CA	CCACGTCATCGGCGCCAAAC	370
CRABP II	CCCAACTTCTCTGGCAAC TGGA	CTCTCGGACGTAGACCCTGGT	410
hAR	CTGATCTGTGGAGATGAAGC	TTCCGAAGACGACAAGATGG	191
hER	TGCAGTGTGCAATGACTATGC	CACTTCGTAGCATTTGCGG	205
β-Actin	GCGCTCGTCGTCGACAACGGC	ATGGGGTAGCTCGTGCCGTAGCAGT	220
MK	CTGCAACTGGAAGAAGGA	TTCCCTTCCCTTTCTTGG	206
TG	GGGCGAACCACCTGAACAAA	GGTCACTACCTAGCATGTTGT	300

sample. If 8 cell lines were to be compared before and after stimulation with retinoic acid, this meant normalizing 16 cDNA samples. Once normalized, however, the concentrations of other cDNAs could be compared to each other as well both before and after stimulation. Fourteen different retinoic acid-responsive genes were assayed in this way. The amplification of each cDNA was performed under conditions optimal for the particular cDNA. For example, hRARa yielded bands within the linear range of the ethidium bromide staining method after 33 amplifications. All amplifications were performed in duplicate, and a separate aliquot was always amplified for β--actin to ensure that the normalization was correct.

3.6. Sequencing of hRARα DNA-Binding Domain cDNA

The hRARαPCR products from all the cell lines were ligated into pCR vector using the TA-cloning kit (Invitrogen, San Diego, CA) and sequenced using the Sequanase kit (Amersham, Arlington Heights, IL) with Sp6 and T7 primers.

4. RESULTS AND DISCUSSION

Figure 2 displays the proliferation response of the different cell lines to both all-trans-retinoic acid and 4-HPR. Three phenotypes appear to be present. One phenotype, comprised of HUC-BC and TCC-Sup, is inhibited by all-trans-retinoic acid but not by 4-HPR. The HUC-PC, SCaBER, 5637, T24 and J82 lines are inhibited by 4-HPR, but not by all-trans-retinoic acid. The third phenotype is exemplified by RT-4 and is resistant to both retinoids. It is possible that the SCaBER cells form a fourth phenotype that is inhibited by both retinoids. These two retinoids function by different mechanisms; aTRA functioning through the set of receptors and 4-HPR acting by mechanisms that are independent of the classical system of receptors.[65] The difference in phenotypic response to inhibition of proliferation emphasizes this difference.

We next asked the questions of what was the status of retinoid signaling in bladder cancer cells and whether there any common patterns of gene expression could be identified that would allow the phenotype of retinoid response to be predicted a priori for bladder cancer cells. Functionality of retinoid signaling in these cell lines was assessed by transient transfection with a plasmid containing a RARE, thymidine kinase promoter, and

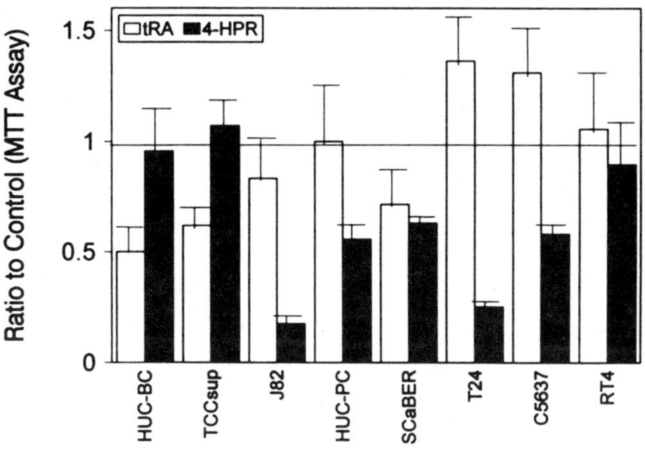

Figure 2. Inhibition of proliferation of bladder uroepithelial cell lines by 10 μM atRA and 4-HPR.

the chloroamphenicol acetyl transferase gene. The results of these studies are shown in Figure 3. RARE signaling was active in all of the cell lines except for the HUC-PC and RT-4 lines.

To assess gene expression, the technique of Relative PCR was used rapidly to assess relative levels of constitutive expression and expression subsequent to stimulation with retinoid compounds. Figure 4 illustrates several of these analyses, which are all summarized in Table 2. Each of the PCR bands was characterized by several techniques to demonstrate that the target cDNA was amplified. First was the expected size. Second was the pattern of degradation with specific nucleases, and third was to Southern blot if specific independent probes were available. Of particular significance is that hRARβ is upregulated weakly in only two cell lines. The list comprises genes encoding the receptors, including both hRAR and hRXR families; elements in other signaling pathways such as EGFR, the angiogenic MK cytokine, androgen receptor (hAR), estrogen receptor (hER) and ICAM-I; transglutaminase, a gene involved with apoptosis and differentiation of a number of epithelial cell types, and CRBP I, and CRABP I and II, three proteins that bind retinol and retinoic acid respectively. All the cell lines expressed hRARα and hRXRα

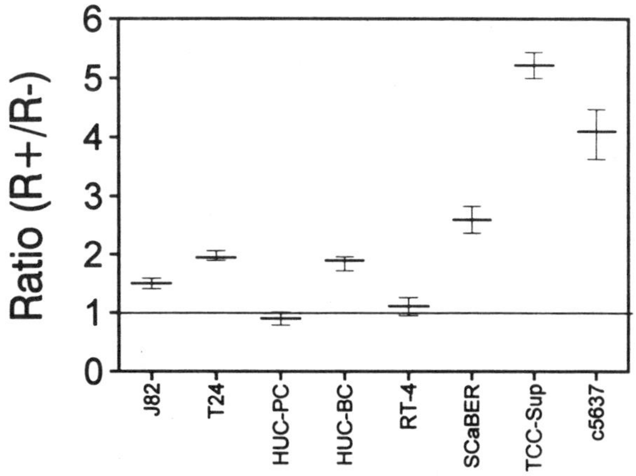

Figure 3. Activity of retinoid signaling assessed with a plasmid containing a RARE element and the CAT gene. The ordinate shows the ratio of CAT activity in the presence and absence of 10 μM atRA. Two cell lines are inactive.

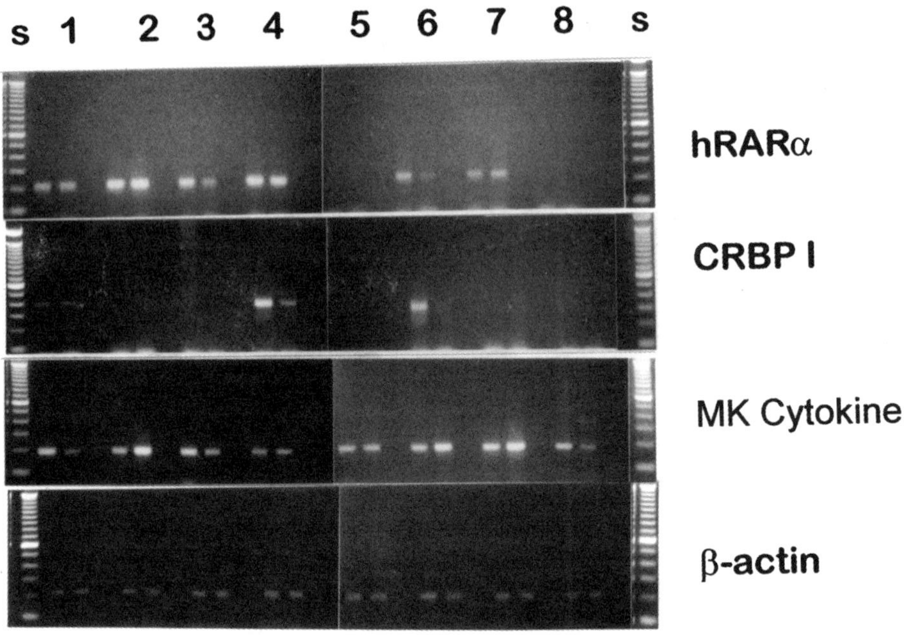

Figure 4. Illustration of Relative PCR anlyses of expression of retinoid-dependent genes in the absence and presence of 10 μM atRA for 48 hours. s = molecular weight standards. Lanes present paired results of constitutive and expression after treatment. Lane 1 = HUC-BC, Lane 2 = TCCSup, Lane 3 = HUC-PC, Lane 4 = J82, Lane 5 = SCaBER, Lane 6 = T24, Lane 7 = 5637, Lane 8 = RT4.

Table 2. Constitutive (C) and stimulated expression (T) of mRNAs of selected retinoic acid receptor genes and binding proteins as determined by Relative RT-PCR. With the exception of cytokine MK, cells would be expected to express all genes. With the exception of EGFR, all should be upregulated by retinoic acid treatment

	HUC-BC		TCC-Sup		HUC-PC		J-82		SCa-BER		T-24		5637		RT-4	
	C	T	C	T	C	T	C	T	C	T	C	T	C	T	C	T
HRARα	M	M	M	H	M	H	H	H	H	L	H	L	M	M	H	L
HRARβ	U	U	U	U	U	M	U	U	U	U	U	L	U	U	U	U
HRXRα	H	H	H	H	H	H	H	H	H	H	H	H	H	H	H	H
HRXRβ	H	H	H	H	H	H	H	H	H	H	H	H	H	H	H	H
HRARγ	N	N	L	H	N	N	L	H	N	N	N	N	N	N	L	H
CRBP I	U	U	U	U	H	L	U	U	L	L	U	U	H	L	U	U
CRABP I	M	L	M	H	M	L	U	U	L	L	U	U	U	U	U	U
CRABP II	L	H	M	H	M	L	L	L	H	H	L	L	H	H	H	H
hAR	U	N	L	H	L	M	U	N	U	N	U	N	U	N	U	N
hER	U	N	U	N	U	N	U	N	U	N	U	N	L	L	U	N
TGase	H	L	L	H	L	M	L	H	L	H	H	H	M	M	U	U
EGFR	H	H	H	L	H	L	H	L	H	H	M	L	M	L	M	L
ICAM-I	M	H	M	L	H	M	M	H	M	H	H	M	H	L	U	U
MK cytokine	L	L	H	H	M	L	M	H	M	H	H	L	H	M	U	U

U = undetectable at limits of assay; L = low: a faint band was evident; M = moderate: a distinct band was present; H = high: an intense band was present. N = not analyzed.

constitutively, but otherwise there was little commonality in the pattern of gene expression. With the exception of cytokine MK, all of these genes would be expected to be expressed, and of those, all except EGFR should be upregulated by retinoid treatment.

The two HUC cell lines showed a single band in the position expected from CRABP I cDNA. TCC-Sup also showed fainter bands above and below the expected product while T24 and SCaBER yielded only the additional bands. The presence of these bands was reproducible with independent cDNA samples but the bands were not further identified. The hRARβ receptor was not expressed constitutively in any cell line in amounts detectable by Northern analysis or RT-PCR, and only in T24 and HUC-PC was it expressed subsequent to retinoid stimulation. Expression of hRARγ was measured in three lines and could not explain the phenotypes because it was expressed in both resistant and sensitive cell lines. The complexity is further illustrated by the comodulation of sex steroid and retinoid signaling. Expression of the binding proteins varied; all the cell lines expressed CRABP II at some level, several did not express CRABP I, and only 3 expressed CRBP I. In general, wide differences of expression were seen. The papilloma-derived RT-4 line was particularly aberrant in comparison to the other cell lines and failed to express many of the genes that were expressed by the other lines. Overall, no correlation could be found between modulation of hRARα mRNA and expression of the other retinoid-responsive genes studied, between exspression of the entire hierarchy of retinoid-responsive genes including hRARα and the phenotypic response of cells, and between retinoid signal transduction as demonstrated by CAT assay and modification of expression of retinoid-responsive genes. The complexity of these systems of signaling is seen in the sex steroid receptor genes, hAR (androgen receptor) and hER (estrogen receptor). Both contain a RARE-element. Yet, most cells failed to express these receptors, and in one, the expected up-regulation is not seen.

Because gene mutation seems to be a common means of inactivating signaling pathways, and because hRARα plays a central role in upregulating many other retinoid-responsive genes, we sequenced the amplimers of hRARα for all the cell lines. A G toT point mutation at position 48 of the cloned sequence, equivalent to position 606 of the hRARα gene was found. This mutation would substitute tyrosine for asparagine in a highly conserved domain. The mutation was found in a second, independently-produced amplimer, confirming that it was not due to a mutation induced by PCR.

Our data indicated that a complex pattern of inactivation or dysregulation of retinoid signaling has occurred in most of the cell lines, including the immortalized lines. In two of the lines, RARE signaling was absent as tested by transactivation of a CAT gene driven by a RARE. The system of receptors activated by retinoic acid is active and functional in the other lines. In the HUC-PC line the lack of signaling could be attributed to a mutation in the hRARα gene. Assuming this mutation in a highly conserved region of the molecule would inactivate RARE transcriptional activation in general, the resistance of this line to all-trans-retinoic acid is explicable. However, no mutation was seen in the RT4 line, indicating that the resistant phenotype can be produced by more than a single mechanism. Interestingly, this papilloma cell line displayed numerous aberrations in expression of genes encoding signaling and retinoid-binding proteins.

Only two of the cell lines showed significant inhibition of proliferation by 10 μM retinoic acid, while five were inhibited by 4-HPR and one, RT4, was resistent to both. Interestingly, cell lines that were inhibited by 4-HPR were not inhibited by aTRA and vice-versa. In two of the cell lines the RARE signaling system of transcription factors was inactive, but even in those cell lines aTRA was capable of altering gene expression. This finding confirms that retinoids can act by mechanisms that are not dependent upon RARE,[64] some of which involve the AP-1 transcription factor.[56,63]

Resistance or sensitivity to retinoids does not seem to be predicted from any of the main genes implicated in their action, namely the receptors or the binding proteins. Most likely this is the result of the dysregulation of gene expression that seems to have occurred in these cells. However, more detailed studies of protein levels will be required to confirm this finding. Refractoriness in the presence of functional receptors has been reported.[62] Examining several cell lines displays a fuller range of the effects of carcinogenic transformation than does examination of a single cell type in depth. The wide range of behavior seen in Table 2 demonstrates clearly that the system responds nonlinearly. In other words, knowledge of the expression of a variety of retinoic acid-responsive genes that supposedly are intimately involved in producing the response of cells to all-trans-retinoic acid provides little ability to predict the sensitivity of the cell line to all-trans-retinoic acid. While one could argue that we simply have not identified the proper set of genes, another explanation should be considered as well. It is distinctly possible that retinoid action, like the weather, represents a complex system. The role of complexity in biological systems is under appreciated, probably because we lack effective tools and theories for studying complex systems.

Components of complex systems cannot be isolated sufficiently to manipulate each variable individually, and such systems are inherently nonlinear, thereby making it difficult to assess causality in the conventional sense. Complexity is inherent to retinoid signaling because of hierarchical links to other signaling pathways, including steroid receptors[79] and others[80]. This complexity is not restricted to bladder-derived cells. The entire hRXR and hRAR homo- and heterodimer system displays both specificity and redundancy,[81–83] and even the activity of retinoids tailored to activate specific receptor systems cannot predict the overall response of lung cancer cells to retinoids.[84] Thus, the phenotypic response will not be readily predictable from a knowledge of gene expression.

An organism is a complex system, not because it contains many components, but because they interact. One assumption underlying much of science is that the components of a system can be isolated and that the operation of the system as a whole can be understood in terms of the functions of the parts. In isolating a system, causality can be deduced; experimental manipulation of the isolated system (cause) will reveal its workings in terms of changes in the isolated system (effect). The assumption that a particular effect always results from a specific cause is often not stated or that other causes, which may not be under investigation, may also produce the effect or may modify it in some fundamental way, often are not stated as well. Great lengths are taken to approximate this causal model, often by simplifying the system. Mechanisms are investigated using only a single, highly inbred strain of animal such as the Sprague-Dawley rat or a single cell line such as the F-9 teratocarcinoma used to investigate retinoid action. Diets are standardized as are cell culture media. Cells are cultured as monolayers on plastic rather than in more complex architectures. While this approach has helped investigators to replicate each others results, it is not clear that it always has been as successful in reaching the goals of understanding biological systems.

Considerable simplification may be achieved by considering phenotypes rather than genotypes. Multiple genotypes can result in a single phenotype. This can be illustrated nicely with a study to identify a phenotype that can predict recurrence of bladder cancer.[26] Previous work in the laboratories of Hemstreet and coworkers have identified changes in the cytoskeleton as representing a phenotype that seems to be common to many clinical tumors.[20, 21, 85–89] The carcinogenic process seems to shift the equilibrium from filamentous actin (F-actin) to more monomeric globular actin (G-actin). The position of the equilibrium reflects the action of promoters, which shift the equilibrium toward G-actin, or reti-

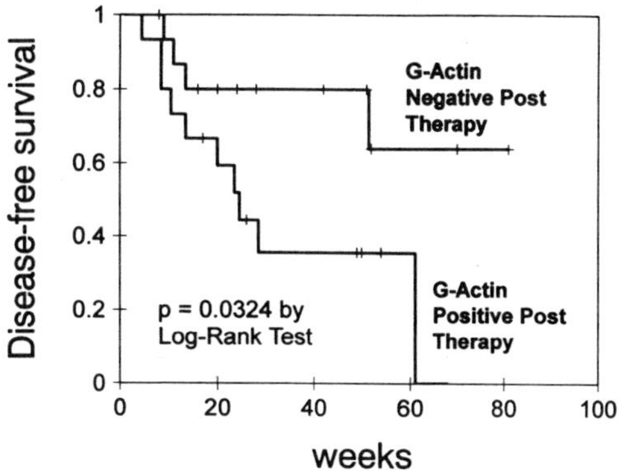

Figure 5. Kaplan-Meyer analysis of recurrence by status of G-actin after completion of therapy. The two curves were compared by the log-rank test. Tick marks indicate right-censored data. The difference between the two survival curves was significant (p=0.0324) by the log-rank test..

noids, which shift the equilibrium towards F-actin.[20, 21, 87,89] The phenotype of the cells of the uroepithelium reflects the risk of having or developing bladder cancer.[88] The alteration of the exfoliated bladder uroepithelial cells from a high F-actin phenotype to a predominately low F-actin (high G-actin) phenotype reflects the interactions illustrated in Figure 1. While transformed cells generally are of the low F-actin phenotype, the fraction of such cells found in urine far exceeds the number of cancer cells in a urine, and so it is highly likely that the presence of a significant proportion of low F-actin cells reflects alterations in the normal phenotype reflecting the presence of residual cancer cells or advanced lesions that then act to alter the cytoplasmic actin equilibrium as do promoters in cell culture.[89] If this is the case, then the presence of such cells subsequent to therapy should be a strong risk factor for recurrence.

The use of the actin phenotype was tested in a study of 25 bladder cancer patients being monitored for recurrence.[26] In this study, patients were subjected to several different therapies including transurethral resection (TUR), BCG immunotherapy, combined TUR-BCG, and DMSO chemoprevention. In all cases, the single most predictive prognostic test was the identification of the low F-actin phenotype as the dominant phenotype within the bladder immediately after completion of therapy, as is shown in Figure 5. Every patient who was positive for the low F-actin phenotype at the completion of therapy recurred. In this study, patients were recycled upon recurrence, so some patients had several recurrences. Even for these high-risk patients, when the F/G-actin phenotype was normalized, they tended not to recur.[26]

Complexity is probably the usual case in biology rather than the exception. While the genetic model has tended to emphasize the system of genes that control the cell cycle, DNA repair and apoptosis, particularly p53 and associated genes. Perhaps it is because this one gene is so central to so many different and intersecting pathways that it is so commonly mutated. Yet, as reviewed by Shackney and Shankey, this model seems active in only a fraction of cancers.[2] Retinoid signaling, which seems to govern equally central pathways involving expression of genes crucial for growth and growth arrest, seems to be dysregulated in bladder cancer by a number of mechanisms of which mutation may prove to be a minority. Genes can be silenced by methylation[90] and by other mechanisms. The complex interactions among linked signaling pathways often can lead to effects that are not easily understood in terms of strict cause and effect.

President Nixon declared a War on Cancer in 1972, in part because his scientific advisors told him that such a "war" could be won; we were on the brink of discovering the causes, and hence cures, of cancer. Expenditures of a very few billion dollars would lead to the conquest of this disease. Although some remarkable advances have been achieved in treating a few cancers, survival rates are little changed from 1972, in spite of 25 years of "war" by a caste of professional scientists supported by expenditures of many billions of dollars. Yet, most of the improvement can be ascribed to prevention and to combined use of drugs, many of which were in the pharmacopoiea prior to 1972 and not to new agents. Moreover, we still are on the brink of discovering the causes of cancer.

Is it possible that the failure to win this "war" is due to an inadequate strategy? What if the causal model of science itself were incomplete? What if in attempting to isolate the system to produce causality the result was instead to obscure those very features of the system that produced cancer? John Cairns proposed the "escape from control hypothesis" many years ago.[91] The essential tenets of this hypothesis are that cancer is the result of the failure of the systems of multiple, redundant controls operating to regulate the processes of cellular proliferation, differentiation, and death. This hypothesis appears in slightly different forms, such as that "subversion" of the set of controls is at the root of carcinogenesis.[12] Mutation is only one of several mechanisms by which cells can escape control. If these hypotheses are correct, then a priori the causal model is limiting. Proponents of this approach argue that once the workings of the parts are discovered, then the overall regulation can be investigated. Intermediary metabolism is an example of how this worked. The elements were discovered between about 1940 and 1970, while subsequent research has focused upon its integration. While this undoubtedly is true, what is not clear is whether an alternate approach might not have brought us to the same point much sooner.

While many cancer biologists seem to believe that mechanistic (i.e. causal) science is the only worthwhile pursuit, ecology and epidemiology represent two examples of sciences that posit that the interactions of the elements of the system are at the core of the effects being observed. After all, the rise and fall of populations of different species within an ecosystem cannot be understood by the behavior of any single species. Diseases may have more than one cause, or multiple risk factors contribute to the appearance of disease. This broader "systems approach" may be better suited to understanding complex systems than is the causal model. After all, can cells be considered in isolation from their milieu representing their own ecosystem? We can ask whether our mechanistic models may simplify many of these interactions to the point that the findings may be irrelevant to the real world of cancer biology.

The neural network provides one means of understanding and modeling complex systems. Figure 6 illustrates a hypothetical network that integrates a number of inputs that could determine the presence of abnormal phenotypes. A neural net functions analogously to a neuron. If the strength of the inputs exceeds some critical threshold level specified by the sum of the inputs, denoted by a_i modulated by a nonlinear transfer function and the weights of all input connections, then the jth neural element will transmit a signal of strength a_j to all elements to which it is attached. In general, the nonlinear transfer function generally means that a_j is zero or 1. These elements actually represent genes, and factors such as genetic susceptibility will alter the weights. Different intrinsic weights could be ascribed to different isoforms to reflect different cancer susceptibility resulting from, for example, slow or fast acetylators.[92] The system carries its memory in these weights, which are altered by experience and exposure through a feedback system indicated by the arrows from the output (phenotype) back to the neural elements.

Whether such more realistic complex models will prove useful will only be determined through further research. They nonetheless can provide a starting point for considering the complex interactions of biological systems *a priori*, and as mechanistic studies elucidate the workings of the internal machinery, more precise modeling can substitute for descriptive models developed on the basis broad studies.

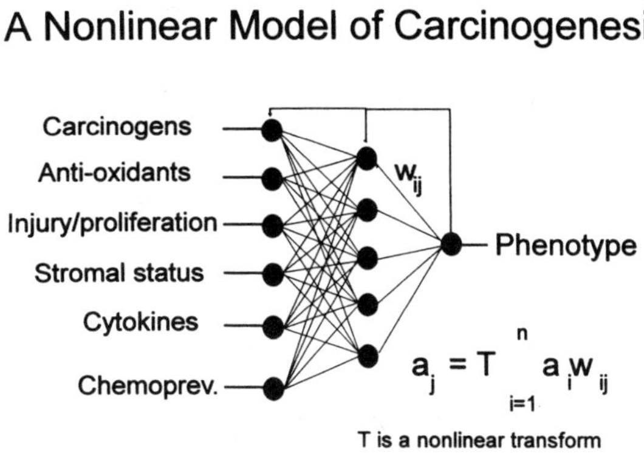

Figure 6. Hypothetical neural network model of carcinogenesis integrating environmental and genetic components. See text for detailed explanation.

ACKNOWLEDGMENTS

The authors wish to thank the many people in their laboratories who contributed to these studies and to the organizers of this conference for providing an exciting venue for the interchange of scientific information.

REFERENCES

1. Kinzler, K. W. and Vogelstein, B.: Lessons from hereditary colorectal cancer. Cell, **87:** 159, 1996.
2. Shackney, S. E. and Shankey, T. V.: Common patterns of genetic evolution in human solid tumors. Cytometry, **29:** 1, 1997.
3. Prehn, R. T.: Cancers beget mutations versus mutations beget cancers. Cancer Res., **54:** 5296, 1994.
4. Weaver, V. M., Petersen, O. W., Wang, F., Larabell, C. A., Briand, P., Damsky, C., and Bissell, M. J.: Reversion of the malignant phenotype of human breast cells in three-dimensional culture and in vivo by integrin blocking antibodies. J. Cell Biol., **137:** 231, 1997.
5. Wingo, P. A., Tong, T., and Bolden, S.: Cancer Statistics, 1995. CA Cancer J. Clin., **45:** 8, 1995.
6. Whelan, S. L. and Ferlay, J.: Cancer Incidence in Five Continents. Age-specific and standardized incidence rates. IARC Sci. Publ. 178, 1992.
7. Cartwright, R. A.: Screening workers exposed to suspect bladder carcinogens. J. Occup. Med., **28:** 1017, 1986.
8. Heney, N. M., Ahmed, S., Flanagan, M. J., Frable, W., Corder, M. P., Hafermann, M. D., and Hawkins, I. R.: Superficial bladder cancer: progression and recurrence. J. Urol., **130:** 1083, 1983.
9. Weinberg, R.: Oncogenes, antioncogenes, and the molecular bases of multistep carcinogenesis. Cancer Res., **49:** 3713, 1989.
10. Pienta, K., Partin, A., and Coffey, D. S.: Cancer as a disease of DNA organization and dynamic cell structure. Cancer Res., **49:** 2525, 1989.
11. Tzen, C., Estervig, D. N., Minoo, P., Filipak, M., Maercklein, P., Hoerl, B., and Scott, R.: Differentiation, cancer, and anticancer activity. Biochem. Cell Biol., **66:** 478, 1988.
12. Heldin, C., Betscholz, C., Claesson-Welsh, l., and Westermark, B.: Subversion of growth regulatory pathways in malignant transformation. Biochim. Biophys. Acta, **907:** 219, 1987.
13. Couture, J. and Hansen, M.: Recessive genes in tumorigenesis. Cancer Bull., **43:** 41, 1991.
14. Kastan, M. B., Onyekwere, O., Sidransky, D., Vogelstein, B., and Craig, R. W.: Participation of p53 protein in the cellular response to DNA damage. Cancer Res., **51:** 6304, 1991.
15. Ruoslahti, E. and Yamaguchi, Y.: Proteoglycans as modulators of growth factors. Cell, **64:** 867, 1991.
16. Nathan, C. and Sporn, M.: Cytokines in context. J. Cell Biol. **113:** 981, 1991.
17. Harris, C. C.: Chemical and physical carcinogenesis: advances and perspectives for the 1990s. Cancer Res., **51:** 5023s, 1991.
18. Trosko, J. E., Chang, C. C., Madhukar, B. V., and Oh, S. Y.: Modulators of gap junction function: the scientific basis of epigenetic toxicology. In Vitro Toxicology, **3:** 9, 1990.

19. Cuthill, S.: Cellular epigenetics and the origin of cancer. BioEssays, **16:** 393, 1994.
20. Hemstreet, G. P., Rao, J. Y., Hurst, R. E., Bonner, R. B., Jones, P. L., Vaidya, A. M., Fradet, Y., Moon, R. C., and Kelloff, G. J.: Intermediate endpoint biomarkers for chemoprevention. J. Cell. Biochem., Suppl. **16I:** 93, 1992.
21. Rao, J. Y., Hemstreet, G. P., Hurst, R. E., Bonner, R. B., Jones, P. L., Min, K. W., and Fradet, Y.: Alterations in phenotypic biochemical markers in bladder epithelium during tumorigenesis. Proc. Natl. Acad. Sci. USA., **90:** 8287, 1993.
22. Koss, L. G.: Tumors of the urinary tract and prostate. In L.G. Koss (Ed.), Diagnostic cytology and its histologic basis. Philadelphia: J.B. Lippincott, 1979. Pp. 749.
23. Norming, U., Nyman, C., and Tribukait, B.: Comparative flow and cytometric deoxyribonucleic acid studies on exophytic tumor and random mucosal biopsies in untreated carcinoma of the bladder. J. Urol., **142:** 1442, 1989.
24. Tsai, Y. C., Simoneau, A. R., Spruck, C. H.,III, Nichols, P. W., Steven, K., Buckley, J. D., and Jones, P. A.: Mosaicism in human epithelium: Macroscopic monoclonal patches cover the urothelium. J. Urol., **153:** 1697, 1995.
25. Spruck, C. H.,III, Ohneseit, P. F., Gonzalez-Zulueta, M., Esrig, D., Miyao, N., Tsai, Y. C., Lerner, S. P., Schm,tte, C., Yang, A. S., Cote, R., Dubeau, L., Nichols, P. W., Hermann, G. G., Steven, K., Horn, T., Skinner, D. G., and Jones, P. A.: Two molecular pathways to transitional cell carcinoma of the bladder. Cancer Res., **54:** 784, 1994.
26. Hemstreet, G. P., Rao, J. Y., Hurst, R. E., Bonner, R. B., Mellott, J., and Rooker, G. M.: Biomarkers in monitoring for efficacy of immunotherapy and chemoprevention of bladder cancer with dimethylsulfoxide. Cancer Detec. Prev., 1998.(in press)
27. Presti, J. C.,Jr., Reuter, V. E., Galan, T., Fair, W. R., and Cordon-Cardo, C.: Molecular genetic alterations in superficial and locally advanced human bladder cancer. Cancer Res., **51:** 5405, 1991.
28. Farrow, G. M.: Urine cytology in the detection of bladder cancer: a critical approach. J. Occup. Med., **32:** 817, 1990.
29. Prehn, R. T.: Many growth factors may not be growth factors. Cancer Res., **52:** 501, 1992.
30. Cohen, S. M. and Ellwein, L. B.: Genetic errors, cell proliferation, and carcinogenesis. Cancer Res., **51:** 6493, 1991.
31. Elliott, B., Ostman, A., Westermark, B., and Rubin, K.: Modulation of growth factor responsiveness of murine mammary carcinoma cells by cell matrix interactions: correlation of cell proliferation and spreading. J. Cell. Physiol., **152:** 292, 1992.
32. Gudas, L. J.: Retinoids, retinoid-responsive genes, cell differentiation, and cancer. Cell Growth Differ., **3:** 655, 1992.
33. Greenwald, P., Kelloff, G., Whitman-Burch, C., and Kramer, B.: Chemoprevention. CA Cancer J. Clin., **45:** 31, 1995.
34. Lipman, S., Heyman, R., Kurie, J., Benner, S., and Hong, K.: Retinoids and chemoprevention: Clinical and basic studies. J. Cell. Biochem., **22:** 1, 1995.
35. McCormick, A. M. and Napoli, J. L.: Identification of 5,6-epoxyretinoic acid as an endogenous retinol metabolite. J. Biol. Chem., **257:** 1730, 1982.
36. Napoli, J. L. and Race, K. R.: Biogenesis of retinoic acid from beta-carotene. Differences between the metabolism of beta-carotene and retinal. J. Biol. Chem., **263:** 17372, 1988.
37. DeLuca, L. H. and Shapiro, S. S.: Modulation of cellular interactions by vitamin A and derivatives (retinoids). Ann. N. Y. Acad. Sci., **359:** 14, 1981.
38. Ong, D. E., Crow, J. A., and Chytil, F.: Radioimmunochemical determination of cellular retinol- and cellular retinoic acid-binding proteins in cytosols of rat tissues. J. Biol. Chem., **257:** 13385, 1982.
39. Sani, B. and Corbett, T.: Retinoic acid-binding protein in normal tissues and experimental tumors. Cancer Res., **27**: 209, 1977.
40. Giguere, V., Lyn, S., Yip, P., Siu, C. H., and Amin, S.: Molecular cloning of cDNA encoding a second cellular retinoic acid-binding protein. Proc. Natl. Acad. Sci. USA., **87:** 6233, 1990.
41. Adamson, P. C., Boylan, J. F., Balis, F. M., Murphy, R. F., Godwin, K. A., Gudas, L. J., and Poplack, D. G.: Time course of induction of metabolism of all-trans-retinoic acid and the up-regulation of cellular retinoic acid-binding protein. Cancer Res., **267:** 21486, 1992.
42. Boylan, J. F. and Gudas, L. J.: The level of CRABP-I expression influences the amounts and types of all-trans-retinoic acid metabolites in F9 teratocarcinoma stem cells. J. Biol. Chem., **267:** 21486, 1992.
43. Giguere, V.: Retinoic acid receptors and cellular retinoid binding proteins: complex interplay in retinoid signaling. Endocrine Reviews **15:** 61, 1994.
44. Sporn, M. B. and Roberts, A. B.: Role of retinoids in differentiation and carcinogenesis. Cancer Res., **43:** 3034, 1983.

45. Pfahl, M., Apfel, R., Bendik, I., Fanjul, A., Graupner, G., Lee, M. O., La-Vista, N., Lu, X. P., Piedrafita, J., Ortiz, M. A., and et al: Nuclear retinoid receptors and their mechanism of action. Vitamins & Hormones ,**49:** 327, 1994.
46. Chambon, P.: The retinoid signaling pathway: molecular and genetic analyses. Semin. Cell Biol., **5:** 115, 1994.
47. Mangelsdorf, D. J., Thummel, C., Beato, M., Herrlich, P., Schutz, G., Umesono, K., Blumberg, B., Kastner, P., Mark, M., Chambon, P., and et al: The nuclear receptor superfamily: the second decade. Cell, **83:** 835, 1995.
48. Giguere, V., Ong, E. S., Segui, P., and Evans, R. M.: Identification of a receptor for the morphogen retinoic acid. Nature, **330:** 624, 1987.
49. Benbrook, D., Lernhardt, E., and Pfahl, M.: A new retinoic acid receptor identified from a hepatocellular carcinoma. Nature, **333:** 669, 1988.
50. Heyman, R. A., Mangelsdorf, D. J., Dyck, J. A., Stein, R. B., Eichele, G., Evans, R. M., and Thaller, C.: 9-cis retinoic acid is a high affinity ligand for the retinoid X receptor. Cell, **68:** 397, 1992.
51. Leid, M., Kastner, P., Lyons, R., Nakshatri, H., Saunders, M., Zacharewski, T., Chen, J. Y., Staub, A., Garnier, J. M., Mader, S., and et al: Purification, cloning, and RXR identity of the HeLa cell factor with which RAR or TR heterodimerizes to bind target sequences efficiently [published erratum appears in Cell 1992 Nov 27;71(5): following 886]. Cell, **68:** 377, 1992.
52. Mangelsdorf, D. J. and Evans, R. M.: The RXR heterodimers and orphan receptors. Cell, **83:** 841, 1995.
53. Pfahl, M.: Vertebrate receptors: molecular biology, dimerization and response elements. Semin. Cell Biol., **5:** 95, 1994.
54. Lehmann, J. M., Zhang, X. K., Graupner, G., Lee, M. O., Hermann, T., Hoffmann, B., and Pfahl, M.: Formation of retinoid X receptor homodimers leads to repression of T3 response: hormonal cross talk by ligand-induced squelching. Mol. Cell. Biol., **13:** 7698, 1993.
55. Salbert, G., Fanjul, A., Piedrafita, F. J., Lu, X. P., Kim, S. J., Tran, P., and Pfahl, M.: Retinoic acid receptors and retinoid X receptor-alpha down-regulate the transforming growth factor-beta 1 promoter by antagonizing AP-1 activity. Mol. Endocrinol., **7:** 1347, 1993.
56. Fanjul, A., Dawson, M. I., Hobbs, P. D., Jong, L., Cameron, J. F., Harlev, E., Graupner, G., Lu, X. P., and Pfahl, M.: A new class of retinoids with selective inhibition of AP-1 inhibits proliferation. Nature, **372:** 107, 1994.
57. Graupner, G., Malle, G., Maignan, J., Lang, G., Prunieras, M., and Pfahl, M.: 6'-substituted naphthalene-2-carboxylic acid analogs, a new class of retinoic acid receptor subtype-specific ligands. Biochem. Biophys. Res. Commun., **179:** 1554, 1991.
58. Mehta, K., McQueen, T., Neamati, N., Collins, S., and Andreeff, M.: Activation of retinoid receptors RARα and RXRα induces differentiation and apoptosis, respectively, in HL-60 cells. Cell Growth Differ., **7:** 179, 1996.
59. Moon, R. C., McCormick, D. L., and Mehta, R. G.: Inhibition of carcinogenesis by retinoids. Cancer Res., **43:** 2469s, 1983.
60. Ponzoni, M., Bocca, P., Chiesa, V., Decensi, A., Pistoia, V., Raffaghello, L., Rozzo, C., and Montaldo, P. G.: Differential effects of N-(4-hydroxyphenyl)retinamide and retinoic acid on neuroblastoma cells: apoptosis versus differentiation. Cancer Res., **55:** 853, 1995.
61. Delia, D., Aiello, A., Formelli, F., Fontanella, E., Costa, A., Miyashita, T., Reed, J. C., and Pierotti, M. A.: Regulation of apoptosis induced by the retinoid N-(4-hydroxyphenyl) retinamide and effect of deregulated bcl-2. Blood, **85:** 359, 1995.
62. Kim, Y. H., Dohi, D. F., Han, G. R., Zou, C. P., Oridate, N., Walsh, G. L., Nesbitt, J. C., Xu, X. C., Hong, W. K., Lotan, R., and et al.: Retinoid refractoriness occurs during lung carcinogenesis despite functional retinoid receptors. Cancer Res., **55:** 5603, 1995.
63. Li, J. J., Dong, Z., Dawson, M. I., and Colburn, N. H.: Inhibition of tumor promoter-induced transformation by retinoids that transrepress AP-1 without transactivating retinoic acid response element. Cancer Res., **56:** 483, 1996.
64. Shao, Z. M., Dawson, M. I., Li, X. S., Rishi, A. K., Sheikh, M. S., Han, Q. X., Ordonez, J. V., Shroot, B., and Fontana, J. A.: p53 independent G0/G1 arrest and apoptosis induced by a novel retinoid in human breast cancer cells. Oncogene, **11:** 493, 1995.
65. Sheikh, M. S., Shao, Z. M., Li, X. S., Ordonez, J. V., Conley, B. A., Wu, S., Dawson, M. I., Han, Q. X., Chao, W. R., and Quick, T.: N-(4-hydroxyphenyl)retinamide (4-HPR)-mediated biological actions involve retinoid receptor-independent pathways in human breast carcinoma. Carcinogenesis, **16:** 2477, 1995.
66. Moon, R. C., McCormick, D. L., Becci, P. J., Shealy, Y. F., Frickel, F., Paust, J., and Sporn, M. B.: Influence of 15 retinoic acid amides on urinary bladder carcinogenesis in the mouse. Carcinogenesis, **3:** 1469, 1982.

67. Moon, R. C., Kelloff, G. J., Detrisac, C. J., Steele, V. E., Thomas, C. F., and Sigman, C. C.: Chemoprevention of OH-BBN-induced bladder cancer in mice by oltipraz, alone and in combination with 4-HPR and DFMO. Anticancer Res. **14:** 5, 1994. 68. Moon, R. C., Detrisac, C. J., Thomas, C. F., and Kelloff, G. J.: Chemoprevention of experimental bladder cancer. J. Cell. Biochem. Suppl., **16I:** 134, 1992.
69. Studer, U. E., Biedermann, C., Chollet, D., Karrer, P., Kraft, R., Toggenburg, H., and Vonbank, F.: Prevention of recurrent superficial bladder tumors by oral etretinate: preliminary results of a randomized, double blind multicenter trial in Switzerland. J. Urol., **131:** 47, 1984.
70. Studer, U. E., Jenzer, S., Biedermann, C., Chollet, D., Kraft, R., Von Toggenburg, H., and Vonbank, F.: Adjuvant treatment with a vitamin A analogue (etretinate) after transurethral resection of superficial bladder tumors - Final analysis of a prospective, randomized multicenter trial in Switzerland. Eur. Urol., **28:** 284, 1995.
71. Decensi, A., Curotto, A., Bruno, S., Costantini, M., Torrisi, R., Gatteschi, B., Cussotto, M., Pizzorno, R., Quattrini, S., Repetto, U., and et al.: DNA flow cytometry as a surrogate end-point in patients with superficial bladder cancer treated with 4-HPR. Eur. J. Cancer Prev., **3:** 377, 1994.
72. Lamm, D. L., Riggs, D. R., Shriver, J. S., VanGilder, P. F., Rach, J. F., and DeHaven, J. I.: Megadose vitamins in bladder cancer: A double-blind clinical trial. J. Urol., **151:** 21, 1994.
73. Bookland, E., Reznikoff, C. A., Lindstrom, M., and Swaminathan, S.: Induction of thioguanine-resistant mutations in human uroepithelial cells by 4-aminobiphenyl and its N-hydroxyderivatives. Cancer Res., **52:** 1615, 1992.
74. Mosmann, T.: Rapid colorimetric assay for cellular growth and survival: application to proliferation and cytotoxicity assays. J. Immunol. Methods, **65:** 55, 1983.
75. Glover, J. F., Irwin, J. T., and Darbre, P. D.: Interaction of phenol red with estrogenic and antiestrogenic action on growth of human breast cancer cells ZR-75–1 and T-47-D. Cancer Res., **48:** 3693, 1988.
76. Chomczynski, P. and Sacchi, N.: Single-step method of RNA isolation by acid guanidinium thiocyanate-phenol- chloroform extraction. Anal. Biochem., **162:** 156, 1987.
77. Maniatis, T., Fritsch, E. F., and Sambrook, J.: Molecular cloning: a laboratory manual, Cold Spring Harbor, New York: Cold Spring Harbor Laboratory, 1990.
78. Rao, J. Y., Bonner, R. B., Hurst, R. E., Qiu, W. R., Reznikoff, C. A., and Hemstreet, G. P.: Quantitative changes in cytoskeletal and nuclear actin levels during cellular transformation. Int. J. Cancer, **70:** 423, 1997.
79. Shibata, H., Spencer, T. E., Onate, S. A., Jenster, G., Tsai, S. Y., Tsai, M. J., and O'Malley, B. W.: Role of co-activators and co-repressors in the mechanism of steroid/thyroid receptor action. Recent Progress in Hormone Research, **52:** 141, 1997.
80. Leid, M., Kastner, P., and Chambon, P.: Multiplicity generates diversity in the retinoic acid signalling pathways. Trends Biochem. Sci. ,**17:** 427, 1992.
81. Forman, B. M., Umesono, K., Chen, J., and Evans, R. M.: Unique response pathways are established by allosteric interactions among nuclear hormone receptors. Cell, **81:** 541, 1995.
82. Zhang, X. K. and Pfahl, M.: Hetero- and homodimeric receptors in thyroid hormone and vitamin A action. Receptor **3:** 183, 1993.
83. Lufkin, T., Lohnes, D., Mark, M., Dierich, A., Gorry, P., Gaub, M. P., LeMeur, M., and Chambon, P.: High postnatal lethality and testis degeneration in retinoic acid receptor alpha mutant mice. Proc. Natl. Acad. Sci. USA., **90:** 7225, 1993.
84. Sun, S.-Y., Yue, P., Dawson, M., Shroot, B., Michel, S., Lamph, W., Heyman, R., Teng, M., Chandraratna, R., Shudo, K., Hong, W., and Lotan, R.: Differential Effects of Synthetic Nuclear Retinoid Receptor-selective Retinoids on the Growth of Human Non-Small Cell Lung Carcinoma Cells. Cancer Res., **57**: 4931, 1997.
85. Hemstreet, G. P., Rao, J. Y., Hurst, R. E., Bonner, R. B., Waliszewski, P., Grossman, H. B., Liebert, M., and Bane, B. L.: G-actin as a risk factor and modulatable endpoint for cancer chemoprevention trials. J. Cell. Biochem. Suppl. , **25S**: 197, 1996.
86. Hemstreet, G. P., Bonner, R. B., Hurst, R. E., and O'Dowd, G. A.: Cytology of bladder cancer. In N.J. Vogelzang, P.T. Scardino, W.U. Shipley, and D.S. Coffey (Eds.), Comprehensive Textbook of Genitourinary Oncology. Baltimore, MD: Williams & Wilkins, 1996. Pp. 338–350.
87. Rao, J. Y., Hemstreet, G. P., Bonner, R. B., Hurst, R. E., Qiu, W. R., and Reznikoff, C. A.: Nuclear actin as a biomarker for bladder cancer risk assessment. J. Urol., **151:** 349A, 1994.(Abstract)
88. Rao, J. Y., Hemstreet, G. P., Hurst, R. E., Bonner, R. B., Min, K. W., and Jones, P. L.: Cellular F-actin levels as a marker for cellular transformation: correlation with bladder cancer risk. Cancer Res., **51:** 2762, 1991.

89. Rao, J. Y., Hurst, R. E., Bales, W. D., Jones, P. L., Bass, R. A., Archer, L. T., and Hemstreet, G. P.: Cellular f-actin levels as a marker for cellular transformation: relationship to cell division and differentiation. Cancer Res., **50:** 2215, 1990.
90. Jones, P. A.: DNA Methylation Errors and Cancer. Cancer Res. ,**56:** 2463, 1996.
91. Cairns, J.: The cancer problem. Sci. Am. **233:** 64, 1975.
92. Rothman, N., Bhatnagar, V. K., Hayes, R. B., Zenser, T. V., Kashyap, S. K., Butler, M. A. X.,Bell DA, Lakshmi, V., Jaeger, M., Kashyap, R., and et al: The impact of interindividual variation in NAT2 activity on benzidine urinary metabolites and urothelial DNA adducts in exposed workers. Proc. Natl. Acad. Sci. USA., **93:** 5084, 1996.

36

TUMOR CELL MOTILITY

A Novel Therapeutic Target in Bladder Carcinoma, Experimental and Clinical Results

T. Otto,[1,3] G. Lümmen,[1,2] A. Be,[1] H. Rübben,[1,3] and A. Raz[4,*]

[1]Department of Urology
[2]Department of Pharmacology
[3]West German Tumor Center
University of Essen, Medical School
Hufelandstr. 55, 45122 Essen, Germany
[4]Karmanos Cancer Institute
110 East Warren Avenue, Detroit, Michigan, 48201

1. SUMMARY

Surgery, chemotherapy and radiotherapy are the current modalities of tumor management. However, in systemic disease, predicted patients' cure can be achieved in but a few tumor diseases. Based on previous basic research we have highlighted that the loss of cell-cell adhesion in association with an increased tumor cell motility is an essential feature of the malignant potential of bladder tumors. Thus, we have attempted therapeutical methods differing from hitherto existing treatments by focusing on a tumor cell function we call cell motility. Characterization of so-called anti-motility drugs was performed biochemically as well analyzed by in vitro by using in established bladder carcinoma cell lines. We evaluated the potential therapeutic benefit in a model of chemically induced bladder carcinoma followed by a phase I/II trial applying antimotility drugs in patients which were chemotherapy-resistant and having metastatic bladder cancer.

Both basic research as well as the results of first translational clinical trials confirmed, that advanced bladder carcinomas can be favorably affected by inhibition of tumor cell motility.

---

* Corresponding author.

Advances in Bladder Research, edited by Baskin and Hayward.
Kluwer Academic / Plenum Publishers, New York, 1999.

2. INTRODUCTION

Urothelial carcinomas are heterogeneous diseases with an aggressive clinical potential. Bladder tumors are distinguished as either superficial or invasive, and the most commonly used prognostic factors of bladder cancer are tumor size, pathological grade, stage, multiplicity, and ploidy. These factors, however, are often not reliable in predicting clinical outcome and prognosis, which depends on the differentiation-dedifferentiation status, depth of invasion and metastases. These biological progresses are usually reflected in the primary tumor by loss of cell-cell contact and acquisition of cell motility, phenomena associated with a decreased expression of the epithelium-specific cell-cell adhesion molecule E-cadherin and increased expression of the autocrine motility factor receptor gp78[1].

Despite advances in the clinical management of bladder cancers, identifying early-stage patients at a high risk for cancer progression is not yet possible. As the prediction of patient outcome solely based on histopathological classifications was found to be inaccurate in approximately 40 % of the cases, it poses an obvious predicament of how to decide which patient may be at risk for recurrence and progression and should undergo aggressive therapy. Thus, the need for a predictable marker(s) for identification of such a group of patients is of an obvious urgent need. Furthermore there is a need for a new therapeutic approach in metastasis bladder cancer disease because long term survival after cystemic therapy in metastatic bladder cancer disease is extremely poor [2].

3. RESULTS OF A PROSPECTIVE MARKER STUDY CONCERNING AUTOCRINE MOTILITY FACTOR RECEPTOR (AMFR, GP78)

Here, we evaluate the clinical outcome of 100 patients with urothelial carcinomas with follow-up information for more than 2 years after surgery in relation to the expression of two cell surface antigens, i.e., E-cadherin and autocrine motility factor receptor (AMF-R, gp78). Frozen bladder tissues were serially cut, stained either with hematoxylin and eosin for grading, with the anti-gp78 antibodies, or with anti-E-cadherin antibodies to determine level of expression.

Of 63 patients presented at the time of diagnosis with pathological loss of E-cadherin associated with increased gp78 expression, 39 (62 %) succumbed to tumor progression or death. Of 37 patients with normal E-cadherin and gp78 expression positive or negative, respectively, 36 (97 %) had favorable disease outcomes ($P < 0.0001$)[3]. The results suggest that in bladder carcinomas abnormal expression of both E-cadherin and gp78 results in poor disease outcome, independent of tumor stage and grade. (Figure 1).

4. SIGNAL TRANSDUCTION PATHWAY OF TUMOR CELL MOTILITY

In order to use tumor cell motility as a therapeutic target, we investigated the signal transduction pathway of cell motility factors and determined drugs which influence, i.e. inhibit motility in highly aggressive bladder cancer cell line J82.

Aim of the present study was to examine the expression and the effects of G protein-coupled, calcium-mobilizing receptors on motility of J82 cells, a cell line originated from

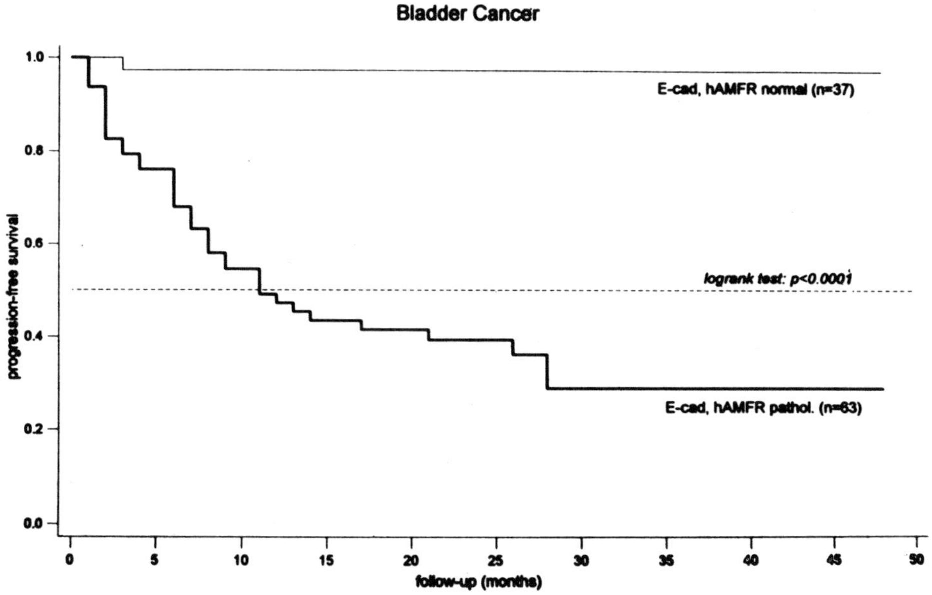

Figure 1. Progression free survival according to E-cadherin / gp78 status of 100 patients with bladder carcinomas.

a human transitional-cell carcinoma. These cells express a variety of such receptors, including those for lysophosphatidic acid (LPA), thrombin, bradykinin, bombesin, and histamine. Agonist activation of these receptors resulted in rapid and transient increase in $[Ca^{2+}]_I$ and inositol phosphate formation. Pertussis toxin (PTX) treatment reduced the LAP-induced $[Ca^{2+}]_I$ increase by 50 %, without affecting the $[Ca^{2+}]_I$ responses to the other agonists. Motility of J82 cells was strongly stimulated by LPA and thrombin, although higher concentrations of either agonist were required than for the Ca^{2+} responses, whereas bradykinin, bombesin and histamine were without effect. Treatment of J82 cells with cytochalasin B and the Rho-inactivating *Clostridum difficile* toxin B prevented LPA- and thrombin-stimulated cell motility. PTX treatment had no effect on thrombin-stimulated motility, whereas the stimulatory effect of LPA was completely abolished (Figure 2).

Half-maximal inhibition was observed at only 10 pg/ml PTX (24 hr), correlating with a similar extent of ADP-ribosylation of endogenous $G_{i/o}$ proteins. Preventing the $[Ca^{2+}]_I$ transients by chelation of intracellular Ca^{2+} with BAPTA/AM blunted the motility response to either agonist. Agonist-stimulated motility was also strongly attenuated by the phorbol ester, phorbol 12-myristate 13-acetate, which by itself had no effect on J82 cell migration, and which prevented the LPA- and thrombin-induced $[Ca^{2+}]_I$ increases. Finally, raising cAMP levels by forskolin and dibutyryl cAMP inhibited the LPA- but not the thrombin-induced J82 cell migration. In conclusion, J82 cells express various G protein-coupled, calcium-mobilizing receptors, out of which only those for LPA and thrombin stimulate cell motility, indicating that phospholipase C-derived second messengers *per se* are not sufficient for increasing cell motility. The complex signal transduction machinery leading to agonist-stimulated motility of these carcinoma cells appears to involve some common, apparently essential factors, such as $[Ca^{2+}]_I$ increase and Rho family G protein-regulated reorganization of the cytoskeleton, as well as some distinct components, most notably distinct subtypes of heterotrimeric G proteins and apparently also distinct cAMP-

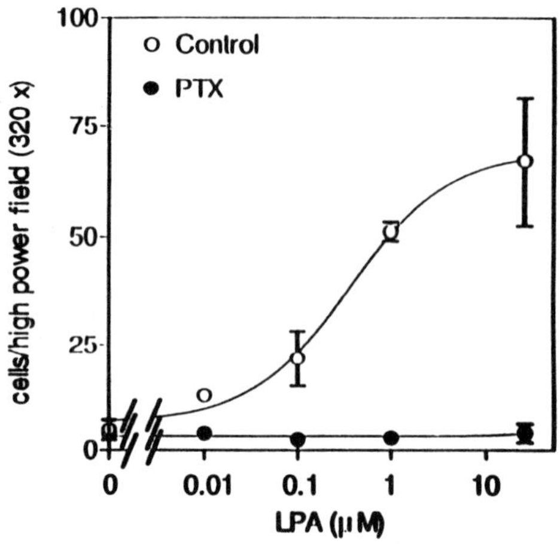

Figure 2. Concentration-dependent effect of PTX on LPA-induced motility and ADP-ribosylation of $G_{i/o}$ proteins in J82 cells. J82 cells were pretreated for 24 hr with the indicated concentrations of PTX. Thereafter, cell migration (A) was determined in the absence (O) and presence of 20 μM LPA (l), and PTX-catalyzed $[^{32}P]$ADP-ribosylation of $G_{i/o}$ proteins was determined in detergent extracts. The inset shows a representative autoradiogram, with the molecular weight markers on the left and the PTX concentrations (ng/ml) used for treatment of intact J82 cells at the bottom.

sensitive targets. From these in vitro data we conclude that PTX is a candidate for an in vivo antimotility therapeutic approach[4].

5. INFLUENCE OF PERTUSSIS TOXIN ON THE DEVELOPMENT OF ADVANCED BLADDER CARCINOMA IN MICE

Since motility factors correlate with the malignant potential of tumor cells and the clinical course of patients with bladder carcinoma[2], we defined antimotility factors potentially influencing tumor cell motility as verified with an in vitro assay (Phagokinetic track motility assay). For the first time antimotility factor (PTX) was applied in a model of chemically induced bladder carcinoma in mice and adverse reactions and therapy effectiveness evaluated.

5.1. Material and Methods

5.1.1. Animals. Female mice of the strain B6D2F1 with a body weight of 25 g and a median age of 11 weeks were selected. The animals were allocated at random with a light-dark cycle of 12 hours. Over a period of 10 weeks 0,05 % BBN was added to the drinking water in light impermeable bottles. Given an average uptake of 6 ml water/day and animal each mouse ingested approximately 0,2 ml BBN (96 %).

5.1.2. Carcinogen. BBN is a yellow oily liquid hardly volatile and soluble due to terminal hydroxylation. BBN was synthesized and provided by the German Cancer Research Center Heidelberg.

5.1.3. Pertussis-Toxin. Already a concentration of 0,5 ug/ml is reported to completely block cell motility and to suspend the effect of LPA and the autocrine motility factor. The effect of pertussis-toxin is dose- and time-dependent and is due to blocking of the G-protein through inhibition of phospholipase A2 and C. The toxin is provided by Takeda.

5.1.4. Application. PTX was injected intra peritoneally without anesthesia or additional medication. From 13 weeks onward therapy was begun and carried out between 2 p.m. and 4 p.m. to avoid influence of quotidian rhythm. All animals were sacrificed at 23 weeks. The mice were allocated to two groups (control group (PBS) versus PTX group).

5.1.5. Vivisection. At the end of the 23rd week all mice were sacrificed followed by a median laparotomy. The bladder is filled with 0,05 saline to facilitate fixation and staining. Apart from three randomly selected mice per group all organs were immediately embedded in liquid nitrogen. Organs of the selected rodents were stored in 5 % formaldehyde.

5.1.6. Macroscopical Evaluation. To assess the clinical tumor stage of the bladder carcinoma a macroscopical examination of the thoracic and abdominal contents was performed. Advanced, bladder wall-penetrating urothelial carcinoma as well as tumor related obstruction of the upper urinary tract was macroscopically verified and classified thus distinguishing locally advanced tumors (> / = T3b).

5.1.7. Light Microscopy. After fixation of the 5 μm frozen sections in ethanol the slides were stained with hematoxylin-eosin and evaluated at 100 and 400 power field (Zeiss microscope).

Since tumor progression is end point of the study assessment of depth of infiltration as classified by the UICC is of supreme importance. In this trial tumors were classified in superficial and early invasive carcinomas (pTa, pT1, pT2) in contrast to extravesical tumor growth (pT3) and infiltration of adjacent organs (pT4). Histopathological examination of the removed organs (kidneys, liver, lung, spleen) was performed to disclose therapy related side-effects and the presence of distant metastases (liver, lung, lymph nodes) though not expected in this animal model.

5.1.8. Statistical Analysis. The results are calculated in P values. Regarding the frequency of tumor development in the animal experiment Fisher ´s exact test, X^2-test or Mantel-Haenszel-test were applied depending on the data being simultaneously obtained. The primary hypothesis of the animal experiment was confirmed when analyzed with a $p < 0{,}05$ being significant. All further analysis was descriptive and explorative. Significance levels were not adjusted in view of the problem of multiple testing. All statistical tests were paired.

5.1.9. Influence on Tumor Development.

5.1.9.1. Control (Group I) animals receiving intraperitoneal injection of phosphate buffered saline (PBS) after exposure to BBN as control developed locally advanced tumors in 41 % (11/27). All Tumors were undifferentiated (Grade IV). five mice died tumor-related before the end of the 23rd week of which 4 underwent a postmortem examination allowing the assessment of tumor-stage in 27 of 28 animals. In all of the 11 mice with locally advanced tumors the macroscopical findings were consistent with histopathological evidence of bladder wall penetrating tumor growth (pT3b) (Table 1)

5.1.9.2. Pertussis toxin (group II) Following a total of 10 applications of 2,4 ug PT / 100 mg body weight locally advanced bladder carcinoma did occur in 3 of 28 mice (11 %). No side effects were observed (Table 1). We conclude from these data, that antimotility therapy with PTX is a promising treatment modality for advanced carcinomas without severe side effects.

Table 1. Results of the therapy with antimotility factor (Pertussis toxin) for locally advanced bladder carcinoma as compared to the sham group (group I)

Group	Drug	Number N	Bladder cancer* > pt3b, G4	P-value
I (control)	phosphate buffered saline (PBS)	27	11 (41%)	0.0001
II	pertussis toxin	28	3 (11%)	

* In both groups we found severe dysplasia in all specimens (p = 1.0).

6. SECOND LINE THERAPY WITH PTX, PACLITAXEL, AND CARBOPLATIN IN METASTATIC BLADDER CANCER DISEASE

6.1. Results of Pertussis Toxin Monotherapy in Bladder Cancer Disease

We performed dose findings studies for PTX in bladder cancer patients and found as well as for intramuscular application as for intravesical application no side, effects WHO grade > 2 (Table 2, 3).

6.2. Patients and Methods

Carboplatin, paclitaxel and PTX were administered to improve the treatment of cisplatin- and methotrexate-resistant metastatic bladder carcinoma. Inclusion criteria for patients were a World Health Organization (WHO) performance status of Grade < 2, progressive disease subsequent to combination chemotherapy with methotrexate and cisplatin, and bidimensionally measurable lesions. After informed consent was obtained from the patients, the treatment cycle was begun on day 1 with an intramuscular injection of 2,9 μg pertussis toxin in 0,5 ml suspension (phosphate-buffered saline). Injections of PTX were repeated on days 4 and 7 (3 intramuscular injections in 1 week, with an interval of 2 days between them). On day 8, 24 hours after the last PTX injection, intravenous paclitaxel 135 mg/m^2 body surface was given via a central venous line for 3 hours. Twelve and 6 hours prior to administration of paclitaxel, dihydrocortisol 20 mg was given. Subsequent to paclitaxel infusion, each patient received carboplatin 400 mg/m^2 body surface. Seven days after chemotherapy, treatment was continued with an intramuscular injection of PTX on days 15 and 19. One cycle lasted 3 weeks; day 22 marked the beginning of the second cycle. A thorough examination, involving assessment of tumor stage and response, was performed prior to a third cycle. Depending on subjective response or complete or partial remission, treatment was to be continued with a third and fourth cycle. Additional staging,

Table 2. Dose finding study (phase I) for intramuscular application of PTX in refractory metastatic cancer disease

	Bladder carcinoma PTX intramuscular M1 retractory disease
Side effect ≥ WHO grade 2	0
N	26

PTX: 9 mcg - 270 mcg.

Table 3. Dose finding study (phase I) for intravesical application of PTX in locally advanced bladder carcinoma prior to radical cystectomy

		Bladder carcinoma PT2 - pT4 PTX, intravesical	
Dose level	N	PTX [mcg]	Side effects WHO grade > 2
I	3	9	0
II	3	18	0
III	3	26	0
IV	3	35	0
V	3	44	0

including chest X-ray, sonography, and computed tomography of the abdomen and pelvis, was performed 3, 6, and 12 months after therapy.

6.3. Results

Since 1994 15 patients had been treated according to the regimen. The median age of the patients was 62 years. Prior to second-line therapy, 11 patients received 2 cycles of methotrexate (40 mg/m^2, days 1 and 15) and cisplatin (70 mg/m^2, day 1), respectively, as first-line chemotherapy. After 28 days the second cycle was begun. Accordingly, 7 patients were each treated with 4 cycles of MVAC as first-line therapy (methotrexate 30 mg/m^2, days 1, 15, and 22; vinblastine 3 mg/m^2, days 2, 15, and 22; doxorubicin 40 mg/m^2, day 2; and cisplatin 70 mg/m^2, day 2). The cycles were repeated after 28 days.

The decision to begin second-line chemotherapy was based on evidence of tumor progression with first-line treatment (Table 4). Tumor progression did occur in 11 of 18 patients receiving chemotherapy with methotrexate and cisplatin and was revealed by staging before the third cycle was to begin. In the other 7 patients, tumor progression was detected after a median of 3 months after MVAC chemotherapy. Second-line chemotherapy was begun a median of 6 weeks (range, 2–8 weeks) after diagnosis of tumor progression. Regarding localization of secondary lesions, all patients presented with retroperitoneal lymph node metastases. Further secondary tumors were detected in the liver in 4 patients, bony lesions in 4 patients, and pulmonary metastases in 7 patients. For second-line therapy with paclitaxel, carboplatin, and PTX, a median of 2,5 cycles was administered. The patients underwent staging examination prior to a third cycle. On examination, 7 patients had stable disease and another 7 had tumor progression. In one patient, progressive disease was already noted after the first cycle.

Objective tumor response occurred in four patients. A female patient age 34 years with histologically confirmed urothelial carcinoma developed a complete remission of a pelvic wall lesion after two courses of paclitaxel, carboplatin, and PTX. Remission was confirmed after a total of four cycles. Four months after therapy, she presented with progressive disease and liver metastases. The second objective response was observed in a male patient age 54 years with retroperitoneal lymph node metastases; however, 2 months later, he again developed progressive disease. A third patient with histologically confirmed urothelial carcinoma and extensive retroperitoneal, mediastinal, and cervical lymph node metastasis achieved complete remission after 4 cycles and 3 months after therapy. Partial response was achieved in a fourth patient with retroperitoneal lymph node metastasis; and

Table 4. Case-oriented presentation of the results of treatment of metastatic urothelial carcinoma of the bladder with PTX, paclitaxel, and carboplatin

				Responses *			
Patient no.	Age	Gender	Cycle	CR	PR	SD	PD
1	34	F	4	4			
2	67	M	4			+8	
3	71	M	1				1
4	62	M	2				2
5	82	F	2			+6	
6	71	F	2				12
7	56	M	3		2		
8	62	M	2				2
9	59	M	2			+3	
10	64	M	2				3
11	68	M	2				3
12	64	M	4	4			
13	56	F	2				4
14	64	F	2			+2	
15	64	M	4		+6		
16	62	M	2			+4	
17	57	F	2				4
18	56	M	2				1

CR: complete response, PR: partial response, SD: stable disease, PD: progressive disease.
* Figures indicate mos.

at follow-up 6 months after treatment, the patient was still in remission. Chemotherapy with paclitaxel and carboplatin caused moderate nausea and vomiting (WHO Grade 1–2) in all patients. In one patient treatment was withdrawn due to continuous vomiting (WHO Grade 4). Myelotoxicity (WHO Grade 3) was detected in 3 patients at the end of the fourth cycle and in 1 patient (WHO Grade 4) at the end of the second cycle. No side effects occurred after administration of PTX.

ACKNOWLEDGMENTS

Supported in part by Deutsche Forschungsgemeinschaft DFG, SFB 354 and by Dr. Mildred Scheel Foundation to T.O. and by NIH grant CA-51714–03 to A.R.

REFERENCES

1. Otto, T., Birchmeier, W., Schmidt, U., Rembrink, K., Schipper, J., Rübben, H., Raz, A. Inverse relation of E-cadherin and autocrine motility factor receptor expression as a prognostic factor in patients with carcinoma of the bladder: Cancer Res. **54**: 3120, 1994.
2. Igawa, M. Limitations of M-VAC chemotherapy for the treatment of advanced bladder carcinomas: J. Urol. **144:** 662, 1994.
3. Otto, T. Be, A., Schmidt, U., Raz, A., Rübben, H.: Improved prognosis assessment for patients with bladder cancer, Am. J. Pathol. **150**: 911, 1997.
4. Otto, T., Be, A., Krege, S., Walz, P.H., Rübben, H.: Paclitaxel based second line therapy for patients with advanced chemotherapy -resistant bladder carcinoma (M1) - A clinical phase II study Cancer **80**: 465, 1997.

DISCUSSION

Cancer of the Bladder

Q & A with Peter Jones, Ph.D.

Avrahim Raz, Detroit.

Q: Why does methylation cause extinction of promoter activity.

A: So, the question is, how does methylation switch off the promoter? Actually, there's a lot of information on that subject. What happens is that when the CPG sites become methylated, they attract a protein to bind to that region, which is called a methylated-DNA binding protein, and there are some very exciting new papers that are going to be coming out very soon that shows what that does when the methyl-binding protein comes there, it attracts a histodeacetylase to that region - keeps the histones deacetylated and causes a heterochromatic state on the promoter. Those papers you will see very soon.

Henry Sun, New York University.

Q: What's the mechanism for this kind of change in the methylation pattern in the tumor formation? Is this a whole cell change for the methylation of many, many genes, or is it relatively selective?

A: I think that's an excellent question. The reason why we developed this screen, this methylation-sensitive APPCR, was to address that question. If you remember, I showed you in some of the bladder tumors that almost 30% of all of the sites that we can assess become abnormally methylated, so it's very profound. The question of why it happens, is we don't know, because actually, we know nothing about how a de novo methylation occurs, either in normal embryonic development or in cancer development.

Chin Minh, UCSF.

Q: How about the methylation in introns? Do you expect to see an increase in expression or decrease in expression?

A: I think the methylation in introns probably also occurs and is probably associated with expression. There are some well-known examples of that, I didn't have time to go into, in normal genes in imprinting. For example, the IgF-2 receptor gene, which is expressed from the maternal copy has methylation in the first intron of the gene on the mother's chromosome, but not the father's. So there is some evidence that intronic sequences show the same behavior.

Simon Hayward, UCSF.

Q: Two questions. The first is, I assume that both the sense and the antisense strands can be methylated, and does it matter which strand is methylated or is it really just the position of methylation relative to the promoter?

A: I actually forgot to bring two key slides. The key slides show you that the methylation is always symmetrical. In other words, because a CG site on the other strand is CG going in the other direction, either both of the Cs are methylated or neither. After DNA replication, there's a new strand of DNA that doesn't have 5-methyl C in it, the enzyme takes its cue from the existing methyl group and applies a new one on the other strand. It's an excellent question. That's how methylation patterns actually get inherited.

Q: The second question is, you told us that Drosophila, for example, survive quite happily without methylating their DNA, and you've told us lots of reasons why we shouldn't methylate our DNA, what's the survival advantage of methylating?

A: For years, those of us who have worked in the field, would sort of cringe under the podium when you ask that question, but we don't do that anymore because the reason we don't is somebody, Rudolph Yanish, knocked out the methyltransferase gene in the mouse. If you do that in the mouse, you don't get a mouse. So, you clearly need this to in order to undergo embryonic development. Why we need it, we don't know, but it's certainly essential.

Q: How far back in the evolutionary chain does methylation go?

A: Yes. I can't remember the exact details, but it seems to be correlated with an increase in the size of the genome. Drosophila tends to have a very efficient genome. We have lots of extra sequences, and maybe that's why we have methylation.

John Park, Boston.

Q: As a novice in the field, I learned a tremendous deal. The question sort of relates to what Simon Hayward just asked. Does DNA methylation have any specific known role during organ development? Your APPCR method, for example, was sort of reminiscent of a differential display, another type of fishing expedition, but much more focused, and differential display when applied to organ development, it sort of pulled out a lot of candidate markers, but really was difficult to really nail down. I wonder if it has a role in gene activation/inactivation during organ development, that you might be able to come away with more specific identification of genes.

A: I think that the main reason for DNA methylation is not to regulate genes, but it's to permanently keep genes off. So, it switches things off. It's a suppression mechanism. And I think what it does is three things. In female mammals, it keeps the X chromosome quiet. In imprinted genes, it keeps either the father's or the mother's gene quiet for good. The third thing, and probably the most important thing it does, is to keep switched off the large number of DNA sequences we have in our genome that we don't want to express. I'm thinking of L1 sequences and transposable elements which populate our DNA and actually make up 30% of our DNA is sequences we never want to express. I think that methylation is a really good way of keeping those things off. Why it goes awry in cancer, I don't know.

Unidentified Questioner.

Q: I'm fascinated by the association between the P16 alterations and the squamous cell. In essence, you've associated a phenotype with a genotype here. The questions I have, relate to what is the status of the second allele where one is methylated?, and what do you see when you look at those squamous cell carcinomas that have an etiology associated with schistosomiasis? Can you tie that all together.?

A: Those of course are the questions we would like to answer. These were non-schistosomiasis patients. We have looked at schistosomiasis patients in the past, but in that case, we were just looking for homozygous deletions. We never looked for methylation changes of that stage. So, I don't really know at this moment. In that particular patient that I showed you, we don't know, unfortunately the status of the other allele. At this moment, we don't know. But, it's conceivable that both alleles can become switched off by methylation.

Henry Sun, New York University.

Q: You mentioned that when you look at the cultured cells, frequently the methylation pattern would change. How rapid is this change? Is this reversible?

A: The change occurs very frequently, and is associated, we think with immortalization, and it's not reversible except if you treat with azacitidine. In fact, that's how we discovered azacitidine, because many of you may not know, but Kathy Reznikoff developed a mutation in a half-cell line, which is a mouse embryo fibroblast, and it's well used in chemical carcinogenesis. We treated it with azacitidine, we got contracting muscle cells. That was a very surprising observation. So we don't actually know.

Q: The reason I was wondering about that, is because I was wondering whether or not this change in the methylation pattern is a reflection of the phenotype. For example, that could potentially explain why in the TCC versus SCC, you actually see different methylation pattern. If so, then it makes you wonder, whatever gene you see, that they have different methylation pattern in the tumor, whether that's cause or effect of the tumor phenotype.

A: Well, I think it's really difficult to answer that at the moment. My own bias is that it's caused, of course.

Unidentified Questioner.

Q: One concern, when you look at bladder cancer, you look at the normal bladder as its control, being a field disease you really wonder how good of a control that is. When you look at the methylation across the various tumors, for a given gene you see increased methylation, unchanged or diminished methylation. What would happen if you looked at controls without bladder cancer, controls with inflammatory diseases of the bladder? What type of methylation do you ? Because it seems so variable per a given tumor, you wonder what that variability would be in other diseases.

A: We have actually compared the normals to all the other normals in these patients, and they're remarkably similar, the patterns as a matter of fact. In that slide that I showed you, remember that each of those dots represents a comparison of the person's normal and their own cancer. It's not the normal of somebody else. It's juxtapositioned. It's clearly an incremental effect that's occurred in the human. But you raise a really, really good question. I think that what we're gonna find is these methylation changes occur very early in carcinogenesis, and I think they might actually participate in some way in the field effect. Maybe the methylation effect is a field effect. So we want to try and address that question.

Jenny Southgate, Leeds, UK.

Q: I was just wondering, do normal cells when they're put into culture, show methylation changes if they adapt to culture or maybe go through to senescence?, and whether all tumor cell lines show these methylation changes, or whether it's particular phenotypes or genotypes?

A: When we've taken mouse cells and put them in culture, we don't see the methylation changes until we get in the mortal line. In every single mouse cell line that's on this earth has completely abnormal methylation patterns. So, it's an acquired defect. Every cancer that we've looked at, with the exception of a few of those patients there, has a methylation change.

Q & A with Fred Waldman, M.D.

Avrahim Raz, Detroit.

Q: The technique is a very limited technique, because it can detect only gross aberration. You are unable to detect whether is a point mutation?

A: Correct. But remember, that point mutations, whether it's in P16 or anything else, the other allele is frequently deleted, is frequently lost. In tumors, that loss is frequently a chromosomal deletion. That's the other side of it.

Bob Hurst, Oklahoma City, Oklahoma.

Q: Looking at your data there on recurrences, actually it looks to me more like sometimes it's clonal and sometimes it's not. It either was high, 75% or more, or down it was 30 to 25%, and it looked almost bimodal. That made me wonder because Peter Jones had shown the same thing, that sometimes it was clonal and sometimes it was not on recurrences.

A: I'm struggling, interpreting these data, and that's why we're doing these statistical convolutions to do that, especially with the superficial ones that have so few changes, that by CGH, the difference between two-thirds and one-third is once change, so it depends on how frequent those changes are. If that change, that 25% is only found in that one tumor, okay, out of all the tumors we've looked at, to me, that's strong evidence that they're clonally related. If it's present in three-quarters of the tumors, then that's very weak evidence that they're clonally related, and that's what that statistic is attempting to do. It's not perfect yet, but that's the goal there.

Monica Liebert, M.D. Anderson Cancer Center, Texas.

Q: Fred, I had two questions. One is, in your pathway, it's possible that there are biological pathways like on chromosome 9 and chromosome 17, that are actually requirements for the development of cancer, and that might be one explanation for some of the commonality that may or may not be, you know, if you don't have that, you're not a tumor. Wouldn't you agree with that? Isn't that possible?

A: Yeah, I mean, again, just being contrary to think of this alternative hypothesis, not that I want to believe it, not that any of my grants describe that in any detail, I'm basing all of our work on the other hypothesis. I'm starting to think that maybe some of these things, like Peter suggested for the gene that he was interested in, they may have two functions.

One function may be for metastasis or for invasion, it may be invasion and metastasis are much more alike than is invasion versus noninvasion. But they also may be selected for clonally in all the tumor cells based on some proliferative event. So if you can define a proliferative event, that gives you an excuse for that aberration to be selected for in the total tumor population.

Q: My second question is actually a technical one, and that is when you compare your normals, you're actually looking at normal leukocyte DNAs, is that correct?
A: For CGH, the normal DNAs, normal genomic DNA from another patient. It's not sensitive to polymorphic differences between patients.
Q: Have you ever considered actually doing a head-to-head comparison, looking at the mets in the primary as the comparisons green, primary red?

A: Sure, that was our elegant suggestion, as you can imagine. Because when you compare the two of them, the differences will now light up as relative gains or losses, and we did that, and it correlates with these other findings. The problem with only doing that is that all of the common changes, which are 80% of the changes, would now disappear. So you can't see that this tumor actually has 20 different changes, all you're seeing are some relative changes higher or lower in a few places. But to point those out, it does correlate very well.

Q: Did you ever find anything hot in that particular comparison, or was there something left out? Apparently not.

Peter Jones, USC.

Q: First of all, this is really stimulating. We also looked at primaries and mets some years ago using loss of heterozygosity studies, and came up with exactly the same observation. Admittedly, we were not looking at so many regions as you were, but we found no differences between the primary and the met. The other very surprising result, maybe you mentioned, you know, you're looking at a very gross level in this kind of approach. Daryl Shibato who is at USC with us did the following experiment. We had a patient where we had a lymph metastasis, and then the patient came back a year later with another metastasis. He looked at 200 micro satellites in the two tumors and could find not a single difference. That's a pretty fine analysis. So I think maybe some of our concepts are going to have to be dramatically changed.

Unidentified Questioner.

Q: Is another hypothesis, maybe naive, is that if the primary and the metastasis are similar, might it be the microenvironment, the angiogenesis, and so forth, so other factors other than the tumor itself that might be dependent upon metastasis?

A: Well of course, methylation wouldn't be picked up here, so there can be lots of methylation differences, if I can answer that for Peter, but any kind of expression differences or individual gene mutations would not be picked up by this procedure, so lots of other things can be going on, that's going to be distinguishing between them.

Q & A with M'Liss Ann Hudson, M.D.

Unidentified Questioner.

Q: There certainly is compelling data that you present with invasiveness and modulating invasiveness by the plasminogen receptor. When you go into the tissues that one obtains clinically, you look at superficial tumors rather than cell line superficial tumors, invasive tumors, and then metastases. Does the receptor density hold up in terms of predicting invasiveness?

A: I'm only gonna quote somebody else's work, it's not mine. None of the data that I had seen actually quantitated the receptor in the metastasis. Many people have commented that when you take tumor specimens from an individual patient, how much of the actual tumor that's representative of the whole tumor do you get? That's where I have a hard time speaking to somebody else's data. They do show higher quantities of an antigen, but its probably not specific. We know that fibroblasts will secrete, UPAR, there may be some other stromal cells and maybe macrophages that are able to contribute when you're talking about a tumor extract. People in colon cancer have even looked at, say, a line similar to the T24 one that expressed a receptor but didn't make the antigen, that the stromal cells that they had in inflammatory exudate that was part of the thing, that those stromal cells would produce the antigen, and that colon cancer would then be able to invade but not without the inflammatory cells.

Q: I guess that some of the exciting therapeutic possibilities that you mentioned in conclusion could well be tested in the animal models that you have.

A: Surely.

Bob Hurst, Oklahoma.

Q: I was wondering, the T24s, is it possible they express another protease?

A: Absolutely. In fact, people have worked with the metaloprotneases and collagenases, and I do believe that, you know, this is hardly the only degrative enzyme that these cells happen to express. So, yes.

Q: So, it's the phenotype that counts? You have to have some sort of protease, and there may be a choice from several.

A: Absolutely.

Q & A with Dr. Monica Leibert.

John Park, Boston.

Q: Is there anything known about the regulation of the PDGH? How does it compare with the inducibility of the COX-2?

A: We haven't actually haven't done the inducibility part of COX-2 in bladder cancers. One little piece of data that I of course glossed over was that in the expression of COX-2 in bladder cancer, since bladder cancers in situ have a lot of inflammation, it may not be constitutive action, but induced action by the presence of inflammatory cells. So, that's something we need to evaluate more critically; however, that some of our bladder cancer cell lines do show constitutive expression may indicate that that's a possibility. PDGH is really interesting because it really hasn't been studied that well in tissues that are not hormonally regulated. It's well understood that it's activity is hormonally regulated in uterine epithelium, and that's been studied very well, and that it's expression is very clearly associated with birth, and so it's very sharply regulated by the hormones that change as birth progresses. But what's controlling and expressing it in bladder cancer, we really don't know. So, that's going to be a whole other area for exploration, I think. Since it's not likely that it's directly hormonally related since many of our specimens come from men, that that will be something interesting to pursue. Similarly, we don't know what's going to be directing the COX-2 expression and why it's constitutive in some of those cells, so I think that that's going to be really important to do.

INDEX